D1692817

Applied Metallomics

Applied Metallomics

From Life Sciences to Environmental Sciences

Edited by Yu-Feng Li and Hongzhe Sun

WILEY VCH

The Editors

Dr. Yu-Feng Li
Chinese Academy of Sciences
19B, Yuquan Road
Beijing
China

Dr. Hongzhe Sun
University of Hong Kong
Pokfulam Road
Hong Kong
China

All books published by **WILEY-VCH** are carefully produced. Nevertheless, authors, editors, and publisher do not warrant the information contained in these books, including this book, to be free of errors. Readers are advised to keep in mind that statements, data, illustrations, procedural details or other items may inadvertently be inaccurate.

Library of Congress Card No.: applied for

British Library Cataloguing-in-Publication Data
A catalogue record for this book is available from the British Library.

Bibliographic information published by the Deutsche Nationalbibliothek
The Deutsche Nationalbibliothek lists this publication in the Deutsche Nationalbibliografie; detailed bibliographic data are available on the Internet at <http://dnb.d-nb.de>.

© 2024 WILEY-VCH GmbH, Boschstraße 12, 69469 Weinheim, Germany

All rights reserved (including those of translation into other languages). No part of this book may be reproduced in any form – by photoprinting, microfilm, or any other means – nor transmitted or translated into a machine language without written permission from the publishers. Registered names, trademarks, etc. used in this book, even when not specifically marked as such, are not to be considered unprotected by law.

Print ISBN: 978-3-527-35144-2
ePDF ISBN: 978-3-527-84037-3
ePub ISBN: 978-3-527-84038-0
oBook ISBN: 978-3-527-84039-7

Cover Design and Image: Schulz Grafik-Design, Fußgönheim, Germany

Typesetting   Straive, Chennai, India
Printing and Binding: CPI Group (UK) Ltd, Croydon, CR0 4YY

Contents

Foreword *xv*
Preface *xvii*

1 **Introduction** *1*
Yu-Feng Li and Hongzhe Sun
1.1 A Brief Introduction to Metallomics *1*
1.2 Key Issues and Challenges in Metallomics *3*
1.3 About the Structure of this Book *4*
References *6*

2 **Nanometallomics** *11*
Hongxin Xie, Liming Wang, Jiating Zhao, Yuxi Gao, Bai Li, and Yu-Feng Li
2.1 The Concept of Nanometallomics *11*
2.2 The Analytical Techniques in Nanometallomics *12*
2.2.1 The Analytical Techniques for Size Characterization of Nanomaterials in Biological System *12*
2.2.1.1 Chromatography-based Techniques for Size Characterization *12*
2.2.1.2 Mass-spectrometry-based Techniques for Size Characterization *13*
2.2.1.3 Laser, X-rays, and Neutron-beam-based Techniques for Size Characterization *13*
2.2.2 The Analytical Techniques for Quantification of Nanomaterials and Metallome in Biological System *14*
2.2.3 The Analytical Techniques for Studying the Distribution of Nanomaterials in Biological System *15*
2.2.4 The Analytical Techniques for Studying the Metabolism of Nanomaterials in Biological System *16*
2.3 The Application of Nanometallomics in Nanotoxicology *17*
2.3.1 Understanding the Size Changes, Uptake and Excretion, Distribution, and Metabolism of Nanomaterials in Biological Systems *17*
2.3.2 Comparative Nanometallomics for Distinguishing Nanomaterials Exposure and Nanosafety Evaluation *20*

2.4	Conclusions and Perspectives 21
	Acknowledgments 22
	List of Abbreviations 22
	References 23

3   Environmetallomics 33

Lihong Liu, Ligang Hu, Baowei Chen, Bin He, and Guibin Jiang

3.1	The Concept of Environmetallomics 33
3.2	The Analytical Techniques in Environmetallomics 34
3.2.1	The Requirements for Environmetallome Analysis 34
3.2.2	Quantitative Analysis for Environmetallomics 35
3.2.3	Metal Distribution and Mapping for Environmetallomics 37
3.2.4	Metal Speciation for Environmetallomics 39
3.2.5	Metalloprotein Analysis 41
3.3	The Application of Environmetallomics in Environmental Science and Ecotoxicological Science and the Perspectives 43
	Acknowledgments 44
	List of Abbreviations 44
	References 45

4   Agrometallomics 49

Xuefei Mao, Xue Li, Tengpeng Liu, and Yajie Lei

4.1	The Concept of Agrometallomics 49
4.1.1	Introduction 49
4.1.2	Agrometallomics and its Concept 51
4.2	Analytical Techniques in Agrometallomics 52
4.2.1	Sensitivity and Multi-elemental Analysis in Agrometallomics 52
4.2.1.1	Mass Spectrometry in Agrometallomics 52
4.2.1.2	Atomic Spectrometry for Agrometallomics 119
4.2.2	Elemental Speciation and State Analysis in Agrometallomics 121
4.2.2.1	Chromatographic Hyphenation for Atomic Spectrometry or Mass Spectrometry 121
4.2.2.2	Synchrotron Radiation Analysis 122
4.2.2.3	Energy Spectroscopy Based on X-ray 123
4.2.3	Spatial Distribution and Micro-analysis Techniques in Agrometallomics 124
4.2.3.1	Laser Ablation Inductively Coupled Plasma Mass Spectrometry 124
4.2.3.2	Electrothermal Vaporization Hyphenation Technique 125
4.2.3.3	Laser-induced Breakdown Spectroscopy 125
4.2.3.4	Single-Cell and Micro-particle Analysis 126
4.3	Application and Perspectives of Agrometallomics in Agricultural Science and Food Science 127
4.3.1	Agricultural Plants and Fungi and Derived Food 127
4.3.2	Agricultural Animal and Derived Food 131

4.3.2.1	Application of Sensitivity and Multielemental Analysis in Agricultural Animals *132*	
4.3.2.2	Application of Elemental Speciation and State Analysis in Agricultural Animals *135*	
4.3.2.3	Application of Spatial Distribution and Micro-analysis in Agricultural Animals *137*	
4.3.3	Soil, Water, and Fertilizer for Agriculture *139*	
	List of Abbreviations *143*	
	References *144*	
5	**Metrometallomics** *153*	
	Liuxing Feng	
5.1	The Concept of Metrometallomics *153*	
5.2	The Analytical Techniques in Metrometallomics *154*	
5.2.1	Analytical Techniques of Protein Quantification in Metrometallomics *154*	
5.2.2	Analytical Techniques of Quantitative *In Situ* Analysis in Metrometallomics *155*	
5.3	The Application of Metrometallomics in Life Science and the Perspectives *159*	
5.3.1	Absolute Quantification of Metalloproteins in Metrometallomics *159*	
5.3.1.1	Naturally Present Elements (P, S, Se, Metals) *159*	
5.3.1.2	Elemental Labeling *160*	
5.3.1.3	Directly Protein Tagging (I, Hg, Chelate Complexes) *162*	
5.3.1.4	Immunological Tagging *164*	
5.3.1.5	Direct Quantification of Proteins by LA-ICP-MS *165*	
5.3.1.6	Calibration for Metalloprotein Quantification by ICP-MS *167*	
5.3.1.7	Perspectives of Absolute Quantification of Metalloproteins *168*	
5.3.2	Calibration Strategies of Quantitative *In Situ* Analysis in Metrometallomics *168*	
5.3.2.1	Internal Standardization *168*	
5.3.2.2	External Calibration *174*	
5.3.2.3	Calibration by Isotope Dilution *182*	
5.3.2.4	Perspectives of Quantitative In Situ Analysis in Metrometallomics *185*	
	Acknowledgments *186*	
	References *186*	
6	**Medimetallomics and Clinimetallomics** *193*	
	Guohuan Yin, Ang Li, Meiduo Zhao, Jing Xu, Jing Ma, Bo Zhou, Huiling Li, and Qun Xu	
6.1	The Concept of Medimetallomics and Clinimetallomics *193*	
6.1.1	Medimetallomics *195*	
6.1.2	Clinimetallomics *195*	
6.2	The Analytical Techniques in Medimetallomics and Clinimetallomics *195*	

6.2.1	Total Analysis of Clinical Elements	*196*
6.2.1.1	Atomic Spectroscopy Detection Technology	*196*
6.2.1.2	Mass Detection Technology	*197*
6.2.1.3	Electrochemical Analysis	*198*
6.2.1.4	Neutron Activation Analysis	*198*
6.2.2	Clinical Element Morphology and Valence Analysis Technology	*199*
6.2.2.1	Atomic Spectroscopy Detection Technology	*200*
6.2.2.2	Mass Spectrometry Detection Technology	*201*
6.2.3	Summary and Outlook	*203*
6.3	The Application of Medimetallomics and Clinimetallomics in Medical and Clinical Science and the Perspectives	*204*
6.3.1	Medimetallomics	*204*
6.3.1.1	Global or National Medimetallomics Research	*204*
6.3.1.2	Standardized Protocol for Medimetallomics Research	*205*
6.3.1.3	The Application of Medimetallomics Results	*207*
6.3.1.4	Next Steps and Opportunities for Medimetallomics	*208*
6.3.2	Clinimetallomics	*208*
6.3.2.1	Diseases Associated with Trace Elements	*208*
6.3.2.2	Toxic-Element-Related Diseases	*221*
6.3.2.3	Combined Toxicity of Multiple Heavy Metal Mixtures	*223*
6.3.2.4	Genetic Diseases Associated with Metallomics	*224*
6.3.2.5	Application of Metallomics in Disease Treatment	*224*
6.3.2.6	Perspectives	*226*
	List of Abbreviations	*226*
	References	*229*
7	**Matermetallomics** *237*	
	Qing Li, Zhao-Qing Cai, Wen-Xin Cui, and Zheng Wang	
7.1	The Concept of Matermetallomics	*237*
7.1.1	Introduction	*237*
7.1.2	Metallic Elements as Dopant	*239*
7.1.3	Metallic Elements as Impurities	*241*
7.1.4	Metallic Elements as Crosslinkers	*242*
7.2	The Analytical Techniques in Matermetallomics	*243*
7.2.1	Element Imaging Analysis	*243*
7.2.1.1	Laser Ablation Inductively Coupled Plasma Mass Spectrometry (LA-ICP-MS)	*246*
7.2.1.2	Laser-Induced Breakdown Spectroscopy (LIBS)	*247*
7.2.1.3	Secondary Ion Mass Spectrometry (SIMS)	*247*
7.2.1.4	TEM/X-EDS	*248*
7.2.1.5	Synchrotron Radiation X-Ray Fluorescence Spectrometry (SR-XRF)	*249*
7.2.2	Quantitative and Qualitative Analysis	*250*
7.2.2.1	Inductively Coupled Plasma Atomic Emission Spectrometry (ICP-AES)	*250*

7.2.2.2	Inductively Coupled Plasma Mass Spectrometry (ICP-MS)	*251*
7.2.2.3	X-Ray Fluorescence (XRF)	*252*
7.2.2.4	GD Optical Emission Spectroscopy (GD-OES) and GD Mass Spectrometry (GD-MS)	*253*
7.2.3	Metal Speciation Analysis	*254*
7.2.3.1	Raman Spectroscopy	*254*
7.2.3.2	X-Ray Photo Electron Spectroscopy (XPS)	*255*
7.2.4	Techniques Providing Depth Information	*255*
7.3	The Application of Matermetallomics in Material Science and the Perspectives	*256*
7.3.1	Matermetallomics in Semiconductor Materials	*256*
7.3.2	Matermetallomics in Artificial Crystal Materials	*257*
	Acknowledgments	*258*
	List of Abbreviations	*258*
	References	*260*

8 Archaeometallomics *265*
Li Li, Yue Zhou, Sijia Li, Lingtong Yan, Heyang Sun, and Xiangqian Feng

8.1	The Concept of Archaeometallomics	*265*
8.2	The Analytical Techniques in Archaeometallomics	*266*
8.2.1	Neutron Activation Analysis (NAA)	*266*
8.2.2	X-Ray Fluorescence Analysis (XRF)	*266*
8.2.3	Laser Ablation Inductively Coupled Plasma Mass Spectrometry (LA-ICP-MS)	*267*
8.2.4	Laser-induced Breakdown Spectroscopy (LIBS)	*267*
8.2.5	Atomic Absorption Spectroscopy (AAS)	*267*
8.2.6	X-Ray Absorption Fine Structure Spectroscopy (XAFS)	*267*
8.2.7	X-Ray Diffraction (XRD)	*268*
8.2.8	Neutron Diffraction	*268*
8.3	The Application of Archaeometallomics in Archaeological Science	*269*
8.3.1	The Application of Archaeometallomics in Ancient Ceramics	*269*
8.3.1.1	Archaeometallomics in Studying the Origin and Dating of Ancient Ceramics	*269*
8.3.1.2	Archaeometallomics in Studying the Color Mechanism and Firing Technology of Ancient Ceramics	*271*
8.3.2	The Application of Archaeometallomics in Metal Cultural Relics	*272*
8.3.2.1	Archaeometallomics in Studying the Origin of Metal Cultural Relics	*273*
8.3.2.2	Archaeometallomics in Studying the Manufacturing Technology of Metal Cultural Relics	*274*
8.3.2.3	Archaeometallomics in Studying the Corrosion of Metal Cultural Relics	*275*
8.3.3	The Application of Archaeometallomics in Ancient Painting	*275*

8.3.3.1	Archaeometallomics in Studying the Aging Mechanism of Painting Cultural Relics *276*	
8.3.3.2	Archaeometallomics in Studying the Authenticity Identification of Painting Cultural Relics *278*	
8.4	Summary and Perspectives *279*	
	Acknowledgments *279*	
	List of Abbreviations *279*	
	References *280*	

9 Metallomics in Toxicology *285*
Ruixia Wang, Ming Gao, Jiahao Chen, Mengying Qi, and Ming Xu

9.1	Metallomic Research on the Toxicology of Metals *285*
9.2	Recent Progresses in Understanding the Health Effects of Heavy Metals *287*
9.2.1	Mercury, Oxidative Stress, and Cell Death *287*
9.2.2	Arsenic and Lung Cancer *291*
9.2.3	Epigenetic Effects of Cadmium *292*
9.2.4	Nephrotoxicity of Uranium in Drinking Water *294*
9.3	Knowledge Gaps, Challenges, and Perspectives *297*
	Acknowledgments *298*
	List of Abbreviations *298*
	References *300*

10 Pathometallomics: Taking Neurodegenerative Disease as an Example *311*
Xiubo Du, Xuexia Li, and Qiong Liu

10.1	Introduction to Pathometallomics *311*
10.1.1	The Concept and Scope of Pathometallomics *311*
10.1.2	Brief Introduction to Methodologies for Pathometallomics *312*
10.2	Application of Pathometallomics in Neurodegenerative Diseases *314*
10.2.1	Pathometallomics in Alzheimer's Disease *314*
10.2.1.1	Dysregulation of Metal Homeostasis in AD *315*
10.2.1.2	Metal-Associated Dysfunction in AD *320*
10.2.1.3	Application of Metallomics in the Prognosis of AD *321*
10.2.1.4	Metal Chelators as AD Therapeutics *322*
10.2.2	Pathometallomics in Parkinson's Disease *324*
10.2.2.1	Dysregulation of Metal Homeostasis in PD *324*
10.2.2.2	Application of Metallomics in the Prognosis of PD *332*
10.2.2.3	Application of Metallodrugs and Metalloproteins in the Treatment of PD *333*
10.2.3	Pathometallomics in Amyotrophic Lateral Sclerosis *333*
10.2.3.1	Dysregulation of Metal Homeostasis in ALS *333*
10.2.3.2	Metal-Associated Dysfunction in ALS *334*
10.2.4	Pathometallomics in Autism Spectrum Disorder *336*

10.3	The Perspectives of Pathometallomics *338*	
	Acknowledgments *338*	
	List of Abbreviations *338*	
	References *340*	

11 Oncometallomics: Metallomics in Cancer Studies *349*
Xin Wang, Chao Li, and Yu-Feng Li

11.1	Introduction to Oncometallomics *349*	
11.2	The Application of Oncometallomics in Cancer Studies *351*	
11.2.1	The Application of Oncometallomics in Cancer Diagnosis *351*	
11.2.1.1	Prostate Cancer *351*	
11.2.1.2	Breast Cancer *351*	
11.2.1.3	Lung Cancer *352*	
11.2.1.4	Gastric Cancer *352*	
11.2.1.5	Colorectal Cancer *353*	
11.2.1.6	Esophageal Cancer *353*	
11.2.1.7	Liver Cancer *353*	
11.2.1.8	Ovarian Cancer *354*	
11.2.1.9	Cervical Cancer *354*	
11.2.1.10	Thyroid Cancer *354*	
11.2.2	The Application of Oncometallomics in Cancer Treatment *354*	
11.3	The Metallome that Involved in the Occurrence and Development of Cancer *355*	
11.4	Conclusions and Perspectives *356*	
	Acknowledgments *358*	
	List of Abbreviations *358*	
	References *358*	

12 Bio-elementomics *363*
Dongfang Wang, Jing Wu, Bing Cao, Lailai Yan, Qianqian Zhao, Tiebing Liu, and Jingyu Wang

12.1	Introduction *363*	
12.1.1	The Concept of Bio-elementomics *363*	
12.1.2	The Development History of Bio-elementomics *363*	
12.1.3	Research Scope *364*	
12.2	Basic Laws of Bio-elementomics *364*	
12.2.1	Review of Bio-elementomics *364*	
12.2.2	Organizational Selectivity of Bio-elements *365*	
12.2.3	Specific Correlation of Bio-elements *365*	
12.2.4	Orderliness of Bio-elements *366*	
12.2.5	Diversity of Bio-elements *366*	
12.2.6	Biological Fractionation *366*	
12.2.7	The Correlation Between the Bio-elementomes and Other "Omes" *367*	
12.3	Rare-Earth Elementome *367*	
12.3.1	Association of Rare-Earth Elements and Related Diseases *367*	

12.3.2	The Mechanism Studies of the Hormesis Effect of REEs Based on the Bio-elementomics	*369*
12.3.3	Beneficial Rebalancing Hypothesis for Hormesis Effect	*370*
12.4	Limitations of Bio-elementomics	*371*
12.4.1	Statistically Higher Level of Some Elements in the Patient's Body	*371*
12.4.2	Environment-independent Biomarkers	*372*
12.4.3	Trace Elements in Immortalized Lymphocytes	*372*
12.5	Perspectives	*373*
12.5.1	Speciation Analysis of Elements	*373*
12.5.2	Bio-elements and Their Interactions with Proteins, Genes, and Small Molecules	*373*
12.5.3	Research Based on the Hormesis "Beneficial Rebalancing" Hypothesis	*374*
12.5.4	Multi-element Analysis of Immortalized Lymphocytes	*374*
12.5.5	Analysis of Bio-elements in Single Cell	*374*
	References	*374*

13 Methodology and Tools for Metallomics *377*
Xiaowen Yan, Ming Xu, and Qiuquan Wang

13.1	Brief Description of Metallomics	*377*
13.1.1	Why Do Research on Biometals?	*377*
13.1.2	What's the Goal of Metallomics?	*378*
13.1.3	How to Perform a Metallomic Study?	*379*
13.2	Methodologic Strategy for Metallomic Research	*380*
13.2.1	In Vivo	*381*
13.2.2	Ex Vivo	*381*
13.2.3	In Vitro	*382*
13.2.4	In Silico	*383*
13.3	Tools for Metallomics	*383*
13.3.1	Tools for Quantitative Metallomics	*383*
13.3.2	Tools for Qualitative Metallomics	*384*
13.3.3	Imaging Tools for Metallomics	*386*
13.4	Concluding Remarks	*387*
	List of Abbreviations	*387*
	References	*388*

14 ICP-MS for Single-Cell Analysis in Metallomics *391*
Man He, Beibei Chen, and Bin Hu

14.1	Introduction	*391*
14.2	ICP-MS Instrumental Optimization for Single-Cell Analysis	*392*
14.2.1	Sample Introduction System	*392*
14.2.1.1	Pneumatic Nebulization	*392*
14.2.1.2	Laser Ablation	*399*
14.2.2	Mass Analyzer and Detector	*400*

14.3	Microfluidic Platform for Single-Cells Analysis	*401*
14.3.1	Droplet-Encapsulation-Based Single-Cell Separation	*403*
14.3.2	Hydrodynamic-Capture-Based Single-Cell Separation	*407*
14.3.3	Magnetic-Separation-Based Single-Cell Capture	*408*
14.4	ICP-MS-Based Single-Cells Analysis in Metallomics	*408*
14.4.1	Endogenous Elements in Single Cells	*409*
14.4.2	Exogenous Metal Exposure to Single Cells	*409*
14.4.3	Nanoparticles Uptake by Single Cells	*415*
14.4.4	Metal-containing Drugs Uptake by Single Cells	*416*
14.4.5	Biomolecular Quantification at Single-Cell Level	*417*
14.4.6	Other Applications	*418*
14.5	Summary and Perspectives	*419*
	List of Abbreviations	*420*
	References	*420*

15 Novel ICP-MS-based Techniques for Metallomics *429*
Panpan Chang and Meng Wang

15.1	Introduction	*429*
15.2	ICP-MS: A Powerful Method in Metallomics	*430*
15.2.1	Solution Introduction System and Plasma Source	*430*
15.2.2	Time-of-flight Mass Analyzer	*431*
15.2.3	Laser Ablation Systems	*432*
15.3	Recent Advances in ICP-MS-based Metallomics	*433*
15.3.1	Single-particle Analysis	*433*
15.3.2	Single-cell Analysis	*435*
15.3.3	Spatial Metallomics	*441*
15.4	Conclusions	*442*
	Acknowledgment	*443*
	List of Abbreviations	*443*
	References	*444*

16 Machine Learning for Data Mining in Metallomics *449*
Wei Wang and Xin Wang

16.1	Data Mining Methods in Metallomics	*450*
16.1.1	Data Preprocessing	*450*
16.1.1.1	Smoothing Process	*450*
16.1.1.2	Normalization	*450*
16.1.1.3	Fourier Transform	*451*
16.1.1.4	Wavelet Transform	*451*
16.1.1.5	Convolution Operation	*452*
16.1.2	Data Dimensionality Reduction	*452*
16.1.2.1	Principal Component Analysis	*453*
16.1.2.2	Independent Component Analysis	*453*
16.1.2.3	Multidimensional Scaling	*454*
16.1.2.4	Local Preserving Projection	*454*

16.1.2.5 T-Stochastic Neighbor Embedding *454*
16.1.3 Sample Set Division *455*
16.1.3.1 Random Sampling *455*
16.1.3.2 Kennard–Stone Sampling *455*
16.1.3.3 Sample Set Partitioning Based on Joint x–y Distances *455*
16.1.3.4 Cross-Validation *456*
16.1.3.5 Leave-One-Out Cross Validation *457*
16.1.4 Predictive Model Building Method *457*
16.1.4.1 Partial Least Squares Regression *457*
16.1.4.2 Support Vector Machine *457*
16.1.4.3 Decision Tree *458*
16.1.4.4 K-means Clustering *458*
16.1.4.5 Deep Learning *459*
16.1.5 Model Evaluation *461*
16.1.5.1 Evaluation Index of the Quantitative Model *461*
16.1.5.2 Evaluation Indicators of the Qualitative Model *462*
16.2 Application of Machine Learning for Data Mining in Metallomics *463*
16.2.1 Applications in Medical Science *463*
16.2.2 Applications in Agricultural Science *466*
16.2.3 Applications in the Environmental Science *467*
References *469*

Index *471*

Foreword

Metallomics is the systematic study of the interactions and functional connections of metallome with genes, proteins, metabolites, and other biomolecules within organisms. It aims to provide a global understanding of metal uptake, trafficking, role, and excretion in biological systems, and potentially to be able to predict all of these in silico using bioinformatics and even artificial intelligence.

With the development for over two decades, metallomics has converged with different research fields to form diverse branches of metallomics, such as nanometallomics, environmentallomics, agrometallomics, clinimetallomics, and metrometallomics, to name a few. Besides, with the development of high throughput analytical techniques, methodologies in metallomics were also formed, such as targeted/non-targeted metallomics, spatial metallomics, temporal metallomics, spatiotemporal metallomics, metalloproteomics, and metametallomics.

In 2010, Prof. Zhifang Chai, Prof. Yuxi Gao, and I co-edited a book entitled *Nuclear Analytical Techniques for Metallomics and Metalloproteomics*, which is the first monographic book to address the key aspects of the application of nuclear analytical techniques to metallomics study. After that, several books on metallomics edited by distinguished scientists were published, such as *Metallomics and the Cell; Metallomics: A Primer of Integrated Biometal Science; Metallomics: Recent Analytical Techniques and Applications; and Metallomics: The Science of Biometals*, which greatly enhanced the development of metallomics.

It my pleasure to introduce this new book of *Applied Metallomics*. This book is a collection of 16 chapters contributed by eminent experts in China. In Chapters 2–12, different branches of metallomics were introduced, while in Chapters 13–15, dedicated methodology and tools were presented. Specifically in Chapter 16, data mining with machine learning and even artificial intelligence for metallomics was introduced, which is highly desired for metallomics studies.

The co-editors of this book, Prof. Yu-Feng Li and Prof. Hongzhe Sun, are co-founders of the "CAS-HKU Joint Laboratory of Metallomics on Health and Environment," who have been dedicated to metallomics and metalloproteomics studies for over 20 years. With their hard work and close collaboration, I am happy to see this book being published, which is suitable for not only scientists already involved in metallomics research but also for young scientists, graduate students, and others who are interested in understanding the vital roles of metals/metalloids in life and many other fields.

Chunying Chen
Member of Chinese Academy of Sciences
National Center for Nanoscience and Technology
Beijing, China

Preface

Since its coinage by Haraguchi in 2002 as "integrated biometal science" and the following definition by International Union of Pure and Applied Chemistry (IUPAC) in 2010, metallomics has seen great development not only in methodology but also in its application in different research fields. Considering the important role of metallome, it is regarded as one of the five pillars of life along with genome, proteome, lipidome, and glycome. With the development of high throughput analytical methods, it is believed that metallomics will evolve to elementomics, which will cover all the elements in the periodic table. However, here in this book, we will keep our eyes mainly on metal/metalloids rather than all the elements.

Over the last two decades, metallomics has converged with nanoscience, environmental science, agricultural science, medical science, toxicological science, metrological science, radiological science, archaeology, and materials science to form different branches of metallomics, such as nanometallomics, environmentallomics, agrometallomics, clinimetallomics, metrometallomics, radiometallomics, archaeometallomics, and matermetallomics. These developments show the diversity and prosperity of metallomics in different research fields.

High throughput analytical techniques are commercially available, such as inductively coupled plasma optical emission spectroscopy (ICP-OES) and inductively coupled plasma optical mass spectrometry (ICP-MS). Besides, large research infrastructures, such as synchrotron radiation facilities, free electron lasers, spallation neutron sources, proton sources, and heavy ion sources, can provide cutting-edge techniques as superb tools for metallomics. Based on these high throughput analytical techniques, targeted/non-targeted metallomics, spatial metallomics, temporal metallomics, spatiotemporal metallomics, metalloproteomics, and metametallomics are formed as dedicated tools for studying the quantification, distribution, speciation, and function of metallomes.

The aim of this book is to present the latest development of metallomics in both its application in different research fields and its methodology development. There are 16 chapters in this book. Chapter 1 gives a brief introduction to metallomics and presents the key issues and challenges in metallomics. From Chapters 2 to 12, different branches of metallomics, such as nanometallomics, environmetallomics, agrometallomics, metrometallomics, medimetallomics, clinimetallomics, matermetallomics, archaeometallomics, metallomics in toxicology, pathometallomics, oncometallomics, and bio-elementomics, are presented, which cover both the analytical techniques and their application in the specific research field. This is also why we call this book *Applied Metallomics*, since we want to show that metallomics is helpful in many fields. From Chapters 13 to 15, the methodology and tools in metallomics, especially ICP-MS, are presented. Last but not least, Chapter 16 presents machine learning in metallomics for data mining. We would like to thank all the contributing authors for their hard work and efforts in forming this book, since we are hoping to present the latest development and application of metallomics in diverse research fields. Besides, through the publication of this book, we are also hoping to attract more and more distinguished researchers, young scientists, graduate students, and other people who are interested in understanding the vital roles of metals/metalloids in life and many other fields to join the metallomics family.

Both Chinese Academy of Sciences (CAS) and the University of Hong Kong (HKU) pay high attention to metallomics study, and a "CAS-HKU Joint Laboratory of Metallomics on Health and Environment" was established in 2019. Besides, the "Beijing Metallomics Facility" was established in 2020, which includes 12 universities, institutes and even hospitals resides in Beijing area. Furthermore, the "National Consortium for Excellence in Metallomics" was also initiated in 2020 to cover all the interested parties around China. We would like to thank all the scientists and colleagues with whom we worked together on metallomics these years.

We would like to acknowledge the financial support from the National Natural Science Foundation of China (NSFC, Nos. 11975247, 11475496, 11205168) and the Ministry of Science and Technology (2022YFA1207300, 2020YFA0710700, 2016YFA0201600). We also thank the support of the joint research scheme from Research Grants Council (RGC) and NSFC (No. 20931160430).

We would like to express our sincere appreciation to Felix Bloeck, Kelly Labrum, Shwathi Srinivasan, Tanya Domeier, Farhath Fathima, and Elizabeth Rose Amaladoss from Wiley. Without their patience and kind help, the publication of this book would be impossible.

March 1, 2024

Yu-Feng Li, Beijing, China
Hongzhe Sun, Hong Kong, China

1

Introduction

Yu-Feng Li[1] and Hongzhe Sun[2]

[1]Chinese Academy of Sciences, Institute of High Energy Physics, CAS-HKU Joint Laboratory of Metallomics on Health and Environment, & CAS Key Laboratory for Biomedical Effects of Nanomaterials and Nanosafety, & Beijing Metallomics Facility, & National Consortium for Excellence in Metallomics, No. 19B, Yuquan Road, Beijing 100049, China
[2]The University of Hong Kong, Department of Chemistry, CAS-HKU Joint Laboratory of Metallomics on Health and Environment, Hong Kong, SAR, China

1.1 A Brief Introduction to Metallomics

Metals and metalloids play vital role in life and even death, acting as catalysis, structural components, signal transmitters or electron donors [1–3], etc. However, metals or metalloids do not work alone; they may interact with each other through synergism or antagonism in biological systems. The systematic understanding on the roles of all the metals and metalloids in biological systems is called "metallomics," which was proposed by Haraguchi in 2002 as "integrated biometal science," [4, 5] while the term "metallome" was first used by Williams in 2001, referring to "an element distribution or a free element content in a cellular compartment, cell or organism."[6] In 2010, metallomics was defined by the International Union of Pure and Applied Chemistry (IUPAC) as a research field focusing on the systematic study of the interactions and functional connections of metallome with genes, proteins, metabolites, and other biomolecules within organisms [7]. Metallomics aims to provide a global understanding of the metal uptake, trafficking, role, and excretion in biological systems, and potentially to be able to predict all of these in silico using bioinformatics [8, 9].

Metallomics aims to understand the biological systems and life process at the atomic level [10], which is similar to ionomics in this regard [11–13]. However, metallomics also works at molecular levels like the study on metalloproteins [14] and speciation study [15, 16]. Therefore, it covers both the atomic and molecular levels of metallome. The metallome has been proposed as a fifth pillar of elemental – vis-à-vis molecular-building blocks alongside the genome, proteome, lipidome, and glycome in life [17] as illustrated in Figure 1.1. With the technical advances, it is expected that metallomics will be evolved to elementomics [18–21], which will cover all the elements in the periodic table.

Applied Metallomics: From Life Sciences to Environmental Sciences, First Edition.
Edited by Yu-Feng Li and Hongzhe Sun.
© 2024 WILEY-VCH GmbH. Published 2024 by WILEY-VCH GmbH.

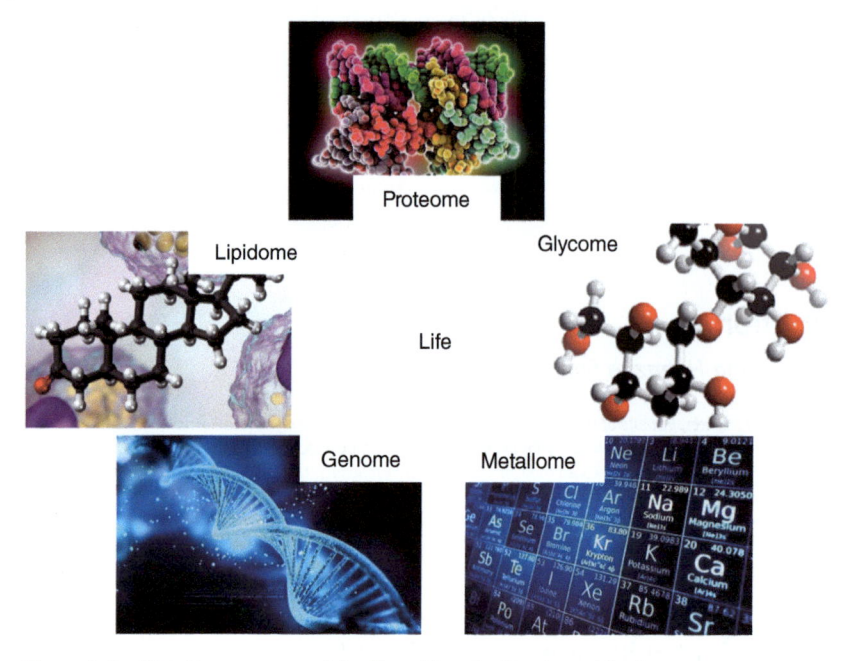

Figure 1.1 Metallome as one of the five pillars in life alongside the genome, proteome, lipidome, and glycome. Source: abhijith3747/Adobe Stock; Yu-Feng Li (Author).

Metallomics is considered both a basic science to understand the chemical structures and biological functions of metallome and an applied science in convergence with many research fields [5, 17]. Metallomics has gained increasing attention among scientists working outside the biological science, such as nanoscience, environmental science, agricultural science, medical science, food science, geoscience, toxicological science, materials science, and metrological science [8–10, 15, 22–28]. The convergence of metallomics with these research field has led to new branches of metallomics such as nanometallomics [22, 29–31], environmentallomics [32–34], agrometallomics [35], metrometallomics [36], clinimetallomics [37], radiometallomics [38], archaeometallomics [39], and matermetallomics [40]. A special issue called "Atomic spectroscopy for metallomics" published in *Atomic Spectroscopy* covered some of these topics [10].

Dedicated analytical techniques are required for the characterization of metallome and their interactions with genome, proteome, metabolome, and other biomolecules in the biological systems. Commercially available instruments like inductively coupled plasma optical emission spectroscopy (ICP-OES) and inductively coupled plasma optical mass spectrometry (ICP-MS) have been widely applied in metallomics for the quantification of metallome [15, 24–26]. On the other hand, there are scientific instruments which are commercially less available but accessible to scientists, i.e. large research infrastructures (LRIs). LRIs are built to solve the strategic, basic, and forward-looking scientific and technological problems in economic and social development [41].

Particle accelerators are such kinds of LRIs, which produce beams of fast-moving, electrically charged atomic or subatomic particles, such as electrons, positrons, protons, or ions [42]. They are used for particle physics and nuclear physics like the structure of nuclei, the nature of nuclear forces, and the properties of nuclei not found in nature [43]. They are also used for radioisotope production, industrial radiography, radiation therapy, sterilization of biological materials, and radiocarbon dating [44, 45], etc. Furthermore, the particle accelerator-derived electrons, positrons, protons, or ions beams have many advantages compared to the commercially available ones, which make them superb tools in many research fields including metallomics.

1.2 Key Issues and Challenges in Metallomics

It is desired to know which and how many elements are there in a biological system, and this includes the high-throughput quantification of elements, their species, and also the metal/metalloid-binding molecules like metalloproteins.

For the quantification of metallome, it is required to know first which metallome is to be quantified. There are over 20 elements like C, N, O, F, Na, Mg, Si, P, S, Ca, Fe, Cu, Zn, and I that have been proved to be essential elements for humans [46]. Plants also require B, while some bacteria need W, La, and Ce instead of Ca [47]. In some marine diatoms, Cd-containing carbonic anhydrase was found [48]. It is also desired to quantify the presence of non-essential elements in the biological system since new technologies and manufacturing practices lead to new industrial emissions, releases, or discharges to the environment. For example, the electronic industry uses rare-earth elements and many transition elements in the periodic table. With nearly 60 million tonnes of electronic waste generated in 2021 alone, it is also desired to know to which extent that humans, plants, animals, and microorganisms are exposed to these elements [49]. Another example is the increased production and application of nanomaterials these years, which inevitably lead to the environmental burden of nanomaterials themselves or their degraded products [31, 50]. Therefore, it is desired to quantify as many elements as possible or even the whole elements in the periodic table in a high-throughput way; however, this indeed is one of the big challenges for the quantification of metallome since huge concentration difference of elements exists in the biological systems [21].

The speciation of elements is also required since different species of the same element may have different biological effects. One example is Cr. Cr^{3+} is positive on glucose on lipid metabolism, while Cr^{6+} in the form of chromate is a carcinogen[46]. Another example is Hg. It is known that mercury selenide (HgSe) is stable and generally not bioavailable, while methylmercury (MeHg) is highly toxic and bioaccumulative in biological systems [51]. For the speciation in metallomics, the challenge is to identify the species in situ. This is also required for the quantification of metalloproteins and other non-protein complexes of metal ions in metallomics study.

Besides knowing the concentration and speciation of metallome in a whole biological system, it is also desired to know how many of them exist in a particular location, i.e. the distribution of metallome. Seeing is believing. This includes knowing the two-dimensional (2D) and three-dimensional (3D) distribution of metallome in a biological system, which is called spatial metallomics [52–54]. Spatial metallomics is the study on the distribution of metallome at the subcellular, cellular, tissue, and whole-body levels including human-sized objects or even larger ones [55], requiring the spatial resolution at the nanometer, micrometer, millimeter, centimeter, and even larger ones [56].

The challenge for mapping the metallome in 2D and 3D is to have a tunable spatial resolution, which will greatly facilitate the study on the distribution of metallome in a biological system through a coarse scan first, following by a fine scan of the samples [57]. For 3D mapping, high-speed data acquisition is highly desired, while non-destructive analysis is highly preferred.

The 2D and 3D spatial distribution of chemical species is also required in the metallomics study. This includes the ex situ and in situ study, while the in situ chemical speciation is always desired. The study on the distribution of metals/metalloids in metalloproteins and other non-proteins complexes, i.e. their 3D structure can also be included in spatial metallomics.

The temporal change of metallome in a biological system forms series of live pictures of metallome in a biological system, i.e. the movie of metallome. The study on this can be called temporal metallomics. Furthermore, the study on the spatial distribution of metallome with a temporal resolution in a biological system can be called the spatiotemporal metallomics [52]. This shows the dynamic changes of metallome in a biological system.

For both the temporal and spatiotemporal metallomics, a challenge lies in the in situ monitoring of the dynamic changes of metallome in the biological system, especially at the single-cell level, which is called single-cell metallomics [17, 52]. Besides, considering the individual variations in mixed communities of organisms like cells, metametallomics was proposed to cover this [17, 26]. Temporal and spatiotemporal metallomics also include the study on the dynamic changes of metals in metalloproteins and other biomolecules.

1.3 About the Structure of this Book

As abovementioned, with the fast development in the last 20 years, metallomics has been converging with different research fields such as nanoscience, environmental science, agricultural science, biomedical science, toxicological science, materials science, archaeology, and analytical science to form new metallomics branches like nanometallomics, environmetallomics, agrometallomics, and medimetallomics. Dedicated tools through synchrotron radiation, neutrons, protons and other commercially available techniques can be applied in metallomics study. Besides, the methodology of metallomics including targeted or non-targeted metallomics, spatial metallomics (including single-particle/single-cell metallomics), temporal

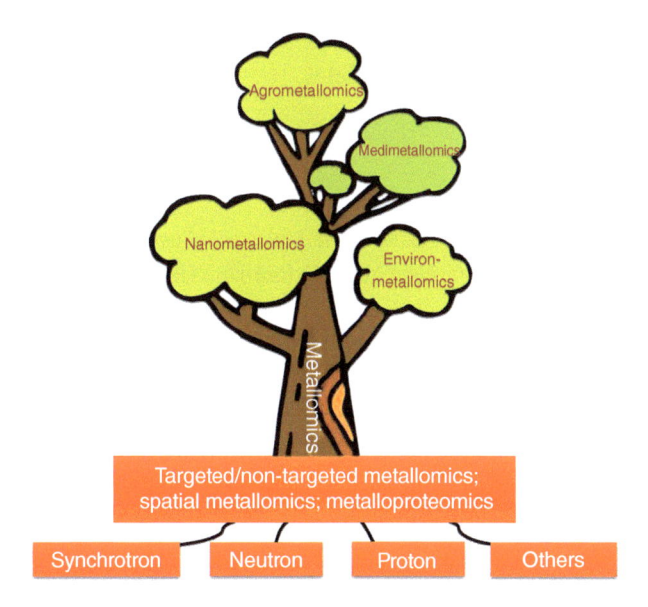

Figure 1.2 The different branches of metallomics and the methodologies and tools. Source: Yu-Feng Li.

metallomics, spatiotemporal metallomics, and metalloproteomics is also formed (Figure 1.2).

Therefore, from Chapters 2–11, these metallomics branches will be introduced: Chapter 2, Nanometallomics (coordinated by Yu-Feng Li); Chapter 3, Environmetallomics (coordinated by Ligang Hu); Chapter 4, Agrometallomics (coordinated by Xuefei Mao); Chapter 5, Metrometallomics (coordinated by Liuxing Feng); Chapter 6, Medimetallomics and clinimetallomics (coordinated by Qun Xu and Huiling Li); Chapter 7, Matermetallomics (coordinated by Zheng Wang); Chapter 8, Archaeometallomics (coordinated by Xiangqian Feng); Chapter 9, Metallomics in toxicology (coordinated by Ming Xu); Chapter 10, Pathometallomics: Taking neurodegenerative disease as an example (coordinated by Qiong Liu); and Chapter 11, Oncometallomics: Metallomics in cancer studies (coordinated by Xin Wang and Chao Li). Each chapter covers both the analytical techniques and the application of metallomics in a definite discipline or more. This is also why we call this book *Applied Metallomics*. Since metallomics is expected to be evolved to elementomics, an introduction and application of bio-elementomics is presented in Chapter 12, Bio-elementomics (coordinated by Jingyu Wang).

In addressing the key issues and challenges of metallomics, methodologies and tools, especially ICP-MS, machining learning and data mining are introduced in Chapter 13, Methodology and tools for metallomics (coordinated by Qiuquan Wang); Chapter 14, ICP-MS for single-cell analysis in metallomics (coordinated by Bin Hu); Chapter 15, Novel ICP-MS-based techniques for metallomics (coordinated by Meng Wang); and Chapter 16, Machine learning for data mining in metallomics (coordinated by Wei Wang).

Through all these efforts, this book is intended to reflect the latest development and application of metallomics in different research fields.

References

1 Fraústo da Silva, J.J.R. and Williams, R.J.P. (2001). *The Biological Chemistry of the Elements: The Inorganic Chemistry of Life*. New York: Oxford University Press.
2 Olwin, J.H. (1977). Metals in the life of man. *J. Anal. Toxicol.* 1: 245–251.
3 Han, N., Li, L.-G., Peng, X.-C. et al. (2022). Ferroptosis triggered by dihydroartemisinin facilitates chlorin e6 induced photodynamic therapy against lung cancer through inhibiting GPX4 and enhancing ROS. *Eur. J. Pharmacol.* 919: 174797.
4 Haraguchi, H. and Matsuura, H. (2003). Chemical speciation for metallomics. Presented at *Proceedings of International Symposium on Bio-Trace Elements 2002 (BITRE 2002)* Wako.
5 Haraguchi, H. (2004). Metallomics as integrated biometal science. *J. Anal. At. Spectrom.* 19: 5–14.
6 Williams, R.J.P. (2001). Chemical selection of elements by cells. *Coord. Chem. Rev.* 216–217: 583–595.
7 Lobinski, R., Becker, J.S., Haraguchi, H., and Sarkar, B. (2010). Metallomics: guidelines for terminology and critical evaluation of analytical chemistry approaches (IUPAC technical report). *Pure Appl. Chem.* 82: 493–504.
8 Li, Y.-F., Sun, H., Chen, C., and Chai, Z. (2016). *Metallomics*. Beijing: Science Press.
9 Mounicou, S., Szpunar, J., and Lobinski, R. (2009). Metallomics: the concept and methodology. *Chem. Soc. Rev.* 38: 1119–1138.
10 Li, Y.-F. and Sun, H. (2021). Metallomics in multidisciplinary research and the analytical advances. *At. Spectrosc.* 42: 227–230.
11 Baxter, I.R., Vitek, O., Lahner, B. et al. (2008). The leaf ionome as a multivariable system to detect a plant's physiological status. *Proc. Natl. Acad. Sci.* 105: 12081–12086.
12 Salt, D.E., Baxter, I., and Lahner, B. (2008). Ionomics and the study of the plant ionome. *Annu. Rev. Plant Biol.* 59: 709–733.
13 Zhang, Y., Xu, Y., and Zheng, L. (2020). Disease ionomics: understanding the role of ions in complex disease. *Int. J. Mol. Sci.* 21: 8646.
14 Zhou, Y., Li, H., and Sun, H. (2022). Metalloproteomics for biomedical research: methodology and applications. *Annu. Rev. Biochem.* 91: https://doi.org/10.1146/annurev-biochem-040320-104628.
15 Michalke, B. (2016). *Metallomics: Analytical Techniques and Speciation Methods*. Berlin: Wiley.
16 Chai, Z., Mao, X., Hu, Z. et al. (2002). Overview of the methodology of nuclear analytical techniques for speciation studies of trace elements in the biological and environmental sciences. *Anal. Bioanal.Chem.* 372: 407–411.

17 Maret, W. (2022). The quintessence of metallomics: a harbinger of a different life science based on the periodic table of the bioelements. *Metallomics*.

18 Li, Y.-F., Chen, C., Qu, Y. et al. (2008). Metallomics, elementomics, and analytical techniques. *Pure Appl. Chem.* 80: 2577–2594.

19 Xiong, Y., Ouyang, L., Liu, Y. et al. (2006). One of the most important parts for bio-elementomics: specific correlation study of bio-elements in a given tissue. *J. Chin. Mass Spectrom. Soc.* 27: 35–36.

20 Jakubowski, N. and Hieftje, G.M. (2008). The broadening scope of JAAS. *J. Anal. At. Spectrom.* 23: 13–14.

21 Liu, J., Peng, L., Wang, Q. et al. (2022). Simultaneous quantification of 70 elements in biofluids within 5 min using inductively coupled plasma mass spectrometry to reveal elementomic phenotypes of healthy Chinese adults. *Talanta* 250: 123720.

22 Chen, C., Chai, Z., and Gao, Y. (2010). *Nuclear Analytical Techniques for Metallomics and Metalloproteomics*. Cambridge: RSC publishing.

23 Banci, L. (2013). *Metallomics and the Cell*. Amsterdam: Springer Netherlands.

24 Maret, W. (2016). *Metallomics: A Primer of Integrated Biometal Science*. London: Imperial College Press.

25 Ogra, Y. and Hirata, T. (2017). *Metallomics: Recent Analytical Techniques and Applications*. Tokyo: Springer Japan.

26 Arruda, M.A.Z. (2018). *Metallomics: The Science of Biometals*. Cham: Springer.

27 Sablok, G. (2020). *Plant Metallomics and Functional Omics: A System-wide Perspective*. Berlin: Springer.

28 Sun, H. and Chai, Z.-F. (2010). Metallomics: An integrated science for metals in biology and medicine. *Annu. Rep. "A" (Inorg. Chem.)* 106: 20–38.

29 Benetti, F., Bregoli, L., Olivato, I., and Sabbioni, E. (2014). Effects of metal(loid)-based nanomaterials on essential element homeostasis: the central role of nanometallomics for nanotoxicology. *Metallomics* 6: 729–747.

30 Li, Y.-F., Gao, Y., Chai, Z., and Chen, C. (2014). Nanometallomics: an emerging field studying the biological effects of metal-related nanomaterials. *Metallomics* 6: 220–232.

31 Wang, L., Zhao, J., Cui, L. et al. (2021). Comparative nanometallomics as a new tool for nanosafety evaluation. *Metallomics* 13: mfab013.

32 López-Barea, J. and Gómez-Ariza, J.L. (2006). Environmental proteomics and metallomics. *Proteomics* 6: S51–S62.

33 Hu, L., He, B., and Jiang, G. (2016). *Metallomics* (ed. Y.-F. Li, H. Sun, C. Chen, and Z. Chai). Beijing: Science Press.

34 Chen, B., Hu, L., He, B. et al. (2020). Environmetallomics: systematically investigating metals in environmentally relevant media. *Trends Anal. Chem.* 126: 115875.

35 Li, X., Liu, T., Chang, C. et al. (2021). Analytical methodologies for agrometallomics: a critical review. *J. Agric. Food. Chem.* 69: 6100–6118.

36 Pan, M., Zang, Y., Zhou, X. et al. (2021). Inductively coupled plasma mass spectrometry for metrometallomics: the study of quantitative metalloproteins. *At. Spectrosc.* 42: 262–270.

37 Song, X., Li, H., Ma, C. et al. (2021). Clinimetallomics: arsenic speciation in urine of arsenism patients by HPLC-ICP-MS. *At. Spectrosc.* 42: 278–281.

38 Liang, Y., Liu, Y., Li, H. et al. (2021). Advances of synchrotron radiation-based radiometallomics for the study of uranium. *At. Spectrosc.* 42: 254–261.

39 Li, L., Yan, L., Sun, H. et al. (2021). Archaeometallomics as a tool for studying ancient ceramics. *At. Spectrosc.* 42: 247–253.

40 Li, Q., Cai, Z., Fang, Y., and Wang, Z. (2021). Matermetallomics: concept and analytical methodology. *At. Spectrosc.* 42: 238–246.

41 Chen, H. (2011). *Large Research Infrastructures Development in China: A Roadmap to 2050*. Berlin, Heidelberg: Springer.

42 Gao, J. (2021). *Physics and Design for Accelerators of High Energy Particle Colliders*. Shanghai: Shanghai Jiao Tong University Press.

43 Yuan, C.-Z. and Karliner, M. (2021). Cornucopia of antineutrons and hyperons from a super j/ψ factory for next-generation nuclear and particle physics high-precision experiments. *Phys. Rev. Lett.* 127: 012003.

44 Oda, H., Akiyama, M., Masuda, T., and Nakamura, T. (2007). Radiocarbon dating of an ancient Japanese document "minamoto no yoritomo sodehan migyosho" by accelerator mass spectrometry. *J. Radioanal. Nucl. Chem.* 272: 439–442.

45 Housley, R.A. (1990). Radiocarbon dating by accelerator mass spectrometry (AMS): an introduction. *Geol. Today* 6: 60–62.

46 Chai, Z. and Zhu, H. (ed.) (1994). *Introduction to Trace Element Chemistry*. Beijing: Atomic Energy Press.

47 Pol, A., Barends, T.R.M., Dietl, A. et al. (2014). Rare earth metals are essential for methanotrophic life in volcanic mudpots. *Environ. Microbiol.* 16: 255–264.

48 Lane, T.W. and Morel, F.M.M. (2000). A biological function for cadmium in marine diatoms. *Proc. Natl. Acad. Sci. U.S.A.* 97: 4627–4631.

49 King, A.H. (2019). Our elemental footprint. *Nat. Mater.* 18: 408–409.

50 Chen, C., Li, Y.-F., Qu, Y. et al. (2013). Advanced nuclear analytical and related techniques for the growing challenges in nanotoxicology. *Chem. Soc. Rev.* 42: 8266–8303.

51 Lin, X., Zhao, J., Zhang, W. et al. (2021). Towards screening the neurotoxicity of chemicals through feces after exposure to methylmercury or inorganic mercury in rats: a combined study using microbiome, metabolomics and metallomics. *J. Hazard Mater.* 409: 124923.

52 Fan, Y., Cui, L., Wang, L. et al. (2022). Large research infrastructure based spatial metallomics and single-cell/single-particle metallomics. *Chin. J. Inorg. Anal. Chem.* 12: https://doi.org/10.3969/j.issn.2095-1035.2022.3904.3021.

53 Zhao, Y., Chen, C., Feng, W. et al. (2022). Professor Zhifang Chai: scientific contributions and achievements. *Chin. Chem. Lett.* 33: 3297–3302.

54 Chang, P.-P., Zheng, L.-N., Wang, B. et al. (2022). ICP-MS-based methodology in metallomics: towards single particle analysis, single cell analysis, and spatial metallomics. *At. Spectrosc.* 43: https://doi.org/10.46770/as.42022.46108.

55 Grüner, F., Blumendorf, F., Schmutzler, O. et al. (2018). Localising functionalised gold-nanoparticles in murine spinal cords by X-ray fluorescence imaging and

background-reduction through spatial filtering for human-sized objects. *Sci. Rep.* 8: 16561.
56 Wang, L., Yan, L., Liu, J. et al. (2018). Quantification of nanomaterial/nanomedicine trafficking in vivo. *Anal. Chem.* 90: 589–614.
57 Chaurand, P., Liu, W., Borschneck, D. et al. (2018). Multi-scale x-ray computed tomography to detect and localize metal-based nanomaterials in lung tissues of in vivo exposed mice. *Sci. Rep.* 8: 4408.

2

Nanometallomics*

Hongxin Xie, Liming Wang, Jiating Zhao, Yuxi Gao, Bai Li, and Yu-Feng Li

Chinese Academy of Sciences, Institute of High Energy Physics, CAS-HKU Joint Laboratory of Metallomics on Health and Environment, & CAS Key Laboratory for Biomedical Effects of Nanomaterials and Nanosafety, & Beijing Metallomics Facility, & National Consortium for Excellence in Metallomics, No. 19B, Yuquan Road, Beijing 100049, China

2.1 The Concept of Nanometallomics

Nanomaterials have at least one dimension within the nanometer size (1–100 nm) possessing magnetic, optical, acoustic, thermal, and electrical properties. This makes them attractive for a wide range of applications, such as optics, chemicals, ceramics, biology, and medicine. On the other hand, safety concerns on nanomaterials were also raised since they may enter biological systems directly during manufacturing processes or indirectly via the food chains [1]. For example, it was showed that nanoparticles could largely escape alveolar macrophage surveillance and gain access to the pulmonary interstitium with greater inflammatory effect than larger particles [2, 3]. A fast translocation of nanoparticles from pulmonary and gastrointestinal epithelium into the systemic circulation through animal studies was also observed [4]. As such, a new research field called nanotoxicology was formed to study the biological effects of nanomaterials [5–7]. Nanosafety evaluation is paramount since it is necessary not only for human health protection and environmental integrity but also as a cornerstone for industrial and regulatory bodies [8].

Metallomics aims to systematically study the metallome and the interactions and functional connections of metallome with genes, proteins, metabolites, and other biomolecules in biological systems [9–14]. Metallomics can act as a useful tool for the nanosafety evaluation in nanotoxicology, which is termed as nanometallomics. Nanometallomics is considered as a branch of metallomics, which focuses on the systematic study of absorption, distribution, metabolism, and excretion (ADME) behavior and their interactions with genes, proteins, and other biomolecules of metal-related nanomaterials in biological systems [15, 16]. Metal-related

---

* This chapter has been modified to feature as Reviews: (1) Li, Y.-F., Gao, Y., Chai, Z., Chen, C. (2014). Nanometallomics: an emerging field studying the biological effects of metal-related nanomaterials. *Metallomics* 6: 220–232; (2) Wang, L., Zhao, J., Cui, L., et al. (2021). Comparative nanometallomics as a new tool for nanosafety evaluation. *Metallomics* 13: mfab013.

Applied Metallomics: From Life Sciences to Environmental Sciences, First Edition.
Edited by Yu-Feng Li and Hongzhe Sun.
© 2024 WILEY-VCH GmbH. Published 2024 by WILEY-VCH GmbH.

nanomaterials include metallic and metal-containing nanomaterials, which are frequently used as catalysts, sensors, or probes due to their unique crystalline forms and superior mechanical, electrical, magnetic, optical, and catalytic properties [17]. Besides, the study on the impacts to elemental homeostasis after nanomaterials exposure is also considered as nanometallomics study [16, 18], where the nanomaterials include the non-metallic nanomaterials like silicon- and carbon-based ones. In this way, all the nanomaterials, either metallic or non-metallic ones, can be included in nanometallomics study. Therefore, nanometallomics systematically study the ADME behavior and their interactions with genes, proteins, and other biomolecules of metal-related nanomaterials in biological systems. It also focuses on the impacts to elemental and/or metallo-biomolecules homeostasis in the biological system when exposed to nanomaterials. Comparative metallomics is the monitoring of the changes of the metallome as a function of time and exposure to external stimuli [19]. Similarly, comparative nanometallomics is the monitoring of the changes of the metallome as a function of time and exposure to nanomaterials, which has been proposed as a tool for screening or predicting the toxicity of nanomaterials [16].

2.2 The Analytical Techniques in Nanometallomics

2.2.1 The Analytical Techniques for Size Characterization of Nanomaterials in Biological System

The size characterization of nanomaterials in biological system is important since the toxicity of nanomaterials may change simply due to the size change. For example, aggregation can reduce the overall surface area-to-volume ratio and accessible active sites of nanomaterials, thereby reducing their toxicity [20, 21].

Electron microscopes like scanning electron microscope (SEM) and transmission electron microscope (TEM) are gold standards for the size characterization of nanomaterials [22]. However, they are more suitable for characterizing pristine ones, which can only look locally rather than globally. Besides, it is often difficult to characterize low concentrations of nanomaterials in biological system [23–25] and to monitor the size changes of nanomaterials after entering the biological system [26]. On the other hand, chromatography-based techniques, mass-spectrometry-based techniques, and laser, X-rays, and neutron-beam-based techniques have been applied to monitor the size of nanomaterials [27].

2.2.1.1 Chromatography-based Techniques for Size Characterization

Chromatography is an effective method to separate components in a complex mixture [28]. Several chromatography-based techniques including liquid chromatography (LC), hydrodynamic chromatography (HDC), field-flow fractionation (FFF), and electrophoresis have been applied in the size characterization of nanomaterials.

LC can be divided into adsorption chromatography, partition chromatography, ion chromatography, size-exclusion chromatography (SEC), bonded-phase chromatography, affinity chromatography, etc., among which SEC is the most widely used in the separation and size characterization of nanomaterials. For example, SEC has been applied to separate and characterize nanomaterials such as Au nanoclusters below 100 nm [29], quantum dots (QDs) [30], single-walled carbon nanotubes (CNTs) [31], and silica NPs [32].

HDC employs non-porous rigid particles to fill the chromatographic column or capillary with different diameters than the chromatographic column. It has the advantages of simple operation, wide particle size range (5–1200 nm), and separation of partially aggregated nanomaterials [33].

FFF can characterize nanomaterials with sizes ranging from 1 nm to 100 µm for fast and efficient separation in liquid medium [34, 35]. It can be divided into several techniques based on the type of the external force employed, among which asymmetric flow-field flow fractionation (AF4) is the most widely applied [36–39]. AF4 in combination with inductively coupled plasma mass spectrometry (ICP-MS) was applied to characterize the size of AgNPs in nutraceuticals with comparable results using TEM [40].

Electrophoresis refers to the phenomenon that charged particles migrate in the opposite direction of their charge under the action of an electric field. Gel electrophoresis (GE) has been applied to separate 5, 15, and 20 nm spherical gold nanoclusters, and to separate a crude suspension of gold spheres, plates, and long rods [41]. Capillary electrophoresis (CE) is a well-established separation technique with advantages of high resolution, speed, and efficiency. CE can be used to identify nanomaterials in different matrices, providing a powerful analytical tool for the isolation and characterization of individual nanomaterials coupled to proteins [42]. It has been connected with UV-Vis and ICP-MS to characterize Au, Ag, and Au@Ag core–shell nanospheres with various particle sizes and morphologies [43].

2.2.1.2 Mass-spectrometry-based Techniques for Size Characterization

ICP-MS is a versatile tool for multielemental quantification, which has been applied as a detector for size characterization of nanomaterials coupled with other techniques like chromatography. It has also been developed to monitor the size distribution of nanomaterials through time-resolved analysis (TRA) mode, which is called single-particle ICP-MS (SP-ICP-MS) [44–48]. For example, the size distribution of gold nanoparticles (AuNPs) in shellfish (clams and oysters) was studied using SP-ICP-MS, which found that it was 35–55 and 30–65 nm in clams and oysters, respectively [49]. Besides size distribution, SP-ICP-MS could also detect the number concentration and mass concentration of AuNPs in soil [50].

Laser ablation (LA)-ICP-MS was also applied for the size characterization of nanomaterials. For example, it was used the size characterization of AuNPs (13, 34, and 47 nm) in water samples in combination with thin-layer chromatography [51].

2.2.1.3 Laser, X-rays, and Neutron-beam-based Techniques for Size Characterization

Dynamic light scattering (DLS) and nanoparticle tracking analysis (NTA) have been widely used for size characterization of nanomaterials from micrometer to nanometer scale [52–54]. They are based on the scattering by nanomaterials after excited by laser. DLS is the most frequently applied size characterization technique [55].

X-rays and neutron beam can also be used as an excitation source for size characterization of nanomaterials. X-rays provided by the synchrotron radiation source are several orders of magnitude higher than the laboratory X-ray tubes in brightness [56].

Synchrotron radiation-based small-angle X-ray scattering (SAXS) is the study of the scattering phenomenon of a sample in a small angle range near the original X-ray beam, which has been used for the size distribution in the region of less than 20 nm of nanomaterials in complex matrices [57].

Similarly, small-angle neutron scattering (SANS) was also applied for size characterization [58, 59]. Due to the extremely strong penetrating ability of neutron beams, SANS can perform in situ, real-time, and non-destructive detection of the internal depth of materials under extreme environments. For example, the size of silica-filled silicone rubber aggregates was found to be about 143 nm, and there were approximately 24–97 silica particles in each aggregate [60].

X-ray fluorescence (XRF) with nano-sized X-ray can be used for the size characterization of nanomaterials. For example, the size distribution of cobalt nanoparticle (CoNP) was found to be approximately 96 ± 42 nm through XRF in *Caenorhabditis elegans*, which was consistent with the results got through DLS [61]. The size characterization of nanomaterials using XRF will highly depend on the beam size of X-rays. Besides, XRF tomography can reconstruct one or more virtual cross sections of the element distribution with depth information, which could be used to construct the 3D structure of nanomaterials [62].

2.2.2 The Analytical Techniques for Quantification of Nanomaterials and Metallome in Biological System

The high-throughput quantification of nanomaterials, especially metallic nanomaterials and metallome in biological system, can be achieved through ICP-MS, neutron activation analysis (NAA), XRF, etc.

ICP-MS is the most applied technique for the quantification of nanomaterials and metallome in biological system with the detection limit down to parts per trillion (ppt) level [63]. ICP-MS can detect most elements in biological systems, but sulfur, phosphorus, and halogens are not efficiently ionized by the ICP owing to their high ionization energies. This can be solved by either using high-resolution (sector field double-focusing) mass spectrometer or quadrupole mass analyzer with the collision/reaction cell techniques [64, 65].

NAA can simultaneously measure more than 30 elements in one sample with detection limit from 10^{-6}–10^{-13} g [66]. The major advantage of NAA is that it is nearly free of any matrix interference effects since it does not need sample digestion or dissolution [67]. However, the radioactive nuclides obtained after neutron activation require professional operation [13], which hinders the wide application of this technique in nanometallomics.

XRF is based on the detection of characteristic fluorescence after being excited by primary X-ray with sufficient energy [68]. If the exciting source is replaced by a proton beam or electron beam, it will be called proton-inducted X-ray emission spectrometry (PIXE) [69], or energy-dispersive spectrometer (EDS) [70], respectively. The absolute detection limit of XRF can be 10^{-12}–10^{-15} g and relative detection limit of several µg/g, even in ng/g levels [71, 72].

2.2.3 The Analytical Techniques for Studying the Distribution of Nanomaterials in Biological System

Studying the spatial distribution of nanomaterials in the biological system provides the information about their trafficking and deposition behavior.

One way to know the distribution of exposed nanomaterials is to quantify the nanomaterials in different tissues and organs with abovementioned quantification techniques after sacrificing the animals.

Direct mapping of the distribution of nanomaterials in biological tissues can be achieved through XRF, LA-ICP-MS, secondary ion mass spectrometry (SIMS), and isotopic tracing.

XRF, especially the synchrotron-radiation-based XRF (SR-XRF) with a beam size in micrometer or nanometer scale, can be used to study the distribution of nanomaterials in cells, tissue slices through raster scanning of the specimen to get the two-dimensional (2D) distribution [73, 74]. To perform depth analysis, the confocal arrangement, which consists of X-ray optics in the excitation as well as in the detection channel, has been set up with a depth resolution from 10 to 40 µm, and minimum detection limits are sub-mg/kg level [75–79]. XRF tomography measurement is also possible for performing three-dimensional (3D) elemental analysis by measuring a series of projected distributions under various angles and back projected using the appropriate mathematical algorithms [80, 81]. Besides, the reconstruction of a series of successive 2D distribution can also achieve the 3D distribution of samples [82].

LA-ICP-MS has become one powerful tool in the quantitative analysis of nanomaterials in situ owing to the very high sensitivity of ICP-MS and direct laser sampling to obtain more information from samples. The spatial resolution of LA-ICP-MS is about 10 µm with a detection limit of sub mg/kg level [83].

SIMS can be used for practically all elements of the periodic table (only the noble gases are difficult to measure because they don't ionize easily) with a detection limit of ng/kg and spatial resolution down to 50 nm, which is called nanoSIMS [70, 84, 85]. During a SIMS measurement, the sample is slowly sputtered (eroded) away and depth profiles can be obtained, which is the 3D measurements [86].

Non-destructive detection of the distribution of nanomaterials can be achieved through isotopic tracing. The isotopes can be radioactive or stable ones. For radioisotope tracing, the radioisotopes are introduced to the nanomaterials, while the emitted α, β, or γ beams by radioisotopes are monitored by different detectors including single-photon emission computed tomography (SPECT) and positron emission tomography (PET) to study the distribution of nanomaterials [7, 13]. The radioisotope tracing has the disadvantage of the generation of radioactive waste and requires strict radioactive protection. Besides, the radiation effects induced by radioisotopes also need to be considered. Stable isotopes can be used in experiments involving living organisms and even humans without concern for radiation hazards. Stable isotope tracing uses the isotopic ratio to monitor the distribution of nanomaterials, which requires a mass spectroscopy technique like isotope ratio mass spectroscopy or multi-collector mass spectrometry [3]. In comparison with radioisotope tracers, a greater amount of the stable isotope tracers with a highly enriched abundance are required.

2.2.4 The Analytical Techniques for Studying the Metabolism of Nanomaterials in Biological System

After entering the body, nanomaterials may dissolve, be decomposed, be oxidized, or be reduced, and this may lead to the electronic and/or ionic transfer either within the nanoparticles lattice or on release to culture medium.

The metabolism of nanomaterials can be studied by X-ray absorption spectroscopy (XAS), which probes the local geometric and electronic structure surrounding a specific element directly in solid, liquid, or gaseous state [13, 87]. XAS can probe the local structure around almost any specific element in the periodic table, which gives information on the number and chemical identities of near neighbors and the average interatomic distances without the requirement for preparation of crystalline samples [88]. For example, long-term retention of gold nanorods in liver and spleen was found in rats; however, XAS showed that no change of the oxidation state of gold was observed, suggesting that gold nanorods are inert after entering the body [89].

XAS requires the concentration of the interested element to be 10 µg/g or higher [90]. With the introduction of high-energy-resolution fluorescence-detected (HERFD) XAS, the detection limit has been lowered to below µg/g or even a few hundred of ng/g [91, 92]. Since HERFD XAS has a better spectral resolution than conventional XAS, the precision for the study of the quantitative transformation of nanomaterials can also be significantly improved [93].

When nanomaterials enter a physiological environment, they can rapidly adsorb a layer of proteins to form the protein corona [94–96]. Researchers found that size, shape, and surface characteristics of nanomaterials could affect protein adsorption and also the structure of the adsorbed proteins. This will affect the toxicity of nanomaterials and determine the route and efficiency of nanomaterial uptake. Circular dichroism (CD) spectroscopy is used for the examination of protein secondary structures, dynamics and folding, monitoring conformational changes when complexed with nanomaterials. Besides, synchrotron-radiation-based CD (SRCD) has higher signal-to-noise ratios than conventional CD, which enable the use of smaller amounts of protein samples and the ability to detect smaller changes of protein structure accurately [97]. On the other hand, quantitative proteomics can be applied to study the protein abundance in the protein corona of nanomaterials [98].

A summary on the selected analytical techniques in nanometallomics is shown in Figure 2.1.

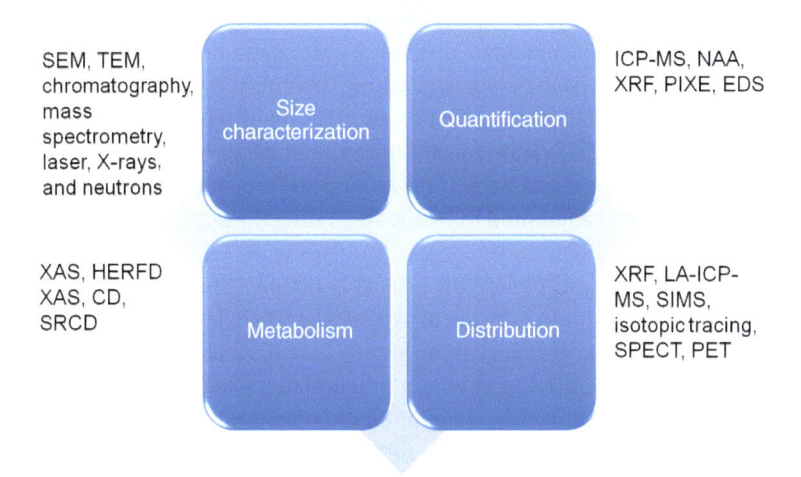

Figure 2.1 The selected analytical techniques in nanometallomics. SEM: scanning electron microscope, TEM: transmission electron microscope, ICP-MS: inductively coupled plasma mass spectrometry, NAA: neutron activation analysis, XRF: X-ray fluorescence analysis, PIXE: proton-inducted X-ray emission spectrometry, EDS: energy-dispersive spectrometer, LA-ICP-MS: laser ablation inductively coupled plasma mass spectrometry, SIMS: secondary ion mass spectrometry, SPECT: single-photon emission computed tomography, PET: positron emission tomography, XAS: X-ray absorption spectroscopy, HERFD XAS: high-energy-resolution fluorescence-detected XAS, CD: circular dichroism, SRCD: synchrotron-radiation-based CD. Source: Yu-Feng Li.

2.3 The Application of Nanometallomics in Nanotoxicology

2.3.1 Understanding the Size Changes, Uptake and Excretion, Distribution, and Metabolism of Nanomaterials in Biological Systems

The size changes of nanomaterials after entering the biological system are important. The size, quantity concentration, and mass concentration of AgNPs in the blood of burn patients were studied. It was noticed that protein corona was formed on the surface of AgNPs, but no AgNPs agglomerates >16 nm were found. Besides, sulfur in the blood played a major role in the degradation of AgNPs [99]. The size of silica NPs in serum was characterized utilizing AF4 and ICP-MS, which agreed with the results obtained through other techniques like NTA [100]. A comparison on the performance between HDC-ICP-MS and AF4-ICP-MS in the separation and characterization of AuNPs was made. It was found that AF4-ICP-MS can better separate and characterize AuNPs at 5, 20, 50, and 100 nm, while HDC-ICP-MS can simultaneously characterize degraded AuNPs [101]. SP-ICP-MS was applied to study the size of TiO_2 NPs in crab sticks after enzymatic hydrolysis, which found the size was between 53.8 and 62.1 nm, suggesting different types of TiO_2 NPs were applied as the additives for preparing the crab sticks [102].

As an in situ technique, SAXS was applied to characterize the size of ZnO and Fe_2O_3 NPs in 1% sodium carboxymethylcellulose solution, which found that the size was much larger than that obtained through TEM, suggesting aggregation occurred [103]. Since neutrons have strong penetrating capability, it can be applied for the in-depth analysis for nanomaterials. For example, lipid nanomaterials in polysorbate-coated sophorolipid-based nanostructured lipid carriers (NLCs) were studied by SANS. Large spherical nanostructures around 100 nm and smaller ellipsoidal micelles around 10 nm were found in NLCs, suggesting the agglomeration of lipid nanomaterials [104].

The uptake of nanomaterials in biological systems is mainly through ingestion and inhalation. A comparison on the uptake of copper nanoparticles (CuNPs) and copper ions found that both were enriched in the kidney 24 hours after oral exposure in mice; however, CuNPs was found to maintain much higher level in the kidney than copper ions after 72 hours, suggesting that the elimination rate of CuNPs is slow [105]. Similarly, another study found that cobalt nanoparticles could be absorbed by earthworm and were retained in the body of earthworm for eight weeks, with less than 20% of the absorbed cobalt nanoparticles being excreted [106].

Inhalation is another pathway of exposure. A comparison on the uptake of different sizes (15 and 80 nm) of ^{192}Ir nanoparticles in adult rats through isotopic tracing found that they were cleared via airways into the gastrointestinal tract and feces one week after inhalation [107]. Besides, it was found that the 15 nm nanoparticles were easier to translocate to secondary organs like liver, spleen, heart, and brain than the 80 nm ones, suggesting the size-dependent behavior of nanoparticles after inhalation.

As foreign objects, nanomaterials will be eliminated through the primary excretory organs/systems. It was found that the particles with sizes over 100 nm will be caught by the reticuloendothelial systems (RES), while particles with sizes below 5 nm can be removed by the kidneys [108]. The formation of protein corona on the surface of nanomaterials was found to reduce their cytotoxicity [94–96]. The coating of nanomaterials was also found to influence their uptake and excretion. For example, a higher clearance was found when quantum dots (QDs) were conjugated to bovine serum albumin than those coated with mercaptoundecanoic acid and cross-linked with lysine [109].

Studying the whole body distribution of nanomaterials may help find the target tissues after nanomaterials exposure. The copper levels were significantly higher in liver, kidneys, olfactory bulb, and blood after instillation of copper nanoparticles for one week in mice than the control, suggesting that the nasal-inhaled copper particles can translocate to other organs and tissues and induce certain lesions [110]. For 20 and 120 nm ZnO exposure in healthy adult mice after oral intake, significant increase of Zn contents was found in the kidney, pancreas, and bone ($p < 0.05$), and slight increase in the liver and heart for the 20 nm ZnO exposure while significantly higher Zn in the bone for the 120 nm ZnO. The liver, spleen, heart, pancreas, and bone were believed to be the target organs for 20 and 120 nm ZnO oral exposure [111]. For silver nanoparticles, predominant accumulation was found in the spleen and liver at 24 hours after intravenous injection in Balb/c mice, while quantum dots (QD,

15.1 ± 7.6 nm) in mice were excreted via renal filtration shortly post-injection and accumulated in the liver [112]. It was found that the exposure to Cu nanoparticles resulted in an obvious elevation of Cu and K levels, and a change of bio-distribution of Cu in *C. elegans* [113]. Accumulation of Cu was distributed in the head and at a location 1/3 of the way up the body from the tail compared to the unexposed control. In contrast, a higher amount of Cu was detected in other portion of worm body, especially in its excretory cells and intestine when exposed to Cu^{2+}. All these studies suggested an extensive distribution of metal-related nanomaterials in the tissues of the RES.

A more detailed study on the distribution of nanomaterials in a specific tissue/organ or even a single cell is also desired. For example, titanium was found to accumulate mainly in the cerebral cortex, thalamus, and hippocampus, especially in the CA1 and CA3 regions of hippocampus 30 days after intranasal instillation. The significantly increased Ti contents in the hippocampus result in the obviously irregular arrangement and loss of neurons in the hippocampus, which indicated that the TiO_2 nanoparticles could enter the brain via the olfactory bulb [114, 115]. For CdSe-QDs, it was found that Cd was almost undetectable, whereas the Se was clearly detected in the cytoplasm of SKOV3 cancer cell and was not present in either the cytoplasm or the nucleus of the control sample [116]. The distribution of unpurified and purified single-walled (SW) and multi-walled (MW) carbon nanotubes (CNT) in macrophages was studied by monitoring the catalyst metal particle employed in most synthesis technique and finally remaining attached to or contained in nanotubes [117]. One or several iron-rich zone(s) was found inside of or close to the cell contours and colocalization of the highest Fe signal was found with the highest *P* signal, suggesting an interaction of SWCNT with the nuclear or perinuclear region.

The metabolism of nanomaterials will affect their biological effects since they may dissolve, be decomposed, be oxidized or reduced, and lead to the electronic and/or ionic transfer either within the nanoparticle lattice or on release to culture medium. A comparison of the toxicity to *Pseudokirchneriella. subcapitata* for nano ZnO, bulk ZnO, and $ZnCl_2$ found that the toxicity aroused solely from the dissolved zinc. An enhanced dissolution of iron oxide nanoparticles in the acidic condition of lysosomes or in a microenvironment containing ligands with a strong affinity was found [118, 119]. The liberation of free Cd^{2+} from CdSe QDs was found and the cytotoxicity of CdSe QDs correlated with the free Cd^{2+} concentration in air-oxidized and UV-exposed samples [120]. Besides, CdSe@ZnS core/shell QDs were degraded in *C. elegans* and Se^{2-} in the CdSe core was oxidized to Se^{4+} [121]. These results confirmed that QDs could be decomposed in the body. On the other hand, long-term retention of gold nanorods in liver and spleen did not induce changes in the oxidation states of gold, suggesting gold nanorods are inert in the body [89].

In all, nanometallomics with a suite of analytical techniques can help understand the size changes, uptake, excretion, distribution, and metabolism of nanomaterials in biological systems, which is important for nanosafety evaluation. The formation of protein corona and other biomolecules corona on the surface of nanomaterials may change their biological effects [1, 98, 122, 123].

2.3.2 Comparative Nanometallomics for Distinguishing Nanomaterials Exposure and Nanosafety Evaluation

Elemental homeostasis exists in a healthy biological system, while the introduction of nanomaterials will inevitably affect the elemental homeostasis. For example, in cells, it was found that CdSe QDs increased intracellular Ca concentration in hippocampal neurons [124], while there was a significant decrease of Mg and Zn contents in mouse fibroblasts Balb/3T3 when exposed to Co nanoparticles [125]. Chronic and acute exposure of rat primary cortical neurons to few-layer pristine graphene (GR) and monolayer graphene oxide (GO) flakes led to interrupted Ca homeostasis in both excitatory and inhibitory neurons [126]. GR could also disrupt intercellular Ca homeostasis and induce a marked increase in membrane cholesterol levels in primary astrocytes [127]. In plants, TiO_2 NPs reduced the bioaccumulation of As in rice seedlings [128], while nanoselenium (SeNPs) reduced the level of mercury (Hg) in garlic plant by preventing the entry of Hg through the formation of HgSe and HgSe NPs [129]. Silicon nanoparticles (SiNPs) significantly increased the concentrations of K, Mg, and Fe in rice seedlings and grains [130, 131]. Silica nanoparticles (SiO_2 NPs) were found to significantly reduce the Hg concentrations in the epidermis and pericycle of the roots and stems, especially the pericycle in soybean seedlings [132]. In animals, it was noticed that Cd/Se/Te QDs increased hepatic Cu, Zn, Mn, and Se levels, in association with higher CTR1, ZIP14, and ZIP8 expression levels in ICR mice [133]. AuNPs caused a decrease in Ca and Zn in liver, lung, heart, kidney tissues, and blood, an increase in Cu and Fe in liver, heart, kidney, and a decrease of Fe in lung by 50 nm AuNPs in male Wistar–Kyoto rats [134]. CNTs were found to disrupt Fe homeostasis and induce anemia of inflammation through inflammatory pathway [135]. In human beings, a 59-year-old woman who ingested 422 ml per day of colloidal silver (AgNPs) for eight months decreased Cu, Se, and caeruloplasmin levels in serum [136].

High-throughput nanosafety evaluation is always desired considering the fact of the emergence of novel nanomaterials at astonishing speed nowadays. Comparative nanometallomics has been proposed as a tool for distinguishing the exposure and nanosafety evaluation of nanomaterials [16]. The working scheme (Figure 2.2) is as follows: (i) The perturbance to elemental and biomolecular homeostasis after exposure to nanomaterials in whole or part (organelles, tissues, bioliquids, biosolids) of cells, plants, animals, and even human beings will be studied, and the nanosafety information will also be collected. (ii) The differences of metallome between nanomaterials-exposed samples and the controls will be distinguished, which can be called "metallome signature" through machine learning techniques [122, 137] and artificial intelligence. (iii) The metallome signature can be used to distinguish the nanomaterials exposure or be linked to their toxicological aspects to construct prediction model for the nanosafety evaluation of emerging nanomaterials.

This working scheme has the merit of high throughput since the analytical techniques like ICP-MS and XRF are high throughput and readily available. Besides, the more nanosafety data available, the more precise prediction will be achieved considering the nature of machine learning and artificial intelligence techniques

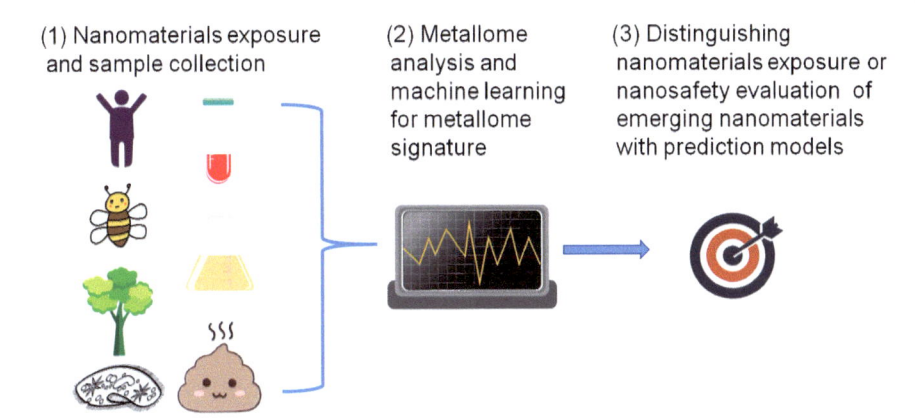

Figure 2.2 The working scheme for distinguishing nanomaterials exposure or nanosafety evaluation through comparative nanometallomics. Source: Yu-Feng Li.

[138]. A successful prediction of the disturbance of metabolic pathways induced by 33 nanomaterials through integrated machine learning models and metabolomics [139], suggesting the promising role of comparative nanometallomics. The functional composition of the protein corona and the cellular recognition of nanoparticles has also been predicted with machine learning [122].

2.4 Conclusions and Perspectives

Nanometallomics aims to systematically study the ADME behavior and their interactions with genes, proteins, and other biomolecules of metal-related nanomaterials in biological system. It also aims to study the impacts to elemental, especially metal(loid) and metallo-biomolecular homeostasis by nanomaterials.

Dedicated analytical techniques are applied in nanometallomics. In this way, nanometallomics helps understand the size changes, uptake and excretion, distribution and metabolism of nanomaterials after entering the biological systems. Besides, comparative nanometallomics can be used for distinguishing nanomaterials exposure or nanosafety evaluation, which is high throughput and will be precise considering the nature of machine learning techniques.

The advancement of high-throughput elemental analytical techniques is always desired in nanometallomics. With the development of laser techniques in recent years [140], laser-induced breakdown spectroscopy (LIBS) is an emerging multielemental technique which may be applied in nanometallomics. Besides, metallome-wide association study (MWAS) is also highly required in nanometallomics, similar to genome-wide association study (GWAS) in genomics [141]. This requires a universal and standard procedure for whole element analysis in the biological systems [142]. Furthermore, huge data volume will be obtained in nanometallomics study; therefore, the big data strategies are necessary [143].

Acknowledgments

This work was financially supported by Ministry of Science and Technology of China (2022YFA1207300) and Natural Science Foundation of China (11975247). We thank staffs at Beijing Synchrotron Radiation Facility (BSRF), Shanghai Synchrotron Radiation Facility (SSRF), National Synchrotron Radiation Laboratory (NSRL), and China Spallation Neutron Source (CSNS).

List of Abbreviations

ADME	absorption, distribution, metabolism, and excretion
CD	circular dichroism
CE	capillary electrophoresis
DLS	dynamic light scattering
EDS	energy dispersive spectrometer
FFF	field-flow fractionation
GE	gel electrophoresis
GWAS	Genome-Wide Association Study
HDC	hydrodynamic chromatography
HERFD XAS	high-energy-resolution fluorescence-detected XAS
ICP-MS	inductively coupled plasma mass spectrometry
LA-ICP-MS	laser ablation ICP-MS
LC	liquid chromatography
LIBS	laser-induced breakdown spectroscopy
MWAS	metallome-wide association study
NAA	neutron activation analysis
NTA	nanoparticle tracking analysis
PET	positron emission tomography
PIXE	proton-inducted X-ray emission spectrometry
SANS	small-angle neutron scattering
SAXS	small-angle X-ray scattering
SEC	size exclusion chromatography
SEM	scanning electron microscope
SIMS	secondary ion mass spectrometry
SPECT	single-photon emission computed tomography
SP-ICP-MS	single-particle ICP-MS
SRCD	synchrotron-radiation-based CD
TEM	transmission electron microscope
UV-Vis	ultraviolet–visible spectroscopy
XAS	X-ray absorption spectroscopy
XRF	X-ray fluorescence

References

1 Chen, C., Leong, D.T., and Lynch, I. (2020). Rethinking nanosafety: harnessing progress and driving innovation. *Small* 16: 2002503.
2 Oberdörster, G. (2000). Pulmonary effects of inhaled ultrafine particles. *Int. Arch. Occup. Environ. Health* 74: 1–8.
3 Oberdörster, G., Sharp, Z., Atudorei, V. et al. (2004). Translocation of inhaled ultrafine particles to the brain. *Inhalation Toxicol.* 16: 437–445.
4 Nemmar, A., Hoylaerts, M.F., Hoet, P.H.M. et al. (2002). Ultrafine particles affect experimental thrombosis in an in vivo hamster model. *Am. J. Respir. Crit. Care Med.* 166: 998–1004.
5 Zhao, Y. and Nalwa, H.S. (2006). *Nanotoxicology – Interactions of Nanomaterials with Biological Systems*. California: American Scientific Publishers.
6 Oberdörster, G., Oberdörster, E., and Oberdörster, J. (2005). Nanotoxicology: an emerging discipline evolving from studies of ultrafine particles. *Environ. Health Perspect.* 113: 823–839.
7 Chen, C., Li, Y.-F., Qu, Y. et al. (2013). Advanced nuclear analytical and related techniques for the growing challenges in nanotoxicology. *Chem. Soc. Rev.* 42, 21: 8266–8303.
8 Liu, S. and Xia, T. (2020). Continued efforts on nanomaterial-environmental health and safety is critical to maintain sustainable growth of nanoindustry. *Small* 16: 2000603.
9 Mounicou, S., Szpunar, J., and Lobinski, R. (2009). Metallomics: the concept and methodology. *Chem. Soc. Rev.* 38: 1119–1138.
10 Haraguchi, H. (2004). Metallomics as integrated biometal science. *J. Anal. At. Spectrom.* 19: 5–14.
11 Ge, R. and Sun, H. (2009). Metallomics: an integrated biometal science. *Sci. Sin. Chim.* 52: 2055–2070.
12 Li, Y.-F., Sun, H., Chen, C., and Chai, Z. (2016). *Metallomics*. Beijing: Science Press.
13 Chen, C., Chai, Z., and Gao, Y. (2010). *Nuclear Analytical Techniques for Metallomics and Metalloproteomics*. Cambridge: RSC Publishing.
14 Li, Y.-F., Gao, Y., Chen, C. et al. (2009). High throughput analytical techniques in metallomics and the perspectives. *Sci. Sin. Chim.* 39: 580–589.
15 Li, Y.-F., Gao, Y., Chai, Z., and Chen, C. (2014). Nanometallomics: an emerging field studying the biological effects of metal-related nanomaterials. *Metallomics* 6: 220–232.
16 Wang, L., Zhao, J., Cui, L. et al. (2021). Comparative nanometallomics as a new tool for nanosafety evaluation. *Metallomics* 13: mfab013.
17 Buzea, C., Pacheco, I.I., and Robbie, K. (2007). Nanomaterials and nanoparticles: sources and toxicity. *Biointerphases* 2: MR17-MR71.
18 Benetti, F., Bregoli, L., Olivato, I., and Sabbioni, E. (2014). Effects of metal(loid)-based nanomaterials on essential element homeostasis: the central role of nanometallomics for nanotoxicology. *Metallomics* 6: 729–747.

19 Szpunar, J. (2005). Advances in analytical methodology for bioinorganic speciation analysis: metallomics, metalloproteomics and heteroatom-tagged proteomics and metabolomics. *Analyst* 130: 442–465.
20 Abbas, Q., Yousaf, B., Amina, M.U. et al. (2020). Transformation pathways and fate of engineered nanoparticles (ENPs) in distinct interactive environmental compartments: a review. *Environ. Int.* 138: 105646.
21 Yung, M.M.N., Wong, S.W.Y., Kwok, K.W.H. et al. (2015). Salinity-dependent toxicities of zinc oxide nanoparticles to the marine diatom *Thalassiosira pseudonana*. *Aquat. Toxicol.* 165: 31–40.
22 Wang, L., Yan, L., Liu, J. et al. (2018). Quantification of nanomaterial/nanomedicine trafficking in vivo. *Anal. Chem.* 90: 589–614.
23 Zhou, X. and Liu, J. (2017). Research advances in separation and determination of metallic nanomaterials in the environment. *Chin. Sci. Bull.* 62: 2758–2769.
24 Lv, J. and Zhang, S. (2012). Methods for separation and analysis of nanomaterials in the environment. *Prog. Chem.* 24: 92–101.
25 Modena, M.M., Rühle, B., Burg, T.P., and Wuttke, S. (2019). Nanoparticle characterization: what to measure? *Adv. Mater.* 31: 1901556.
26 Hoo, C.M., Starostin, N., West, P., and Mecartney, M.L. (2008). A comparison of atomic force microscopy (AFM) and dynamic light scattering (DLS) methods to characterize nanoparticle size distributions. *J. Nanopart. Res.* 10: 89–96.
27 Xie, H., Wei, X., Zhao, J. et al. (2022). Size characterization of nanomaterials in environmental and biological matrices through non-electron microscopic techniques. *Sci. Total Environ.* 835: 155399.
28 Khatun, Z., Bhat, A., Sharma, S., and Sharma, A. (2016). Elucidating diversity of exosomes: biophysical and molecular characterization methods. *Nanomedicine* 11: 2359–2377.
29 Wilcoxon, J.P., Martin, J.E., and Provencio, P. (2000). Size distributions of gold nanoclusters studied by liquid chromatography. *Langmuir* 16: 9912–9920.
30 Krueger, K.M., Al-Somali, A.M., Falkner, J.C., and Colvin, V.L. (2005). Characterization of nanocrystalline CDSE by size exclusion chromatography. *Anal. Chem.* 77: 3511–3515.
31 Ziegler, K.J., Schmidt, D.J., Rauwald, U. et al. (2005). Length-dependent extraction of single-walled carbon nanotubes. *Nano Lett.* 5: 2355–2359.
32 Truillet, C., Lux, F., Tillement, O. et al. (2013). Coupling of HPLC with electrospray ionization mass spectrometry for studying the aging of ultrasmall multifunctional gadolinium-based silica nanoparticles. *Anal. Chem.* 85: 10440–10447.
33 Li, J.-J., Liu, P., and Geng, X.-D. (2009). Hydrodynamic chromatography and slalom chromatography and their applications. *Chin. J. Anal. Chem.* 37: 1082–1087.
34 Giddings, J.C. (1966). A new separation concept based on a coupling of concentration and flow nonuniformities. *Sep. Sci.* 1: 123–125.
35 Giddings, J.C., Yang, F.J., and Myers, M.N. (1976). Flow-field-flow fractionation: a versatile new separation method. *Science* 193: 1244–1245.

36 Gigault, J., Pettibone, J.M., Schmitt, C., and Hackley, V.A. (2014). Rational strategy for characterization of nanoscale particles by asymmetric-flow field flow fractionation: a tutorial. *Anal. Chim. Acta* 809: 9–24.

37 Liang, Q., Wu, D., Qiu, B., and Han, N. (2017). Present situation and development trends of asymmetrical flow field-flow fractionation. *Chin. J. Chromatogr.* 35: 918–926.

38 Laborda, F., Bolea, E., Ceprià, G. et al. (2016). Detection, characterization and quantification of inorganic engineered nanomaterials: a review of techniques and methodological approaches for the analysis of complex samples. *Anal. Chim. Acta* 904: 10–32.

39 Li, B., Chua, S.L., Chng, A.L. et al. (2020). An effective approach for size characterization and mass quantification of silica nanoparticles in coffee creamer by AF4-ICP-MS. *Anal. Bioanal.Chem.* 412: 5499–5512.

40 Ramos, K., Ramos, L., Cámara, C., and Gómez-Gómez, M.M. (2014). Characterization and quantification of silver nanoparticles in nutraceuticals and beverages by asymmetric flow field flow fractionation coupled with inductively coupled plasma mass spectrometry. *J. Chromatogr. A* 1371: 227–236.

41 Xu, X., Caswell, K.K., Tucker, E. et al. (2007). Size and shape separation of gold nanoparticles with preparative gel electrophoresis. *J. Chromatogr. A* 1167: 35–41.

42 Zarei, M. and Aalaie, J. (2019). Profiling of nanoparticle–protein interactions by electrophoresis techniques. *Anal. Bioanal.Chem.* 411: 79–96.

43 Liu, F.-K., Tsai, M.-H., Hsu, Y.-C., and Chu, T.-C. (2006). Analytical separation of Au/Ag core/shell nanoparticles by capillary electrophoresis. *J. Chromatogr. A* 1133: 340–346.

44 Degueldre, C. and Favarger, P.Y. (2004). Thorium colloid analysis by single particle inductively coupled plasma-mass spectrometry. *Talanta* 62: 1051–1054.

45 Degueldre, C., Favarger, P.Y., and Bitea, C. (2004). Zirconia colloid analysis by single particle inductively coupled plasma–mass spectrometry. *Anal. Chim. Acta* 518: 137–142.

46 Degueldre, C., Favarger, P.Y., and Wold, S. (2006). Gold colloid analysis by inductively coupled plasma-mass spectrometry in a single particle mode. *Anal. Chim. Acta* 555: 263–268.

47 Laborda, F., Bolea, E., and Jiménez-Lamana, J. (2014). Single particle inductively coupled plasma mass spectrometry: a powerful tool for nanoanalysis. *Anal. Chem.* 86: 2270–2278.

48 Krause, B., Meyer, T., Sieg, H. et al. (2018). Characterization of aluminum, aluminum oxide and titanium dioxide nanomaterials using a combination of methods for particle surface and size analysis. *RSC Adv.* 8: 14377–14388.

49 Zhou, Q., Liu, L., Liu, N. et al. (2020). Determination and characterization of metal nanoparticles in clams and oysters. *Ecotoxicol. Environ. Saf.* 198: 110670.

50 Gao, Y.-P., Yang, Y., Li, L. et al. (2020). Quantitative detection of gold nanoparticles in soil and sediment. *Anal. Chim. Acta* 1110: 72–81.

51 Yan, N., Zhu, Z., Jin, L. et al. (2015). Quantitative characterization of gold nanoparticles by coupling thin layer chromatography with laser ablation inductively coupled plasma mass spectrometry. *Anal. Chem.* 87: 6079–6087.

52 Caputo, F., Clogston, J., Calzolai, L. et al. (2019). Measuring particle size distribution of nanoparticle enabled medicinal products, the joint view of EUNCL and NCI-NCL. A step by step approach combining orthogonal measurements with increasing complexity. *J. Controlled Release* 299: 31–43.

53 Bhattacharjee, S. (2016). DLS and zeta potential – what they are and what they are not? *J. Controlled Release* 235: 337–351.

54 Lim, J., Yeap, S.P., Che, H.X., and Low, S.C. (2013). Characterization of magnetic nanoparticle by dynamic light scattering. *Nanoscale Res. Lett.* 8: 381.

55 Sotelo-Boyás, M.E., Correa-Pacheco, Z.N., Bautista-Baños, S., and Corona-Rangel, M.L. (2017). Physicochemical characterization of chitosan nanoparticles and nanocapsules incorporated with lime essential oil and their antibacterial activity against food-borne pathogens. *LWT* 77: 15–20.

56 Pan, D., Roessl, E., Schlomka, J.-P. et al. (2010). Computed tomography in color: Nanok-enhanced spectral CT molecular imaging. *Angew. Chem. Int. Ed.* 49: 9635–9639.

57 Pauw, B.R., Kastner, C., and Thunemann, A.F. (2017). Nanoparticle size distribution quantification: results of a small-angle X-ray scattering inter-laboratory comparison. *J. Appl. Crystallogr.* 50: 1280–1288.

58 Feigin, L.A. and Svergun, D.I. (1987). *Structure Analysis by Small-angle X-ray and Neutron Scattering*, 263–269. New York: Plenum Press.

59 Halamish, H.M., Trousil, J., Rak, D. et al. (2019). Self-assembly and nanostructure of poly(vinyl alcohol)-graft-poly(methyl methacrylate) amphiphilic nanoparticles. *J. Colloid Interface Sci.* 553: 512–523.

60 Liu, D., Chen, J., Song, L. et al. (2017). Parameterization of silica-filled silicone rubber morphology: a contrast variation sans and TEM study. *Polymer* 120: 155–163.

61 Cagno, S., Brede, D.A., Nuyts, G. et al. (2017). Combined computed nanotomography and nanoscopic X-ray fluorescence imaging of cobalt nanoparticles in *Caenorhabditis elegans*. *Anal. Chem.* 89: 11435–11442.

62 Pashkova, G.V., Smagunova, A.N., and Finkelshtein, A.L. (2018). X-ray fluorescence analysis of milk and dairy products: a review. *TrAC Trends Anal. Chem.* 106: 183–189.

63 Thompson, M. and Walsh, J.N. (1983). *Handbook of Inductively Coupled Plasma Spectrometry*. Glasgow: Blackie.

64 Feldmann, I., Jakubowski, N., Thomas, C., and Stuewer, D. (1999). Application of a hexapole collision and reaction cell in ICP-MS part II: analytical figures of merit and first applications. *Fresenius J. Anal. Chem.* 365: 422–428.

65 Koppenaal, D.W., Eiden, G.C., and Barinaga, C.J. (2004). Collision and reaction cells in atomic mass spectrometry: development, status, and applications. *J. Anal. At. Spectrom.* 19: 561–570.

66 Chai, Z., Sun, J., and Ma, S. (1992). *Neutron Activation Analysis in Environmental Sciences, Biological and Geological Sciences*. Beijing: Atomic Energy Press.

67 Chai, Z. and Zhu, H. (ed.) (1994). *Introduction to Trace Element Chemistry*. Beijing: Atomic Energy Press.

68 Jenkins, R. (1999). *X-ray Fluorescence Spectrometry*. New York: Wiley.

69 Johansson, E. (1989). PIXE: a novel technique for elemental analysis. *Endeavour* 13: 48–53.

70 Schaumlöffel, D., Hutchinson, R., Malherbe, J. et al. (2016). *Metallomics: Analytical Techniques and Speciation Methods* (ed. B. Michalke), 83–116. Weinheim: Wiley-VCH.

71 Gama, E.M., Nascentes, C.C., Matos, R.P. et al. (2017). A simple method for the multi-elemental analysis of beer using total reflection X-ray fluorescence. *Talanta* 174: 274–278.

72 Wang, L.L., Yu, H.S., Li, L.N. et al. (2016). The development of TXRF method and its application on the study of trace elements in water at SSRF. *Nucl. Instrum. Methods Phys. Res., Sect. B* 375: 49–55.

73 Paunesku, T., Vogt, S., Maser, J. et al. (2006). X-ray fluorescence microprobe imaging in biology and medicine. *J. Cell. Biochem.* 99: 1489–1502.

74 Iida, A. (1997). X-ray spectrometric applications of a synchrotron X-ray microbeam. *X-Ray Spectrom.* 26: 359–363.

75 Kanngießer, B., Malzer, W., and Reiche, I. (2003). A new 3D micro X-ray fluorescence analysis set-up – first archaeometric applications. *Nucl. Instrum. Methods Phys. Res. Sect. B.* 211: 259–264.

76 Muradin, A.K. (2000). Capillary optics and their use in X-ray analysis. *X-Ray Spectrom.* 29: 343–348.

77 Vincze, L., Vekemans, B., Brenker, F.E. et al. (2004). Three-dimensional trace element analysis by confocal X-ray microfluorescence imaging. *Anal. Chem.* 76: 6786–6791.

78 Kouichi Tsuji, K.N. (2007). Development of confocal 3D micro-XRF spectrometer with dual Cr-Mo excitation. *X-Ray Spectrom.* 36: 145–149.

79 Janssens, K., Proost, K., and Falkenberg, G. (2004). Confocal microscopic X-ray fluorescence at the HASYLAB microfocus beamline: characteristics and possibilities. *Spectrochim. Acta B.* 59: 1637–1645.

80 Vincze, L., Vekemans, B., Szaloki, I., et al. (2001). High resolution X-ray fluorescence micro-tomography on single sediment particles. Presented at *46th SPIE Annual Meeting International Symposium of Optical Science and Technology*, San Diego.

81 Gan, H., Gao, H., Zhu, H. et al. (2006). X-ray fluorescence tomography. *Laser Optoelectr. Prog.* 43: 56–64.

82 Xie, H., Tian, X., He, L. et al. (2023). Spatial metallomics reveals preferable accumulation of methylated selenium in a single seed of the hyperaccumulator *Cardamine violifolia*. *J. Agric. Food. Chem.* 71: 2658–2665.

83 Motelica-Heino, M., Le Coustumer, P., Thomassin, J.H. et al. (1998). Macro and microchemistry of trace metals in vitrified domestic wastes by laser ablation ICP-MS and scanning electron microprobe X-ray energy dispersive spectroscopy. *Talanta* 46: 407–422.

84 Briggs, B.Y.D., Brown, A., and Vickerman, J.C. (1988). Handbook of static secondary ion mass spectrometry (sims). *Anal. Chem.* 60: 1791–1799.

85 Moore, K.L., Lombi, E., Zhao, F.-J., and Grovenor, C.R.M. (2012). Elemental imaging at the nanoscale: nanosims and complementary techniques for element localisation in plants. *Anal. Bioanal.Chem.* 402: 3263–3273.

86 Chandra, S. (2004). 3d subcellular sims imaging in cryogenically prepared single cells. *Appl. Surf. Sci.* 231-232: 467–469.

87 Li, Y.-F., Chen, C., Li, B. et al. (2008). Mercury in human hair and blood samples from people living in Wanshan mercury mine area, Guizhou, China: an XAS study. *J. Inorg. Biochem.* 102: 500–506.

88 Mai, Z. (2013). *Synchrotron Radiation Light Source and Its Application*. Beijing: Science Press.

89 Wang, L., Li, Y.-F., Zhou, L. et al. (2010). Characterization of gold nanorods in vivo by integrated analytical techniques: their uptake, retention, and chemical forms. *Anal. Bioanal.Chem.* 396: 1105–1114.

90 Li, Y.-F., Wang, X., Wang, L. et al. (2010). Direct quantitative speciation of selenium in selenium-enriched yeast and yeast-based products by X-ray absorption spectroscopy confirmed by HPLC-ICP-MS. *J. Anal. At. Spectrom.* 25: 426–430.

91 Bauer, M. (2014). HERFD-XAS and valence-to-core-XES: new tools to push the limits in research with hard X-rays? *Phys. Chem. Chem. Phys.* 16: 13827–13837.

92 Xu, W., Du, Z., Liu, S. et al. (2018). Perspectives of XRF and XANES applications in cryospheric sciences using Chinese SR facilities. *Condens. Matter* 3: 29.

93 Proux, O., Lahera, E., Del Net, W. et al. (2017). High-energy resolution fluorescence detected X-ray absorption spectroscopy: a powerful new structural tool in environmental biogeochemistry sciences. *J. Environ. Qual.* 46: 1146–1157.

94 Walkey, C.D. and Chan, W.C.W. (2012). Understanding and controlling the interaction of nanomaterials with proteins in a physiological environment. *Chem. Soc. Rev.* 41: 2780–2799.

95 Ge, C., Du, J., Zhao, L. et al. (2011). Binding of blood proteins to carbon nanotubes reduces cytotoxicity. *PNAS* 108: 16968–16973.

96 Yang, S.-T., Liu, Y., Wang, Y.-W., and Cao, A. (2013). Biosafety and bioapplication of nanomaterials by designing protein–nanoparticle interactions. *Small* 9: 1635–1653.

97 Laera, S., Ceccone, G., Rossi, F. et al. (2011). Measuring protein structure and stability of protein-nanoparticle systems with synchrotron radiation circular dichroism. *Nano Lett.* 11: 4480–4484.

98 Baimanov, D., Wang, J., Zhang, J. et al. (2022). In situ analysis of nanoparticle soft corona and dynamic evolution. *Nat. Commun.* 13: 5389.

99 Roman, M., Rigo, C., Castillo-Michel, H. et al. (2016). Hydrodynamic chromatography coupled to single-particle ICP-MS for the simultaneous characterization of AGNPS and determination of dissolved Ag in plasma and blood of burn patients. *Anal. Bioanal.Chem.* 408: 5109–5124.

100 Bartczak, D., Vincent, P., and Goenaga-Infante, H. (2015). Determination of size- and number-based concentration of silica nanoparticles in a complex biological matrix by online techniques. *Anal. Chem.* 87: 5482–5485.

101 Gray, E.P., Bruton, T.A., Higgins, C.P. et al. (2012). Analysis of gold nanoparticle mixtures: a comparison of hydrodynamic chromatography (HDC) and asymmetrical flow field-flow fractionation (AF4) coupled to ICP-MS. *J. Anal. At. Spectrom.* 27: 1532–1539.

102 Taboada-López, M.V., Herbello-Hermelo, P., Domínguez-González, R. et al. (2019). Enzymatic hydrolysis as a sample pre-treatment for titanium dioxide nanoparticles assessment in surimi (crab sticks) by single particle ICP-MS. *Talanta* 195: 23–32.

103 Wang, B., Jing, L., Feng, W. et al. (2007). Characterization of size and morphology of ZnO and Fe_2O_3 nanoparticles in dispersive media by SAXS. *He Jishu/Nucl. Techn.* 30: 576–579.

104 Kanwar, R., Gradzielski, M., Prevost, S. et al. (2019). Experimental validation of biocompatible nanostructured lipid carriers of sophorolipid: optimization, characterization and in-vitro evaluation. *Colloids Surf., B* 181: 845–855.

105 Meng, H., Chen, Z., Xing, G. et al. (2007). Ultrahigh reactivity provokes nanotoxicity: explanation of oral toxicity of nano-copper particles. *Toxicol. Lett.* 175: 102–110.

106 Oughton, D.H., Hertel-Aas, T., Pellicer, E. et al. (2008). Neutron activation of engineered nanoparticles as a tool for tracing their environmental fate and uptake in organisms. *Environ. Toxicol. Chem.* 27: 1883–1887.

107 Kreyling, W.G., Semmler, M., Erbe, F. et al. (2002). Translocation of ultrafine insoluble iridium particles from lung epithelium to extrapulmonary organs is size dependent but very low. *J. Toxicol. Environ. Health A* 65: 1513–1530.

108 Li, Y.-F. and Chen, C. (2011). Fate and toxicity of metallic and metal-containing nanoparticles for biomedical applications. *Small* 7: 2965–2980.

109 Fischer, H.C., Liu, L., Pang, K.S., and Chan, W.C.W. (2006). Pharmacokinetics of nanoscale quantum dots: in vivo distribution, sequestration, and clearance in the rat. *Adv. Funct. Mater.* 16: 1299–1305.

110 Liu, Y., Gao, Y., Zhang, L. et al. (2009). Potential health impact on mice after nasal instillation of nano-sized copper particles and their translocation in mice. *J. Nanosci. Nanotechnol.* 9: 1–9.

111 Wang, B., Feng, W., Wang, M. et al. (2008). Acute toxicological impact of nano- and submicro-scaled zinc oxide powder on healthy adult mice. *J. Nanopart. Res.* 10: 263–276.

112 Tu, C., Ma, X., House, A. et al. (2011). PET imaging and biodistribution of silicon quantum dots in mice. *ACS Med. Chem. Lett.* 2: 285–288.

113 Gao, Y., Liu, N., Chen, C. et al. (2008). Mapping technique for biodistribution of elements in a model organism, *Caenorhabditis elegans*, after exposure to copper nanoparticles with microbeam synchrotron radiation X-ray fluorescence. *J. Anal. At. Spectrom.* 23: 1121–1124.

114 Wang, J., Liu, Y., Jiao, F. et al. (2008). Time-dependent translocation and potential impairment on central nervous system by intranasally instilled tio$_2$ nanoparticles. *Toxicology* 254: 82–90.

115 Wang, J., Chen, C., Liu, Y. et al. (2008). Potential neurological lesion after nasal instillation of TiO$_2$ nanoparticles in the anatase and rutile crystal phases. *Toxicol. Lett.* 183: 72–80.

116 Corezzi, S., Urbanelli, L., Cloetens, P. et al. (2009). Synchrotron-based X-ray fluorescence imaging of human cells labeled with CdSe quantum dots. *Anal. Biochem.* 388: 33–39.

117 Bussy, C., Cambedouzou, J., Lanone, S. et al. (2008). Carbon nanotubes in macrophages: imaging and chemical analysis by X-ray fluorescence microscopy. *Nano Lett.* 8: 2659–2663.

118 Zhu, M., Li, Y., Shi, J. et al. (2012). Exosomes as extrapulmonary signaling conveyors for nanoparticle-induced systemic immune activation. *Small* 8: 404–412.

119 Wang, B., Yin, J.-J., Zhou, X. et al. (2013). Physicochemical origin for free radical generation of iron oxide nanoparticles in biomicroenvironment: catalytic activities mediated by surface chemical states. *J. Phys. Chem. C* 117: 383–392.

120 Derfus, A.M., Chan, W.C.W., and Bhatia, S.N. (2003). Probing the cytotoxicity of semiconductor quantum dots. *Nano Lett.* 4: 11–18.

121 Qu, Y., Li, W., Zhou, Y. et al. (2011). Full assessment of fate and physiological behavior of quantum dots utilizing *Caenorhabditis elegans* as a model organism. *Nano Lett.* 11: 3174–3183.

122 Ban, Z., Yuan, P., Yu, F. et al. (2020). Machine learning predicts the functional composition of the protein corona and the cellular recognition of nanoparticles. *Proc. Natl. Acad. Sci. U.S.A.* 117: 10492–10499.

123 Cai, R., Ren, J., Ji, Y. et al. (2020). Corona of thorns: the surface chemistry-mediated protein corona perturbs the recognition and immune response of macrophages. *ACS Appl. Mater. Interfaces* 12: 1997–2008.

124 Tang, M., Xing, T., Zeng, J. et al. (2008). Unmodified CDSE quantum dots induce elevation of cytoplasmic calcium levels and impairment of functional properties of sodium channels in rat primary cultured hippocampal neurons. *Environ. Health Perspect.* 116: 915–922.

125 Sabbioni, E., Fortaner, S., Farina, M. et al. (2014). Cytotoxicity and morphological transforming potential of cobalt nanoparticles, microparticles and ions in Balb/3T3 mouse fibroblasts: an in vitro model. *Nanotoxicology* 8: 455–464.

126 Bramini, M., Sacchetti, S., Armirotti, A. et al. (2016). Graphene oxide nanosheets disrupt lipid composition, Ca^{2+} homeostasis, and synaptic transmission in primary cortical neurons. *ACS Nano* 10: 7154–7171.

127 Bramini, M., Chiacchiaretta, M., Armirotti, A. et al. (2019). An increase in membrane cholesterol by graphene oxide disrupts calcium homeostasis in primary astrocytes. *Small* 15: 1900147.

128 Wu, X., Hu, J., Wu, F. et al. (2020). Application of TiO$_2$ nanoparticles to reduce bioaccumulation of arsenic in rice seedlings (*Oryza sativa* l.): a mechanistic study. *J. Hazard. Mater.* 405: 124047.

129 Zhao, J., Liang, X., Zhu, N. et al. (2020). Immobilization of mercury by nano-elemental selenium and the underlying mechanisms in hydroponic-cultured garlic plant. *Environ. Sci.: Nano* 7: 1115–1125.

130 Wang, S., Wang, F., and Gao, S. (2015). Foliar application with nano-silicon alleviates Cd toxicity in rice seedlings. *Environ. Sci. Pollut. Res.* 22: 2837–2845.

131 Chen, R., Zhang, C., Zhao, Y. et al. (2018). Foliar application with nano-silicon reduced cadmium accumulation in grains by inhibiting cadmium translocation in rice plants. *Environ. Sci. Pollut. Res.* 25: 2361–2368.

132 Li, Y., Zhu, N., Liang, X. et al. (2020). Silica nanoparticles alleviate mercury toxicity via immobilization and inactivation of Hg(II) in soybean (*Glycine max*). *Environ. Sci.: Nano* 7: 1807–1817.

133 Lin, C.-H., Yang, M.-H., Chang, L.W. et al. (2011). Cd/Se/Te-based quantum dot 705 modulated redox homeostasis with hepatotoxicity in mice. *Nanotoxicology* 5: 650–663.

134 Abdelhalim, M.A.K., Siddiqi, N.J., Alhomida, A.S., and Alayed, M.S. (2012). Size effect of gold nanoparticles on various trace elements levels in different tissues of rats. *Afr. J. Microbiol. Res.* 6: 2246–2251.

135 Ma, J., Li, R., Liu, Y. et al. (2017). Carbon nanotubes disrupt iron homeostasis and induce anemia of inflammation through inflammatory pathway as a secondary effect distant to their portal-of-entry. *Small* 13: 1603830.

136 Stepien, K.M. and Taylor, A. (2012). Colloidal silver ingestion with copper and caeruloplasmin deficiency. *Ann. Clin. Biochem.* 49: 300–301.

137 He, L., Lu, Y., Li, C. et al. (2022). Non-targeted metallomics through synchrotron radiation X-ray fluorescence with machine learning for cancer screening using blood samples. *Talanta* 245: 123486.

138 Luo, Y., Chen, S., and Valdes, G. (2020). Machine learning for radiation outcome modeling and prediction. *Med. Phys.* 47: e178–e184.

139 Peng, T., Wei, C., Yu, F. et al. (2020). Predicting nanotoxicity by an integrated machine learning and metabolomics approach. *Environ. Pollut.* 267: 115434.

140 Duan, Y. and Lin, Q. (2016). *Laser-induced Breakdown Spectroscopy and Its Applications*. Beijing: Science Press.

141 Donnelly, P. (2008). Progress and challenges in genome-wide association studies in humans. *Nature* 456: 728–731.

142 Li, Y.-F., Chen, C., Qu, Y. et al. (2008). Metallomics, elementomics, and analytical techniques. *Pure Appl. Chem.* 80: 2577–2594.

143 Wang, C., Steiner, U., and Sepe, A. (2018). Synchrotron big data science. *Small* 14: 1802291.

3

Environmetallomics*

Lihong Liu[1], Ligang Hu[1], Baowei Chen[2], Bin He[1], and Guibin Jiang[1]

[1]*Chinese Academy of Sciences, State Key Laboratory of Environmental Chemistry and Ecotoxicology, Research Center for Eco-Environmental Sciences, 8 Shuangqing Road, Haidian District, Beijing 100085, China*
[2]*Sun Yat-Sen University, School of Marine Sciences, Southern Marine Science and Engineering Guangdong Laboratory, No. 135, Xingang Xi Road, Guangzhou 510275, China*

3.1 The Concept of Environmetallomics

Metals (and metalloids) play crucial roles to lives as essential or beneficial components and meanwhile exhibit toxic effects to organisms as contaminants. Toxic metals have become one of the most concerning contaminants in the field of environment science worldwide, among which mercury, lead, arsenic, cadmium, and chromium have been listed as class one carcinogen by International Agency for Research on Cancer (IARC) and announced to be the top 10 priority pollutants by the World Health Organization (WHO). These toxic metals could transfer from the environment to organisms and bioaccumulate in human through food chain even at low exposure levels, posing severe adverse effects to human health. For example, approximately 94–200 millions of people worldwide are chronically exposed to high arsenic concentrations in groundwater [1], which could increase the risks of malignancies of skin, bladder, and lung and adverse health effects, such as skin lesions, hypertension, cardiovascular effects, and respiratory disease [2].

Currently, there is still a lack of comprehensive cognition and understanding of the environmental health risks and ecological toxicity of toxic metals, which is basically considered the essential aim of environmental research. Numbers of researches on the environmental fate and behavior of toxic metals and their potential ecological and health risks have been carried out. However, systematic studies on the mechanisms of metal absorption, transport, transformation, and function in complex organisms or human are very limited. This is primarily due to the insufficiency and limitation of methodologies in the research of environmental science and toxicology, which hampers the understanding of the key issues and

---

* This chapter has been modified to feature as Reviews: (i) Liu, L., Yin, Y., Hu*, L., et al. (2020). Revisiting the forms of trace elements in biogeochemical cycling: analytical needs and challenges. *Trends Anal. Chem. 129*: 115953; (ii) Chen, B., Hu* L., He, B., et al. (2020). Environmetallomics: systematically investigating metals in environmentally relevant media. *Trends Anal. Chem. 126*: 115875.

Applied Metallomics: From Life Sciences to Environmental Sciences, First Edition.
Edited by Yu-Feng Li and Hongzhe Sun.
© 2024 WILEY-VCH GmbH. Published 2024 by WILEY-VCH GmbH.

risks of toxic metals and meantime limit the improvement of effective management strategies. The comprehensive and systematic methodologies are in great demand for further clarifying the toxicology mechanisms and health impact of metals.

As an emerging field, metallomics emphasizes the systematic understanding of metal-dependent life processes by using the comprehensive techniques and novel approaches combining both theory and experiment. From an environmental perspective, the environmental metallomics (*environmetallomics*), which was described as "an environmental branch in metallomics" [3], was initially proposed by Lopez-Barea et al. in 2006 [4] and its explicit definition and related terms was established by Baowei Chen et al. in 2020 [5]. The term "environmetallome" has been defined as follows: (i) the entirety of metal and metalloid species that can induce toxic effects on living organisms at an environmentally relevant concentration range; (ii) all metals and metalloids within a biological system which homeostasis could be directly or indirectly interfere and regulated to induce potential adverse effects [5]. Environmetallomics, accordingly, is the study of an environmetallome and aims to elucidate the metal's molecular aspects, potential health, and ecosystems risks through exploration of occurrence, distribution, speciation, transformation, and toxicity of metal pollutants by using systematic and comprehensive methods. This also includes the relative investigation of interactions of metals with other compounds or pollutants in the natural environment as well as the effects of metals on living organisms, in terms of proteins, genes, metabolites, and other biomolecules. Distinct from the traditional metallomics, the scope of environmetallomics is extended from the interior of living organisms to a local habitat or ecosystem (Figure 3.1), constituting an independent interdisciplinary research area combining metallomics and environmental science. Generally, it emphasizes the entire range of potential implications of metal contamination (including the non-biogenic compounds) in the natural environment, which are usually outside the range of traditional metallomics studies or "metallome" (the entirety of metals and metalloid species in the living systems). Robust methods capable of separation, identification, localization, quantification, and in silico prediction of multi-metal species are critically needed for characterizing the metals in the environment and evaluating their interactions with the genomes, proteomes, and metabolomes. Therefore, this chapter focuses on the review of the potential metallomics techniques and approaches that may be promising in environmetallomics research.

3.2 The Analytical Techniques in Environmetallomics

3.2.1 The Requirements for Environmetallome Analysis

Traditional analytical approaches in metallomics have been utilized for environmetallomics analysis, such as quantitative analysis, metal distribution and mapping, metal speciation and structure, as well as metalloprotein analysis. Nevertheless, the range of potentially suitable analytical methods for environmetallomics are still

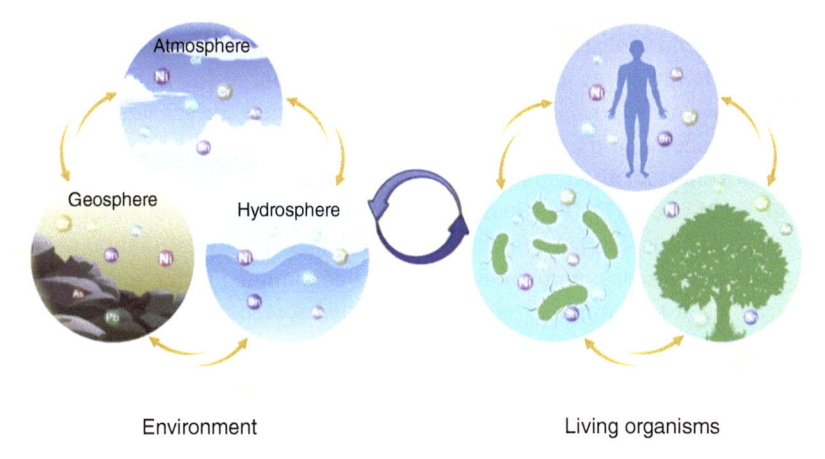

Figure 3.1 The schematic figure for the scope of environmetallomics. Source: From Ref. [5] with permission from Elsevier.

considerably restricted due to its typical characteristics. On the one hand, the concentrations of different metals in the external environment and biological organisms distribute within a wide range, differing by orders of magnitude. Generally, the toxic metals (e.g. Hg, Pb and Cd) are present at low concentrations and one or several order of magnitudes lower than essential elements (e.g. Fe, Cu, and Zn). Therefore, high sensitivities of analytical instruments are required for environmetallomics analysis to achieve accurate measurement of trace metals. On the other hand, the environmental samples commonly contain much more complex and variable matrices than those samples involved in the traditional metallomics studies. To be specific, the environmental samples usually contain a variety of types of inorganic and organic compounds, complex biological materials as well as geogenic matter, which poses a notable challenge to environmetallome analysis. The complexity of matrix in environmental samples would cause serious interferences for metal analysis, which hampers the accurate quantification and thus requires more robust analytical approaches to overcome this problem for further environmetallomic studies.

3.2.2 Quantitative Analysis for Environmetallomics

For quantitative analysis of trace metals in different samples, the most commonly used techniques are plasma-based techniques, including inductively coupled plasma optical emission spectrometry (ICP-OES) and inductively coupled plasma mass spectrometry (ICP-MS), as well as neutron activation analysis (NAA) technique [6]. In general, both ICP-MS and NAA are considered as powerful tools for element quantification, with comparative advantages of high sensitivity, capacity of simultaneous quantitative detection, wide ranges of elements from very low to high concentrations, and good performance even in complicated environmental medium [7]. Among these advantages, the sensitivity of the method is the primary concern for successful detection of ultra-trace metals and extensive applications in natural environmental and biological samples. Generally, the detection limits (LOD) for ICP-MS, which can be as low as ng/l or pg/g level, are about 2 to 3 orders of magnitudes lower than the LOD for ICP-OES [8]. The liquid and gas samples can be directly introduced into the ICP-MS, while for solid-based samples, laser-ablation (LA)-ICP-MS offers an alternative and competitive choice for direct in situ analysis [9]. As for the NAA, the LOD are in the range of 10^{-7} µg/g to µg/g level for more than 30 elements, and this technique has obvious advantages of non-destructive analysis, little matrix interference, and barely laboratory contamination for solid samples without the need for sample extraction [10]. Rapid screening and quantification of multi-elements in environmental samples by the above-mentioned methods is an efficient way to find the relationships between metals and biological processes, such as uptake, distribution, translocation, and bioaccumulation. For instance, the relationships of 15 elements (As, Ba, Cd, Co, Cs, Cu, Mn, Mo, Ni, Pb, Se, Sr, Tl, W, Zn) in the human urine and the environment were investigated by rapid analysis of ICP-MS, revealing that the potential sources of metal and metalloid exposure was groundwater (As, Mo, and W) and betel (Cd) in the investigated population [11].

In addition to the average total concentration of elements in the integrated sample, the metal contents in a single cell are much more essential. After the metal uptake into cells, it is considered that the distribution of metals between individual cells is heterogeneous for unicellular and multicellular organisms. For instance, the existence of heterogeneity in cells has been demonstrated in the metal uptake process and the metal content differs according to the cell cycle status [12], although this has not been comprehensively assessed. These dynamics process and intracellular homeostasis of toxic metals in individual cells are scarcely reported in previous literatures due to method limitations. Therefore, the development of analysis techniques for detecting various elements in a single cell is of great importance and considered as a significant challenge at the present period [13]. Recently, a mass-spectrometry-based technique coupled with a certain introduction system, named single-cell ICP-MS (SC-ICP-MS), has been developed and has become an innovative and emerging tool for the quantification of metals in a single cell of a population [14]. The SC-ICP-MS allows one single-metal analysis in a single cell during one test with notable advantages of fast detection, high sensitivity, simple operation, and wide application. The method has been developed rapidly and successfully applied in environmetallomics studies for toxic metals, obtaining more and more research interest. For instance, a quantitative method for Hg in single cell was developed based on capillary SC-ICP-MS online system and was used to monitor Hg in individual unicellular *tetrahymena*, discovering the accumulation and discerning differences in the uptake of Hg among different cells, especially under the exposure of higher concentration Hg [15]. Nevertheless, the SC-ICP-MS has limitations because the detector allows detection of only one element in a single run. Mass cytometry, also named as cytometry time of flight (CyTOF), is an emerging technique based on ICP time-of-flight MS and could promise real simultaneous detection of various elements, which can provide both the information of selected metals and biological signals (such as cell viability,

DNA content, and relative cell size) by using labels antibodies tagged with isotopes of rare-earth metals [16]. By using the CyTOF, analysis of Pb content in single erythrocytes (anucleate cell) was established, and research found the high heterogeneity of Pb concentration in individual mature erythrocytes (m-erythrocytes), which dynamically decreased with the lifetime of m-erythrocytes and were fit to the pattern of gamma distribution [17]. Aside from the advantages of this method, it also has some distinct technical limitations in the aspect of signal acquiring, reliance on antibody specificity, complicated and costly sample preparation, as well as metal coverage, making the detection of elements with m/z less than 80 impossible. On the contrary, laser ablation coupled with ICP-MS (LA-ICP-MS) systems is capable of simultaneously monitoring multiple elements (without restrictions of element types) in an individual single cell by coupling with a time-of-flight mass analyzer, without the use of expensive labels and additional sample preparation [18]. But the limitations of sensitivity and feasibility of this method proposed further optimization and practical validation to achieve better performance. Overall, the application of all these single-cell analysis techniques in the field of environmetallomics needs further research.

3.2.3 Metal Distribution and Mapping for Environmetallomics

Apart from the total concentration of metals, metal distribution in the organisms at the level of cells, tissues, and subcellular organelles is also important for better understanding the transportation and function of metals and the identification of possible toxicological targets of toxic metals. After being exposed to toxic metals, the biological organisms were pretreated by physical dissection or by destroying the cells to obtain organ samples and subcellular fractions, which were then digested and analyzed to get the information of metal distribution profiles [19]. Generally, mapping techniques for metals in environmental and biological samples include X-ray fluorescence-based technique, such as synchrotron radiation X-ray fluorescence (SR-XRF) and synchrotron radiation X-ray absorption spectroscopy (SR-XAS), and mass-spectrometry-based technique, such as laser ablation inductively coupled plasma mass spectrometry (LA-ICP MS) and secondary ion mass spectrometry (SIMS).

The above techniques have been applied extensively in metal imaging with unique features of relatively non-destructive analysis, high sensitivity (LOD less than mg/kg), and good spatial resolutions (about the scope of μm) [20]. For instance, the Hg distribution in zebrafish larvae was imaged by SRXRF, and the accumulation of most Hg was found to occur in the rapidly dividing lens epithelium, demonstrating that the impairment of visual processes was Hg-related as previously reported. By using SR-XAS, the distribution and identification of arsenic species, including the oxidative states (trivalent and pentavalent arsenic species), and possible complexes (e.g. As(V) bound to ferric sulfate precipitates and As(III) complex bound to triglutathione) can be depicted in biological specimens [21]. The speciation, spatial distribution, and abundance of different elements (including iron, arsenic, and sulfur speciation) using SR-XRF and SR-XAS then proved the phytostabilization mechanism of apparent arsenic detoxification, which had been reported as a hypothesis without direct evidence prior to the confirmation of that study [21]. In general, the interaction of arsenic with root cellular metabolism was prevented by the transformation process of arsenic species, in which pentavalent arsenate immobilized on the root epidermis is reduced to trivalent species and bound to ferric sulfate precipitates in root vacuoles [21]. The multiple energy XRF-based in situ imaging techniques provided powerful tool for elemental research associated with their distribution, localization, and accumulation, which consequently improved the thorough understanding of the biological processes of different elements at the levels of tissue, cell, and even subcell [22].

Compared with XRF techniques, mass-spectrometry-based imaging instruments showed higher sensitivities and provided the determination of both metal concentrations and isotope ratios [23]. For instance, LA-ICP-MS was applied for the co-localization of Hg deposits and macrophage centers (representing the biomarker of contaminant toxicity) in the fish tissue collected from a natural lake, and accordingly the causative relationship between Hg exposure and the immune responses was established [24]. Except for imaging analysis of trace metals (Fe, Gd, C, P, S, Ni, Cu, Zn) in biological samples, LA-ICP-MS can also be applied in metal nanoparticles analysis [25]. The quantitative imaging of gold nanoparticles (AuNPs) in different organs of mice was obtained by LA-ICP-MS methods, and the results illustrated that the suborgan distribution of AuNPs in spleen, kidney, and liver was dominated by the surface charge after the mice was intravenously injected with functionalized AuNPs [26]. Time-of-flight SIMS (TOF-SIMS), another powerful imaging technique based on mass spectrometry, was proposed for imaging and depth profiling of metals with spatial resolution in the μm range and depth resolution in the ~nm range, allowing the profile of subcellular imaging of As and Hg species in bacteria and investigation of their uptake and transformation processes inside cells [27]. The imaging result showed that in the periplasmic space mercury ions were transformed to methylmercury, while pentavalent arsenate ions were reduced to trivalent arsenite species inside the cells, which was nearby the cell membrane [27]. However, SIMS usually suffers from bad interferences caused by atomic or molecular ions and a variety of matrices which may affect the information depth and quantification of SIMS images. Moreover, except for these techniques, an emerging scattered light imaging (SLI) technique based on confocal microscope with high spatial resolution was developed for imaging analysis of both metal nanoparticles and fluorescent biomolecules in live cells, allowing the dynamic observation of uptake, distribution, and transformation of silver nanoparticles in single, living cells [28].

To date, great efforts are still needed to improve the performance and efficiency of imaging techniques. Presently, the analysis duration of metal mapping is too long for routine application, especially for the heterogeneous environmental samples and it is necessary to reduce it [29]. Moreover, it is not practically possible to obtain imaging profiles for different metal species which could be detected by speciation analysis technique. For in vivo metal imaging, there is great need to develop non-invasive and effective methods enabling dynamic localization and transportation analysis of metals inside the organisms. Recently, due to the good permeability, small molecular probes have been designed for visualization of metal distribution and have been used for mapping of both essential and biological metals (e.g. Zn, Cu, Fe, Mn, Ni, Co) and toxic metals (Pb, Cd and Hg) in living systems by specific and stoichiometric metal-binding targets [30]. For example, a rhodamine-based mercury probe was developed for the detection of mercury in living cells and zebrafish and the probes, which irreversibly reacted with Hg, showed advantages in terms of selectivity and sensitivity over other reversible ones [31].

3.2.4 Metal Speciation for Environmetallomics

Speciation analysis of metals is essential because the metals exist in the form of various chemical species and different sizes in the natural environment and organisms. The chemical species of metals have been categorized by different valence state, isotopic composition, and molecular structure, while species of metals from the macroscopic perspective include metal ions or molecules, metal clusters, macromolecular complexes, nanoparticles, micro-sized particle as well as bulk materials (e.g. minerals) in the order of smallest to largest. These metal forms varied constantly and showed quite distinct chemical activities and consequently diversified mobility, bioavailability, and toxicity, all of which are expressly included in the environmetallomics studies [32]. For example, under saturation condition free ionic trivalent chromium (Cr(III)) in water solution will firstly turn into the polymeric Cr(III) species, accounting for approximately 60% of the total chromium in wood leachates and then gradually transit to large species of chromic particles and precipitates [33]. Bivalent cadmium (Cd(II)) ions could in vivo interact with metallothionein (MT) coexisting with Cu and Zn, and more than half of them form a complex of Cd1-7MT in liver, preventing its reaction with other biological targets and leading to detoxification of Cd exposure [34]. Accordingly, in the environmetallomics studies of metals associated with their transformation, bioavailability, and toxicity, appropriate extraction and analysis techniques for various metal species are the essential prerequisite. The reliable and efficient methods to extract labile metal species from the original environment without interconversion are the first key step for accurate speciation analysis. Many methods have been developed for extracting ionic or molecular species of metals by using both fierce extraction (such as microwaves, Soxhlet extraction, and sonication) and mild extraction (shaking, heating) with water, dilute acid, and organic solvents (e.g. methanol) as the extraction solvents [35]. In addition, novel functionalized miniaturized membrane for highly selective adsorption of trivalence arenite (As(III)) was developed and applied for arsenic speciation analysis by energy-dispersive X-ray fluorescence spectrometry in high salinity water [36]. For macromolecules or particles, which were usually removed from the extracts by centrifugation or filtration, mild digestion methods have been achieved with acidic, alkaline, and enzymatic as the extraction solvents [37].

For speciation analysis of various forms of metals in terms of both chemical species and different sized forms, the primary requirements include separation, recognition, characterization, and quantification. Presently, the most effective techniques for metal speciation are based on sensitive detectors (such as ICP-MS and NAA techniques) hyphenated with effective separation techniques which enables the effective separation and determination of metal species which have been identified previously in different types of samples (e.g. environmental, biological, food, and clinical samples) [38]. A number of techniques have been utilized for separating element species, such as gas chromatography (GC), liquid chromatography (LC), and capillary electrophoresis (CE). Nevertheless, it is still a challenge to achieve the separation of different dimensional metals, such as free metal ions, metal nanoparticles, and macromolecular binding with metals. Typically, off-line separation ways have been used for separation of these metal forms by centrifugation and ultrafiltration as an important pretreatment procedure. Meanwhile, several on-line techniques have been recently developed and exhibit remarkable application potential in the elemental separation. Generally, the most effective separation techniques can be categorized as two types: the ones with stationary phase (e.g. LC) and that without stationary phase (e.g. CE and different modes of field-flow fractionation [FFF]) based on the different column structures and corresponding separation principles [35c]. The separation techniques with stationary phase are, by definition, performed by a column filling with solid phases of different sizes and structure. Its separation was derived by the different physicochemical interactions of the target metal species with the stationary and mobile phases, leading to different elution orders and separately retention times. So far, several modes of LC techniques, such as hydrodynamic chromatography (HDC), size exclusion chromatography (SEC), and reversed-phase high-performance liquid chromatography (RP-HPLC), have been applied for separating metal species (nanoparticles, metalloproteins, and metal ions) according to their molecular sizes [39]. For example, with the use of hyphenated technique of HPLC and ICP-MS or NAA, arsenic species were quantified in hypoallergenic formulas and grain porridges samples [40]. The SEC separation coupled with metal specific detector allowed the separation and detection of nanoparticles with different sizes [41] and metalloproteins with different molecular weights [42]. The separation techniques without stationary phase, on the other hand, are usually performed on a hollow column, and the separation of different species is achieved under high-voltage electric field by different mass-to-charge ratio of analytes, such as CE and gel electrophoresis (GE) [43], or by the different flow field effect, such as asymmetric flow (a cross flow field), sedimentation (a centrifugal field), and hollow fiber flow (a flow field) FFF [44]. For instance, the CE technique has been utilized for highly effective separation of arsenic and selenium species [45] as well as different sized metal nanoparticles [43b].

Currently, new analytical strategies combining elemental and molecular mass spectrometric techniques have been applied for metal speciation analysis and have shown especially to be capable of determination and identification of unknown metal species. Take arsenic as example, the unrecognized thiol-containing methylated arsenicals were quantified by the ICP-MS with high sensitivity and jointly qualified by electrospray ionization mass spectrometry (ESI-MS) to profile its structure [46]. Similarly, the combination of ESI-MS and ICP-MS was synergically applied for arsenic speciation analysis to verify the presence of monomethylarsonous acid (MMA^{III}) in As-contaminated groundwater [47], and to investigate the arsenic metabolites in chicken liver after being fed with roxarsone, in which three new methylated phenylarsenical metabolites were finally identified with highly toxic effect [48]. Moreover, synchrotron-based micro-based XRF mapping can also be utilized for the direct analysis of arsenite species and other metal contents (e.g. Sb, Fe, Mn, and Zn) in rice samples without laborious pretreatment [36].

3.2.5 Metalloprotein Analysis

The generic strategy for environmental metalloproteomics is based on the complementary coupling of separation techniques (on-line or off-line 1D or 2D GE, LC or multiple if necessary, and CE) with metal or molecule-related detectors (ICP-MS, ESI-MS, XFS, and XAS). The development of a column-type GE coupled with ICP-MS allowed the comprehensive measurement of both metals and their associated proteins and showed excellent performance in rapid profiling of the metal-associated proteins in the fields of proteome and metallome [43c]. This technique was further utilized for investigating and further characterizing the metal-binding protein profiles in *Helicobacter pylori* under treatment with a bismuth-based drug [43c]. Besides, in a bacterial sample of *Serratia* Se1998 which showed very high tolerance ability to Pb, flagellin was discovered to be a novel super Pb-binding protein by the GE-ICP-MS method [49]. The wide application of this technique provided novel perspectives for understanding the function and bioaccumulation mechanism of metals in bacteria. Even though GE offers an attractive tool for protein separation, the separation relied on the preservation of the labile bonds of metals with proteins, which may possibly hinder the recovery and identification of metalloproteins. Given this, several precautions should be considered: firstly, the impacts of separation protocols should be taken into account, such as the denaturation and electrophoresis voltage that would affect the stability of metalloproteins and cause metal dissociation from proteins [43c]; and secondly, for the metal contamination from laboratory, gels and staining should be avoided by using ultra-clean and high-purity reagents [50].

Compared with the gel separation systems, chromatographic techniques have been utilized for separation of metalloproteins with some advantages. The chromatographic separation system is much more moderate than that of GE, and it is much easier to monitor the recovery of metalloproteins and to collect the fractionation of metal-binding complexes for further analysis. With the coupling of ICP-MS and the combination of high matrix introduction (HMI) mode, a size exclusion column (SEC) was used to separate four iodine-labeled proteins with different molecular weights, and the hyphenated system was then used in rapid separation and quantitative detection of Pb-binding proteins as well as Pb ions in neutrophil of mice treated with lead acetate [42]. The SEC-ICP-MS system was also used for analysis of brain cytosol and the Hg-containing protein fraction displayed a significant difference between the maternal and infant rats, which was helpful for the study of metal toxicology [51]. Besides, multi-dimensional chromatography or orthogonal separation configurations have also been approved to be superior methods to obtain good separation of metalloproteins [52]. For example, two-dimensional separation technique combining weak anion exchange chromatography (WAX) and GE has been developed for identification of metalloproteins by coupling with ICP-MS, and proteins containing metallic elements of Mg, Cu, Zn, Ag, and I were efficiently analyzed in *E. coli* by this method [52]. Another recent research used another 2D-HPLC separation system (SEC and WAX) through the on-line heart-cutting way to hyphenate with ICP-MS and locate the Hg-containing protein, unequivocally confirming that Hg binding with serum albumin was the primary form in human plasma [53].

Still, the improvement of proteomics analysis is needed for metalloproteins in environmental and biological systems due to the following reasons. On the one hand, compared with the genomics and transcriptomics which can increase the sensitivity by amplification of polymerase chain reaction (PCR), proteomics has limitations of sensitivity, requiring the improvement and evolution of sensitive proteomics analytical instruments. On the other hand, despite the identification of proteins in proteomics, the existence of metals in metalloproteins requires an additional dimensional analysis in the scope of metallomics [54]. In consideration of the low concentration and complex matrix of metalloproteins, the metallomics approaches particularly need to ensure the detection and quantification of metal-containing proteins with good specificity and ultra-high sensitivity. Besides, the requirement for characterization of metals in metalloprotein also directs the metallomics analysis to transfer the focus from classical protein-based purification to metal-based purification [55]. To characterize the structure of the metal-free components of metalloproteins, ESI-MS provides additional information as an optimal and alternative partner to ICP-MS [56]. For example, with the synergistic application of ESI-MS and ICP-MS after CE and LC separation, identification and quantitative determination of metallothionein isoforms were achieved [57].

To date, there are significant technical challenges remaining in the comprehensive profiling of metalloproteins in complicated samples (e.g. organisms), which can bring incomparable insight into the aspect of metal metabolism. However, the limitations of separation capacity, sensitivity, and matrix interferences have dramatically hindered the analysis of all metalloproteins in the organisms containing numerous accompanying proteins. To improve the separation performance for proteins, the combined utilization of chromatographic fractionation as well as gel electrophoresis is considered as an effective way. The immobilized metal affinity chromatography (IMAC) has relatively lower demands for the separation and detection of proteins in terms of resolution and interference elimination and has been applied for identifying the target metalloproteins from plenty of non-specific proteins in biological samples [58]. For instance, a copper(II)-loaded IMAC column combined with 2-D GE and mass spectrometry was used to identify the copper-binding proteins in Hep G2 cells and finally confirmed 48 cytosolic proteins and nineteen microsomal proteins, in which 52 were found to contain putative or known metal-binding domains [59]. Another research reported the development of Hg-associated IMAC for identification of thirty-eight mercury-binding proteins in SK-N-SH cells, and the method exhibited excellent performance at capturing proteins under very low abundance (i.e. less than 50 ppm) without obvious abundance discrimination [60]. In addition, the development of a recent technology named proteolysis targeting chimera (PROTAC) has been proposed, and its application for identification of targeted proteins in chemical biology has showed great potential for analysis of unrecognized metalloproteins [61].

3.3 The Application of Environmetallomics in Environmental Science and Ecotoxicological Science and the Perspectives

The emergence of environmetallomics provides powerful tool for comprehensively elucidating the environmental health risks of metal pollutants, which covers the interdisciplinary research in the field of environmental science, biology, inorganic chemistry, and toxicology. The revelation of source, existence, and transformation of metals in different cycles is the basis for the environmental health research of metals. Metals in biosphere, including the biological and exogenous metals, are closely related with those in geosphere through anthropogenic and natural way. There was a quite good correlation between the elemental concentrations (including major-to-ultratrace elements) in human blood serum and their concentrations in seawater, significantly explaining the origin of metal concentrations in organisms was from metals in environment [62]. To find the associations between the environmental metals and its health effects as well as the pathway into organism is one key step for the environmental health research. Except for the serious health problems of heavy metals (e.g. minamata disease incident and arsenic contamination of groundwater), the potential health risk of long-term and low-dose exposure of toxic metals should be considered although there is still lack of conclusive evidence. More specifically, the innovative environmetallomics approaches have been applied for investigating the environmental health and toxicology of metals by in situ imaging of metals in living cell or tissue and profiling toxic-metals-associated proteins. Study of the distribution and localization of metals in cells and tissues (or even cellular compartments) after metals entry into the organisms is the first important step to clarify the biotransformation and toxicology mechanism of metals at molecular level [63]. In situ imaging techniques, such as the XRF and LA-ICP-MS, could provide the most straightforward information. With the use of synchrotron XRF, the distribution of organic mercury in zebrafish larvae treated with MeHg-L-cysteine was mapped in micrometer resolution and it was found that mercury was heterogeneously distributed and strikingly accumulated in the rapidly dividing lens epithelium. The high-resolution imaging of metals showed that mercury was preferentially concentrated at the outer single-cell layer of fish eyes epithelial tissue, which implied that mercury-impaired visual process may arise not only from previously reported neurological effects, but also from direct effects on the ocular tissue [64]. Profiling metals-associated proteins at the proteome level, which is still a big challenge, is the basis for integrally elucidating the biotransformation and molecular mechanisms of metals in biological systems. The high-resolution separation techniques combined with highly sensitive element-specific detector provides efficient approach for analysis of the metalloproteins. There have been limited reports that investigated the metalloproteome in specific biological samples, such as microorganisms (*Pyrococcus furiosus*, *Escherichia coli*, and *Sulfolobus solfataricus*), in which metalloproteomes were largely uncharacterized [55]. Even though most works focused on limited numbers of metal-containing proteins, those findings still provided important information on behaviors and toxicity of metals. For example, silver-regulated and -binding proteins in *P. aeruginosa* treated with both silver nanoparticles and silver ions were identified by using the metalloproteomic approaches and clarified the antimicrobial activity of silver nanoparticles by disrupting the functions of cell membrane and inducing oxidative stress [65].

Currently, the environmetallomics techniques have brought not only novel insights but also challenges to the metal-related research in the aspects of both methodological development and functional studies. Firstly, except for the chemical species of metals, "larger-scale" forms of metals, including metal particles and metal–macromolecule complexes ubiquitously present in the environment, play important role in the environmental fate, biotransformation, and ecological toxicity of metals. New requirements have arisen for simultaneous separation, identification, and quantification of these various metal species, including metal ionic species, clusters, macromolecular complexes as well as nanoparticles, for comprehensive investigation and better understanding their environmental fate and health risks. Secondly, in consideration of the particular features of the target compounds in environmental samples, which is in very low concentration levels and suffers from complex matrices, great efforts are still needed to build up analytical methods with high sensitivity, high throughput, wide coverage, and good tolerance to interferences for environmetallomics studies. Thirdly, the combination of metallomic studies with multidisciplinary approaches, such as bioinformatics, genomic, proteomic, and metabolomic analysis will provide a comprehensive view and significantly promote the research on environmental health and toxicological studies of metals.

Acknowledgments

This work was financially supported by National Natural Science Foundation of China (22193052, 22076200, 22276201).

List of Abbreviations

CE	capillary electrophoresis
CyTOF	mass cytometry
ESI-MS	electrospray ionization mass spectrometry

FFF	field-flow fractionation
GC	gas chromatography
GE	gel electrophoresis
HDC	hydrodynamic chromatography
ICP-MS	inductively coupled plasma mass spectrometry
ICP-OES	inductively coupled plasma optical emission spectrometry
IMAC	immobilized metal affinity chromatography
LA-ICP-MS	laser-ablation inductively coupled plasma mass spectrometry
LC	liquid chromatography
NAA	neutron activation analysis
PCR	polymerase chain reaction
PROTAC	proteolysis targeting chimera
RP-HPLC	reversed-phase high-performance liquid chromatography
SC-ICP-MS	single cell inductively coupled plasma mass spectrometry
SEC	size exclusion chromatography
SIMS	secondary ion mass spectrometry
SLi	scattered light imaging
SR-XAS	synchrotron radiation X-ray absorption spectroscopy
SR-XRF	synchrotron radiation X-ray fluorescence
TOF-SIMS	time-of-flight secondary ion mass spectrometry
WAX	weak anion exchange chromatography

References

1 Podgorski, J. and Berg, M. (2020). *Science* 368: 845.
2 (a)Huang, H.-W., Lee, C.-H., and Yu, H.-S. (2019). *Int. J. Env. Res. Pub. Health* 16: 2746; (b)Yoshida, T., Yamauchi, H., and Sun, G.F. (2004). *Toxicol. Appl. Pharmacol.* 198: 243.
3 Hu, L., He, B., Wang, Y. et al. (2013). *Chin. Sci. Bull.* 58: 169.
4 López-Barea, J. and Gómez-Ariza, J.L. (2006). *Proteomics* 6: S51.
5 Chen, B., Hu, L., He, B. et al. (2020). *Trends Anal. Chem.* 126: 115875.
6 Meermann, B. and Nischwitz, V. (2018). *J. Anal. At. Spectrom.* 33: 38.
7 (a)Rodríguez-Moro, G., Ramírez-Acosta, S., Arias-Borrego, A. et al. (2018). *Metallomics: The Science of Biometals* (ed. M.A.Z. Arruda). Cham: Springer International Publishing; (b)Szpunar, J. (2005). *Analyst* 130: 442.
8 Pröfrock, D. and Prange, A. (2012). *Appl. Spectrosc.* 66: 843.
9 Neilsen, J.L., Abildtrup, A., Christensen, J. et al. (1998). *Spectrochim. Acta, Part B* 53: 339.
10 Yuan, J. T. and Lin, T. C. (2009). Presented at *MILCOM 2009 – 2009 IEEE Military Communications Conference*. 18–21 Oct. 2009.
11 Sanchez, T.R., Slavkovich, V., LoIacono, N. et al. (2018). *Environ. Int.* 121: 852.
12 (a)Kim, J.A., Aberg, C., Salvati, A., and Dawson, K.A. (2012). *Nat. Nanotechnol.* 7: 62; (b)Zhou, Y., Wang, H.B., Tse, E. et al. (2018). *Anal. Chem.* 90: 10465.
13 Yin, L., Zhang, Z., Liu, Y.Z. et al. (2019). *Analyst* 144: 824.

14 (a)Mavrakis, E., Mavroudakis, L., Lydakis-Simantiris, N., and Pergantis, S.A. (2019). *Anal. Chem.* 91: 9590; (b)Sun, H.Z., Tsang, C.N., Ho, K.S., and Chan, W.T. (2011). *J. Am. Chem. Soc.* 133: 7355.

15 Shi, J.B., Ji, X.M., Wu, Q. et al. (2020). *Anal. Chem.* 92: 622.

16 Bendall, S.C., Simonds, E.F., Qiu, P. et al. (2011). *Science* 332: 687.

17 Liu, N., Huang, Y., Zhang, H. et al. (2021). *Environ. Sci. Technol.* 55: 3819.

18 Theiner, S., Schweikert, A., Van Malderen, S.J.M. et al. (2019). *Anal. Chem.* 91: 8207.

19 Feng, W.Y., Meng, W., Ming, G.A. et al. (2011). *J. Anal. At. Spectrom.* 26: 156.

20 (a)Becker, J.S. (2013). *J. Mass Spectrom.* 48: 255; (b)Gao, Y.X., Chen, C.Y., and Chai, Z.F. (2007). *J. Anal. At. Spectrom.* 22: 856.

21 Hammond, C.M., Root, R.A., Maier, R.M., and Chorover, J. (2018). *Environ. Sci. Technol.* 52: 1156.

22 Bongiovanni, G.A., Perez, R.D., Mardirosian, M. et al. (2019). *Appl. Radiat. Isot.* 150: 95.

23 (a)Becker, J.S., Zoriy, M., Matusch, A. et al. (2010). *Mass Spectrom. Rev.* 29: 156; (b)Qin, Z.Y., Caruso, J.A., Lai, B. et al. (2011). *Metallomics* 3: 28.

24 Barst, B.D., Gevertz, A.K., Chumchal, M.M. et al. (2011). *Environ. Sci. Technol.* 45: 8982.

25 Hsieh, Y.K., Jiang, P.S., Yang, B.S. et al. (2011). *Anal. Bioanal.Chem.* 401: 909.

26 Elci, S.G., Jiang, Y., Yan, B. et al. (2016). *ACS Nano* 10: 5536.

27 Nygren, H., Dahlen, G., and Malmberg, P. (2014). *Basic Clin. Pharmacol.* 115: 129.

28 Wang, F.B., Chen, B.L., Yan, B. et al. (2019). *J. Am. Chem. Soc.* 141: 14043.

29 Hare, D.J., New, E.J., de Jonge, M.D., and McColl, G. (2015). *Chem. Soc. Rev.* 44: 5941.

30 (a)Carter, K.P., Young, A.M., and Palmer, A.E. (2014). *Chem. Rev.* 114: 4564; (b)Singh, V.K., Singh, V., Yadav, P.K. et al. (2019). *J. Photochem. Photobiol. A.* 384.

31 Yang, Y.K., Ko, S.K., Shin, I., and Tae, J. (2007). *Nat. Protoc.* 2: 1740.

32 (a)Tan, Q.G., Zhou, W.T., and Wang, W.X. (2018). *Environ. Sci. Technol.* 52: 484; (b)Gupta, V.K., Nayak, A., and Agarwal, S. (2015). *Environ. Eng. Res.* 20: 1.

33 Hu, L.G., Cai, Y., and Jiang, G.B. (2016). *Chemosphere* 156: 14.

34 Jara-Biedma, R., Gonzalez-Dominguez, R., Garcia-Barrera, T. et al. (2013). *Biometals* 26: 639.

35 (a)Yuan, C.G., He, B., Gao, E.L. et al. (2007). *Microchim. Acta* 159: 175; (b)Leufroy, A., Noel, L., Dufailly, V. et al. (2011). *Talanta* 83: 770; (c)Liu, L.H., Yin, Y.G., Hu, L.G. et al. (2020). *Trend Anal. Chem.* 129.

36 Lukojko, E., Talik, E., Gagor, A., and Sitko, R. (2018). *Anal. Chim. Acta* 1008: 57.

37 (a)Dan, Y., Zhang, W., Xue, R. et al. (2015). *Environ. Sci. Technol.* 49: 3007; (b)Loeschner, K., Brabrand, M.S.J., Sloth, J.J., and Larsen, E.H. (2014). *Anal. Bioanal. Chem.* 406: 3845; (c)Deng, Y., Petersen, E.J., Challis, K.E. et al. (2017). *Environ. Sci. Technol.* 51: 10615.

38 Marcinkowska, M. and Baralkiewicz, D. (2016). *Talanta* 161: 177.

39 (a) Zhou, X.-x., Liu, J.-f., and Jiang, G.-b. (2017). *Environ. Sci. Technol.* 51: 3892; (b) Pergantis, S.A., Jones-Lepp, T.L., and Heithmar, E.M. (2012). *Anal. Chem.* 84: 6454; (c) Soto-Alvaredo, J., Montes-Bayon, M., and Bettmer, J. (2013). *Anal. Chem.* 85: 1316; (d) Chen, L.Q., Guo, Y.F., Yang, L.M., and Wang, Q.Q. (2007). *J. Anal. At. Spectrom.* 22: 1403.

40 Chajduk, E. and Polkowska-Motrenko, H. (2019). *Food Chem.* 292: 129.

41 Zhou, X.X., Liu, R., and Liu, J.F. (2014). *Environ. Sci. Technol.* 48: 14516.

42 Tang, Y., Liu, L., Nong, Q. et al. (2022). *J. Chromatogr. A* 1677: 463303.

43 (a) Timerbaev, A.R. (2013). *Chem. Rev.* 113: 778; (b) Liu, L., He, B., Liu, Q. et al. (2014). *Angew. Chem. Int. Ed.* 53: 14476; (c) Hu, L.G., Cheng, T.F., He, B. et al. (2013). *Angew. Chem. Int. Ed.* 52: 4916.

44 (a) Schmidt, B., Loeschner, K., Hadrup, N. et al. (2011). *Anal. Chem.* 83: 2461; (b) Dubascoux, S., Le Hecho, I., Hassellov, M. et al. (2010). *J. Anal. At. Spectrom.* 25: 613; (c) Tan, Z.Q., Liu, J.F., Guo, X.R. et al. (2015). *Anal. Chem.* 87: 8441.

45 Liu, L., Yun, Z., He, B., and Jiang, G. (2014). *Anal. Chem.* 86: 8167.

46 Chen, B.W., Lu, X.F., Arnold, L.L. et al. (2016). *Chem. Res. Toxicol.* 29: 1480.

47 McKnight-Whitford, A., Chen, B.W., Naranmandura, H. et al. (2010). *Environ. Sci. Technol.* 44: 5875.

48 Peng, H.Y., Hu, B., Liu, Q.Q. et al. (2017). *Angew. Chem. Int. Ed.* 56: 6773.

49 Chen, B.W., Fang, L.C., Yan, X.T. et al. (2019). *J. Hazard. Mater.* 363: 34.

50 Wittig, I., Braun, H.-P., and Schägger, H. (2006). *Nat. Protoc.* 1: 418.

51 Wang, M., Feng, W.Y., Wang, H.J. et al. (2008). *J. Anal. At. Spectrom.* 23: 1112.

52 Yan, X.T., He, B., Wang, D.Y. et al. (2018). *Talanta* 184: 404.

53 Yun, Z.J., Li, L., Liu, L.H. et al. (2013). *Metallomics* 5: 821.

54 Calderon-Cells, F. and Encinar, J.R. (2019). *J. Proteomics* 198: 11.

55 Cvetkovic, A., Menon, A.L., Thorgersen, M.P. et al. (2010). *Nature* 466: 779.

56 Fu, D. and Finney, L. (2014). *Expert Rev. Proteomics* 11: 13.

57 (a) Mounicou, S., Połeć, K., Chassaigne, H. et al. (2000). *J. Anal. At. Spectrom.* 15: 635; (b) Coufalíková, K., Benešová, I., Vaculovič, T. et al. (2017). *Anal. Chim. Acta* 968: 58.

58 Yan, H.M., Wang, N., Weinfeld, M. et al. (2009). *Anal. Chem.* 81: 4144.

59 Smith, S.D., She, Y.M., Roberts, E.A., and Sarkar, B. (2004). *J. Proteome Res.* 3: 834.

60 Li, Y.L., He, B., Hu, L.G. et al. (2018). *Talanta* 178: 811.

61 Schapira, M., Calabrese, M.F., Bullock, A.N., and Crews, C.M. (2019). *Nat. Rev. Drug Discovery* 18: 949.

62 Haraguchi, H. (2004). *J. Anal. At. Spectrom.* 19: 5.

63 McRae, R., Bagchi, P., Sumalekshmy, S., and Fahrni, C.J. (2009). *Chem. Rev.* 109: 4780.

64 Korbas, M., Blechinger, S.R., Krone, P.H. et al. (2008). *Proc. Natl. Acad. Sci. U.S.A.* 105: 12108.

65 Yan, X.T., He, B., Liu, L.H. et al. (2018). *Metallomics* 10: 557.

4

Agrometallomics

Xuefei Mao, Xue Li, Tengpeng Liu, and Yajie Lei

Institute of Quality Standard and Testing Technology for Agro-products, Chinese Academy of Agricultural Sciences, and Key Laboratory of Agro-food Safety and Quality, Ministry of Agriculture and Rural Affairs, NO. 12 Zhongguancun South Street, Haidian District, Beijing 100081, P.R. China

4.1 The Concept of Agrometallomics

4.1.1 Introduction

Agriculture has been in existence for more than thousands of years in human history, is characterized with cultivating and breeding animals, plants, and fungi for the raw material of food, medicine, industrial production, and other products [1]. Modernized agriculture (Figure 4.1) is aimed to satisfy the developing demands of human beings, including not only agricultural products and the related by-products, such as vegetable, grain, fruit, tea, cotton, oilseed, farmed animals, and nontoxic fungi, but also agricultural inputs such as seed, fertilizer, feedstuff, pesticide, and other cultivating additions, as well as environmental and ecological media such as water, soil, and atmosphere, and even microbiological species such as plant pathogens and some microorganisms [2].

Metals and metalloids always play very important roles in agricultural products and related media. Firstly, macro- and micro-elements including Mg, Cu, Fe, Zn, Ca, Na, and K are essential to agricultural organisms due to their influence in their breeding, planting, growth, and fruiting processes; on the other hand, as the main pollutants worldwide, toxic metals including Cd, As, Pb, and Hg are receiving more concern as regards living organisms and human health. The World Health Organization (WHO) and many countries' organizations have built the threshold values for metallic or metalloid elements normally existing in nature or living organisms. So far, a number of researches have been performed to assess the behaviors and effects of beneficial and toxic elements in agricultural biology in order to reveal the interaction among elements, compounds, and agricultural organisms. For instance, it was reported that foliar application of Zn and Se could alleviate the Cd and Pb poisoning of water spinach through bioavailability/cytotoxicity in human cell lines; furthermore, this research established a feasible way for biofortification of Zn and Se in agri-food [3]. For another example, by using knockout biotechnology, researchers found the key genes (*hgcAB*) could reduce the bacteria ability for

Applied Metallomics: From Life Sciences to Environmental Sciences, First Edition.
Edited by Yu-Feng Li and Hongzhe Sun.
© 2024 WILEY-VCH GmbH. Published 2024 by WILEY-VCH GmbH.

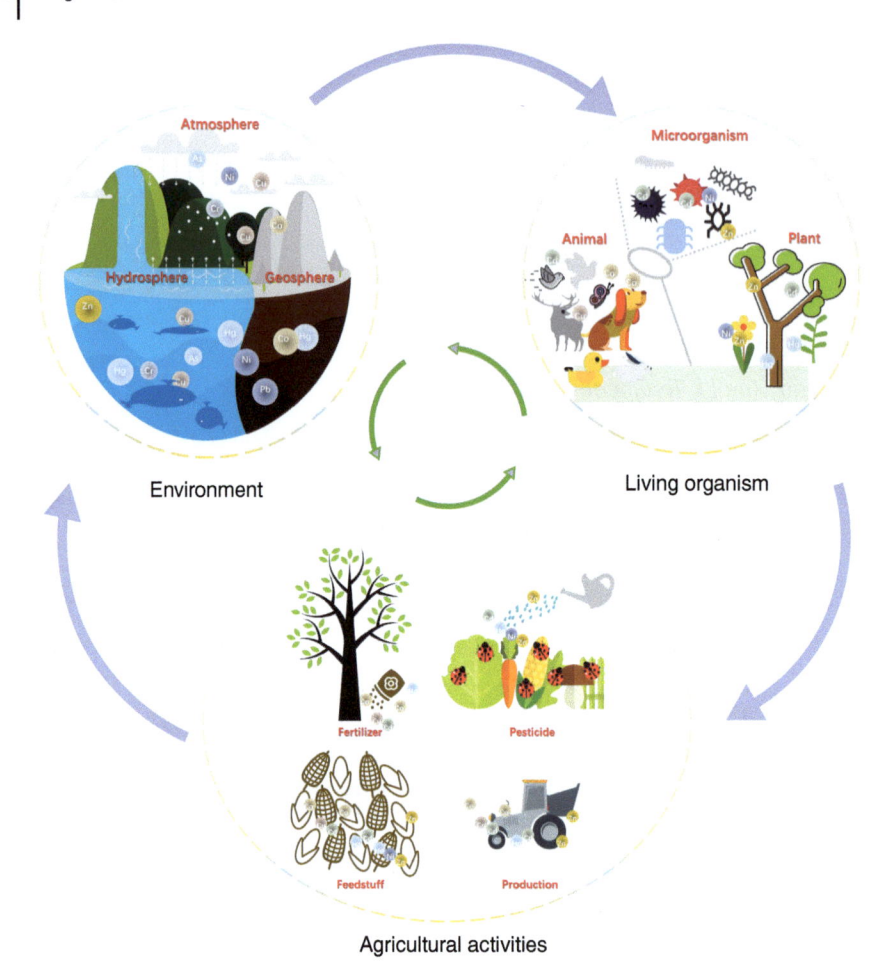

Figure 4.1 Logical diagram of agricultural system in agrometallomics. Source: Xuefei Mao.

yielding methylmercury without impairing cellular growth [4], which was indirectly related with agricultural organisms' health and agri-food safety. However, it must be noticed that a comprehensive conception about biological benefits and health risks caused by metal exposure to agricultural products and media is still unclear and requires a further investigation based on comprehensive methods.

Williams first introduced the term "metallome" in 1990 with an analogy "proteome," indicating the soluble fraction of metals in cells [5]. In 2004, Japanese scientist Haraguchi first used "metallomics" to describe "integrated biometal science" [6] in order to provide a systematic comprehension for uptaking, trafficking, inter-reaction, and excretion of the metals in biological systems [6, 7]. Metallomics is aimed at systematically studying metallomes and the interactions and functional connections of metal or metalloid species for genes, proteins, metabolites, and other biomolecules within lives [8]. There is no doubt that understanding the metallic roles is highly demanded for ensuring the yield, quality, and safety of agricultural products, media, and environment in modernized agriculture. Therefore,

4.1 The Concept of Agrometallomics | 51

Macro scale	
Agricultural system	Cropping, livestock, environment, etc.
Agricultural lives and media	Livestock, crop, pest, fertilizer, feedstuff, etc.
Biological tissue	Root, leaf, stem, grain, animal offal, spore, etc.
Biological cell and micro particle	Bacteria, alga, single cell, nanoparticle, etc.
Biological molecules chemical species	Molecular mechnism of heavy metal, etc.
Macro scale	

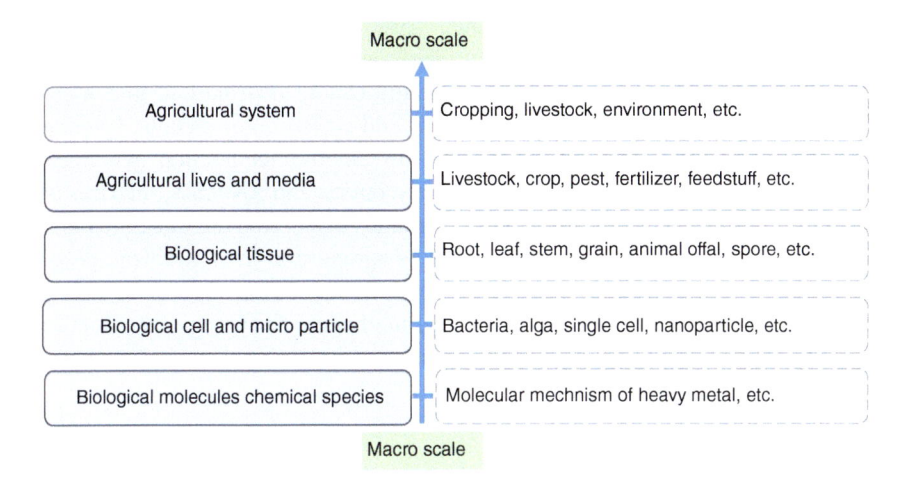

Figure 4.2 Different scales of agrometallomics. Source: Xuefei Mao.

the metallomic research in agricultural science, namely "**agrometallomics**," is necessary and innovative during this period of modernized agriculture. In this work, the concept of agrometallomics is first defined and then the advances of metallic analytical technologies for agrometallome are showcased.

4.1.2 Agrometallomics and its Concept

As a branch of metallomics, agrometallomics is first rendered here as follows: an independent inter-discipline involving an agricultural metallome (*agrometallome*) and metallomics (*agrometallomics*), as well as the potential metallomic techniques and approaches that can be promising in agricultural science and production applications. Referring to the previous term of "environmetallome" [9], the closest concept of "agrimetallome" can be considered as: (i) the entirety of metal and metalloid species that can induce healthy or toxic effects on living organisms at agriculturally relevant levels and ranges; (ii) all metals and metalloids within a biological or non-biological system in which homeostasis could be directly or indirectly interfered and regulated to induce possible beneficial or adverse effects. In Figure 4.2, from the macro-scale perspective, agrometallomics mainly involves cropping, livestock, fishing, and their environmental and ecological systems; secondly, agricultural products and media are individuals containing metallome in livestock, crop, pest, fertilizer, feedstuff, etc.; for biological tissues, agrometallomics can provide solution to the metal distribution, speciation, and changing processes in target organs *via* elemental concentration, speciation, and mapping analysis; for biological cells and micro-particles, agrometallomics can show metal information in target cells and particles via single-cell and nanoparticle analysis and other methods; for biological molecules, agrometallomics is capable of revealing molecular mechanism in target DNA, RNA, protein, lipid, glycogen, and so on *via* labeled elemental analysis.

"**Agrometallomics**" involves elemental existence, level, distribution and speciation in agricultural lives, and relevant media, so a holistic methodology of metallic analysis is thereby indispensable. These analytical approaches mainly cover ultra-sensitive and high-throughput analysis, speciation and state analysis, spatial and micro-analysis for elemental quantification, distribution, and mapping, single-cell and micro-particle, speciation and structure, or metallic labeling in agrometallome. In addition, compared with environmetallomics, genomics, metabolomics, proteinomics, lipidomics and glycomics, and RNomics, matrices of agricultural sample are more complex and variable considering the integrated and comprehensive agriculture system consisting of agricultural lives, fertilizer, pesticide, feedstuff, and soil, water, atmosphere, etc. Therefore, this chapter is mainly focused on the advances of metallic analytical methodologies for agrometallomics to summarize this emerging field of metallomics referring to element-dependent agricultural lives and media.

4.2 Analytical Techniques in Agrometallomics

4.2.1 Sensitivity and Multi-elemental Analysis in Agrometallomics

4.2.1.1 Mass Spectrometry in Agrometallomics

Inductively Coupled Plasma Mass Spectrometry Inductively coupled plasma mass spectrometry (ICP-MS) is the most powerful approach for multi-elemental analysis and hyphenation analysis for solid sampling analysis, micro-particle analysis, micro-zone analysis, elemental speciation, and metal labeling analysis, which benefits from its high sensitivity, less spectral interference, wide linear dynamic range, and isotopic detection [10, 11]. Multi-elements and isotopes in agricultural animals and plants, soil, air and water, fertilizers, feedstuff, and other agricultural media can be measured by various ICP-MS approaches under different conditions [12, 13]. And the detailed information about analytical methodologies related with agrometallomics in this review have been all shown in Table 4.1 involving instrumental method, sample matrix and preparation, target element or isotope, detection capacity, calibration, application, and technical characteristics. Elemental detection capacity of ICP-MS is commonly ng/l level or lower using mathematical correction, collision/reaction cell, or mass-shift mode to alleviate molecular ion interference. The ultratrace multi-elemental information enables the investigation of agrometallome.

For non-hyphenated ICP-MS, nebulizer is the most commonly used liquid sample introduction system after sample digestion or extraction treatment. On the basis of mass analyzer, ICP-MS is commonly classified into inductively coupled plasma quadrupole mass spectrometry (ICP-Q-MS) or tandem quadrupole mass spectrometry (ICP-Q-MS/MS or ICP-QQQ), inductively coupled plasma sector field mass spectrometry (ICP-SF-MS), inductively coupled plasma time of flight mass spectrometry (ICP-TOF-MS), and multi-collector inductively coupled plasma mass spectrometry (MC-ICP-MS) [14]. Considering the complicated matrices of various agricultural media, composition separation and purification *via* ion-exchange resin [15] and other means should be further investigated to eliminate matrix interference and isotopic fractionation effect [16].

Table 4.1 Summarization and comparison of analytical methodologies for agrometallomics from literatures.

No.	Instrumental	Matrix	Sample preparation	Element or isotope	LOD or LOQ	Calibration	Application and technical characteristics	DOI of references
1	LA-ICP-MS	Human eyes	Tissue sections are fixed on glass slides	Metallothionein 1/2 (MT1/2); complement factor H (CFH); amyloid precursor protein (APP) in retina	/	/	Use LA-ICP-MS to simultaneously locate three proteins related to neurodegenerative diseases (such as AMD) in the same eye tissue. The multiplex methodology allows the amplification of the protein detection. Moreover, knowing the number of tag atoms and available antibodies per labeled immunoprobe, semi-quantification can be achieved.	https://doi.org/10.1016/j.talanta.2020.121489
2	MC-ICP-MS	Blend ore BCR 027; rye grass standard BCR 281; basalt; soil	Microwave digestion; Anion-Exchange Chromatography.	Zinc stable isotope ratios	Blend ore BCR 027: $0.25 \pm 0.06‰$ (2SD); ryegrass BCR 281: 0.40 ± 0.09; Imperial Zn: $0.10 \pm 0.08‰$; 2 SD; London Zn: 0.08 ± 0.04	Double-spike sample analysis protocol; Standard sample bracketing protocol	Characterize the Zn isotope pool available to plants in a natural setting. Developed a DS protocol for the accurate and precise determination of Zn isotope ratios.	https://doi.org/10.1007/s00216-010-4231-5

(Continued)

Table 4.1 (Continued)

No.	Instrumental	Matrix	Sample preparation	Element or isotope	LOD or LOQ	Calibration	Application and technical characteristics	DOI of references
3	ICP-MS	Arctic ice core samples	Acid washing and water washing	As, Cd, Co, Cu, Ni, Pb, Zn, U	As: 1.1 pg/g Cd: 0.9 pg/g Co: 0.1 pg/g Cu: 0.4 pg/g Ni: 0.6 pg/g Pb: 0.1 pg/g Zn: 3.0 pg/g U: 0.1 pg/g	External standard method, 0.1 (U)–673.3 (Zn) pg/g	Measurement and Evaluation of Trace Heavy Metals in Arctic Ice Cores. An all-plastic device was developed and used together with ceramic knives to remove the contaminated surface of ice cores.	https://doi.org/10.1016/S0039-9140(01)00509-4
4	ICP-MS/OES	Tea	Microwave digestion	More than 80 elements such as Al, B, Cu, Fe, Mn, Na, ^{137}Ba, ^{138}Ba, ^{79}Br, ^{81}Br, ^{52}Cr, ^{53}Cr, ^{58}Ni, ^{60}Ni, ^{85}Rb, ^{86}Sr, ^{88}Sr, ^{46}Ti, ^{47}Ti, ^{48}Ti, ^{64}Zn, ^{66}Zn	/	External standard method, $R^2 > 0.999$	Used for tea traceability. Simultaneous determination of various mineral elements and stable isotopes by ICP-MS and ICP-OES.	https://doi.org/10.1016/j.foodcont.2020.107735
5	ICP-MS/MS	Fruit wines	Acidified with nitric acid.	Al, V, Cr, Mn, Fe, Co, Ni, Cu, Zn, As, Se, Cd, Hg, Pb	0.41–58.1 ng/l	External standard method, liner range: 0.1–500 µg/l	Establishing comprehensive standards is necessary for aiming at the quality control and food safety of fruit wines. Spectral interferences in the complex matrix composition of different fruit wine samples, in the MS/MS mode, were eliminated using mixed reaction gases of O_2/H_2 and $NH_3/He/H_2$ through the mass shift and on-mass methods.	https://doi.org/10.1016/j.foodchem.2019.125172

#	Method	Sample	Sample pretreatment	Analyte	Detection limit	Internal standard	Conclusion	DOI
6	CE/ICP-MS/MS	Liposomal doxorubicin formulations	Dilute with 5% glucose	Intra-liposomal and external sulfate	Sulfate: 0.8 μg/ml	Internal standard method (25.0 μg/ml of arsenate in 5% dextrose)	Simultaneously determine the content of liposulfate in adriamycin liposome preparations in vivo and in vitro. CE method is more reliable than the filtration method for separation of external sulfate and intra-liposomal sulfate in nanoparticle drug carriers such as liposomes.	https://doi.org/10.1016/j.ijpharm.2019.03.003
7	ICP-QQQ-MS	A mercury resistant bacterial strain (SA2) in soil sample	Cultivation and isolation under low phosphate conditions	Hg	/	/	Potential of SA2 in the bioremediation of mercury-contaminated waste. Rapid removal of Hg from experimental medium and the presence of mercuric reductase enzyme show the potential of SA2 in the bioremediation of Hg contaminated wastes.	https://doi.org/10.1016/j.chemosphere.2015.08.061
8	ICP-QQQ-MS	/	Digestion	As and Se	As: 0.001 μg/l Se: 0.002 μg/l	/	Accurate determination of rare-earth elements in food. ICP-QQ adopts the mass transfer mode, which reduces the double charge effect, and accurately determines As and Se without correction equations.	https://doi.org/10.1039/C4JA00310A

(Continued)

Table 4.1 (Continued)

No.	Instrumental	Matrix	Sample preparation	Element or isotope	LOD or LOQ	Calibration	Application and technical characteristics	DOI of references
9	ICP-TOF-MS	Natural water samples"	Acidification and refrigeration	Lead and lead isotope ratios	6 ng/l	Internal standard method	Aimed at the development of a suitable on-line preconcentration method for the determination of precise isotopic ratios in complex matrices. ICP-TOF-MS provides a very useful technique for simultaneous determination of traces lead and lead isotope ratios in natural water samples.	https://doi.org/10.1016/j.aca.2003.11.007
10	ICP-TOF-MS	Seawater	/	Transition and rare-earth elements	20–50 pg/l	/	Determination of trace elements and rare-earth elements in seawater. On-line separation and preconcentration was achieved using a timed flow-injection FI system incorporating a column.	https://doi.org/10.1016/S0584-8547(01)00263-4
11	ICP-oa-TOF-MS	Tomato	Microwave digestion	Li, Be, Al, V, Cr, Mn, Co, Cu, Zn, Ga, As, Rb, Sr, Ag, Cd, In, Cs, Ba, Tl, Pb, Bi, U		External standard method	Food traceability. Features of ICP-oa-TOF-MS simultaneous measurement.	https://doi.org/10.1016/j.foodcont.2014.04.027

12	ICP-SF MS	Surface and subsurface waters	Filter	Rare-earth elements (REE)	0.05–0.2 ng/l	External standard method Line range: 0.5–5 μg/l	Use of REEs as diagnostic tools for rock–water interactions and for monitoring surface and subsurface waters. Use two atomizers to compare the ability to determine rare-earth metals.	https://doi.org/10.1016/j.sab.2009.06.013
13	ICP-MS/MC-ICP-MS	Soil	Crushing and digestion with acid	Pb, Cd, Pb isotope	/	/	Elucidate the occurrence and accumulation of Pb and Cd in the mining region. MC-ICP-MS can accurately determine lead isotopes.	https://doi.org/10.1016/j.jhazmat.2019.121528
14	TIMS ICP-MS	Branches, leaves and olives of olives trees.; Soil; Irrigation and rainwaters; fertilizers	Nitric acid for Sr separation and purification on the Eichrom Sr-resin	Sr isotope	Sr contents in the Leaves: 14.3–41.4 μg/g; Sr contents in the irrigated soils: 4.52–8.28 μg/g; Sr contents in the soil: 17.2–41.7 μg/g; Water: 0.708539 (±9E–6); Rainwaters: 0.70877 (±1E–5)	/	The sources of strontium absorption by trees in different cultivation environments were discussed. Collection and determination of strontium isotopes in different environments.	https://doi.org/10.1016/j.apgeochem.2017.05.010
15	TIMS	Dolomite basalt peridotite	HF + HNO$_3$ digestion, drying and dissolve with HCl. Separate the matrix with LN resin column and Ln Spec resin column	Cr isotope ratios	Internal precisions: (2 SE) of 0.017 ± 0.037 Reproducibility: 0.113454 ± 0.000004	Cr double spike	To enhance Cr ionization efficiency and develop a highly sensitive and precise analytical protocol to determine Cr isotopes using TIMS. Propose a Nb$_2$O$_5$ emitter and improve the traditional two-step anion-exchange resin column procedure.	https://doi.org/10.1039/C6JA00265J

(Continued)

Table 4.1 (Continued)

No.	Instrumental	Matrix	Sample preparation	Element or isotope	LOD or LOQ	Calibration	Application and technical characteristics	DOI of references
16	LIMS	Soil	/	Fe, Mg, As, Na, K, Ca, Ga, Ge, Ba, Be, Bi, Cd, Ce	/	The molar concentration of the element is approximately equal to the signal intensity divided by the total ion current.	Analyze geochemical standard reference soil samples to demonstrate the rapid analytical capability of this technique. Owing to the introduction of buffer gas in the ion source, the LIMS system proves to be a suitable technology for rapid detection of multiple elements in soil with little or no sample preparation.	https://doi.org/10.46770/AS.2020.04.001
17	MC-ICP-MS TIMS ICP-AES	Solid carbonate samples	Soak in H_2O_2, dissolve in HCl, and purify by Amberlite IRA 743 resin column and mixed resin column	Boron isotopes	The average ($n^1/_410$) internal analytical precision and the external precision of the measured 11B/10B ratios of NIST of NIST 951 are 70.05‰ and 0.09‰, respectively.	Sample-standard bracketing procedure (SSB) PTIMS-Cs_2BO_2 +-static double collection method	To achieve high precision and high accuracy measurements of boron isotopic compositions in natural geological samples, improving in the separation/purification process of boron from complex matrices. Establishing of static double-collection PTIMS-Cs_2BO_2 method without special requirements on instrumental hardware.	https://doi.org/10.1016/j.talanta.2014.02.009

#	Instrument	Sample	Preparation	Analyte	Precision	Method	Description	DOI
18	TIMS MC-ICP-MS	Ni standard solution	/	Ni Isotope	Internal precision (2 RSE) of ±0.03 to 0.05	Isotope dilution method	Improved the Ni ionization efficiency and precise measurement. Proposed a zirconium hydrogen phosphate emitter, combined with phosphate acid.	https://doi.org/10.46770/AS.2020.212
19	Q-ICP-MS ICP-MS AAS	Wine	Microwave digestion and Dowex 50W-X8 ion-exchange resin to separate Sr and Rb	$^{87}Sr/^{86}Sr$ Isotope Ratio and Mineral elements	ICP-MS: 0.1–0.6 µg/l; AAS: 0.05–5.0 mg/l	External standard method linear range: 2.5–50 µg/l R^2: 0.9991–0.9999	Geographical traceability of Wines. The combination of mineral elements and isotopes strengthens the feasibility of traceability.	https://doi.org/10.1007/s12161-016-0550-2
20	TIMS LA-MC-ICP-MS	Columbite-tantalite minerals	HF immersion, ion-exchange chromatography column separation of Pb and U	U-Pb	LA-ICP-MS precise <0.5%	/	To determine reliable crystallization ages for rare-element pegmatites. LA-MC-ICP-MS technique as a method of directly analyzing primary zones in columbite-tantalite, avoiding inclusions and alteration.	https://doi.org/10.1007/s00410-003-0538-y
21	Compact quadrupole mass spectrometer	Water	/	Fe isotope ratio	/	Isotope dilution mass spectrometry	Determined Fe traces in different water. Using boric acid in addition to silica gel, was developed to determine Fe isotope ratios with thermal ionization mass spectrometry.	https://doi.org/10.1016/0168-1176(88)80036-3

(Continued)

Table 4.1 (Continued)

No.	Instrumental	Matrix	Sample preparation	Element or isotope	LOD or LOQ	Calibration	Application and technical characteristics	DOI of references
22	TIMS	Fecal samples; blood; urine	Dissolve with HCl and purify by Ion-exchange chromatography	Zn, P, and Fe stable isotope	/	/	Enriched stable isotopes were used in nutrition studies. After obtaining baseline values for zinc, copper, and Fe absorption from diets adequate in all nutrients, the effects of age, pregnancy, and several dietary variables were studied.	https://doi.org/10.1007/BF02796684
23	GD-QMS/ICP-OES/XRF	Hf metal	Hf metal was cleaned with iso-propanol, dried; Hf metal dissolved ultrapure water	Hf	/	Non-matrix matched standard	Determinated impurities Hf metal. Utilizes a non-matrix matched standard and an internal reference element for the quantification of more than 70 elements.	https://doi.org/10.46770/AS.2020.03.002
24	ICP-MS/GD-QMS	High purity cadmium	Dowex-50Wx8 ion-exchange resin purification; HNO_3 is digested, then dissolved in HCl	Trace element impurities	^{49}Ti: 0.47 ng/g ^{51}V: 0.30 ng/g ^{52}Cr: 1.1 ng/g ^{55}Mn: 0.4 ng/g ^{59}Co: 0.05 ng/g ^{62}Ni: 0.47 ng/g ^{63}Cu: 1.1 ng/g ^{64}Zn: 6 ng/g ^{86}Sr: 0.3 ng/g	/	Developed a separation procedure suitable for the determination of trace and ultra-trace level impurities using ICP-QMS. More elements could be reported using GD-QMS.	https://doi.org/10.1016/S0003-2670(99)00860-0

25	PGD-TOFMS	Cu//NiCu nanometric multilayers	Samples were prepared by using Si//Cr(5 nm)//Cu(20 nm) substrates (cathode)	^{28}Si, ^{52}Cr, ^{56}Fe, ^{59}Co, ^{62}Ni	/	Multi-matrix calibration approach	Focused towards the depth profile analysis of Cu/NiCu nanolayers and multilayers electrodeposited on Si wafers. Using the afterglow region for all the sample components except for the major element (Cu) that was analyzed in the plateau.	https://doi.org/10.1016/j.sab.2017.06.016
26	GD-MS	Soil; sediment and vegetation	The sample is compacted with the silver powder.	Trace radioisotopes	Detection limits in the ng/g range	Optimizing the mass resolution	Determination of some radioisotopes of cesium, strontium, plutonium, uranium and thorium in soils, sediments and vegetations. Reasonable sample preparation.	https://doi.org/10.1007/s00216-3550642
27	PGD-TOF-MS/ICP-OES	Geological sample of rare-earth elements enriched ore 65-1B-ATK-97	Grinding and pressing	Major and Trace Elements	$2-4 \times 10^{-6}$ mass %	Using relative sensitivity factors	Direct Quantification of Major and Trace Elements in Geological Samples; geological analysis. The method requires a minimal sample pretreatment and is applicable for the determination of wide range of elements of the periodic table in a single analytical procedure without sample dissolution.	https://doi.org/10.1080/00032719.2018.1485025]

(Continued)

Table 4.1 (Continued)

No.	Instrumental	Matrix	Sample preparation	Element or isotope	LOD or LOQ	Calibration	Application and technical characteristics	DOI of references
28	GD-MS	Certified reference materials	HNO_3 and water washing, argon drying	B, C, Al, Si, P, S, V, Ti	/	Universal RSFs (u-RSFs)	Using universal RSFs (u-RSFs) to establish a direct quantification method by GDMS. Based on the optimum parameters, a set of u-RSFs were obtained by averaging the RSFs in certified reference materials (CRMs) covering most common matrixes.	https://doi.org/10.1016/j.sab.2019.01.004
29	GD-MS	Chromium (III) oxide (Cr_2O_3) and chromium (VI) oxide (CrO_3)	Grinding and pressing of graphite powder or silver powder	Cr speciation	Chromium (III) oxide and chromium (VI) oxide were found to be 2.89% and 1.97%	Normalize	Direct speciation of chromium in solid-state samples. The impact of glow discharge operating conditions on the appearance of these characteristic cluster ions is discussed.	https://doi.org/10.1039/B803358G
30	Pulsed millisecond radio frequency-GD-TOF-MS	Solid-state samples	Sample mixed with silver powder	Fe(II) monoxide (FeO), Fe(III) oxide (Fe_2O_3) and Fe(II, III) oxide (Fe_3O_4) speciation	/	/	Direct Fe_xO_y speciation in solid-state materials. Adjustment of parameters such as sampling distance, temporal regime, discharge gas pressure, pulse frequency, and duty cycle is essential to enable such speciation.	https://doi.org/10.1039/C0JA00229A

31	APGD-MS	Apple, cranberry, grape and orange juices	Spray the fruit juice containing pesticides on the filter paper strip	Pesticides	Metolcarb: 1 µg/l; Arbofuran and dinoseb: 2 µg/l	Internal standard method	A fast and sensitive analysis of pesticides. No sample pretreatment was necessary to analyze these pesticides by direct desorption/ionization using APGD-MS.	https://doi.org/10.1002/rcm.3677
32	Nano SIMS	Flooded rice soil	freeze-dried and ground.	Active diazotrophs	11.33 ± 1.90 kg N/ha in the rice-planted soil and 3.55 ± 1.18 kg N/ha in the nonplanted soil	/	Investigate the effects of rice-planting on the BNF and its associated diazotrophic communities and discover the active diazotrophs in a rice-planted soil. The biological nitrogen fixation in planted and nonplanted paddy soils was quantified using a chamber-based $^{15}N_2$-labeling technique, and the active diazotrophs of soil were assessed by $^{15}N_2$-DNA-stable isotope probing (SIP).	https://doi.org/10.1007/s00374-020-01497-2
33	Nano SIMS ICP-MS	Radish	Freeze in liquid nitrogen, fix the cells with acrolein, embed the sections with Araldite 502 resin	Zn, Ca	^{70}Zn 180 ± 24 and Cd 352 ± 11	/	Examine the in situ Zn and Cd distribution in the root apices of metal-sensitive radish (Raphanus sativus) plants after short-term (24 h) exposure to low concentration (2.2 µM) of each Cd and Zn. Using Nano-SIMS, we mapped the accumulation of 70 Zn and 114 Cd in various root apical tissues in situ for the first time.	https://doi.org/10.1016/j.ecoenv.2019.01.021

(Continued)

Table 4.1 (Continued)

No.	Instrumental	Matrix	Sample preparation	Element or isotope	LOD or LOQ	Calibration	Application and technical characteristics	DOI of references
34	Nano SIMS	Pea seeds	Protein blot analysis	Fe- Ferritin	/	/	Fe bioavailability in pea. Using Nano-SIMS analysis, we provide unprecedented detail on the subcellular distribution of Fe and ferritin.	https://doi.org/10.1038/s41598-018-25130-3
35	Cryo-TOF-MS	Sepal tissues of *Hydrangea macrophylla*	Low temperature freezing treatment	Hydrangea blue-complex	/	/	Studied the mechanisms of blue flower coloration and focused on the color variation of hydrangea. Imaging mass spectrometry is a powerful tool in mapping of inorganic ions and organic molecules in tissues.	https://doi.org/10.1038/s41598-019-41968-7
36	ToF-SIMS	Breast tissues in broiler chicken	Liquid Nitrogen Freeze Cutting	Fatty acid	/	/	Observe fatty acid profiles of breast tissues in broiler chicken subjected to varied vegetable oil diet. ToF-SIMS method allows to obtain useful information in the comparative analysis of animal tissue and facilitates the acquisition of new and previously inaccessible information helpful in understanding the biology and chemistry of lipid-related processes.	https://doi.org/10.1002/jms.4486

37	ICP-MS ICP-OES DMA	Echinodermata and Chordata species	The samples were digested by a microwave system	Pb, Cd	Pb: 0.002 mg/kg Cd: 0.001 mg/kg As: 0.002 mg/kg Hg: 0.002 mg/kg Al: 0.004 mg/kg	External standard method	Analyzing concentrations of heavy metals including arsenic, lead, cadmium, aluminum and mercury in commonly consumed seafood species. ICP-MS and ICP-OES can quickly monitor high-throughput pollutant elements.	https://doi.org/ 10.1080/ 19393210.2014 .932311
38	LIBS ICP-OES	Five species of the medicinal plants	Wash, dry, grind and press and Microwave digestion	Carbon, Fe, magnesium, silicon, calcium, oxygen, hydrogen, nitrogen and sodium	/	Intensity ratio method	Rapid analysis of the major and trace elements in five medicinal plant samples. The intensity ratio method was adopted for the relative profiling of the elements qualitatively present in the five medicinal plant samples. The chemometric multivariate method PLS-DA was also used on the LIBS data of plant samples for a better understanding of their characteristic variations.	https://doi.org/ 10.46770/AS .2020.06.003
39	MP-AES	Soil, sediment, water reference materials, particulate matter, and real-life samples	Water: acidification Soil: HF: +HNO$_3$: +HClO$_4$: +HCl digestion RSPM:HNO$_3$ digest	As, Cr, Ni, Zn, Pb, Cu	0.05–5 ng/g	Matrix matching	Environmental monitoring of contaminated sites. By exploring the performance of MP-AES under different conditions, determine the optimal conditions to measure pollutants.	https://doi.org/ 10.1007/s10661- 014-3913-4

(Continued)

Table 4.1 (Continued)

No.	Instrumental	Matrix	Sample preparation	Element or isotope	LOD or LOQ	Calibration	Application and technical characteristics	DOI of references
40	MP (Microwave Plasma)-AES	Seed cake and husk	Seeds were dehulled, milled and processed	Zn, As, Cu, Fe, Pb	/	/	The usefulness of seed cake and husk as industrial wastes generated from Jatrophacurcas in biotechnology. Nutritive components, macro and micro-elements were investigated following standard protocols.	https://doi.org/10.22161/ijeab/3.4.52
41	MIP (Microwave-Induced Plasma)-OES	Fish	Freeze well and then digest with nitric acid	Mercury, Cadmium, Lead, Arsenic, Copper, Fe, and Zinc	Hg: 0.01 mg/kg Pb: 0.07 mg/kg Cd: 0.01 mg/kg As: 0.02 mg/kg Fe: 0.14 mg/kg Cu: 0.12 mg/kg Zn: 0.05 mg/kg	External standard method $R^2 > 0.995$	Determination of heavy metal elements in fish. The method described can be considered adequate for the simultaneous determination and quantification of the chosen heavy metals in fish matrices.	https://doi.org/10.1007/s12161-017-0908-0
42	MP-OES/ ICP-MS	Toenails	Digest with nitric acid	Al, Ba, Ca, Cr, Cs, Cu, Fe, Mg, Mn, Ni, P, Pb, Rb, S, Sb, Se, Sn, Sr, V and Zn	Al: 10 ng/g, Ba: 4 ng/g, Ca: 0.08 ng/g, Cr: 10 ng/g, Cs: 0.2 ng/g, Cu: 0.001 ng/g, Fe: 0.04 ng/g, Mg: 30 ng/g, Mn: 3 ng/g, Ni: 0.007 ng/g, P: 0.009 ng/g, Pb: 0.04 ng/g, Rb: 5 ng/g, S: 0.004 ng/g, Sb: 0.003 ng/g, Se: 3 ng/g, Sn: 0.001 ng/g, Sr: 0.5 ng/g, V: 0.03 ng/g, Zn: 0.1 ng/g	External standard method	Classification of type-2 diabetes. Combining elemental analysis of toenails and machine learning techniques as a non-invasive diagnostic tool for the robust classification of type-2 diabetes.	https://doi.org/10.1016/j.eswa.2018.08.002

#	Method	Sample	Sample preparation	Analyte	LOD	Calibration	Description	DOI
43	Matrix-Assisted Microwave Induced Plasma Surface Sampling Atomic Emission Spectrometry	Water, soil, marine bottom sediment, gold ore	For solid powder samples was sandwiched between two sheets of the whole filter paper.	Ag, Au, Ba, Cd, Cr, Cu, Eu, La, Mn, Ni, Pb, Sr, y	1.0–88 ng/ml	/	Determination of Metallic Elements in Environmental Samples. The proposed method provided several advantages, including fast analysis speed, little sample consumption, and simple instrument design and system operation.	https://doi.org/10.1016/S1872-2040(15)60857-X
44	HG-DBD-AFS	Surface Water	Filter paper filtration	As	1.0 ng/l	External standard method linear range: 0.05–5 µg/l $R^2 > 0.995$	Ultra-trace Arsenic Determination in Surface Water. A novel DBD reactor was designed for arsenic trap/release coupled to HG-AFS.	https://doi.org/10.1021/acs.analchem.6b00506
45	In Situ-DBDT-AFS	Tap water, river water, and lake water	The samples were cleaned using filter paper prior to usage, respectively.	As	2.8 pg	External standard method Linear range: 0.1–8 µg/l $R^2 > 0.998$	Rapid determination of arsenic on site. The in situ DBD trap is applicable to both AAS and OES as well as a wide array of atomic spectrometers capable of on-site and fast analysis.	https://doi.org/10.1021/acs.analchem.8b01199
46	UVG-DBD	Soil Surface Water	Soil digestion, membrane filtration of water samples	Se	4 pg	External standard method Linear range: 0.05–50 µg/l $R^2 > 0.995$	Detection of Se in water and soil samples. UVG was combined with DBD for the gas-phase preconcentration of volatile selenium species to enhance the analytical sensitivity and eliminate matrix interference.	https://doi.org/10.1021/acs.analchem.0c00878

(Continued)

Table 4.1 (Continued)

No.	Instrumental	Matrix	Sample preparation	Element or isotope	LOD or LOQ	Calibration	Application and technical characteristics	DOI of references
47	HG-PD (point discharge)-OES	Soil	HF+HNO$_3$+HCl digestion; HClO$_4$ + HCl+ ascorbic acid for subsequent analysis	As, Bi, Sb, Sn	As: 7 µg/l Bi: 1 µg/l Sb: 5 µg/l Sn: 2 µg/l	External standard method RSD < 5%	Simultaneously determine As, Bi, Sb and Sn in soil samples. The hydrogen produced by the HG process is used to effectively minimize the background spectrum emission of the micro-plasma, the analyte is converted into volatile species, and separated from the sample solvent and matrix by the HG process to eliminate the interference of the sample matrix.	https://doi.org/10.1039/C6JA00341A
48	PEVG -AFS	Simulated natural water sample GSB 07-1184-2000; soil standard; rice standard	The water sample is diluted to a pH of 3.2; Digestion of solid samples with HCl+HNO$_3$	Zn, Cd	Zn: 0.3 µg/l Cd: 0.003 µg/l	External standard method linear range: 0.05–0.5 µg/l RSD = 2.4%	Solution anode glow discharge for the determination of Cd and Zn by atomic fluorescence spectrometry. The overall efficiency of the PEVG system was much higher than the conventional electrochemical hydride generation (EcHG) and HCl-KBH$_4$system.	https://doi.org/10.1021/acs.analchem.7b00126

#	Method	Sample	Pretreatment	Analyte	LOD	Method/Linear range	Description	DOI
49	HG-SCGD-AES	Rice	Microwave digestion with HNO_3 followed by membrane analysis	Inorganic arsenic valence	0.3 μg/l	External standard method Linear range: 2–100 μg/l RSD = 2.5%	Separate As(III) and As(V). Arsenic was converted into volatile hydrides by the reaction with $NaBH_4$ and HCl, after which they were delivered and subsequently injected into the near-anode region of the SCGD.	https://doi.org/10.1039/C7JA00228A
50	MWP-PTR-MS	Water	/	Volatile organic compounds	Acetone: $2.7 \times 10{-12}$ mol/l Toluene: $12 \times 10{-12}$ mol/l Acetonitrile: $1.9 \times 10{-12}$ mol/l Acetaldehyde: $2.3 \times 10{-12}$ mol/l	External standard method linear range: 4.5×10^{-11} – 4.5×10^{-9} mol/l	Volatile Organic Compound Monitoring. MWP has high potential in improving the performance and reliability of current PTR-MS.	https://doi.org/10.1016/j.talanta.2019.120468
51	MIPDI	Drug tablets, capsules, and ointments	No preprocessing required	Volatile Organic Compound Monitoring	60 pg	/	Experiments have been demonstrated that this method can be used to produce positive and negative ion mass spectra with a wide range of organic compounds. The simple construction, easy operation, stability, reproducibility, the possibility of producing plasmas with alternative gases, and its high efficiency for desorbing/ionizing analytes directly from sample surfaces in an ambient atmosphere.	https://doi.org/10.1021/ac400296v

(Continued)

Table 4.1 (Continued)

No.	Instrumental	Matrix	Sample preparation	Element or isotope	LOD or LOQ	Calibration	Application and technical characteristics	DOI of references
52	DLAAS	Chlorine and fluorine.	/	CCl_2F_2	Detection limits of 400 ppt and 2 ppb for CCl_2F_2 in He were found using the Cl 837 nm and the F 685 nm line, respectively.	/	Evaluate the results of diode laser absorption spectrometry obtained with the dielectric barrier discharge and DBD measurement. Complete dissociation and efficient excitation of metastable atoms.	https://doi.org/10.1016/S0584-8547(00)00286-X
53	HG-DBD-AAS	SRM 2670Urine SRM 1571 orchard leaves	Add water directly to the freeze-dried urine	Arsenic Speciation	As(III): 1.0 µg/l As(V): 11.8 µg/l MMA: 2.0 µg/l DMA: 18.0 µg/l	External standard method arsenic linear range: 20–500 µg/l MMA, DMA linear range: 25–500 µg/l	Arsenic Speciation Analysis. A novel hydride atomizer based on atmospheric pressure DBD plasma.	https://doi.org/10.1021/ac051022c
54	Atmospheric pressure dielectric barrier discharge plasma	Stock solutions of 1000 mg/l of Se, Sn and Sb	/	Se, Sb and Sn	Se: 13.0 µg/l Sb: 0.6 µg/l Sn: 10.6 µg/l	External standard method Sb linear range: 25–500 µg/l Se linear range: 25–100 µg/l Sn linear range: 25–100 µg/l $R^2 > 0.9999$	Determination of Se, Sb, and Sn with Ar as plasma gas to investigate its applicability to other hydride-forming elements. A new atomizer sensitive to hydride.	https://doi.org/10.1016/j.sab.2006.06.012
55	DBD-AFS	Lake water, River water	/	As(III)	As(III): 0.04 µg/l	External standard method Linear range: 0.5–50 µg/l	On-site arsenic analysis. The most attractive characteristic of the present DBD atomizer is the low operation temperature.	https://doi.org/10.1016/j.aca.2007.11.041

56	DBDI-MS	/	Explosive chemicals	Pg-ng	/	Used for the fast detection of explosives in the field of antiterrorism and conservation. Demonstrated to be highly sensitive for the detection of explosives which is comparable with that of DART and DESI.	https://doi.org/10.1002/jms.1243
57	DBDI-MS	Monosodium glutamate	Filtered and placed on a glass slide.	Amino acids	L-alanine: 3.5 pmol	/	For rapid monitoring in the field. A new ion source based on dielectric barrier discharge was developed as an alternative ionization source for ambient mass spectrometry.	https://doi.org/10.1016/j.jasms.2007.07.027
58	UVG-DBD	Soil, surface water	Soil digestion, membrane filtration of water samples	Se	4 pg	External standard method Linear range: 0.05–50 μg/l $R^2 > 0.995$	Detection of Se in water and soil samples. For the first time, UVG was combined with DBD for the gas-phase preconcentration of volatile selenium species to enhance the analytical sensitivity and eliminate matrix interference.	https://doi.org/10.1021/acs.analchem.0c00878
59	ETV-DBD-AFS	Aquatic food	solid sampling	Hg	0.5 μg/kg	External standard method Linear range: 2–12 mg/kg $R^2 > 0.996$	On-line microplasma decomposition of gaseous-phase interference for solid sampling mercury analysis in aquatic food samples. DBD was first utilized to eliminate volatile matrix interference for Hg analysis in aquatic food samples. On-line DBD reactor succeeds in replacing catalytic pyrolysis furnace for decomposing VOCs.	https://doi.org/10.1016/j.aca.2020.04.057

(Continued)

Table 4.1 (Continued)

No.	Instrumental	Matrix	Sample preparation	Element or isotope	LOD or LOQ	Calibration	Application and technical characteristics	DOI of references
60	Electromagnetic heating-microplasma AES	Soil	Solid sampling	Hg, Cd, Pb	Hg: 8.0 µg/kg Cd: 17.8 µg/kg Pb: 3.5 mg/kg	External standard method Hg linear range: 0.015–0.590 mg/kg Cd linear range: 0.060–0.450 mg/kg Pb: 14.0–314.0 mg/kg	Developed portable, miniaturized atomic spectrometers for solid sample analysis. Electromagnetic heating was firstly explored as sample introduction approach in portable DBD-AES to directly analyze soil.	https://doi.org/10.1016/j.talanta.2020.121348
61	INAA	Cigarettes	Grinding, sieving and drying	Toxic heavy metals	/	NAA k0-standardization method	Analysis of toxic heavy metals in cigarettes. The samples were irradiated at the core of the Second Research Egyptian Reactor ET-RR-2, and the Induced activities were counted by <gamma>-ray spectrometry using an efficiency-calibrated High Purity Germanium (HPGe) detector.	https://doi.org/10.1016/j.jtusci.2017.01.007
62	INAA	Beef	Freeze-drying,	Br, Ca, Cs, La, Sc, Se, Sr, Th, Zn	/	/	Discriminate between beef cattle diets. The feeding method implemented in the finishing stage of beef cattle was investigated as a case study, applying NAA and data mining techniques for assessing the differences between two types of diets commonly used.	https://doi.org/10.1007/s10967-019-06874-2

#	Method	Sample	Preparation	Elements	Results	Standard	Description	DOI
63	INAA	Biological materials	Irradiation samples	Al	/	Internal standard method	Determination of aluminum in various biological materials. In order to determine the interference factors from Si and P, SiO$_2$ and (NH$_4$)$_2$HPO$_4$ were used.	https://doi.org/10.1016/j.legalmed.2009.02.046
64	IBA, INAA	Drug products	Crushed into granules	Sildenafil; Na, Mg, Si, P, S, Ca, Ti, Fe, Ni, Zn;	/	/	Identify drugs. It is proposed for the first time a combined approach based on IBA (both PIXE and MeV-SIMS) and INAA, allowing characterization of both authentic and illegal pharmaceuticals containing sildenafil.	https://doi.org/10.1016/j.talanta.2020.121829
65	INAA	Stream sediments	Filtering drying grinding	As, Ba, Mn, Sb, V, Zn	As: 0.94 mg/kg Ba: 43.0 mg/kg Mn: 7.35 mg/kg Sb: 0.14 mg/kg V: 34.9 mg/kg Zn: 5.43 mg/kg	Uncertainty sources can be defined in terms of a gamma-ray counting, weighing and the uncertainty of a reference material.	Determination of pollutant elements in river sediments. INAA was applied for the determination of the elemental contents in the sediment samples by using the NAA irradiation holes of the HANARO research reactor and HPGe gamma-ray spectrometers.	https://doi.org/10.1016/j.aca.2008.03.064
66	INAA	Algerian Artemisia Plant	Grind	As, Ba, Br, Ca, Ce, Co, Cr, Cs, Eu, Fe, Hf, K, La, Na, Rb, Sb, Sc, Sm, Sr, Yb, and Zn	/	Internal standard method	Used to offer scientific basis for an optimum usage of the studied plants and so enriches the database of medicinal herbs. A sensitive nuclear analytical approach	https://doi.org/10.1007/s12011-020-02358-7

(Continued)

Table 4.1 (Continued)

No.	Instrumental	Matrix	Sample preparation	Element or isotope	LOD or LOQ	Calibration	Application and technical characteristics	DOI of references
67	AES XRF ICP-MS AFS	Soil	Digestion	As, Pb, Cu, Zn	AES: Cd : 0.01 mg/kg; PXRF: As, Pb, Cu, Zn: 7 mg/kg, 8 mg/kg, 15 mg/kg, 12 mg/kg	External standard method RSD<20%	Rapid risk assessment of heavy metals in agricultural soils. Rapid heavy metal analysis using PXRF and AES for accurately assessing ecological risk of agricultural soils in greenhouses.	https://doi.org/10.1016/j.ecolind.2019.01.069
68	ICP-MS PXRF	Rice	Digestion	As, Mn, Fe, Ni, Cu, Zn	/	External standard method	Use of portable XRF for the assessment of trace elements in rice and rice products. Portable XRF has the advantage of being a fully mobile technique which is rapid, cost effective, and requires little sample preparation. Since different elements of interest emit characteristic X-rays with different photon energies.	https://doi.org/10.1016/j.apradiso.2015.07.014
69	ED-XRF	Soybean	Digestion	Mg, K, Ca, Mn, Fe, Ni, Cu, Zn, and Rb	/	The certified samples were employed as matching matrix for XRF calibration; External standard method	Soybean Traceability. A novel technical combination of a quick testing method with ED-XRF and MLP data analysis.	https://doi.org/10.46770/AS.2020.01.003

70	TXRF ICP-MS	Whole blood reference materials	Digestion	Pb	0.28 µg/dl	External standard method	Developed, validated, and applied a method to analyze Pb in DBS samples using TXRF for use in human biomonitoring studies. Developed a method and demonstrated that TXRF analysis of processed DBS filter paper samples yields results that are accurate and precise; By purposefully applying the method to study two populations that typify divergent conditions.	https://doi.org/ 10.1016/j.envres .2020.110444
71	IC-HG-AFS	Underground water, urine	Urine: acid extraction	As speciation	As(III): 0.02 ng, As(V): 0.166 ng, MMA: 0.043 ng, DMA: 0.045 ng	External standard method	Separation of arsenic species. A new gas–liquid separator system in IC–HG-AFS.	https://doi.org/ 10.1016/j.talanta .2007.04.026
72	HPLC-(CV)-AFS	Fish muscle tissue, water	Acid leaching	Hg speciation	Hg^{2+}: 0.7 µg/l, MeHg: 1.1 µg/l, EtHg: 0.8 µg/l, PhHg: 0.9 µg/l	External standard method Correlation coefficients: 0.9974–0.9982	High sensitivities analysis of both CV inactive and active mercury species. Post-column oxidation method using Fe_3O_4 magnetic nanoparticles.	https://doi.org/ 10.1039/ C3AN00010A
73	IC-ICP-MS	Soil	KOH extraction	AMPA, phosphate, glyphosate	AMPA: 1.0 µg/l, phosphate: 1.0 µg/l, glyphosate: 1.5 µg/l	External standard method Linear range: 5.0–1000 µg/l, Correlation coefficient $s > 0.999$	Detection of complexes in the soil extract without pre-treatment. Separate phosphate from glyphosate and AMPA by anion-exchange chromatography.	https://doi.org/ 10.1016/j.talanta .2008.12.052

(Continued)

Table 4.1 (Continued)

No.	Instrumental	Matrix	Sample preparation	Element or isotope	LOD or LOQ	Calibration	Application and technical characteristics	DOI of references
74	HPLC-ID-ICP-MS	Soil	Citric acid extraction	Inorganic Sb	Sb(V): 20 ng/l, Sb(III): 65 ng/l	Sb species:isotope dilution, Linear range: 0.2–50 µg/l	The first application of HPLC species unspecific spike isotope dilution ICP-MS to separate and quantify inorganic Sb species in real soil samples. Online isotope dilution concentration determination after a chromatographic separation.	https://doi.org/10.1007/s00216-005-0049-y
75	HILIC-ICP-MS	Human plasma	Protein precipitation by acetonitrile addition	Total Pt/Pt species	Cisplatin: 9.8 µg/l, Oxaliplatin: 40 µg/l, Carboplatin: 28.4 µg/l	External calibration	The determination of the free intact CPC in human plasma. The potential of a novel HILIC-ICP-MS approach in combination with simplified sample preparation procedures.	https://doi.org/10.1039/B907011G
76	LC-ICP-OES	absorbing liquid samples	Denitrification system	Fe(II)-EDTA, Fe(II)-EDTA	Fe(III)-EDTA: $1.2*10^{-5}$ mol/l, Fe(II)-EDTA: $1.6*10^{-5}$ mol/l	External standard method Typical concentration: 0.025–0.1 mol/l, $R > 0.9991$	Speciation of Fe(II)-EDTA and Fe(III)-EDTA. The hyphenated technique of LC-ICP-OES based on ultrasound-assisted replacement reaction.	https://doi.org/10.46770/AS.2020.06.004

77	IC-ICP-MS	Rain, snow, aerosols	Filtered	Iodine speciation	Iodate: 0.2 nmol/l,	Calibration curve	Aqueous-phase iodine chemistry.	https://doi.org/10.5194/acp-8-6069-2008
78	HPLC-(UV)-HG-AFS	Biological reference materials	Water-methanol mixture extraction	Arsenic species	Anion-exchange columns: 0.2 ng, cation-exchange columns: 0.4 ng	External standard method, linear range: 0–20 μg/g, $R^2 = 0.998$	Identification and quantification of arsenic compounds. Anion-exchange chromatography and cation-exchange chromatography.	https://doi.org/10.1016/S0039-9140(99)00048-X
79	HPLC-HG-AAS	Edible *Boletus badius*	Phosphoric acid and Triton-X-100 extraction	Arsenic species	1.00 mg/kg	External standard method	The arsenic species in edible Boletus badius with ambient substrate contamination. Total arsenic concentration was determined by GFAAS, As(III) and As(V) were determined by HPLC-HG-AAS.	https://doi.org/10.1080/03601234.2016.1159459
80	LC-HG-in situ DBDT-AFS	Rice	1% (v/v) HNO_3 and 1% (v/v) H_2O_2 extraction	As	0.05 μg/l,	External standard method, linear range: 0.5–50 μg/l, $R > 0.997$,	Elemental speciation analysis, LC atomic spectrometric instrumentation. LC combined with DBDT for the gas-phase preconcentration of iAs species to enhance the analytical sensitivity.	https://doi.org/10.1039/D0JA00222D

(Continued)

Table 4.1 (Continued)

No.	Instrumental	Matrix	Sample preparation	Element or isotope	LOD or LOQ	Calibration	Application and technical characteristics	DOI of references
81	AEC-HPLC-ICP-MS	Edible seaweed	In vitro digestion procedure and Microwave assisted alkaline digestion	Iodine/bromine species	Iodine: 24.6 ng/g; Bromine: 19.9 ng/g; Iodide: 26.9 ng/g; Iodade: 4.8 ng/g; MIT: 1.11 ng/g; DIT: 1.35 ng/g; bromide: 124 ng/g; bromate: 174 ng/g	External standard method,	Assess the bioavailable fractions, positive correlation between bioavailability and protein contents was found. AEC hyphenated with ICP-MS.	https://doi.org/10.1016/j.aca.2012.07.035
82	HPLC-ICP-MS	Lotus seed	HNO_3 extraction	As-species; Hg-species; Pb-species	As-species: 0.036–0.20 µg/l; Hg-species: 0.023–0.041 µg/l; Pb-species: 0.0076–0.14 µg/l	External standard method, Linear range: As: 1–500 µg/l, $R^2 > 0.9965$; Hg: 0.2–100 µg/l, $R^2 > 0.9929$; Pb: 0.5–100 µg/l, $R^2 > 0.9942$	A promising tool for studying the toxic, metabolic and bioavailable behaviors of arsenic, mercury and lead. Simultaneous speciation of arsenic, mercury and lead was analyzed by HPLC-ICP-MS; four As- and Hg-, and three Pb species were simultaneously eluted within 8 min.	https://doi.org/10.1016/j.foodchem.2019.126119
83	SPME-GC-ICP-MS	Plants, Brassica juncea seedlings	SPME, thermally desorb	Volatile alkyl-selenides and their sulfur analogues	Se-species: 1–10 ppt; Volatile S species: 30–300 ppt	External standard method, linearity up to 0.5 ppm	Investigate Se metabolism in Brassica juncea seedlings, identify unknown species in the standards. coupled technique of HS-SPME/GC/ICP-MS has proven suitable for the speciation of volatile selenium species in plants.	https://doi.org/10.1021/ac020285t

#	Method	Sample	Extraction	Analytes	LOD	Calibration/Linearity	Remarks	DOI
84	SPME–MC–MIP-AES	Lupine; Yeast; Indian mustard; Garlic	SPME	Organo-selenium species	DMSe: 0.57 ng/ml, DEtSe: 0.47 ng/ml, DMDSe: 0.19 ng/ml	External standard method; linearity up to 100 ng/ml; DMSe: $R^2 > 0.991$, DEtSe: $R^2 > 0.975$, DMDSe: $R^2 > 0.967$	Element specific detection of volatile species; inorganic selenium transformation research during metabolism. SPME-MC-MIP-AES for organo-selenium detection in selenium accumulating biological matter.	https://doi.org/10.1016/j.aca.2003.09.027
85	HS-SPME-GC-MS/MS	Surface water; wastewater	HS-SPME, home-made polydimethylsiloxane fiber	Mercury speciation	Alkyl mercury: 0.03 ng/l; inorganic mercury: 6 ng/l	External standard method; Linear range: Hg^{2+}: 25.0–200 ng/l, alkyl mercury: 0.12–80 ng/l; correlation coefficients $(R) > 0.994$	Speciation of trace and ultra-trace mercury in some surface water and wastewater samples. A home-made PDMS fiber was applied for the extraction of mercury species coupled with GC-MS/MS.	https://doi.org/10.1016/j.microc.2019.104459
86	GC-ICP-MS	Seafoods	ExtrFaction (KOH, HCl, KBr/$CuSO_4$)	Mercury species	MeHg: 0.5 pg; EtHg: 1.0 pg	External standard method; Linear range: 1–1000 pg	Mercury speciation in seafoods. GC-ICP-MS.	https://doi.org/10.1016/S1872-2040(08)60042-0
87	CGC-pyro-AFS	Fish reference materials	Closed-vessel microwave-assisted extraction (methanolic tetramethylammonium hydroxide)	Mercury species	MMHg: 2 pg; inorganic mercury: 1 pg	External standard method; Linear range: 5–200 µg/l; determination coefficients (R^2): MeHgEt: >0.995, inorganic mercury: >0.996	Mercury speciation in fish materials, simultaneous determination of methylmercury and inorganic mercury. Closed-vessel microwave-assisted extraction coupled with ethylation and analysis by capillary GC with AFS.	https://doi.org/10.1016/j.chroma.2005.07.054

(Continued)

Table 4.1 (Continued)

No.	Instrumental	Matrix	Sample preparation	Element or isotope	LOD or LOQ	Calibration	Application and technical characteristics	DOI of references
88	P-CT–GC–ICP-MS	Natural waters	Cryogenic traps ash desorption	Me_2Se, Me_2Se_2, Me_2Hg, Et_2Hg, Me_4Sn, Et_4Sn, Me_4Pb, Et_4Pb	Se, Hg, Sn and Pb volatile species: 10.0(0.8), 1.0(0.2), 0.4(0.05) and 0.4(0.08) pg/l	External standard method; relative isotopic abundance measurement	New assessment on the occurrence and speciation of dissolved volatile compounds of trace metals and metalloids in coastal envFements; Overall distribution of volatile metal and metalloid species in estuarine waters. Purge and cryogenic trapping method for the pre-concentration; ash desorption, cryofocusing gas chromatography system hyphenated to an ICP/MS.	https://doi.org/10.1016/S0003-2670(98)00425-5
89	CE-VSG-AFS	Certified biological reference Material (dogfish muscle), lake water and river water	CE (capillary electrophoresis)	Mercury species	6.8–16.5 µg/l;	External standard method; correlation coefficients (R) > 0.9918	Analysis of a certified biological reference material. CE-VSG-AFS technique for mercury speciation in biological materials.	https://doi.org/10.1021/ac026272x
90	CE-VSG-QICP-MS	Certified biological reference material (dogfish Liver)	CE	Mercury species	Hg_2^+: 0.2 pg; CH_3Hg^+: 7 pg;	External standard method; correlation coefficients (R) = 0.9949	Analysis of a certified biological reference material. Capillary electrophoresis coupled to volatile species generation-inductively coupled plasma mass spectrometry.	https://doi.org/10.1039/B101638P

91	CE-ICP-OES/MS	Certified biological reference material (dogfish Liver)	CE	Elemental species	ICP-OES: 0.1–100 ppb ICP-MS: 0.0001–1 ppb	External standard method; Linear over 4–7 orders of magnitude	CE-ICP spectrometry for elemental speciation. Combination of capillary electrophoresis and inductively coupled plasma optical emission or mass spectrometry holds great promise for rapid elemental speciation at concentrations as low as part per million to part per billion levels.	https://doi.org/10.1021/ac00097a003
92	CE-ICP-MS	Rice	Extraction (protease, lipase, water); CE	Se species	0.1–0.9 ng/ml	External standard method; Linear range: 10–400 ng/ml; Correlation coefficient: 0.990–0.998	The nutritional and toxic evaluation of different selenium compounds in nutritional supplements. CE-ICP-MS and a enzyme-assisted extraction used to extract all species of selenium in rice sample.	https://doi.org/10.1016/j.talanta.2011.03.004
93	CE-ICP-MS	Human Plasma	CE	Free Calcium and Calcium-Containing Species	Ca^{2+}: 25 µg/l	External standard method;	A new method for the determination of free calcium concentration in human plasma. Determination of free calcium and concentration estimation of calcium for other calcium-containing species in human plasma have been developed using CE-ICP-MS.	https://doi.org/10.1021/ac800715c
94	CE-HG-AFS	/	CE	Speciation of Organotin Compounds	1–10 µmol/l	External standard method	Speciation analysis of organotin compounds. Hydride-generation atomic fluorescence spectrometric detector was employed to detect organotin compounds after eluted from the outlet of capillary.	https://doi.org/10.1080/00032710701384667

(Continued)

Table 4.1 (Continued)

No. Instrumental	Matrix	Sample preparation	Element or isotope	LOD or LOQ	Calibration	Application and technical characteristics	DOI of references
95 PT-LLLME-LVSS-CE/UV	Biological and envFemental samples	PT-LLLME (phase transfer based liquid–liquid microextraction)	Mercury species	1.40–5.21 ng/ml	External standard method; Linear range: $EtHg^+$ 0.5–50 ng/ml, $R^2 = 0.9957$; $MeHg^+$ 0.5–50 ng/ml, $R^2 = 0.9976$; $PhHg^+$ 0.25–25 ng/ml, $R^2 = 0.9971$; Hg^{2+} 2.5–100 ng/ml, $R^2 = 0.9993$	The CE–UV-based approach is applicable for Hg speciation in real sample analysis. A phase transfer-based liquid–liquid–liquid microextraction technique with high sensitivity and fast extraction kinetics; A new approach for the simultaneous speciation of inorganic/organic Hg.	https://doi.org/10.1016/j.chroma.2011.10.071
96 CE-ICP-MS, CE-ESI-MS/MS	Goji berries	Grounded, homogenizated; Tris-HCl	Zinc speciation	/	/	Zinc binding ligands identification using CE-ICP-MS and CE-ESI-MS/MS; identification of those compounds in fruits extracts. Detection and identification of Zn-binding ligands based on using CE-ICP-MS and CE-ESI-MS/MS.	https://doi.org/10.1016/j.talanta.2018.02.040

97	CE-ICP-MS,	Liver	CE	Speciate Ag$^+$, AgNPs, Ag-metallothionein species	/	Investigated applicability of different CE methods hyphenated to ICP-MS for speciation of AgNPs in biological systems. CE hyphenated to the ICP-MS as promising and elegant technique to study AgNPs in biological systems.	https://doi.org/10.1016/j.chroma.2018.08.031
98	CE-ICP-MS	Liver	CE	Mn speciation	1.1 µg/l	Standard addition; linear range: 10–500 µg/l; coefficients (R^2) = 0.9998	Speciate the compounds quickly and with minimal risk of species alteration. Develop a speciation method for manganese species in liver extracts.	https://doi.org/10.1016/j.chroma.2004.05.076
99	CE-ESI-MS	Bottled, tap, spring and well water samples	CE	Organic and inorganic arsenic	0.02–0.04 µg/l	External standard method;	The validated method CE-ESI(−)-MS for arsenic speciation was applied to the analysis of the weakly mineralized water sample (dry residue < 50 mg/l) of both groundwater and bottled water. Ammonia and hexafluoro-2-propanol mixtures as efficient media for CE-ESI(−)-MS; CE-ESI(−)-MS as a sensitive and suitable alternative to speciation of inorganic and organic As species; Coupling of DLLME to CE-ESI(−)-MS for determination of total inorganic As.	https://doi.org/10.1016/j.talanta.2020.120803

(Continued)

Table 4.1 (Continued)

No.	Instrumental	Matrix	Sample preparation	Element or isotope	LOD or LOQ	Calibration	Application and technical characteristics	DOI of references
100	SR-m-XRF	Rat brain	Brain specimens were cut into 80 mm	Ca, Fe, Cu, Zn	Ca: 1.15 µg/g Fe: 0.53 µg/g Cu: 0.21 µg/g Zn: 0.20 µg/g	The Compton scattering is used as an internal standard	Accurate and precise imaging of the element variations in the brain section of a transgenic mouse model of Alzheimer's disease; quantitative imaging of trace elements in sections of bio-tissues. Quantitative imaging of element spatial distribution in the brain section of a mouse model of Alzheimer's disease using synchrotron radiation X-ray fluorescence analysis.	https://doi.org/10.1039/B921201A
101	FPXRF	Soil	Dry	K, Ca, Ti, Cr, Mn, Fe, Co, Ni, Cu, Zn, As, Se, Sr, Zr, Mo, Hg, Pb, Rb, Cd, Sn, Sd, Ba, Ag	Cr: 295 mg/kg Mn: 1010 mg/kg Co: 1160 mg/kg Ni: 350 mg/kg Cu: 137 mg/kg Zn: 204 mg/kg As: 134 mg/kg Se: 59 mg/kg Sr: 72 mg/kg Zr: 44 mg/kg Mo: 13 mg/kg Hg: 150 mg/kg Pb: 66 mg/kg Rb: 79 mg/kg Cd: 110 mg/kg Sn: 67 mg/kg Sd: 52 mg/kg Ba: 58 mg/kg Ag: 85 mg/kg.	NIST SRM 2709	In situ analysis of metals in soils and sediments, thin films/particulates, and lead in paint. FPXRF spectrometry has become a common analytical technique for on-site screening and fast turnaround analysis of contaminant elements in environmental samples.	https://doi.org/10.1016/S0304-3894(00)00330-7

#	Method	Sample	Preparation	Elements	Concentrations	Standards	Description	DOI
102	SXRF	Target cells	Homogenizing medium	Si, Mn, Fe, Ni, Zn	Si: 7.0×10^{-16} mol/μm^2; Mn, Fe, Ni, Zn: 5.0×10^{-20} – 3.9×10^{-19} mol/μm^2	NIST thin-film standards (SRM 1832 and 1833), Si: $R^2 = 0.993$, Fe: $R^2 = 0.9998$	Identify and quantify the elemental composition of a wide variety of geological, biological, and manufactured targets; synoptic quantitative analyses of cellular element contents. Enables the accurate and precise measurement of trace metals in individual aquatic protists collected from natural environments; This technique distinguishes between different types of cells in an assemblage and between cells and other particulate matter.	https://doi.org/10.1021/ac034227z
103	SRXRF	Fresh green leaves	Cut into sections	Pb, As, Cd, K, Ca, Mn, Ni, Cu, Zn	Pb: 1215 µg/g, As: 708.9 µg/g, Cd: 255.6 µg/g, K: 2046 µg/g, Ca: 2977 µg/g, Mn: 212.9 µg/g, Ni: 105.1 µg/g, Cu: 34.61 µg/g, Zn: 389.5 µg/g.	/	The cellular distributions of Pb and As in the leaves of co-hyperaccumulator Viola principis H. de Boiss. In vivo cellular localization of Pb and As in the leaves provides insight into the physiological mechanisms of metal tolerance and hyperaccumulation in the hyperaccumulators.	https://doi.org/10.1016/j.chemosphere.2008.04.084
104	m-SRXRF	Caenorhabditis elegans (C. elegans)	Homogenizing medium	La, Cu, Cd, Pb	Cd: 14.4–19.4 mg/l, Cu: 2.84–4.07 mg/l, Pb: 7.54–9.76 mg/l	/	Developed as a sensitive technique to illustrate the response of C. elegans to toxicant. Life-cycle endpoints were chosen along with elemental assay to evaluate the aquatic toxicity of lanthanum (La), a representative of REEs.	https://doi.org/10.1039/c0mt00059k

(Continued)

Table 4.1 (Continued)

No.	Instrumental	Matrix	Sample preparation	Element or isotope	LOD or LOQ	Calibration	Application and technical characteristics	DOI of references
105	SRXRF	Seeds	Elements excess treatments	Cu, Fe, Zn, Mn	/	/	Distribution characters of Cu and other metals in root growth zones were investigated by SRXRF.	https://doi.org/10.1007/s12011-010-8710-5
106	SRXRF	Seeds; Roots and shoots	Elements soak treatments	Cr, Ca, Mn, Fe, Cu, Zn, Pb	Pb: 200 mg/l Cr: 5 mg/l	/	Determination of the distribution of elements in heavy metal stressed plants by SRXRF. Differential centrifugation and SRXRF microprobe were used to study the distribution of the elements in tissue cross sections of pakchoi (Brassica chinensis L.) under stress of elevated Pb and Cr.	https://doi.org/10.1021/jf4005725
107	SRXRF	Rice seeds (Oryza sativa L.)	Elements treatments	Hg, Se	2.5 µmol/l	Peak areas of Hg and Se were normalized by the Compton scattering's peak count.	Better understand the molecular mechanism of Hg tolerance as well as the molecular antagonism between Hg and Se in rice plants. We investigated the effect of Se on Hg containing and Hg-responsive proteins in rice using 1, 2-dimensional electrophoresis combined with SR-XRF techniques.	https://doi.org/10.1016/j.jtemb.2017.10.006

108	m-SRXRF	180 leaves from 60 *Citrus sinensis* (L.) Osbeck, variety Valencia/Swingle	PCR amplification	K, Ca, Fe, Cu, Zn	/	A calibration model was constructed using soft independent modelling of class analogy (SIMCA)21 with the whole spectral region and the matrix with the mean data ($R^2 = 0.995$).	Investigated the mineral constituents of healthy leaves and leaves infected with citrus greening (or Huanglongbing). Investigate the mineral composition of citrus leaves infected by citrus greening using spectra obtained by m-SR-XRF.	https://doi.org/10.1039/B920980H
109	SRXRF	*Elsholtzia splendens*	Elements soak treatments; Cut into sections	P, S, Cl, K, Ca, Mn, Fe, Cu, Zn	<0.01 pg	/	Synchrotron radiation X-ray fluorescence spectroscopy (SRXRF) microprobe was used to study the Cu and other elements distribution in E. splendens. There was a significant correlation between Cu and P, S, Ca in distribution, which suggested P, S, and Ca played an important role in Cu accumulation of E. splendens. Based on the significant correlation between Cu and elements Mn, Fe, and Zn in distribution, it seemed that Cu, Mn, Fe, and Zn could be transported by the same transporters with a broad substrate range.	https://doi.org/10.1016/j.micron.2004.02.011

(Continued)

Table 4.1 (Continued)

No.	Instrumental	Matrix	Sample preparation	Element or isotope	LOD or LOQ	Calibration	Application and technical characteristics	DOI of references
110	SRXRF and LA-ICP-MS	*C. chinensis*	Elements expose treatments	Cr	LA-ICP-MS: 0.01 μg/g, XRF: 0.1–1 μg/g	SRM NIST 1547 Peach Leaves was used for calibration and used SRM NIST 1570 to validate the quantification procedure. $R^2 = 0.9961$ (Mn) and $R^2 = 0.9991$ (Ca)	Spatially locate Cr, analyze Cr speciation, detect Cr subcellular concentration. We utilized synchrotron radiation microscopic SRXRF and LA-ICP-MS to spatially locate Cr, XANES to analyze Cr speciation, and ICP-MS to detect Cr subcellular concentration.	https://doi.org/10.1038/s41598-018-26774-x
111	SEM-EDS, XRD, FT-IR	The magnetic biochar	Pyrolyze	Cd, As	/	/	Elucidate the relevant processes and mechanisms of As (III) and Cd (II) co-adsorption on the magnetic biochar. Elucidates mutual effects of Cd (II) and As (III) on their adsorption. Both competition and synergistic effects exist between Cd (II) and As (III). The synergistic effect is controlled by formation of type B ternary complex.	https://doi.org/10.1016/j.jhazmat.2018.01.011

#	Method	Sample		Elements	Findings	DOI
112	XANES, μ-XRF	Soils	/	Cd, As	Arsenic was detected over the whole area with hotspots highly correlated to the Fe hotspots ($R^2 = 0.789$); The distribution pattern of Cd was more similar to that of Ca with a relatively higher correlation ($R^2 = 0.498$)	Evaluate the effect of Ca-MBC on speciation, spatial distribution and stabilization mechanisms of As and Cd using synchrotron-based techniques. The stabilization mechanisms were explored through synchrotron-based micro-XRF and XANE.	https://doi.org/10.1016/j.jhazmat.2019.122010
113	XPS	Soil	/	Cr	/	XPS for qualitative analysis of elemental speciation in the soils before and after the GATP amendment. Prepared green-tea impregnated attapulgite promoted the conversion of weak acid extractable Cr to the residual state such as stable Cr (III) and Cr-Si oxide in soil.	https://doi.org/10.1016/j.jclepro.2020.123967
114	XPS	Tailing soil	/	C, O, S	/	XPS testified the thiol group that was easily oxidized. /	https://doi.org/10.1016/j.chemosphere.2020.127403
115	XPS	Fertilizer	/	P, F and heavy metals	/	Verify the mechanism underlying P, F and heavy metals in phosphogypsum (PG) leachates adsorbed onto biochar. Showing that heavy metal precipitation and complexation was caused by surface ions and different functional groups and application of the combination to loam or sand soil significantly increased crop yield.	https://doi.org/10.1016/j.jclepro.2020.124052
116	XPS	Aromatic Compounds Polymerization	/	Aromatic Compounds Polymerization; C1s, O1s	/	XPS spectra indicate the presence of oxygen containing groups in the chemical composition of plasma-polymerized films and oxidation reactions are pointed. Identify the potential intermediary reactions during plasma polymerization process.	https://doi.org/10.1007/s11090-014-9555-z

(Continued)

Table 4.1 (Continued)

No.	Instrumental	Matrix	Sample preparation	Element or isotope	LOD or LOQ	Calibration	Application and technical characteristics	DOI of references
117	XPS	Polythiophene	/	O, C, S	/	/	Adsorption of ozone was examined on PTh surfaces by angle resolved XPS. /Chemical shift or peaks could suggest the structure changes of samples.	https://doi.org/10.1016/S0169-4332(01)00316-6
118	XPS	PANI@TiO$_2$ composites.	/	C	/	/	XPS was employed to analyze the chemical components of PANI@TiO$_2$ composites. XPS peaks originate from aromatic compounds. The XPS results further confirm that PANI was successfully coated onto the surface of TiO$_2$ NPs.	https://doi.org/10.1080/00405000.2020.1848113
119	XPS	CDs (carbon dot)	/	C, O	/	/	The elemental compositions of CDs were researched by XPS. /	https://doi.org/10.1016/j.dyepig.2020.108831
120	XPS	A-CDs	/	Cu	/	/	Fluorescence quenching mechanism. The fluorescence quenching mechanism of the A-CDs by Cu^{2+} is investigated; XPS and XAES are further used to probe the valence state of Cu element.	https://doi.org/10.1016/j.saa.2020.118531
121	XPS	Biological samples	/	C1s	/	/	Imaging XPS, surface chemical mapping and blood cell visualization. XPS combined with photoemission electron microscopy (PEEM) to fulfil the surface chemical mapping and blood cell visualization.	https://doi.org/10.1116/1.4982644

122 XRD	Adsorption of toxic Pb (II) on agriculture waste (Mahogany fruit shell)	/	Pb(II)	/	/	Derived adsorbent material properties and plausible adsorption mechanism. XRD pattern confirm the amorphous and graphitic nature of MFSAC adsorbent.	https://doi.org/10.1080/03067319.2020.1849648
123 XRD	Phosphate Fertilizer	/	Magnesium Silicate Nanocomposites Coating	/	/	The formation of struvite in a calcareous soil was evaluated in three incubation periods by XRD. The XRD patterns showed the transformation of sepiolite to amorphous fibrous silica.	https://doi.org/10.1080/00103624.2020.1845352
124 XRD	Agriculture waste	/	Au–Pt nanoparticles	/	/	XRD analyze synthesized Au-Pt NPs structurally characterized. The crystalline nature of the bimetallic nano fluid was determined by XRD; XRD data confirmed the polycrystalline nature of the Au-Pt, NPs.	https://doi.org/10.1007/s11356-020-11435-2
125 XRD	Rapeseed	/	Nano zinc	/	/	Characterization of Zn nanoparticles. XRD authenticates the crystalline nature of nanoparticles and provides the average size of the particle.	https://doi.org/10.1371/journal.pone.0241568
126 XRD	Cotton fabrics	/	CS/ZnO nanocomposite	/	/	Reveal wurtzite crystalline structure of CS/ZnO nanocomposite for cotton fabrics. The XRD analysis revealed wurtzite crystalline structure of CS/ZnO nanocomposite.	https://doi.org/10.1016/j.ijbiomac.2020.08.047
127 XRD	Agricultural by-products (hydroxyl-eggshell), aqueous solutions	/	P recovered	/	/	On-firm the primary process of P recovered by agricultural by-products named hydroxyl-eggshell was via precipitation as hydroxyapatite to highly recover phosphorus from aqueous solutions. Fourier transform infrared spectroscopy (FTIR), scanning electron microscopy with energy dispersive X-ray spectrometry (SEM-EDS), and XRD were used to characterize the material before and after P reaction.	https://doi.org/10.1016/j.jclepro.2020.123042

(Continued)

Table 4.1 (Continued)

No.	Instrumental	Matrix	Sample preparation	Element or isotope	LOD or LOQ	Calibration	Application and technical characteristics	DOI of references
128	XRD	Starch gel samples	/	Corn starch gels	/	/	A novel method of preparing starch gels by alcohol soaking was developed; showed that the crystallinity of starch gel increased from 3.8% to 8.9% after alcohol soaking for rapid production of corn starch gels. The mechanical properties, crystallization behaviors, and structural characteristics of starch gels were investigated by XRD; gels and improves the starch concentration in the gels, thus promoting the retrogradation of side chains of the amylopectins and forming more crystals.	https://doi.org/10.1016/j.ijbiomac.2020.08.042
129	LA-ICP-MS	Brain, Gels, plants	Cut into sections	Fe, Zn, Cu, C, P, Mg	Cu: 0.34 mg/g, Zn: 0.14 mg/g, Pb: 12.5 ng/g g, U: 6.9 ng/g	standard reference material (NIST SRM 1515 apple leaves	Apply as a powerful imaging (mapping) technique to produce quantitative images of detailed regionally specific element distributions in thin tissue sections of human or rodent brain, investigate metal distributions in plant and animal sections to study. Using quadrupole-based LA-ICP-MS in comparison to the more sensitive double-focusing sector field LA-ICP-MS to perform imaging analysis of metals and non-metals on soft biological tissues for selected life science studies.	https://doi.org/10.1002/mas.20239

130	ICP-MS	Concentrations	Coprecipitation	Rare-earth and other trace elements (Y, Sc, Zr, Ba, Hf, Th)	/	Nine certified reference materials were analyzed to validate	The Na$_2$O$_2$-NaOH fusion is suitable for the determination of several trace elements. The determination of concentrations of rare-earth (REE) and other trace elements (Y, Sc, Zr, Ba, Hf, Th) in geological samples.	https://doi.org/10.1111/j.1751-908X.2008.00880.x
131	GC/QqQ-MS	Anabolic agents in human urine	Solvent extraction	Clenbuterol, 19-norandrosterone, 19-Noretiocholanolone, 17-Methyl-5-androst-1-ene-3, 17-diol, 17-Methyl-5-androstane-3, 17-diol, 3-Hydroxystanozolol, 16-Hydroxyfurazabol	Clenbuterol: 0.01 ng/ml; 19-norandrosterone: 0.3 ng/ml; 19-Noretiocholanolone: 0.2 ng/ml; 17-Methyl-5-and rost-1-ene-3, 17-diol: 0.8 ng/ml; 17-Methyl-5-androstane-3,17-diol: 0.2 ng/ml; 3-Hydroxystanozolol: 0.3 ng/ml; 16-Hydroxyfurazabol: 1 ng/ml	Methyltestosterone used as internal standard	The use of a complementary GC/QqQ qualitative method for the detection of a selected number of anabolic agents with special sensitivity requirements (2 ng/mL) including 17-methyl-5-androstane-3, 17-diol was investigated. A rapid, sensitive and robust gas chromatography–triple quadrupole mass spectrometry method was developed for the determination of seven anabolic agents in human urine.	https://doi.org/10.1016/j.jchromb.2012.03.037
132	NF-LA-ICP-MS	A 20 nm Au film deposited onto a Si substrate	/	Au	2.7×10^{-5} cps per ablated Au atom		Applying for measurements of thin Au film deposited onto a Si substrate. A near-field laser ablation inductively coupled plasma mass spectrometric (NF-LA-ICP-MS) procedure was created for element analysis in the nm resolution range.	https://doi.org/10.1016/j.ijms.2008.03.008

(Continued)

Table 4.1 (Continued)

No.	Instrumental	Matrix	Sample preparation	Element or isotope	LOD or LOQ	Calibration	Application and technical characteristics	DOI of references
133	LA-MC-ICP-MS	Sagittal otoliths	Cut into sections	Sr, Ba, Mg, Ca	/	Internal standard: ^{83}Kr, ^{84}Sr, ^{85}Rb, ^{86}Sr, ^{87}Sr	Measurements of Sr isotopic compositions together with elemental abundances (Ca, Ba, Sr, Mg). Using *in situ* LA MC-ICP-MS measurements of Sr isotopic compositions together with elemental abundances (Ca, Sr, Ba and Mg), characterize the various types of habitats encountered throughout the lifecycle history of individual barramundi.	https://doi.org/10.1071/MF04184
134	LA-MC-ICPMS	The PSPT-2 and PSPT-3 samples are pyrite and sphalerite powders	In situ analysis	Sulfur isotope	/	The external bracketing standards for in situ sulfur isotope analysis comprise some natural sulfides	In situ sulfur isotope analysis by LA-MC-ICPMS has an advantage in tracing sources of sulfur in an ore deposit that has complex mineral types and multiple ore-forming processes. Matrix effects of different sulfides were examined by matrix-unmatched tests, in which the δ34S values of different types of sulfides were calibrated against a laboratory pyrite standard.	https://doi.org/10.1016/j.jseaes.2019.02.017
135	LA-ICP-TOF-MS	Rat kidney	Cut into sections	Platinum	4.5 mg	External calibration	A cisplatin perfused rat kidney sample was analyzed using a high-throughput and quasi-simultaneous full spectral imaging approach using LA-ICP-TOF-MS. The platinum concentration results of a previous study could be consumed with additional all-elemental information revealing interesting elemental distributions for copper, nickel and tungsten.	https://doi.org/10.1039/C8JA00288F

#	Method	Sample	Preparation	Element	Internal standard	LOD	Remarks	DOI
136	LA-ICP-TOF-MS	Rat brain tissues	Homogenize	/	/	/	Developing a calibration strategy procedure in bioimaging trace elements in rat brain tissue by the LA ICP-TOF-MS method. Discussed the steps of a developed calibration procedure in the determination of trace elements in rat brain tissues by LA ICP-TOF-MS method.	https://doi.org/10.1016/j.talanta.2013.04.055
137	LA-ICP-MS, SR-μ-XRF, XANES EXAFS	Mushrooms	Cut into sections	Hg	MeHg: 4 μg/kg	Ge, Ga, Y, Sc (5 mg/l) were used as internal standards	Provide novel information on the localization and chemical speciation of Hg in caps of B. aereus, B. edulis and S. pescaprae by LA-ICP-MS, μ-XRF and XANES/EXAFS. EXAFS has opened new possibilities to study the chemical speciation and ligand environment in biological materials, without the use of extraction procedures that can alter the "in-vivo" ligand state; EXAFS has opened new possibilities to study the chemical speciation and ligand environment in biological materials, without the use of extraction procedures that can alter the "in-vivo" ligand state.	https://doi.org/10.1016/j.ecoenv.2019.109623
138	LA-ICP-MS	Soil	Homogenize, tablet compressing	Rare-earth elements	Ce: 0.63 μg/g; Dy: 0.34 μg/g Er: 0.48 μg/g EU: 0.23 μg/g Gd: 0.21 μg/g Ho: 0.34 μg/g La: 0.87 μg/g Lu: 0.15 μg/g Nd: 0.99 μg/g Pr: 0.75 μg/g Sm: 0.70 μg/g Tb: 0.25 μg/g Tm: 0.15 μg/g Yb: 0.96 μg/g	Internal standard	Evaluated the concentration and distribution of REEs in cultivated and non-cultivated soil. Establish a method for direct solid sample analysis by LA-ICP-MS.	https://doi.org/10.1016/j.chemosphere.2018.01.165

(Continued)

Table 4.1 (Continued)

No.	Instrumental	Matrix	Sample preparation	Element or isotope	LOD or LOQ	Calibration	Application and technical characteristics	DOI of references
139	MALDI-FT-ICR-MS	Yeast mitochondrial proteins	Microwave digestion	Sulfur	P: 0.18 mg/g, S: 1.3 mg/g, Cu: 6.4 mg/g, Zn: 17.6 mg/g, Fe: 9.5 mg/g	Internal standard element	Protein spots in two-dimensional gels were screened with respect to P, S, Fe, Cu and Zn content, and multielement determination was investigated in separated protein spots by LA-ICP-MS as the microlocal analytical technique. A new screening technique using two-dimensional gels was developed in order to rapidly identify various elements in well-separated protein spots.	https://doi.org/10.1039/B404797B
140	LA-ICP-MS/LC-MS	Roots of *Elsholtzia splendens*	Cut into smaller pieces	Cu, Zn, Na, Mg, K, Ca, Mn, Fe, P, S, Ag	/	Internal standards	LA-ICP-MS is used to screen metal-containing proteins on the dried 2D gels. LA-ICP-MS imaging technique, via the images of Ag, was able to distinguish almost all the protein spots in 2D gels, even those with relative low abundance. The metal images were correlated very well with those obtained by Ag staining.	https://doi.org/10.1016/j.ijms.2011.01.018
141	LA-ICP-MS	Wheat seeds	/	^{70}Zn	/	The Zn: ^{13}C ratio (^{13}C as internal standard) of the calibration standards	Using LA-ICP-MS, the spatial distribution of Zn within the grains was studied to elucidate the Zn transport pathway within the developing grain. The spatial distribution of Zn within the grains was studied by LA-ICP-MS.	https://doi.org/10.1111/j.1469-8137.2010.03489.x

142	HPLC-ID-ICP-MS	Standard proteins	Filter	S, Fe	/	Sulfur isotopic reference material with a natural abundance	The quantification of proteins by ICP-MS offers an alternative method for quantitative proteomics. Developed a method based on HPLC coupled to ICP-MS via post-column isotope dilution for the quantification of transferrin (Tf) and albumin (Alb) in human serum using enriched 34S and 54Fe isotopic solutions.	https://doi.org/10.1039/C4AY00907J
143	LA-ICP-MS	Human Serum Transferrin	Homogenize	^{56}Fe, ^{13}C	/	A serum certified reference material (ERM-DA470k/IFCC)	The absolute quantification of a metalloprotein separated by nondenaturing gel electrophoresis using LA-ICP-MS in combination with species-specific IDMS. Absolute Quantification of Human Serum Transferrin by Species-Specific Isotope Dilution Laser Ablation ICP-MS.	https://doi.org/10.1021/ac200780b
144	LA-ICP-MS	Brain tissues	Homogenize, cut into 40 mm sections	Fe, Cu, Zn	/	Homogeneous in-house standard;^{54}Fe/^{56}Fe, ^{63}Cu/^{65}Cu and ^{64}Zn/^{67}Zn are 0.11313, 0.11584 and 0.02486	A quantitative imaging strategy of intact brain section. Spatial resolution capacity and precise means for quantification; quantitative imaging of trace elements in biological samples.	https://doi.org/10.1016/j.aca.2017.07.003
145	LA-ICP-MS	Moss	Cut into sections	^{242}Pu, ^{243}Am	Pu: 3.6 fg/g	Calibration against the NIST-500 SRM using TIMS	The direct determination of long-lived radionuclides at ultra-trace concentration levels in solid environmental samples. LA-ICP-MS was combined with isotope dilution for the ultra-trace level determination of ^{239}Pu, ^{240}Pu and ^{241}Am on the surface of targets after electroplating.	https://doi.org/10.1016/S1387-3806(03)00024-1

(Continued)

Table 4.1 (Continued)

No.	Instrumental	Matrix	Sample preparation	Element or isotope	LOD or LOQ	Calibration	Application and technical characteristics	DOI of references
146	SC-IDLA–ICP–MS	Macrophage cells	Digest	Silver NPs	0.2 fg Ag per cell	NIST 612 glass standard RM	LA–ICP–MS is an emerging method for the analysis of metal nanoparticles (NPs) in single cells. Multiplexed imaging of proteins and protein modifications in tumor tissues with a subcellular resolution (~1 μm) is achieved using immunohistochemical techniques coupled to the new generation of LA–ICP–MS.	https://doi.org/10.1021/acs.analchem.0c01775
147	W-coil ET-AFS/AAS	Rice and water	Cloud point extraction (Triton X-114)	Cd	W-coil ET-AFS: 0.01 μg/l; W-coil ET-AAS: 0.03 μg/l	External standard method; linear range: W-coil ET-AFS: 0.01–0.5 μg/l; W-coil ET-AAS: 0.03–2 μg/l $R^2 > 0.994$;	The accurate determination of trace amounts of cadmium in rice and water samples by W-coil ETVAFS and ET-AAS after CPE preconcentration. Use of CPE with W-coil atomization/vaporization, high sensitivities and improved LODs were easily obtained for cadmium.	https://doi.org/10.1016/j.aca.2009.01.053

148	SPE-ICP-OES	Water and food samples	SPE	Cd(II), Cu(II), Ni(II), Pb(II)	0.012, 0.098, 0.056 and 0.14 µg/l for Cd(II), Cu(II), Ni(II) and Pb(II) enrichment factor: 100	Calibration curve; linear range: 0.5–4 µg/ml for Cd(II), 0.5–4 µg/ml for Cu(II), 0.5–4 µg/ml for Ni(II) and 0.5–4 µg/ml for Pb(II)	A simple and efficient AEDHB-SG column solid-phase extraction method was used for the separation, preconcentration and simultaneous determination of Cd(II), Cu(II), Ni(II) and Pb(II) ions in water and food samples using ICP-OES. Silica gel was functionalized with N-(2-aminoethyl)-2,3-dihydroxy-benzaldimine for solid-phase extraction of trace elements. This study allows separation and preconcentration at µg/l level of trace elements; Simultaneous determination of Cd(II), Cu(II), Ni(II) and Pb(II) in water and food samples using ICP-OES.	https://doi.org/10.1016/j.jiec.2014.12.041
149	ETV-ICP-AES	Seawater, tap water	On-site electrodeposition	Pb	25 pg/ml (3 min)	External standard method; Linear range: 100–1000 pg/ml $R^2 > 0.9974$	An electrodeposition method for preconcentration and separation of labile Pb from liquid samples with a complex matrix is described. Electrothermal, near torch vaporization sample introduction for ICP-AES was used. Liquid samples with a complex matrix (e.g. seawater, tap water) were tested; On-site electrodeposition was used to test for compliance with EPA's Pb–Cu rule; Detection limit for Pb by NTV-ICP-AES was 25 pg/ml after 3 min electrodeposition.	https://doi.org/10.1016/j.microc.2012.10.013

(Continued)

Table 4.1 (Continued)

No.	Instrumental	Matrix	Sample preparation	Element or isotope	LOD or LOQ	Calibration	Application and technical characteristics	DOI of references
150	ETV-AAS	Solid envFemental and biological samples	Soil sampling	Hg	0.12 ng/g	Solid CRMs calibration graph; linear range: 0–50 ng, $R^2 = 0.9802$; 0–1600 ng, $R^2 = 0.9992$;	A new, rapid technique for the determination of total mercury in envFemental and biological samples. The pyrolysis of the sample in a combustion tube at 750°C under an oxygen atmosphere, collection on a gold amalgamator and subsequent detection by AAS using a silicon UV diode detector.	https://doi.org/10.1016/S0003-2670(99)00742-4
151	SS-ETV-ICP-MS	Geological Samples (rock, soil and sediment)	Sodium citrate was used in the pyrolysis process to improve the transfer and the ionization efficiency of iodine	I	Rock: 11 ng/g Soil: 9 ng/g Sediment: 8 ng/g	Matrix-matched external calibration curves	Developing a robust and high-throughput method based on SS-ETV-ICP-MS for the quantitative determination of trace iodine in geological materials. $Pd(NO_3)_2$ and ascorbic acid were chemical modifier.	https://doi.org/10.46770/AS.2020.02.006
152	QT-ETV-QTAT-AFS	Food samples	Solid sampling	Pb	2 pg	External standard method; linear range: 0.04–4.0 ng; correlation coefficient (R) was >0.998	A novel QT-ETV-QTAT-AFS method was developed for the direct determination of trace Pb in food samples. A quartz tube was first used to trap Pb after SS-ETV and its relative mechanism was investigated.	https://doi.org/10.1039/C6JA00316H

#	Method	Sample	Technique	Analyte	LOD	Linear range	Comments	Reference
153	ETAAS	Water	Solid sampling	Cd, Co, Cr, Mn, Ni, Pb	0.06–2.3 µg/l	Cd: 0–2.0 µg/l; Ni/Pb: 0–100 µg/l; Co, Cr, Mn: 0–25.0 µg/l	The analytical figures of merit obtained with the TCA using a simple homemade power unit, proved the potential of this unenclosed TCA as a simple electrothermal atomization accessory to supplement AA instruments for trace element analysis of waters. The analytical performance of an unenclosed tungsten coil atomizer operating in Ar-H$_2$ was assessed by ETAAS for the determination of pg to ng levels of Cd, Co, Cr, Mn, Ni and Pb in liquid samples.	https://doi.org/10.1016/0003-2670(95)00002-H
154	d-CPE-ETAAS	Water and freshwater fish samples.	d-CPE	Hg(II) CH$_3$Hg	0.23 µg/l	External standard method; linear range: 0.40–15.0 µg/l correlation coefficient (R) = 0.9997	A useful tool for speciation of mercury in water and freshwater fish samples and an attractive method for routine quality control laboratories especially for food applications. A d-CPE was developed for mercury speciation in freshwater fish samples by ETAAS. Method was applied to the speciation of Hg in water and freshwater fish samples.	https://doi.org/10.1016/j.sab.2019.105685

(Continued)

Table 4.1 (Continued)

No.	Instrumental	Matrix	Sample preparation	Element or isotope	LOD or LOQ	Calibration	Application and technical characteristics	DOI of references
155	Thermolysis-AAS	Traditional Chinese Medicines	Solid sampling	Hg	3.9 ng/g	External standard method; Linear range: ng/g to mg/g; correlation coefficient (R) > 0.994	Mercury conversion from cinnabar to biological matrices-bound Hg could occur because of the aid of other ingredients in the formulated drug. Develop and apply a mercury analyzer system capable of quantitative analysis of mercury in Traditional Chinese Medicines (TCM) drugs in the concentrations range from ng/g to mg/g.	https://doi.org/10.1016/j.talanta.2005.05.014
156	Electromagnetic heating -OES	Soil	Solid sampling	Hg, Cd and Pb	Hg: 0.008 mg/kg; Cd: 0.0178 mg/kg; Pb: 3.54 mg/kg	Hg: 0.015–0.590 mg/kg, $R^2 = 0.996$; Cd: 0.060–0.450 mg/kg, $R^2 = 0.9917$; Pb: 14.0–314 mg/kg, $R^2 = 0.9918$	/ /	https://doi.org/10.1016/j.talanta.2020.121348

157	ETV-TC-ICP-MS	Food	Solid sampling	Cd	LOQ:0.5 pg	Matrix-matching calibration strategy 0.16 pg–50 ng $R^2 > 0.995$	The proposed SS-ETV-ICP-MS method is extremely suitable for rapid Cd detection for food sample screening, which maybe a good option for laboratories that use the ICP-MS. A tungsten coil trap was firstly employed to ICP-MS for Cd analysis in food. The modified gas circuit showed an effective performance for plasma stability. The minimum sample mass of fresh food samples well ground proved <30 mg. This SS-ETV-ICP-MS method is fast, accurate, easy, green and safe.	https://doi.org/10.1016/j.sab.2016.02.017
158	SS-ETV-TC-ICP-MS	Wheat, corn and rice samples	Solid sampling	Zn, Cd	Zn: 1 pg; Cd: 0.1 pg	Increasing masses of CRM or grain samples; $R^2 > 0.995$	ETV-TCT-ICP-MS method is highly suitable for rapid Zn and Cd detection when screening food samples, which is a good choice for some laboratories equipped with an ICP-MS. A method in which a TCT captured Zn and Cd atoms at the same time at room temperature after SS-ETV was used for the first time.	https://doi.org/10.1039/C6RA03524H
159	SS-ETV-AFS	Rice	Solid sampling	Cd	40 ng/l	Matrix-matching calibration strategy linear range: 0.3 μg/l–1.0 mg/l $R^2 > 0.995$	The SS-ETV-AFS method was a good solid microsample analysis technique to determine trace Cd in rice and avoided troublesome digestion and unnecessary dilution procedures. SS-ETV-AFS was used to detect Cd in rice samples for the purpose of assessing the homogeneity and minimum sample mass.	https://doi.org/10.1021/jf3045473
160	hydride generation (HG)-DBD-AFS	Water	/	As	1.0 ng/l	External standard method; linear range: 0.05 μg/l–5 μg/l, linear regression coefficient $(R^2) > 0.995$	A novel dielectric barrier discharge reactor (DBDR) was utilized to trap/release arsenic coupled to HGAFS. A coaxial DBDR installed between HG and AFS to enrich arsine under ambient temperature and atmospheric pressure, novel HG-DBD-AFS system was applied to measure ultra-trace arsenic in surface water.	https://doi.org/10.1021/acs.analchem.6b00506

(Continued)

Table 4.1 (Continued)

No.	Instrumental	Matrix	Sample preparation	Element or isotope	LOD or LOQ	Calibration	Application and technical characteristics	DOI of references
161	ETV-ICP-OES	Avian bone and slag	Solid sampling	Al, Ca, Fe, Mg, S, Ag, As, Au, Ba, Cd, Ce, Co, Cr, Cs, Cu, Dy, Eu, Ga, Hf, Hg, Ho, In, K, La, Mn, Mo, Na, Nd, Ni, P, Pb, Rb, Sb, Sc, Se, Si, Sm, Sr, Ti, U, V, W, Yb, Zn, Zr	/	External calibration curves, which were constructed using DUWF-1, NIST 8433, NIST 2711, and SL-1, with a mass range of 1–3.5 mg; $R^2 = 0.97–0.99$,	Use the ETV-ICP-OES method to perform an initial investigation on the effect of slag ingestion on the elemental composition of avian bones, elemental ratios may be used as a fingerprint for slag, allowing further confirmation of slag ingestion. Developing a method of direct analysis of solid avian bone and slag samples to determine bulk (Al, Ca, Fe, Mg, S) and trace (Ag, As, Au, Ba, Cd, Ce, Co, Cr, Cs, Cu, Dy, Eu, Ga, Hf, Hg, Ho, In, K, La, Mn, Mo, Na, Nd, Ni, P, Pb, Rb, Sb, Sc, Se, Si, Sm, Sr, Ti, U, V, W, Yb, Zn, Zr) element concentrations, without the need for sample digestion, using ETV-ICPOES.	https://doi.org/10.1039/D0JA00288G
162	ICP-oa-TOF-MS	Food and beverages	Microwave digestion	Multi-element analysis	0.04–1360 ng/g	Internal standardization (^{103} Rh) 0–50 mg/l	The advantages of ICP-TOF-MS for food control procedures. Accurate results are obtained for some elements by ICP-oaTOF-MS which was difficult for Q-ICP-MS.	https://doi.org/10.1016/j.foodchem.2011.05.047

#	Technique	Sample	Preparation	Analytes	LOD/Values	Notes	DOI
163	ICP-oa-TOF-MS	*Indocalamus tesselatus* samples	Microwave digestion	Pb, Cd, Mn, Ni, Cr, As, Hg, Cu, Zn	Pb: 0.011 µg/g, Cd: 0.047µg/, Mn: 0.016 µg/g, Ni: 0.019 µg/g, Cr: 0.036 µg/g, As: 0.139 µg/g, Hg: 0.017 µg/g, Cu: 0.013 µg/g, Zn: 0.034 µg/g	Cd, Ni, As, Hg: 0–20 µg/l; Cu, Zn, Cr: 5–200 µg/l; Mn: 0–1000 µg/l, Pb: 0–40 µg/l; $R^2 > 0.992$	Showed *I. tesselatus* can be unreservedly used as food packing materials without any health risk. Proposed a method for the determination of heavy metals by ICP-oa-TOF-MS.	https://doi.org/10.1016/j.foodchem.2013.04.103
164	SF-ICP-MS	Grape (*Vitis vinifera* sp.) seeds and skins, green tea (*Camellia sinensis*) leaves and Limousin oak (*Quercus robur*) heartwood	Microwave digestion	Multi-element analysis	/	/	The ICP-MTS elemental profile confirmed the levels of potentially toxic contaminants. MALDI-TOF-MS and UV-vis informed on food-grade tannins composition and authenticity. ICP-MS highlighted the elemental composition of tannin additives.	https://doi.org/10.1016/j.jfca.2017.01.014
165	SF-ICP-MS	River water	Cidifiead	^{226}Ra	0.46 fg/l or 0.02 mBq/l	/	The naturally existed Ba was used as a yield tracer to quantify ^{226}Ra concentration. A novel and simple method to measure ultra-trace ^{226}Ra in river water samples at fg/l (mBq/l) levels.	https://doi.org/10.1016/j.jenvrad.2020.106305
166	SF-ICP-MS	Japanese coastal seawater	100-fold with 0.5%	I	0.23 ng/m	Internal standard (Te)	Applied for the study of total iodine concentrations in Japanese coastal waters for the estimation of sediment–seawater distribution coefficient (Kd) and concentration ratios (CRs). Reported a sensitive analytical method for direct analysis of total iodine in seawater using SF-ICP-MS.	https://doi.org/10.1016/j.microc.2011.08.007

(*Continued*)

Table 4.1 (Continued)

No.	Instrumental	Matrix	Sample preparation	Element or isotope	LOD or LOQ	Calibration	Application and technical characteristics	DOI of references
167	SF-ICP-MS	Milk	Muffle furnace; high-pressure washer; Microwave digestion	Multi-element analysis	/	External calibration method; standard addition calibration or matrix matching calibration	Developed screening procedure allows the detection of differences in elemental mass fractions of milk, feed and water samples from different locations. Developed an analytical procedure for multi-element screening of 40 elements in milk and feed samples.	https://doi .org/10.1016/ j.foodchem .2010.07.050
168	MC-ICP-MS	Plants	Dry ashing in combination with cation-exchange chromatography and micro-sublimation	Boron isotope analysis	/	/	The $\delta^{11}B$ variation in different plants, the fractionation among different parts within plants and the underlying reasons were explained, will replenish the fractionation mechanisms of boron isotopes, and expand the application of $\delta^{11}B$ in biogeochemistry and atmospheric B cycling. A valid chemical and mass spectrometric method was developed to separate and determine $\delta 11B$ of plant tissues including dry ashing, cation-exchange chromatography and micro-sublimation.	https://doi .org/10.1016/ j.talanta .2018.12.087
169	LA-MC-ICP-MS	Individual planktonic foraminifera	/	Boron isotopes	/	/	Individual foraminifera (*O. universa*) boron isotope analyses by LA-MC-ICP-MS without the need for matrix matched standards.	https://doi .org/10.1016/ j.chemgeo .2019.119351

170	LIBS	Soil samples	/	Ba, Ca, Cr, Cu, Fe, Li, Mg, Mn, Pb, Sr, Ti and V	<33 ppm; a precision lower than 10%	Single-point calibration Approach, $R^2 > 0.90$	Applied to the analysis of bulk soil samples for discrimination between specimens. Developed a method for the quantitative elemental analysis of surface soil samples using LIBS, employed a LIBS with 266 nm laser to soil analysis for forensic purposes, and limits of detection that are sufficiently low to conduct forensic analysis.	https://doi .org/10.1016/ j.surfrep .2012.09.001
171	LIBS	Water (mineral salts)	/	Cd, Ni, Cr	Cd, Ni, Cr: 532 nm: 9.61, 8.49, 71.6 µg/l 1064 nm: 22.5, 20.4, 83.8 µg/l	Calibration curves, Ni: 0.118–2.36 µg/g, Cr: 0.0975–1.95 µg/g, Cd: 0.246–4.92 µg/g	Aluminum oxide nanoparticles to enrich Cd, Ni, Cr in water sample to improve LIBS detection limit. Al_2O_3 nanoparticles adsorption enrichment of target elements was verified, and the detection sensitivity was improved by increasing the amount of sample solutions.	https://doi .org/10.1007/ s00216-011- 4869-7
172	LIBS	Red wine	Gel formation	Mg, Ca, K, Na	/	/	Applied to the identification of geographical origins of agri-food. Wine samples belonging to different Spanish geographical regions have been analyzed by LIBS. A novel liquid-to-solid matrix conversion by gel formation have been applied, reported on a simple and fast classification procedure for the quality control of red wines with protected designation of origin (PDO) by LIBS technique. Neural Networks were selected as classification methods to discriminate between wines.	https://doi .org/10.1177/ 000370281982 95

(Continued)

Table 4.1 (Continued)

No.	Instrumental	Matrix	Sample preparation	Element or isotope	LOD or LOQ	Calibration	Application and technical characteristics	DOI of references
173	LIBS	Animal histological sections	/	Trace elements	/	/	Distinguished trace elements between normal and malignant tumor cells from animal histological sections by LIBS. The first LIBS application to our knowledge to explore the possibility of using LIBS for cancer detection.	https://doi.org/10.1016/j.talanta.2016.05.059
174	LIBS	Fruit and Vegetables	/	P, S	P: 0.009–0.029 mg/kg; Cd: 1.6 ng/g	Calibration curves P: 0–0.25 mg/kg; $R^2 > 0.89$ Cd: 3–60 ng/g; $R^2 = 0.917$	Used the nanoparticle-enhanced LIBS technique to study the distributions of harmful chemicals in vegetable leaves. Employed nanoparticle-enhanced LIBS technique to enhance the analytical sensitivity of pesticide residue and Cd in fruit and vegetables, and achieved Cd mapping analysis on edible plant leave revealing the elemental uneven distribution.	https://doi.org/10.1364/AO.43.005399
175	LIBS	Sugar cane leaves	/	Ca, Mg, K, P, Cu, Mn, Zn	Ca 0.01 g/kg, Mg 0.01 g/kg, K 1.4 g/kg, P 0.03 g/kg, Mn 0.8 mg/kg, Zn 1.0 mg/kg and Cu 0.6 mg/kg	Matrix matching calibration standards	Proposed a calibration strategy for LIBS analysis of pellets of plant materials. Demonstrated for quantitative determination of Ca, Mg, K, P, Cu, Mn and Zn in pellets of sugar cane leaves by LIBS and for estimating the detection limits based on the analysis of the corresponding blank. Blanks and/or low concentration standards of plant materials were produced for LIBS analysis.	https://doi.org/10.1016/j.sab.2013.03.009

176	ICP-AES	Individual biological cells	Homogenize	Ca	0.01 pg	Using monodisperse aerosols	Measure the distribution of calcium content of a large number of cells; show distributions of different groups of cells. Continuously determining the calcium content in individual biological cells by inductively coupled plasma atomic emission spectrometry in real time.	https://doi.org/10.1021/ac00091a004
177	time-resolved ICP-MS	Unicellular microbes	Cell homogenize	C, Mg, Al, P, S, K, Ca, Cr, Mn, Fe, Zn	/	$R^2 = 0.9989$	Localization of elements in cells. Realize highly efficient single-cell analysis of microbial cells by time-resolved inductively coupled plasma mass spectrometry (ICP-MS).	https://doi.org/10.1039/C4JA00040D
178	time-resolved ICP-MS	A unicellular alga *Chlorella vulgaris*,	Acid digestion	Na, Mg, Ca, K, Mn, Fe, Cu, Co, Zn, Mo	10 ng/ml	The filtrate containing MgO particles of average diameter of approximately 100nm; $R^2 = 0.999$	Quantitative determination of the metal contents of the algal cells using metal oxide particles for calibration is feasible. Semi-quantitative measurement is also possible using aqueous standards for calibration. Simultaneous cell counting and determination of constituent metals in single cells using time-resolved ICP-MS.	https://doi.org/10.1039/C002272A
179	IEC-SP-ICPMS	Chemicals	Dilute	Silver nanoparticles	17 nm (0.05 µg/l added Ag$^+$) 21 nm (0.1 µg/l added Ag$^+$) 36 nm (0.5 µg/l added Ag$^+$) 45 nm (1 µg/l added Ag$^+$) 57 nm (2 µg/l added Ag$^+$)	Standard ICP-MS reference materials	SP-ICPMS is becoming a very promising technique for nanoparticle detection and characterization, especially at very low concentration. A new approach to attaining lower particle size detection limits has been developed by the online coupling of an IEC with SP-ICPMS	https://doi.org/10.1021/ac5004932

(*Continued*)

Table 4.1 (Continued)

No.	Instrumental	Matrix	Sample preparation	Element or isotope	LOD or LOQ	Calibration	Application and technical characteristics	DOI of references
180	SC-ICP-MS	Single yeast, green alga, or red blood cell	Homogenize	P, S, Mg, Zn, Fe	Yeast: Mg: 8.79 fg/cell Zn: 1.34 fg/cell P: 115 fg/cell S: 32.6 fg/cell Fe: 0.566 fg/cell Green alga: Mg: 293 fg/cell Zn: 3.88 fg/cell P: 2044 fg/cell S: 223 fg/cell Fe: 14.5 fg/cell RBC: Mg: 3.43 fg/cell Zn: 0.497 fg/cell P: 40.1 fg/cell S: 177 fg/cell Fe: 70.2 fg/cell	Comparing the integrated intensity of the transient signals from a cell (IC) with the signal intensity of an ionic standard of known concentration	Study both the physiological and nutritional importance and toxicological effects of elements within a nano-sized sample. The elemental composition of a single yeast, green alga, or red blood cell was precisely determined by using ICP-MS operating in fast time-resolved analysis (TRA) mode.	https://doi.org/10.1002/cbic.202000358

#	Method	Sample	Sample pretreatment	Analyte	LOD/LOQ	Quantitative method	Remark	Reference
181	GC-MS	Essential oils	Dry	The aromatic Monoterpenes; the thujanes	/	The evaluation of the biological activity of the essential oil	GC-MS Analysis of Essential Oils from Some Greek Aromatic Plants and Their Fungitoxicity on *Penicillium digitatum*. The isolated essential oils from seven air-dried plant species were analyzed by GC-MS.	https://doi.org/10.1021/jf990835x
182	LC/IC-MS/MS	Proteins	Digestion	Peptides	10 fmol	Known yeast ORFs (6139 proteins) and a second database containing 6139 bogus proteins	Highly complex protein mixtures can be directly analyzed. We have utilized the combination of strong cation-exchange (SCX) and reversed-phase (RP) chromatography to achieve two-dimensional separation prior to MS/MS. One milligram of whole yeast protein was proteolyzed and separated by SCX chromatography (2.1 mm i.d.).	https://doi.org/10.1021/pr025556v
183	SC-ICP-MS	Fetal bovine serum; plasmocin	Homogenize	Cisplatin	10 μmol/l	External standard method	Total Pt analysis in single cells has been implemented using a total consumption nebulizer coupled to ICP-MS. Address the number of cells in the suspension and the efficiency of the sample introduction system; Quantitative uptake studies of a nontoxic Tb containing compound by individual cells were conducted.	https://doi.org/10.1021/acs.analchem.7b02746
184	SC-ICP-MS	Cells	Homogenize	Cu, Fe, Zn Mn, Co, P, S	^{55}Mn: 0.1 fg, ^{56}Fe: 1.9 fg, ^{59}Co: 0.1 fg, ^{65}Cu: 0.5 fg, ^{66}Zn: 6.6 fg, ^{31}P^{16}O^{+}: 0.6 pg, ^{32}S^{16}O^{+}: 0.4 pg	/	The single-cell analysis of Mn, Fe, Co, Cu, Zn, P, and S in human cancer cell lines and normal human bronchial epithelial cell line. ICP-MS equipped with a high efficiency cell introduction system (HECIS) was developed as a method of single-cell ICP-MS (SC-ICP-MS). The method was applied to the single-cell analysis of Mn, Fe, Co, Cu, Zn, P, and S in human cancer cell lines (HeLa and A549) and normal human bronchial epithelial cell line (16HBE).	https://doi.org/10.1007/s00216-016-0075-y

(Continued)

Table 4.1 (Continued)

No.	Instrumental	Matrix	Sample preparation	Element or isotope	LOD or LOQ	Calibration	Application and technical characteristics	DOI of references
185	ICP-SF-MS	Yeast; Cells	Homogenize, dilute	Na, Mg, Fe, Cu, Zn, Se	Na: 0.91 fg, Mg: 9.4 fg, Fe: 5.9 fg, Cu: 0.54 fg, Zn: 1.2 fg, Se: 72 fg	Internal standard: Indium	Investigate the uptake of essential elements in selenized yeast cells; develop quantification procedures for single-cell analysis. Has the detection power to measure elements, in particular metals, in time resolved single cells.	https://doi.org/10.1039/C3JA30370E
186	SC-ICP-MS	Human epithelial lung adenocarcinoma cells	Homogenize	As	0.35 fg per cell	External calibration, liner range: 0–75 mg/l, $R^2 = 0.9998$	Evaluate the elemental composition at a single-cell level; determine minerals such as calcium and magnesium; assess the cellular uptake of TiO_2 and Ag. the cellular bioavailability of arsenite (incubation of 25 and 50 μm for 0–48 h) has been successfully assessed by SC-ICP-MS/MS for the first time directly after re-suspending the cells in water.	https://doi.org/10.1039/c7mt00285h
187	SC-ICP-MS	Freshwater algae: Cyptomonas ovate	Homogenize	Au Nanoparticle	3 ppb	Internal standard: Ce^{140}, Eu^{151}, Eu^{153}, Ho^{165}, and Lu^{175}; NIST Au Standards	Quantify the mass of Au metal strongly associated with fresh water algae (Cyptomonas ovate) for both nanoparticle and dissolved metal exposures; human toxicology and nanomedicine. Quantification of metal concentrations on an individual cell basis down to the attogram (ag) per cell level; capable of delivering intact individual cells into the ICP-MS plasma and quantifying the metal content associated with each cell down to a few attograms (ag; 10–18 g) per cell.	https://doi.org/10.1021/acs.est.7b04968

188	ICP-TQ-MS	Yeast cells	Homogenize	Se, P	Se: 0.16 fg per cell	Citrate-stabilized gold nanoparticles	Distinguish between nanoparticle-bound elements and their corresponding low-molecular species. The use of complementary analytical strategies that enabled the detection and characterization of selenium-containing nanoparticles in selenized yeast.	https://doi.org/10.1039/C9AN01565E
189	SC/SP-ICP-TQ-MS	Organisms, S. aureus and E. coli	Homogenize	Tellurium nanoparticles (Te NPs)	Aureus: 0.5–1.9 fg Te/cell; E. coli: 0.08 to 0.88 fg Te/cell	A concentration range from 0 to 50 mg L1 with a tellurium standard for ICP obtained from Sigma Aldrich	Quantify the uptake of Te NPs in single bacterial cells; investigate, quantitatively, nanoparticle uptake in bacterial cells and to estimate the dimensions of biogenic Te nanorods. SC-ICP-MS was used to quantify the uptake of Te NPs in single bacterial cells.	https://doi.org/10.1016/j.aca.2020.06.058
190	time-resolved ICP-MS	Yeast suspension	Homogenize, dilute	Mg, P, Ca, Mn, Fe, Cu, Zn	/	External calibration: P and Zn (correlation factor 0.69); P and Mg (0.63), Mg and Zn (0.63); R^2 = 0.9939	Using the HECIS and described method of multielement analysis with algae and bacteria cells, along with extracted organelles, is underway in our laboratory. Qua-simultaneous multi-element detection for multielement correlation analysis within cells.	https://doi.org/10.2116/analsci.29.597

(Continued)

Table 4.1 (Continued)

No.	Instrumental	Matrix	Sample preparation	Element or isotope	LOD or LOQ	Calibration	Application and technical characteristics	DOI of references
191	time-resolved ICP-MS	Unicellular alga, *Chlorella vulgaris*	Filtration, homogenize	Mg	10 ng/ml	Aqueous standards (10 mg/l each of Li, Y, Co, Ce and Tl). For calibration; calibration with MgO particles ($R^2 = 0.999$)	A method of simultaneous cell counting and determination of constituent metals in single cells using time-resolved ICP-MS. Quantitative determination of the metal contents of the algal cells using metal oxide particles for calibration. Semi-quantitative measurement is also possible using aqueous standards for calibration.	https://doi.org/10.1039/C002272A
192	LA-ICP-MS	Marine microalgae *Scrippsiella trochoidea* (Von Stein) Loeblich III	Degradation	Cu	0.5–100 μg/l	External standards 10, 20, 50, or 200 μg/g of Cu; an oxide ratio ($^{238}U^{16}O^+/^{238}U^+$) below 1.0% and $^{232}Th^+/^{238}U^+ \approx 1$	Detailed analysis of (sub) cellular microenvironments. The accumulation of Cu in the model organism *Scrippsiella trochoidea* resulting from transition metal exposure (ranging from 0.5 to 100 μg/l) was evaluated.	https://doi.org/10.1021/acs.analchem.6b00334
193	SP-ICP-MS	Plastic particles	Filter, dilute. cut,	Carbon-13 isotope	1.2 μm for PS microparticles	Aqueous dissolved carbon standards	The analysis of a microplastic suspensions by ICP-MS operated in single-particle mode using microsecond dwell time. The analysis of a microplastic suspensions by ICP-MS operated in single-particle mode using microsecond dwell times. Enables the detection of metallic nanoparticles below 10 nm.	https://doi.org/10.1016/j.talanta.2020.121486

194	SP-ICP-MS	Cat; Spleen samples	Sonication-homogenize	Au NPs	100 ng/l	An external calibration curve using Rh as internal standard	SP-ICPMS is a promising method for the detection of metal-containing nanoparticles (NPs) and the quantification of their size and number concentration. Use of alkaline or enzymatic sample pretreatment prior to characterization of gold nanoparticles in animal tissue by single-particle ICPMS.	https://doi.org/10.1007/s00216-013-7431-y
195	SP–ICP–MS	Plant tissues	Homogenize, digest	Au NPs	1000 NPs/ml	AuNP standards with sizes of 30, 50, 80, and 100 nm served as particle calibration standards	Develop for simultaneous determination of gold NP (AuNP) size, size distribution, particle concentration, and dissolved Au concentration in tomato plant tissues. Simultaneously detect both the particle analyte and the dissolved analyte.	https://doi.org/10.1021/es506179e
196	SP–ICP–MS	Environmental samples	Homogenize	Nanoparticles	/	Environmental samples, detection and determination of native particles in waters	Cope and solve the challenges and problems related to the analysis of ENMs in environmental samples. Qualitative information about the presence of particulate and/or dissolved forms; quantitative information as particle number as well as mass Concentrations; characterization information about the mass of elements per particle and particle size.	https://doi.org/10.1016/j.teac.2016.02.001

(Continued)

Table 4.1 (Continued)

No.	Instrumental	Matrix	Sample preparation	Element or isotope	LOD or LOQ	Calibration	Application and technical characteristics	DOI of references
197	SP–ICP–MS	Drinking water	Disinfection.	Titanium dioxide, silver, and gold nanoparticles	Ti: 0.75 µg/l Ag: 0.10 µg/l Au: 0.10 µg/l	Citrate-capped Au NPs and Ag NPs were used as particle calibration standards	Rapid tracking of nanoparticles during simulated drinking water treatment; investigate the fate and transportation of NPs during drinking water treatments. Single-particle-ICP-MS methods for analysis for TiO2, Ag, and Au nanoparticles. Rapid tracking of nanoparticles during simulated drinking water treatment. Nanoparticles are removed during lime softening and alum coagulation. Ti-containing particles present in source water were removed by water treatment.	https://doi.org/10.1016/j.chemosphere.2015.07.081
198	SP–ICP–MS	Radish	Lyophilization, grind	Cerium oxide and copper oxide nanoparticles	0.39 and 4.3 µg/l for Ce and Cu	External calibration, liner ranges: 0.0–20 µg/l	A powerful tool for the analysis of NPs in environmental samples. The presented methodology was developed to study the bio-accessibility of cerium oxide (CeO_2) and copper oxide (CuO) NPs from radish after the in vitro simulation of gastro intestinal digestion using single-particle inductively coupled plasma mass spectrometry (SP-ICP-MS).	https://doi.org/10.1002/jsfa.10558

199	SP–ICP–MS	Plant tissues	Homogenize, digest	Au NPs	1000 NPs/ml	AuNP standards with sizes of 30, 50, 80, and 100 nm served as particle calibration standards	Develop for simultaneous determination of gold NP (AuNP) size, size distribution, particle concentration, and dissolved Au concentration in tomato plant tissues. Simultaneously detect both the particle analyte and the dissolved analyte.	https://doi.org/10.1021/es506179e
200	SP–ICP–MS	Arabidopsis plants	Homogenize	Silver nanoparticles (Ag NPs)	0.02 mg/l	The 30 nm Au NP standard was used for particle calibration	The measurement of the size distribution of gold nanoparticles (Au NPs) in NP-exposed tomato plants; the analysis of water, food matrices, materials, biological tissues, body fluids for trace metals, and toxic elements. A powerful tool to directly quantify single-particle size, concentration, and size distribution.	https://doi.org/10.3389/fpls.2016.00032
201	SP–ICP–MS	Chicken meat	Sonication-homogenize, enzymatic digestion	Silver nanoparticles (Ag NPs)	0.05 mg/kg	Standard of the 60-nm gold NPs; ionic silver standards in the concentration range of 0.2–5 µg/l for calibration	Using SP-ICP-MS for the sizing and quantification of silver NPs in a food matrix. Sizing and quantifying nano-silver in chicken meat using single-particle ICP-MS.	https://doi.org/10.1007/s00216-013-7571-0

Laser Ionization Mass Spectrometry Laser ionization mass spectrometry (LIMS) is a micro-analysis technique employing a laser ionization source, in which TOF-MS is well matched to laser pulse nature and thereby used to analyze molecular mass or and its signal intensity. Compared with the former, LIMS obtains chemical information of almost all elements on sample surface with 1–5 µm laser spot size and typical 0.01–0.5 µm sampling depths. It was proven that LIMS is a reliable semi-quantitative method for the rapid micro-analysis of metallic and non-metallic elements in solids, which is valuable to supervise the harmful substances for evaluation of soil contamination. Considering the current limitation of LIMS, it is not popular in agricultural research field, but it could investigate metallomics for solid agricultural media such as soil, feedstuff, fertilizer, animal and plant tissues, or other related materials that may have a significant reward anticipation [17, 18].

Thermal Ionization Mass Spectrometry Thermal ionization mass spectrometry (TIMS) is a highly sophisticated tool for measuring precise isotope ratio. The measurement precision could be down to a few parts per million due to an arrangement of two or even three heating filaments, simultaneous multicollection of different isotopes using an array of Faraday collectors, and a moveable electron multiplier for different elements [19]. Though the LOD of TIMS was higher than ICP-MS, TIMS has better sensitivity and precision analysis for isotope ratio with small size samples (low to sub-picogram to nanogram). However, one should also pay attention to the measurement accuracy of TIMS that is still confronted with mass discrimination (e.g. ion optical system or ion detector) and mass fractionation effects during sample evaporation. Till now, TIMS has been frequently used for nuclear materials, geochemistry, environment [20]; and other applications of TIMS for agri-food, water, and soil are related to authentication, safety, bioavailability, and nutrient studies.

Glow Discharge Mass Spectrometry Glow discharge mass spectrometry (GD-MS) is a very powerful and effective analytical approach for direct analysis and depth profiling of elements and isotopes. During the last decades, GD-MS has received increasing attention in various research fields. In fact, GD-MS instrument should be one of the oldest MS forms, in which an argon gas glow discharge is usually used as an ionization source with direct current (dc) or radiofrequency (RF) power. The GD-MS instrument is successfully used for metals, alloys, and conductive solid samples. However, it must be known that most of agricultural media are not available as conducting materials, which cause relatively poor application of GD-MS in agrometallomics. Therefore, the related application. Even so, GD-MS displays versatility, e.g. pulsed GD-TOF-MS could analyze elemental oxidation states from the solid material [21, 22] and GD-MS/MS to analyze pesticide in food sample [23]. Hence, one has to believe that GD-MS owns a potential of elemental speciation and organic composition analysis in agrometallome.

Secondary Ion Mass Spectrometry Secondary ion mass spectrometry (SIMS) instrumentation is a sensitive technique for surface and depth profiling analysis to provide in situ micro-analysis of elements information; as well, detailed elemental and isotopic information on parts-per-million level in a variety of solid samples [24] such as soil, agricultural animal and plant, and so on. In addition to normal SIMS, high-resolution secondary ion mass spectrometry (NanoSIMS) and time-of-flight secondary ion mass spectrometry (TOF-SIMS) are more popular for micro-analysis (even subcellular imaging scale) and composition analysis.

4.2.1.2 Atomic Spectrometry for Agrometallomics

Optical Emission Spectrometry Inductively coupled plasma optical emission spectrometry (ICP-OES) is a kind of multipurpose analytical tool for multiple metallic and some non-metallic elements [25, 26], and is suitable for variety of agricultural samples. OES-MS instrumentation with liquid nebulizer is the most frequently used sampling system, while chemical vapor generators (CVG) such as ultraviolet vapor generation (UVG), hydride generation (HG), and photochemical vapor generation (PVG, non-ultraviolet) or other assistant sampling techniques are designed to improve the analytical sensitivity and eliminate matrix effect for special elemental analysis. Furthermore, LA, ETV, and LC are also hyphenated with ICP-OES for various application requirements. ICP-OES has a wide linear dynamic range for multi-elements almost suitable for all of agricultural samples [27]. Compared with ICP-MS, ICP-OES is especially suitable to micro- or macro-elements analysis, as well as non-metallic elements such as S, N, P, and Si; as well, it is immune to molecular ions interference from mass spectrometer, while it mainly depends on instrumental spectrum resolution and background correction for spectral and matrix interferences from complicated agricultural samples. Microwave plasma optical emission spectrometry (MP-OES), originating from 1963 [28], usually used nitrogen gas generated from nitrogen generator as a self-sustained atmospheric pressure microwave plasma source to replace the ICP mentioned above. Compared with ICP-OES, MP-OES shows several advantages such as smaller footprint, low price, and maintenance cost besides multi-elemental analysis and wide linear range. However, due to relatively lower absolute temperature of excitation source, MP-OES shows it is difficult to emit more atomic spectrum compared with ICP-OES, which leads to the weaknesses for anti-interference and LOD [29]. Even so, MP-OES is completely superior to flame atomic absorption spectrometry (FAAS) and partly comparable to ICP-OES, and has been applied in various agricultural research fields such as environmental [30], plant [31], agri-food [32], health [33], and biotechnology [34].

Low-Temperature Plasma Atomic Spectrometry Temperature is one of the important parameters for plasma. Most of plasma perform at high temperature and is widely used, such as ICP or MP. On the contrary, low-temperature plasma (LTP) is receiving more attention for elemental analysis because of ambient-temperature operation, miniaturization, simplicity, and low expense/energy consumption. Typically, solution anode/cathode glow discharge [35–37], point discharge (PD) [38], corona discharge [39], glow discharge (GD) [40], dielectric barrier discharge (DBD) [41–43], and microwave or radio frequency discharge [44, 45] are commonly used to generate LTP, namely non-thermal plasma (NTP). The above LTP techniques were popular for developing portable and miniature atomic spectrometers mainly using liquid injection, CVG, or ETV sampling system coupled with atomic fluorescence spectrometry (AFS), AAS, OES, and even mass spectrometer as detectors.

DBD is a typical micro-plasma and with great versatility and popularity in agriculture-related elemental analysis. DBD is defined as a discontinuous plasma using at least one dielectric layer such as quartz, ceramics, or Teflon placed between the electrodes, which can run under atmospheric pressure with different working gases (Ar, He, N_2, air, and their mixtures) [46]. The reactive free radicals, energetic electrons and ions, excited-state species, and ultraviolet light generated in plasma enable elemental analysis. The commonly used structures of DBD can be identified as planar and cylindrical configuration regardless of surface DBD (surface discharges) and capillary DBD [47, 48]. In 2001, DBD was firstly utilized for diode laser spectrometry [49]; in 2006, Zhang et al. used atmospheric pressure DBD as an atomizer for AAS and AFS [50–52]; up to now, DBD has been utilized as ionization source for mass spectrometer [53], excitation source for OES or induced source for CVG [54, 55], gas-phase enrichment (GPE) device for elemental analytes [41–43]. However, considering the vulnerability of micro-plasma, CVG such as HG and UVG is the most frequently used sampling method for DBD atomization or excitation to analyze liquid or slurry sample. To direct analyze solid sample, Liu and Mao employed a quartz tube ETV to introduce Hg and a DBD tube to degrade gaseous organic interferents for Hg analysis in food samples by AFS [56]; in another study, Jiang et al. fabricated an ETV coupled with DBD-OES for the direct solid sampling analysis of Hg, Cd, and Pb in soil [57]. Like LA-ICP-MS, the micro-plasma by DBD can be used for elemental ablation in solid surface following ICP-MS (Au, Ni, and Cu) or AFS (Hg) [58, 59]. Although the excitation power of DBD is obviously weaker than that of LA, solid sampling DBD technique is still capable of giving elemental mapping information to some extent in solid surface.

Instrumental Neutron Activation Analysis Instrumental neutron activation analysis (INAA) is a radiometric analysis technique, which uses neutrons to irradiate target elements to yield a nuclear reaction and change these elements into radionuclides called activation; then, characteristic half-life and rays are determined for elemental quantification with high sensitivities in the 10^{-13}–10^{-6} g/g ranges [60]. INAA is nondestructive for solid, liquid, and gas samples and is relatively free of serious interferences and without sample preparation. So, INAA is a favorable tool to rapidly screen multiple elements in agrometallome to explore the uptake, bio-accumulation, and translocation of different elements in agri-processes [61]. Agri-food, soil, plastics, biological tissue, medicine, or even atmospheric particulates samples were reported to be analyzed by INAA [62–65]. The limitation of INAA instrumentation is it is too expensive and too complicated for operation. At the same time, the analysis time of traditional reactor INAA (Re-INAA) is too long due to elemental decay period.

X-ray Fluorescence Spectrometry X-ray fluorescence spectrometry (XRF) is a non-destructive approach for solid sampling analysis of multi-elements without chemical or digestion treatment. XRF uses an X-ray tube or radioisotopic source to excite electrons causing these atoms to emit secondary X-rays, that is, to fluoresce, in which these fluorescence emissions give specific energy spectra or signal peak height for elemental quantification. XRF can be roughly separated into two categories: wavelength-dispersive XRF (WDXRF) and energy-dispersive XRF (EDXRF). WDXRF is governed by using diffraction crystals monochromatic device in each characteristic spectrum to sequentially scan certain elements. Unlike WDXRF, conventional EDXRF allows detector to collect all of the X-rays emitted by the sample at the same time, thereby giving great speed in acquisition and display of data [66]. However, the analytical sensitivity and LODs of EDXRF are always inferior to those of WDXRF. Both WDXRF and EDXRF are widely applied to multi-elemental analysis ranging from sub-part-per-million to percent levels in various agricultural media including soil, agri-food, feedstuff, fertilizer, plant and animal tissues, and atmospheric particle samples.

4.2.2 Elemental Speciation and State Analysis in Agrometallomics

Element analysis technique provides information about element types, but it might be uninformative and even misleading to any positive or negative effects on living organisms if there is no elemental speciation information. Elemental speciation studies are critical in evaluating agri-food nutrition, animal and plant growth, and agricultural water and soil, because the bioavailability, mobility, and toxicity are greatly dependent on the physical and chemical forms of an individual element. Therefore, the purpose of speciation analysis is to identify and determine individual physical–chemical forms or existing state of elements in real samples. Chromatography coupled with atomic spectrometry, synchrotron radiation, and X-ray energy spectroscopy can be used to analyze the elemental speciation or state analysis in agrometallome.

4.2.2.1 Chromatographic Hyphenation for Atomic Spectrometry or Mass Spectrometry

It is well known that the function of liquid chromatography (LC) is compounds separation. The application choice for elemental speciation composition for LC depends on the element of interest and the chemical environment in which it is located. LC includes ion chromatography (IC), high-performance liquid chromatography (HPLC), monolithic column, and low-pressure liquid chromatography [67, 68]. Among them, HPLC is the most commonly used one with a range of hyphenation applications for AAS, AFS, ICP-MS, and ICP-OES instruments; and, allows organic or inorganic elemental species analysis of Sb, P, Br, I, Cr, Se, Hg, As, platinum group elements, and rare-earth elements in agricultural samples [69–74]. This hyphenation technique provides a promising strategy for investigating the toxic, metabolic, and bioavailable elements behaviors in agricultural lives.

From the analytical point of view, gas chromatography (GC) is one of the most powerful separation means for the volatile inorganic and organic composition analysis due to its high separation efficiency and sensitivity, wide dynamic ranges, and good precision. The mobile phase of GC is inert gases, which is different with LC instrument. During the experimental operation, the gaseous analyte separated and was transferred into the atomizer or the excitation/ionization source following ICP-MS, MS, AAS, or AFS detection [75]. Considering gaseous analyte introduction, GC separation techniques are suitable to Pb, As, Sn, Se, Hg elements, etc. [76]. For example, volatile alkyl-selenides and their sulfur analogues could be determined by solid-phase microextraction (SPME) GC-ICP-MS to investigate Se metabolism in agricultural lives. However, non-volatile selenium compounds such as some seleno-amino acids are hard to be separated by GC. Furthermore, Hg speciation including inorganic or organic Hg in agricultural sample, environment, seafood, and food package can be measured by GC-ICP-MS [77], GC-MS/MS [78], or GC-AFS [79]. Besides, it has already been reported that GC-ICP-MS, GC-AFS, or GC-MS/MS could be adopted for elemental speciation analysis such as of Sn, As, and Pb in agrometallomics-related samples; further, GC-ICP-MS with or without high-resolution mass spectrometer to analyze non-traditional elements (Cl, Br, Si, S, and P) speciation can also be attempted which also extend the GC coupling techniques' application [76, 80, 81].

Compared with LC and GC, capillary electrophoresis (CE) is one kind of a more high-efficiency separation technique. The application of CE as an atomic spectrometric hyphenation technique for speciation analysis has been growing in the past decades [82, 83]. During the operation of CE, the separation process takes place inside capillaries containing no stationary phase. Compared with GC or LC, CE contains several unique advantages such as fast and low cost, minimized reagent consumption, high resolving ability, and the possibility of separations with only minor disturbances of the existing equilibrium among different elemental species [84, 85]. Up till now, elemental speciation including Mg, Ag, Zn, Cr, Sb, As, Hg, and Se [86–89] has been separated and analyzed by CE-ICP-MS for several agricultural samples such as animal tissue, plant, medicine, agri-food, soil, and water, and which is in order to further investigate the related metabolic mechanism in agricultural environment and biology. However, several problems should be noticed, that CE technique followed with insufficient sensitivity and separation instability, which was mainly due to inevitable small volumes of injected sample and inside the structure and electrical properties of capillary. So, these characteristics render CE unable to perform real-world speciation analyses to some extent.

4.2.2.2 Synchrotron Radiation Analysis

Synchrotron-radiation (SR)-based techniques have been utilized in agricultural fields with increasing frequency in the past few decades. SR-related techniques possess various strong imaging or speciation analysis methods with label-free, non-destructive, high-sensitive, and rapid characteristics based on X-ray fluorescence, absorption, diffraction, etc. [90], because the SR resource could produce high brilliance and photon flux, tunable wavelength, and bandwidth energy spectrum [91]. Meanwhile, the SR techniques can be mostly performed without sample pre-treatment, which is pretty suitable for in situ analysis and elemental mapping in agricultural samples in the investigation of the uptake, bio-accumulation, and translocation of various elements in agricultural sample or processes [92, 93]. Typically, SR X-ray fluorescence (SRXRF) and SR X-ray absorption spectroscopy (SRXAS) are the most commonly used for micro-analysis and speciation discrimination especially for the researches of agricultural samples.

In principle, SRXRF is a kind of atoms fluorescence analysis excited by X-ray from SR resource, of which most elements in sample can be qualitatively and quantitatively analyzed via detecting specific fluorescent photons [94]. From the view of application, SRXRF analysis is particularly suitable for studying microscopic distribution [95] (to several micrometer) of trace elements in samples, because of high spatial resolution and the ability of simultaneous multi-elemental analysis with 50–100 ng/g LOD [96]. SRXRF has already been reported for element types, elemental spatial distributions, and concentration analysis in various agricultural samples such as soil samples [97], microorganisms [98], plants [99], and animal [100].

As mentioned before, chromatography is the classic power technique for determination of elemental speciation, but a complicated and time-consuming sample preparation is inevitable to ensure sufficient extraction of all target speciation and no speciation transformation during this process. By comparison, SR-based analytical techniques such as XAS display a powerful ability of in situ elemental speciation analysis at micro-scale level; especially, non-destructive analysis is more appropriate for molecular and processing mechanism study of agricultural lives and activities [11]. XAS is able to measure the transition from the core electronic state of a metal to the excited electronic states (LUMO) and continuum, in which the former is X-ray absorption near-edge structure (XANES), and the latter is extended X-ray absorption fine structure (EXAFS). XANES renders complementary structural information of samples such as symmetry of the metal site and electronic structure, types, EXAFS numbers, and distances to ligands and neighboring atoms from the absorbing element [101]. Due to the use of SR X-ray as the excitation source, SRXAS owns higher sensitivity *vs.* common XAS to elucidate the oxidation state and coordination number, the identities of its nearest neighbors, and the length of the bond for tested elements [102–104].

4.2.2.3 Energy Spectroscopy Based on X-ray

In agricultural field, another commonly used technique based on SR is X-ray photoelectron spectroscopy (XPS), well known as electron spectroscopy for chemical analysis (ESCA). XPS is a semi-quantitative or quantitative tool for elemental composition in surface within a depth of ~10 nm of a substance under ultra-high vacuum via measuring the binding states of elements. XPS has extensive use in the investigation of agrometallomic mechanism for agricultural samples such as fertilizer, cotton textile, soil, fertilizer, and even cell.

XRD is a kind of analytical technique to measure the structural characteristic of composition to evaluate the crystal size, crystalline structure, the ratio of crystalline to non-crystalline (amorphous), the distance between planes of crystal, and the crystal arrangement pattern. XRD is the most commonly used for inorganic material and composite samples and cooperates with XPS to render more elemental and crystal information. So, soil, fertilizer, agri-food and agricultural by-product, textile material, waste and nanomaterial can be measured by XRD to explore the composition speciation and structure of samples [105–109]. As a versatile approach of crystalline analysis, XPS coupled with other analysis approaches such as ICP-MS, XPS, SEM, SIMS may play an extraordinary role in providing elemental speciation and composition structure information in agrometallome.

4.2.3 Spatial Distribution and Micro-analysis Techniques in Agrometallomics

4.2.3.1 Laser Ablation Inductively Coupled Plasma Mass Spectrometry

Technology with spatial distribution or micro-analysis features was also one of an important demand in agricultural field. Laser ablation inductively coupled plasma mass spectrometry (LA-ICP-MS) is capable of providing spatially resolved distribution and concentration information for a large number of elements simultaneously [110–112], involving both solid sampling ultraviolet (UV) nanosecond (ns) or femtosecond (fs) laser ablation system made of solid-state Nd:YAG (266 and 213 nm), ArF mixture (193 nm), and Ti-sapphire (femtosecond laser) [113], and high sensitivity mass spectrometric detection system such as ICP-Q-MS, ICP-SF-MS (single-ion collection) or MC-ICP-MS (multiple ions collection) [114]. Compared with SRXRF, LA-ICP-MS is significantly cheaper and provides isotope ratios information under lower elemental level [114]. Compared with LIMS, the LODs of LA-ICP-MS are lower around 1 or 2 orders of magnitude but with better reproducibility and much faster analysis speed. As a direct solid sampling technique of elemental analysis, LA-ICP-MS does not consume HNO_3, HCl, HF, or other reagents, extremely diminishing the risk of contamination and elemental loss caused by sample pretreatment [115, 116]. In terms of application, LA-ICP-MS has been extensively used to perform the *in situ* distribution or mapping investigation of metal/metalloids, some non-metals (C, P, and S) elements and their isotopic information at trace level in raw specimen to provide the localization features and molecular mechanism of element-labeled compounds [10], metallobiomolecules [11], etc. According to laser source, shorter wavelengths are beneficial to reduce fractionation effects that is the sum of all non-stoichiometric effects occurring in the processes of ablation, transport, and ICP ionization as well as improve the quantification [113, 117]. In terms of mass spectrometric detector, analytical sensitivity of LA-ICP-SF-MS is one order of magnitude higher than that of LA-ICP-QMS and could provide a higher sensitivity for imaging of interested elements [118]; as well, another advantage of SF-MS is its superior signal-to-noise ratio and even measurement in nanometer scale [119]. MC-ICP-MS could obtain the accurate and precise isotopic ratio [120], which could be used for tracing elements and the life history such as individual or mineral [121, 122]. Additionally, LA-ICP-TOF-MS is capable of simultaneously detecting all targeted elements in small sample size within dozens of microseconds mass to give more elemental information than sequentially scanned elements mentioned above; so, TOF-MS is especially suitable to couple with LA to save experimental time in practice but also for simultaneous multi-elements analysis [123, 124]. However, LA-ICP-MS is still in its early stages for agricultural application, although some excellent achievements have already been demonstrated.

Even so, the fundamental problem of LA-ICP-MS quantification is the signals that are not representative of elemental composition of sample because of some limited problem such as matrix effects, elemental fractionation, and laser plasma. So far, the most credible method for the accurate quantification of LA-ICP-MS is external calibration utilizing matrix-matching certified reference materials (CRMs) [125]. However, it is too hard to find all matrix-matching CRMs with good homogeneity and quantified elements we need in practice. To solve this problem, the matched matrix can be utilized via homogenizing target elemental standard with solids [114, 126]. In addition, isotope dilution mass spectrometry (IDMS) is internationally regarded as an absolute quantification method [127]. Compared with other calibration strategies, IDMS can correct for some element losses, fractionation, and matrix effect [128]. So, in agricultural field, ID-LA-ICP-MS is suitable for in situ quantitative imaging of trace elements in biological samples, such as animals [129], plants [130], single cell and even protein scale [131].

4.2.3.2 Electrothermal Vaporization Hyphenation Technique

Direct sampling technique is of great benefit for rapid detection of agricultural field. Electrothermal vaporization (ETV), always being made from some high-melting point metals, such as graphite, quartz, tungsten, and rhenium [132–135], is a high-efficiency sampling manner coupled with ICP-MS (ETV-ICP-MS), ICP-OES (ETV-ICP-OES), AAS (ETV-AAS), and AFS (ETV-AFS), and to detect metal or non-metal elements in various samples [132, 136–138]. According to the material and structure of electrothermal vaporizer, ETV hyphenation atomic spectrometry allows directly introducing sub-milligram to gram levels sample size, so it can also be classified into micro-analysis. Hence, ETV has been extensively used to fabricate fast and miniature analyzer for heavy metals in agricultural samples, such as biological samples, soil, agri-food, and plant and animal [56, 57, 132, 134, 139–141]. Direct sampling Hg analyzer consisting of ETV catalytic pyrolysis atomic spectrometer with or without amalgamation has become one of the most successful commercial instruments for solid sampling elemental analysis [134]. Rapid detection of toxic heavy metals such as Zn, Cu, Cr, As, Pb, and Cd in agricultural food and soil is very helpful to control heavy metal contamination in advance [136, 142, 143]. However, due to the limitation of sampling apparatus, ETV hyphenation is difficult for applying to elemental mapping or depth profiling for agrometallome, but there is no denying its application potential in the field and rapid detection.

4.2.3.3 Laser-induced Breakdown Spectroscopy

Laser-induced breakdown spectroscopy (LIBS) has been employed in the past two decades as a promising technique for analyzing the composition of a broad variety of objects. LIBS is an atomic emission spectrometry using focused pulsed laser beam to generate plasma on sample surface, which could have been used for the analysis of metal or non-metal elements and even organic composition characterization and identification [144]. For the agricultural applications, LIBS can replace time-consuming analytical methods for assessing the quality and composition of agricultural samples involving authenticity control, sample discrimination, and traceability, bacteria contamination, elemental surface mapping, and depth profiling, because of their characteristics of efficient and reagent-free operation, in situ or on-line solid sampling tool [27, 145]. Considering rapid detection using laser excitation, no or minimal sample preparation can be achieved. So, LIBS is always designed to fast analyze agricultural soils, fertilizers, animal tissues, fruits, seeds, grains, pasture vegetables, and crop plant leaves; further, the application range involves animal or plant nutrients and phytotoxic elements, pesticide residues, impurities, and other pollutants, as well as the quality control of food derivatives, the identification of early disease diagnosis [144, 146]. Moreover, LIBS could also be used for elemental mapping in various biological matrices including the spatially resolved analysis of plant and animal samples, which is of great and increasing interest to many agronomists and biochemists in order to establish compositional interrelationships of the elemental constituents in agricultural samples [145]. The function between LA-ICP-MS and LIBS is quite similar, but LA mapping involves sample transport and inevitable sample dispersion in the ablation chamber and in the tubing; while LIBS demonstrates a full, quantitative, fast-scanning on-line mapping, although the analytical sensitivity is obviously inferior to that of LA-ICP-MS [145].

4.2.3.4 Single-Cell and Micro-particle Analysis

The recent maturation of single-cell and micro-particle analysis technologies belongs to transformative new method. Traditional analyses usually display overall average information and ignore the elemental behavior under single-cell or micro-particle level. However, researchers gradually realized that cells or particles have individual differences, which is crucial in various agricultural fields such as metallomics, medical diagnosis, toxicology, or clinical tests [147]. To fulfil the discrimination of individual cell(s) or particle(s), high time-resolved analyzers must be used as detectors, e.g. Nomizu et al. [148] measured Ca signal in single cells by time-resolved ICP-OES; by comparison, time-resolved ICP-MS shows higher resolution and lower LOD, e.g. detecting Mn, Zn, K, S, P, Mg, and other elements in single cell [149, 150]. Therefore, nanoparticles and cells analyses based on MS, such as single-particle ICP-MS (SP-ICP-MS) and single-cell ICP-MS (SC-ICP-MS), have been regarded as a powerful tool for micro-analysis at cellular and nano-levels [151].

Single-cell analysis is gaining more and more attention for simultaneously measuring and characterizing nanoparticles in single cell, considering the differentiation of elemental composition at cellular or sub-cellular level. The single-cell analysis techniques mainly based on ICP-MS [152, 153] are suitable for various cell types [154] involved in agricultural application; however, they suffer from analytical sensitivity, resolution, and throughput due to absolutely small size sample. Sampling system or mode is very critical to separate or introduce individual cell(s); flow cytometer [155], GC [156], LC [157], microfluidic system [131], and LA are thereby frequently utilized. All these sampling systems allow cells introduction directly or indirectly into plasma resource for atomization and excitation; and the next analysis with high time resolution MS for elemental content in individual cell or cell population [158–160]. These methods also have been successfully and commonly used to investigate the uptake of elements contents by single cell from animal [161], algae [162], fungi [163], bacteria [164], etc.

On the other hand, considering the widespread application of engineered nanomaterials leading to their release into the environment, the method for reliable and fast detection, quantification, and characterization of nanomaterial becomes critical for the eventual understanding of their implication to human health and environment quality. Like single-cell analysis with various sampling techniques mentioned above, SP-ICP-MS also fulfils the measurement on a "particle-by-particle basis" via counting and sizing particles at lowest concentrations; as well as simultaneously distinguishes between dissolved and particulate analytes. SP-ICP-MS has been widely used in the particle size and quantification of particles in various substrates, such as consumer products [165], animal and plant tissues [166, 167], and environmental samples [168].

4.3 Application and Perspectives of Agrometallomics in Agricultural Science and Food Science

4.3.1 Agricultural Plants and Fungi and Derived Food

HPLC is the most commonly used one with a range of hyphenation applications for AFS, AAS, ICP-OES, and ICP-MS instruments; and, allows inorganic or organic elemental species analysis of As, Hg, Se, Cr, I, Br, P, Sb, platinum group elements, and rare-earth elements in agricultural samples [69–74]. For AFS and AAS hyphenation, HPLC-(UV)-HG-AFS using anion-exchange chromatography or cation-exchange chromatography was employed to measure eight arsenic species [169]; HPLC-HG-AAS was also reported to analyze As(III) and As(V) in Boletus badius sample [170]. Herein, HG is designed as a hyphenation unit to connect LC with AAS or AFS, as well as to eliminate interference and enhance sensitivity. Furthermore, Liu and Mao designed a DBD reactor to preconcentrate As speciation following LC separation to overcome the weakness of insufficient analytical sensitivity of LC-AFS [171]. For reversed-phase high-performance liquid chromatography (RP-HPLC) with UV–Vis detection, the RP-HPLC technique was used for testing elements such as Pb, Cd, Hg, Ni, Cu, Zn, Sn in Chinese herbal medicine [171]. However, these LC hyphenating AAS or AFS techniques are only limited to hydride-forming elements such as As, Hg, Se, Sb, and Sn. By comparison, ICP-MS demonstrates a stronger multi-elemental analysis with better sensitivity and linearity range, and other elements (e.g. P, Cr, I, Br, V, Mo, Te, Tl, W) amenable to ICP-MS were reported recently [68, 172]. For instance, iodide, iodate, 3-iodo-tyrosine (MIT), 3,5-diiodo-tyrosine (DIT), bromide, and bromate were separated by anion-exchange chromatography followed with ICP-MS to assess the bio-available fractions of iodine and bromine species from edible seaweed [173]. In addition, due to multi-elemental detection, HPLC-ICP-MS method was established for simultaneous speciation analysis of As, Hg, and Pb in lotus seed sample including four arsenicals (As(III), DMA, MMA, and As(V)), four mercurials (Hg(II), MeHg, EtHg, and PhHg), and three lead compounds (Pb(II), TML, and TEL) within only eight minutes [174]. HPLC-ICP-MS method was also established for simultaneous speciation analysis of As in Cordyceps sinensis (C. sinensis) sample including four arsenicals (As(III), DMA, MMA, and As(V)) [175]. This hyphenation technique provides a promising strategy for investigating the toxic, metabolic, and bioavailable elements behaviors in agricultural lives.

The application of capillary electrophoresis (CE) as an atomic spectrometric hyphenation technique for speciation analysis has been growing in the past decades [82, 83]. For CE, the separation process takes place inside capillaries containing no stationary phase. Compared with GC or LC, CE offers several unique advantages such as high resolving ability, minimized reagent consumption, fast and low cost, and the possibility of separations with only minor disturbances of the existing equilibrium among different elemental species [84, 85]. Since Olesik et al. [176] first rendered an idea of interfacing CE with ICP-OES and ICP-MS, many studies on the hyphenation of CE with atomic spectrometer have been performed [177–180]. So far, elemental speciation including Se, Hg, As, Sb, Cr, Zn, Ag, and Mg [86–89] has been separated and analyzed by CE-ICP-MS for water, soil, agri-food, medicine, plant and animal tissue samples to further investigate the related metabolic mechanism in agricultural environment and biology. Besides, Lena Ruzik intended an approach for examining speciation analysis of Zn in goji berries using CE-ICP-MS and CE-ESI-MS/MS, furthermore, identifying nine compounds with zinc isotopic profile [87]. Lars Bendahl separated selenium compounds by CE-ICP-MS applied to selenized yeast samples [181]. LihongLiu successfully applied CE-ICP-MS method for arsenic speciation in two herbal plants (flower of *Chrysanthemum morifolium* Ramat) [182]. However, compared to HPLC, it should be noticed that insufficient sensitivity and separation instability are the most serious shortcomings of CE due to inevitably small volumes of injected sample and inside structure and electrical properties of the capillary. So, these characteristics render CE unable to perform real-world speciation analyses to some extent.

Synchrotron radiation (SR) techniques possess various strong imaging or speciation analysis methods with non-destructive, rapid, and unmarked characteristics based on X-ray fluorescence, absorption, diffraction, etc. [90], because the SR resource could produce high brilliance and photon flux, tunable wavelength, and bandwidth energy spectrum [91]. Meanwhile, the SR techniques can be mostly performed without sample pre-treatment, which is pretty suitable for in situ analysis and elemental mapping in agricultural samples to investigate the uptake, bio-accumulation, and translocation of different elements in agricultural processes [92, 93]. Typically, SR X-ray fluorescence (SRXRF) and SR X-ray absorption spectroscopy (SRXAS) are most commonly used for micro-analysis and speciation discrimination.

SRXRF is a kind of atoms fluorescence analysis excited by X-ray from SR resource, of which most elements in sample can be qualitatively and quantitatively analyzed via detecting specific fluorescent photons [94]. SRXRF analysis is particularly suitable for studying microscopic distribution [95] (to several micrometers) of trace elements in samples, because of high spatial resolution and the ability of simultaneous multi-elemental analysis with 50–100 ng/g LOD [96]. SRXRF has already been reported for elemental concentration and spatial distributions analysis in various agricultural samples such as soil [97], microorganisms [98], plants [99], and animal [100]. For example, Chen Xu used SRXRF to measure the distribution of Cu in Elsholtzia splendens' (E. splendens) root, finding the transport of Cu from cortex to xylem in roots [183]. Shen, Ya-Ting used micro-SRXRF to determine element distribution characteristics of K, Ca, Mn, Fe, Cu, Zn, Pb in an Arabidopsis thaliana seedling [184]. Wang et al. [185] revealed the response mechanism of pakchoi under stress of Pb and Cr by studying the distribution of other elements related using SRXRF; Gao et al. [186] investigated the effect of Se on Hg containing and Hg-responsive protein in rice and confirmed the antagonism of Se against Hg toxicity by using SRXRF. SRXRF can also be used to predict and diagnose plant diseases, e.g. Milori et al. [187] investigated the constituents and contents of mineral elements (K, Ca, Fe, Cu, and Zn) in leaves with and without infection of citrus greening by micro-SR-XRF (spot size: 200 μm), and could distinguish the infected group from the normal group with 90% accuracy by combining chemometric analysis. In addition, the high resolution of SRXRF could also be used to study the elemental accumulation or even the tolerance mechanism of plant on cellular level [188]. However, SRXRF cannot provide the elemental speciation and form information like SRXAS.

X-ray photoelectron spectroscopy (XPS), well known as electron spectroscopy for chemical analysis (ESCA), is a semi-quantitative or quantitative tool for elemental composition in surface within a depth of ~10 nm of a substance under ultra-high vacuum via measuring the binding states of elements. XPS has been proved to have extensive use in the investigation of agrometallomic mechanism for agricultural samples such as soil, fertilizer, cotton textile, and even cell. In terms of heavy metal pollution abatement in soil, XPS was utilized to measure Cr composition to prove that prepared green-tea-impregnated attapulgite promoted the conversion of weak acid extractable Cr to the residual state such as stable Cr(III) and Cr–Si oxide in soil in order to reduce the Cr contamination in planted crops [189]. Guangqun Tan validated that wheat stems can be utilized as bioadsorbent to bind lead(II) from aqueous solution by XPS analysis [190]. Jian Liu found Si–Cd co-complexation can explain the inhibition of Cd ion in rice (Oryza sativa) cells uptake with XPS analysis [191].

LA-ICP-MS is still in its early stages for agricultural application, although some excellent achievements have already been demonstrated. For example, a 193 nm ArF excimer LA-ICP-QMS was utilized to measure and image the elemental distribution of Se and Hg in wild growing mushrooms with 50 μm beam size. The results showed that Hg mainly localized in the hymenium of the cap and the concentration of Hg in mushroom could be eliminated by more than 50%

after removing this part; furthermore, the antagonism of Se and Hg co-localized, which is of great importance for consumers [192]. Esmira Alirzayeva analyzed Artemisia fragrans tissue-specific Cd, Cu, Zn distribution with LA-ICP-MS, and revealed ecotype-specific mechanism in internal distribution of metals within leaf tissues [193]. Michael Moustakas revealed spatial heterogeneity of Cd effects on Salvia sclarea leaves by LA-ICP-MS, which can monitor heavy metal effects and plant tolerance mechanisms [194]. Anetta Hanć determined long-distance root to leaf transport of lead in Pisum sativum plants by LA-ICP-MS [195]. On the other hand, a 266 nm LA system equipped with a larger spot size is more suitable for the rapid solid sampling quantification that does not require high resolution, e.g. heavy metals in soil sample [196]. In addition to two-dimensional (2D) analysis, LA-ICP-MS also has an excellent mapping performance in three-dimensional (3D) scale. Michaela Galiová mapped accumulation and distribution of Pb, K, Mn in leaves of Capsicum annuum L with LA-ICP-MS [197]. Due to high sensitivity of multi-elemental analysis, LA-ICP-MS possesses another application for detecting metal or semimetal-containing protein bands or protein spots [198], e.g. Xiaoxia Yu separated and identified cadmium-binding proteins from different parts (root, stem, leaf, and grain) of rice (*Oryza sativa* L.) with LA-ICP-MS. The results indicated that these plant proteins can bind cadmium to reduce heavy metal toxicity [199]. As a powerful isotopic analysis means, LA-ICP-MS is able to render isotopic labeling information to trace the distribution and movement of labeled composition in agricultural lives or media, e.g. [41]. Zn isotope labeling and distribution in grains was measured by LA-ICP-MS in an ear culture system to reveal Zn transport barriers during grain filling in wheat [200].

Electrothermal vaporization (ETV), always being made from graphite, quartz, tungsten, rhenium, and other high-melting point metals [132–135], is a high-efficiency sampling manner coupled with AAS (ETV-AAS), AFS (ETV-AFS), ICP-OES (ETV-ICP-OES), and ICP-MS (ETV-ICP-MS) to detect metal or non-metal elements in various samples [132, 136–138]. According to the material and structure of electrothermal vaporizer, ETV hyphenation atomic spectrometry allows directly introducing sub-milligram to g levels sample size, so it can also be classified as micro-analysis. Hence, ETV has been extensively used to fabricate a fast and miniature analyzer for heavy metals in agri-food, soil, plant and animal, biological samples, and so on [56, 57, 132, 138, 139, 141]. Direct sampling Hg analyzer consisting of ETV catalytic pyrolysis atomic spectrometer with or without amalgamation has become one of the most successful commercial instruments for solid sampling elemental analysis [138]. Rapid detection of toxic heavy metals such as Cd, Pb, As, Cr, Cu, and Zn in agricultural food and soil is very helpful to control heavy metal contamination in advance [136, 142, 143]. Peizhe Xing rapidly detected Cd in grain samples with ETV-AAS [201]. Luisa Šerá used ETV-ICP-OES for determination of Ba, Ca, Co, Cu, Fe, K, Mg, Mn, P, Sr, Zn, S, and Se in freeze-dried and homogenized samples of broccoli (*Brassica oleracea convar. italica*) [202]. However, due to the limitation of sampling apparatus, ETV hyphenation is difficult to apply to elemental mapping or depth profiling for agrometallome.

Laser-induced breakdown spectroscopy (LIBS) is an atomic emission spectrometry using focused pulsed laser beam to generate plasma on sample surface, which recently has become an attractive and promising analytical technique for metal or non-metal elements and even organic composition characterization and identification [144]. Elif Ercioglu use LIBS to discriminate aromatic plants, e.g. black pepper (*Piper nigrum*), ginger (*Zingiber officinale*) [203]. As an efficient and reagent-free, in situ or on-line solid sampling tool, LIBS can replace time-consuming analytical methods for assessing the quality and composition of agricultural samples involving authenticity control, sample discrimination and traceability, bacteria contamination, elemental surface mapping, and depth profiling [27, 145]. Considering rapid detection using laser excitation, no or minimal sample preparation can be achieved. So, LIBS is always designed to fast analyze crop plant leaves, pasture vegetables, grains, seeds, fruits, animal tissues, and agricultural soils and fertilizers; further, the application range involves plant or animal nutrients and phytotoxic elements, pesticide residues, impurities and other pollutants, as well as the quality control of food derivatives, the identification of early disease diagnosis [144, 146]. Microchemical imaging applications by LIBS are also presented in a broader selection of plant compartments (e.g. leaves, roots, stems, and seeds) [204]. Anielle *C. Ranulfi* evaluated comparatively the content of the three macronutrients Ca, K, and Mg by LIBS to better distinguish soybean leaves collected from healthy and sick plants [205]. On the other hand, like LA-ICP-MS, use of LIBS for elemental mapping in various biological matrices including the spatially resolved analysis of plant and animal samples is of great and increasing interest to many agronomists and biochemists in order to establish compositional interrelationships of the elemental constituents in agricultural samples [145]. L. Krajcarová mapped the spatial distribution of silver nanoparticles in root tissues of Vicia faba by LIBS [206]. Moreover, LIBS could be used to determine contaminants in plants, so Zhao et al. [207] employed nanoparticle-enhanced LIBS technique to enhance the analytical sensitivity of pesticide residue and Cd in fruit and vegetables, and finally achieved Cd mapping analysis on edible plant leaves to reveal the elemental uneven distribution. Notwithstanding, calibration and background correction are still critical issues for both LIBS and XRF, which can be properly improved *via* matrix-matched standards, chemometrics, and computerized algorithm [208]. Meri Barbafieri rapidly on-site detected Pb pollution concentration of leaf mustard (*Brassica juncea*) with LIBS [209].

4.3.2 Agricultural Animal and Derived Food

Agricultural animals generally refer to artificially raised animals, including livestock (pigs, cattle, sheep, etc.), poultry (chicken, ducks, geese, etc.), aquatic animals (fish, shrimp, crab shells, etc.), special economic animals (silkworm, bees, etc.), and agricultural laboratory animals (mice, earthworms, etc.). According to the different effects of metal elements on agricultural animals, they can be divided into beneficial metal elements and harmful metal elements. Among them, beneficial metal elements, such as mineral elements, are an important part of the body tissue of agricultural animals. In addition to maintaining animal life and reproduction, the lactation of female animals also depends on mineral elements. In addition, according to the amount of minerals in animals, minerals are divided into two types. One is the element accounting for more than 0.01% of the body weight of animals, known as the major elements, including calcium, phosphorus, magnesium, sodium, potassium, chloride, and sulfur. The other accounts for less than 0.01% of the body weight of animals, known as trace elements, mainly iron, copper, manganese, zinc, iodine, selenium, cobalt, molybdenum, chromium, and so on. The correct use of major elements and trace elements can improve the animal health level, promote animal growth, improve production performance, improve the quality of livestock and poultry products. On the other hand, harmful metal elements mainly include mercury, arsenic, lead, cadmium, and other heavy metal elements, and can enter agricultural animals through bioenrichment and cause serious threats to their life and health. Effective monitoring of the content, form, and distribution of metal elements in agricultural animals is of great significance to the health of agricultural animals and human beings [210].

Based on agrometallomics, this section illustrates the application of advanced techniques of metal analysis methods in agricultural animals in combination with ultra-sensitive and high-throughput analysis, speciation and state analysis, spatial and micro-analysis.

4.3.2.1 Application of Sensitivity and Multielemental Analysis in Agricultural Animals

The methods commonly used for high sensitivity and multi-element detection of metal elements include mass spectrometry, atomic spectrometry, instrumental neutron activation analysis, and X-ray fluorescence spectrometry. Among them, inductively coupled plasma mass spectrometry (ICP-MS), inductively coupled plasma emission spectrometry (ICP-OES), X-ray fluorescence spectrometry (XRF), and instrumental neutron activation analysis (INAA) have been widely used in agricultural animals.

Application of ICP-MS in Agricultural Animals In general, icp-ms was applied to detect heavy metals in agricultural animal researches. In terms of the whole process of breeding, heavy metals in livestock and poultry mainly come from feed, while heavy metals in aquatic agricultural animals mainly come from polluted water and are easily absorbed and enriched by fish, shrimps, and other organisms, causing serious damage to the whole food chain. Nargis Jamila et al. [211] determined the concentrations of toxic metals including As, Cd, Hg, and Pb in commonly consumed crustaceans collected from South Korea. For As, Cd, and Pb, inductively coupled plasma–mass spectrometry (ICP-MS) was used. Hg content was determined using gold amalgamation direct mercury analysis (DMA). The distribution of toxic metals was determined in whole muscles and digestive tracts of crustaceans. Mielcarek et al. [212] assessed arsenic (As), cadmium (Cd), lead (Pb), and mercury (Hg) contamination of freshwater fish from Poland. Selected species of raw, smoked, and pickled freshwater fish ($n = 212$) were evaluated by atomic absorption spectrometry (AAS) and inductively coupled plasma – mass spectrometry (ICPMS).

Assessment of health risk associated with intake of investigated elements present in fish was performed. Contamination of fish with As, Cd, Hg, and Pb ranged as follows: As 23.3–59,290.1 µg/kg, Cd 0.02–97.0 µg/kg, Hg 9.04–606.3 µg/kg, and Pb 0.04–171.4 µg/kg. Consumption of selected species of freshwater fish, especially smoked fish products, may pose a non-carcinogenic and also carcinogenic health risk.

The use of free-range animals for monitoring environmental health offers opportunities to detect exposure and assess the toxicological effects of pollutants in terrestrial ecosystems. Ogbomida et al. [213] evaluated potential human health risk of dietary intake of metals and metalloid via consumption of offal and muscle of free range chicken, cattle, and goats by the urban population in Benin City. Muscle, gizzard, liver, and kidney samples were analyzed for Cr, Mn, Fe, Co, Ni, Cu, Zn, As, Cd, and Pb concentrations using inductively coupled plasma mass spectrometer (ICPMS), while Hg was determined using Hg analyzer. The hazard index of the results suggests that chicken liver and gizzard may have a significant impact on dietary heavy metal exposure in adults and children. Chijioke et al. [214] determined the persistence of metal elements in poultry giblets in the environment and food chain by ICP-MS, in order to evaluate its potential non-carcinogenic and carcinogenic risks to human health. Considering the non-degradability of toxic metals and their potential accumulation in animal tissues, reduction in metal supplementation in animal feed should be introduced and periodic monitoring of chicken giblets may help to mitigate non-essential metal toxicity to public health. What's more, animal manure is one of the diffusion routes of heavy metals and metalloids into the environment, where the soil can accumulate them. Heavy metals and metalloids can then be released into groundwater sources, be absorbed by crops, and enter the food chain with negative effects for human and animal health. Hejna et al. [215] evaluated the concentration of heavy metals and mineral nutrients in modern animal feeding systems. Feed and feces samples were taken from pigs and cattle. The following elements were detected by ICP-MS after pretreatment: Na, Mg, K, Ca, Cr, Mn, Fe, Co, Ni, Cu, Zn, As, Se, Mo, Cd, and Pb. The results indicate that manure is the source of Zn and Cu exported to the environment. Zn and Cu contents should be strictly monitored according to agro-ecological principles.

In addition, origin identification of agricultural products is of great significance for food traceability and prevention of adulteration. Bandoniene et al. [216] developed a method for labeling lamb meat and goat milk by selective enrichment of terbium and thulium in the feed for the animals. Detection of rare-earth element labels was carried out using ICP-MS after acid digestion. Alternatively, laser ablation ICP-MS (LA-ICP-MS) was applied, allowing direct analysis of bone samples and analysis of meat and milk samples after dry ashing and pressing pellets. The method can effectively distinguish labeled and unlabeled animal products, while the rare-earth element content in all labeled products remains low enough to avoid any health risks to consumers.

Application of ICP-OES in Agricultural Animals Monitoring of toxic heavy metals in fish samples is a matter of a great importance from the nutritional and toxicological points of view. Bozorgzadeh et al. [217] designed a dispersive micro-solid-phase extraction (dμSPE) for preconcentration of trace Pb, Cd, Hg, Co, Ni ions using pectin-coated magnetic graphene oxide (pectin/Fe3O4/GO). Inductively coupled plasma optical emission spectroscopy (ICP/OES) was utilized for analyzing the samples. Detection and quantification limits were between 0.01–0.21 μg/g and 0.04–0.67 μg/g for fresh fish sample, respectively. Concentration of the toxic heavy metals was successfully determined in 11 different fish samples using the proposed method. Eduardo et al. [218] sampled 20 individuals of demersal (*Caulolatilus princeps* and *Mycteroperca olfax*) and pelagic (*Thunnus albacares* and *Seriolella violacea*) species. The levels of the toxic elements (Al, B, Ba, Cd, Ni, Pb, and Sr), and the macroelements, microelements, and trace elements (Ca, Cr, Fe, K, Li, Mn, Mo, Mg, Na, V, and Zn) of species muscle tissue were analyzed by ICP-OES. The results showed that Cr K, Mg, and Mo concentrations were higher in deep-sea species, while Zn content was higher in pelagic species. Alberto et al. [219] similarly measured the concentrations of 22 metals and nonmetals in the intestinal and edible parts of brown trout after chemical digestion by ICP-OES, and the results showed that the accumulation of metal and nonmetal elements in the edible parts and guts at different levels, following the series Cu > Zn > Ba > Al > Sr > Fe > Pb and Fe > Al > Hg > As > Mn > Cu > Ba > B > Zn > Pb, respectively. In the study of Enrique et al. [220], 963 specimens of pelagic fish have been collected, of which 345 are *Scomber colias*, 294 are *Trachurus picturatus*, and 324 are *Sardina pilchardus*, the development and ontogeny in the three species were studied by observing if there were variations in their metallic content. The concentration of 11 anthropic metals was determined in each sample using ICP-OES technique. Milenkovic et al. [221] investigated the radioactivity levels and heavy metal concentrations in fish and seafood commercially available in Serbian markets. Concentrations of heavy metals (Cd, Hg, and Pb) were determined using ICP-OES method. The health risks associated with human ingestion of heavy metals and radionuclides through fish consumption were assessed.

Application of XRF and INAA in Agricultural Animals Rajib et al. [222] presented the detection and propagation of heavy metals, particularly chromium (Cr), in poultry in the Rajshahi area using the X-ray fluorescence (XRF) technique. The investigation was done to assess the possible transfer of heavy metals from poultry feeds to chicken meat. Among the four most widely used feeds in the Rajshahi region, "Adorsho Feed (Pabna)" shows a maximum Cr concentration of 17.3 ppm. Six different parts of the chicken of different ages which were grown by "Adorsho Feed" were considered for further study. About 4.3 ppm of Cr was found in yolk, whereas a slightly lesser amount of about 2.7 ppm was found in the albumin. The average value of Cr in six parts of chicken remains almost the same, which is 3.74 ppm, irrespective of age. It is found that about 21.6% of Cr was propagated from the poultry feed to chicken flesh and also about 90% of Cr was propagated from the chicken to egg. The experimental results indicate that the investigation of the transmission of heavy metals like Cr from feed to egg and poultry is efficiently evaluated by using X-ray Fluorescence Technique. Baldassini et al. [223] described a metalloproteomics study of bovine muscle tissue with different grades of meat tenderness from animals of the Nellore breed (Bos indicus) based on protein separation by two-dimensional gel electrophoresis, the identification of calcium ions in protein spots by X-ray fluorescence (SR-XRF), and the characterization of proteins by electrospray ionization mass spectrometry. The procedures were efficient and preserved the metal–protein structure, enabling calcium detection in protein spots by SR-XRF at a given molecular weight range of 14–97 kDa.

In the study of Veado et al. [224], instrumental neutron activation analysis (INAA) and inductively coupled plasma mass spectrometry high resolution (ICP-MSHS) were applied to determine Al, As, B, Ba, Co, Cr, Cs, Cu, Fe, K, Mg, Mn, Mo, Na, P, Pb, Rb, Zn, and Ti in Nile tilapia fish, *Oreochromis niloticus*. The organs analyzed were intestine, spleen, heart, testicle, kidney, liver, gills, and muscle. The results demonstrated relatively high concentrations of Al, Co, Cu, Fe, P, and Ti in gills, Al and Cu in liver, Al in intestine, and Fe in muscle and spleen. CHAFF et al. [225] studied zinc- and cadmium-binding proteins in bovine kidney by chromatography and electrophoresis combined with INAA, and found that bovine kidney contained about 78 ppm zinc and 0.78 ppm Cd, and about 45% zinc and 60% cadmium existed in cytoplasm. More than 95% of these two metals were bound with macromolecules. Shinichi et al. [226] used instrumental neutron activation analysis and concentrated acid digestion method to measure the concentrations of trace metals zinc (Zn), chromium (Cr), manganese (Mn), and iron (Fe) in arable soil applied with animal feces. The 5-year application of animal manure did not result in a considerable increase in Cr, Mn, and Fe concentrations in soils, whereas longer-term application may lead to a gradual accumulation of Cr and Fe depending on the soil and manure types. The types of soil and manure affect trace metal accumulation in arable soils, and elements in the manure relate to the ability of soils to retain trace metals.

4.3.2.2 Application of Elemental Speciation and State Analysis in Agricultural Animals

It is difficult to separate and determine different morphology of elements, which requires the combination of efficient separation technology and sensitive detection method. The combination of chromatography and spectrum/mass spectrometry has the advantages of high sensitivity of chromatographic technology and accurate qualitative of mass spectrometry, which is the mainstream method for the separation and detection of elemental morphology. At present, the combined techniques used for elemental morphology analysis mainly include gas chromatography-inductively coupled plasma mass spectrometry (GC-ICP-MS), gas chromatography-atomic fluorescence spectrometry (GC-AFS), liquid chromatography-atomic fluorescence spectrometry (LC-AFS), and high-performance liquid chromatography-inductively coupled plasma mass spectrometry (HPLC-ICP-MS). Examples of application of combined techniques for elemental morphology analysis in agricultural animals are given below.

Identification and quantification of the selenium species in biological tissues is imperative, considering the need to properly understand its metabolism and its importance in various fields of sciences, especially nutrition science.

Selenium is an essential element for mammals with diet being the major source of intake. In the study of Oliveira et al. [227], cattle feed was enriched with combinations of selenium enriched yeast, canola oil, and/or vitamin E in order to evaluate the accumulation and metabolism of selenium in beef cattle. A method to identify and/or quantify the selenium species: selenocystine (SeCys2), selenomethionine (SeMet), selenomethionine-Seoxide (SeOMet), and inorganic selenium species, selenate (Se(VI)), and selenite (Se(IV)), was developed and applied to cattle feed and beef samples. Gawor et al. [228] present a study aimed at examining speciation analysis of Se in tissues of livers, muscles, and hearts obtained from lambs. The studied lambs were fed with the diet enriched with an inorganic (such as sodium selenate) and organic chemical form of Se (such as Se-enriched yeast) compounds with simultaneous addition of fish oil (FO) and carnosic acid (CA). First, selenium compounds were extracted from animal tissues, then hyphenated high-performance liquid chromatography and inductively coupled plasma mass spectrometry (HPLC–ICP–MS) was used for the identification of five seleno-compounds—Se-methionine (SeMet), Se-cystine (SeCys2), Se-methyl-Se-cysteine (SeMetSeCys), Se(IV), and Se(VI). Verification of the identified seleno-compounds was achieved using triple-quadrupole mass spectrometer coupled to high-performance liquid chromatography (HPLC–ESI–MS/MS). The developed analytical protocol is feasible for speciation analysis of small molecular seleno-compounds in animals samples. In the study of Jorge et al. [229], in vitro bioavailability of total selenium and selenium species from different raw seafood has been assessed by using a simulated gastric and intestinal digestion/dialysis method. ICP-MS was used to assess total selenium contents after a microwave-assisted acid digestion, and also to quantify total selenium in the dialyzable and non-dialyzable fractions. Selenium speciation in the dialyzates was assessed by high-performance liquid chromatography (HPLC) coupled with ICP-MS detection. Major Se species (selenium methionine and oxidized selenium methionine) from dialyzate were identified and characterized by HPLC coupled to mass spectrometry (HPLC–MS). Low bioavailability percentages for total selenium (6.69 ± 3.39 and 5.45 ± 2.44% for fish and mollusk samples, respectively) were obtained. Similar bioavailability percentage was achieved for total selenium as a sum of selenium species (selenocystine plus oxidized selenium methionine and selenium methionine mainly).

Speciation analysis of mercury elements is of more critical importance in nutritional and toxicological assessments. Zhang et al. [230] proposed a sample preparation method based on dispersed solid-phase extraction combined with LC-AFS to detect different mercury species in aquatic animal samples. The spiked recoveries of Hg^{2+}, MetHg, and EtHg in aquatic animal samples were in the 88% to 104% range, and the method could achieve desirable limits of detection below 3 mg/kg (sample mass = 2 g). The method was suitable for the determination of Hg^{2+}, MetHg, and EtHg in aquatic animal samples, and had the advantages of being simple, rapid, accurate, and having high sensitivity. In the study of Deng et al. [231], a simple solid sampling platform using multi-wall carbon nanotubes (MWCNTs) assisted matrix solid-phase dispersion (MSPD) was constructed for online coupling to high-performance liquid chromatography inductively coupled plasma mass spectrometry (HPLC-ICP-MS) for the high accuracy and sample throughput mercury speciation in fish samples. The limits of detection of 9.9 and 8.4 ng/g were obtained for Hg^{2+} and CH_3Hg^+, respectively, based on 1 mg of fish sample. The proposed method was applied for two fresh fish samples for Hg speciation.

Human exposure to high concentrations of arsenic from water and food is a main health concern. Determination of potential arsenic metabolites present at trace concentrations is an analytical challenge, requiring efficient separation and sensitive detection. Mao et al. [232] established a new method for the determination of benzene-arsenic compounds and their possible transformation products in chicken tissues by stirring rod adsorption extraction (SBSE) combined with HPLC and ICP-MS. The proposed method was successfully applied to the speciation of arsenic in chicken meat/liver samples, and the recoveries for the spiked samples were in the range of 78.5–120.4%. Peng et al. [233] developed a method that enabled the identification and quantification of various arsenic species in chicken liver. They described a method of HPLC separation with both ICP-MS and electrospray ionization tandem mass spectrometry (ESI-MS/MS) detection. Detection with both ICPMS and ESI-MS/MS allowed for identification and quantification of eight arsenic species in chicken liver, including arsenobetaine, inorganic arsenite, dimethylarsinic acid, monomethylarsonic acid, inorganic arsenate, 3-amino-4-hydroxyphenylarsonic acid, N-acetyl-4-hydroxyphenylarsonic acid (N-AHPAA), and roxarsone.

4.3.2.3 Application of Spatial Distribution and Micro-analysis in Agricultural Animals

Application of LA-ICP-MS in Agricultural Animals Bioimaging using laser ablation inductively coupled plasma mass spectrometry (LA-ICP-MS) offers the capability to quantify trace elements and isotopes within tissue sections with a spatial resolution ranging about 10–100 μm. Distribution analysis adds to clarifying basic questions of biomedical research and enables bioaccumulation and bioavailability studies for ecological and toxicological risk assessment in humans, animals, and plants. Major application fields of mass spectrometry imaging (MSI) and metallomics have been in brain and cancer research, animal model validation, drug development, and plant science [234].

Copper is essential for eukaryotic life, and animals must acquire this nutrient through the diet and distribute it to cells and organelles for proper function of biological targets. Ackerman et al. [235] combined metal imaging and optical imaging techniques at a variety of spatial resolutions to identify tissues and structures with altered copper levels in the Calamity^{gw71} zebrafish model of Menkes disease. Rapid profiling of tissue slices with LA-ICP-MS identified reduced copper levels in the brain, neuroretina, and liver of Menkes fish compared to control specimens. In the study of Matusch et al. [236], an imaging LA-ICP-MS technique for Fe, Cu, Zn, and Mn was developed to produce large series of quantitative element maps in native brain sections of mice subchronically intoxicated with 1-methyl-4-phenyl-1,2,3,6-tetrahydropyridin (MPTP) as a model of Parkinson's disease. Images were calibrated using matrix-matched laboratory standards. The

LA-ICP-MS technique yielded valid and statistically robust results in the present study on 39 slices from 19 animals. The findings underline the value of routine micro-local analytical techniques in the life sciences and affirm a role of Cu availability in Parkinson's disease.

Tzadik et al. [237] evaluated Atlantic Goliath Groupers, *Epinephelus itajara*, in their nursery habitats via microchemical analyses of fin rays. Trace metal constituents in the fin rays were quantified with LA-ICP-MS. The results highlight the importance of spatial scale for interpreting microchemical analyses on calcified structures in fishes. LA-ICP-MS has also been developed in the generation of quantitative images of metal distributions in thin tissue sections of brain samples (such as human, rat, and mouse brains), with applications in research related to neurodegenerative disorders. Becker et al. [114] described sample preparation by cryo-cutting of thin tissue sections and matrix-matched laboratory standards, mass spectrometric measurements, data acquisition, and quantitative analysis. Specific examples of the bioimaging of metal distributions in normal rodent brains are provided. What's more, isotope dilution mass spectrometry (IDMS) is internationally regarded as an absolute quantification method [127]. Compared with other calibration strategies, IDMS can correct for some matrix effect, fractionation, and element losses [128]. Feng et al. [129] reported a novel quantitative imaging strategy for biological thin section based on ID-LA-ICP-MS. Quantitative imaging of Fe, Cu, and Zn in real mouse brain of Alzheimer's Disease (AD) was measured by the improved methodology. The similar distributional patterns demonstrated that the proposed methodology has potential to investigate the correlation of biomarker heterogeneity and elements distribution, and may be useful to understand such complex brain mechanisms in the future.

In addition, the application of LA-ICP-MS in biological imaging could further extend its capability in drug development, and Lum et al. [238] used this bioimaging tool to visualize deposition behavior of chemically different metallo-complexes, including two that differed only subtly in structure. ICP-MS analysis was performed on six organs of the treated mice to study the bioavailability of the complex. In LA-ICP-MS bioimaging analysis of paraffin-embedded mouse liver and kidney sections, a spatial resolution of 50 m was adopted. Deposition trends matched the findings obtained in elemental analysis. This work demonstrates that LA-ICP-MS imaging is a valuable tool for therapeutic drug development, especially in assisting molecular modification of metal-containing complexes.

Application of ETV and LIBS in Agricultural Animals Ultrasonic slurry sampling electrothermal vaporization inductively coupled plasma mass spectrometry (USS-ETV-ICP-MS) has been applied to the determination of mercury in several fish samples. Chen et al. [239] determined mercury in fish samples by using USS-ETV-ICP-MS method and palladium as modifier to delay mercury evaporation. This method has been applied to the determination of mercury in dogfish muscle reference material (DORM-1 and DORM-2) and dogfish liver reference material (DOLT-1). Accuracy was better than 4% and precision was better than 7% with the USS-ETV-ICP-MS method. Tormen et al. [240] proposed a fast method for the determination of As, Co, Cu, Fe, Mn, Ni, Se, and V in biological samples by ETV-ICP-MS, after a simple sample treatment with formic acid. The detection limits in the samples were between 0.01 (Co) and 850 g/kg (Fe and Se), and the precision expressed by the relative standard deviations (RSD) were between 0.1% (Mn) and 10% (Ni). Accuracy was validated by the analysis of four certified reference biological materials of animal tissues (lobster hepatopancreas, dogfish muscle, oyster tissue, and bovine liver).

Quantitative analysis of heavy metal uptake in different organs and tissues of organisms is of great value in environmental biology. Laser-induced breakdown spectroscopy (LIBS) is a practical tool for quantitative analysis. Kumar et al. [241] first distinguished trace elements between normal and malignant tumor cells from animal histological sections by LIBS, and the results showed good agreement with ICP-OES analysis. Wan et al. [242] proposed an optimized LIBS approach to analyze the contents of heavy metal elements in various tissues of a contaminated fish sample. Experimental results showed that heavy metals deposit selectively in certain fish tissues such as liver and gill rather than evenly distribute all over the fish body. The optimized quantitative LIBS approach can be promoted for the environmental biology.

In a word, agrometallomics-related analytical technique has been widely used in agricultural animals. On the one hand, it studies the distribution, transport, and metabolism of beneficial metal elements in agricultural animals from the perspective of nutrition science. On the other hand, it monitors the content and accumulation of toxic metal elements in agricultural animals from the perspective of toxicology science. Agrometallomics plays an important role in the sustainable development of agricultural animals.

4.3.3 Soil, Water, and Fertilizer for Agriculture

The environmental and ecological composition of soil, water, fertilizer, and so on are also an important part of agrometallomics [2]. The metals and metalloids in these components directly affect agricultural products and their by-products, and have a direct impact on the reproduction, planting, growth, and fruiting of agricultural organisms and the enrichment of toxic metals. Related metal analysis methods for soil, water, and fertilizer mainly include ultrasensitive and high-throughput analysis, speciation, and state analysis, spatial- and microanalysis for elemental quantification, distribution and mapping, single cell and microparticle, speciation and structure, or metallic labeling.

In terms of sensitivity and multi-element analysis methods, it mainly includes mass spectrometry, atomic spectroscopy, instrumental neutron activation analysis, and X-ray fluorescence spectrometry. Inductively coupled plasma mass spectrometry (ICP-MS) has become a powerful multi-element analysis and equipment combination analysis method due to its high sensitivity, less spectral interference, wide linear range, and isotope detection. For nonhyphenated ICP-MS, it is often introduced into the system by nebulizer after sample digestion or extraction. Studies have used two atomizers to compare the ability to determine rare-earth metals in surface water and subsurface waters [243]. The ICP-QQQ, designed with high sensitivity and immunity using tandem quadrupole mass spectrometry, has also been put into the analysis of trace metal Hg in soil, demonstrating the potential for bioremediation in mercury-contaminated waste [244]. SF-MS, on the other hand, has better sensitivity and higher resolution than Q-MS, and is extremely suitable for rare-earth and other elemental analyses with severe spectral interference [243, 245]. In addition, MS-ICP-MS's ability to perform high-precision analysis of multi-isotope ratios with mass resolutions of more than 10 000 at the ultra-trace level makes it a powerful analytical tool for determining isotope ratios in various agricultural applications. MC-ICP-MS analysis of the isotope composition of lead has been studied to achieve tracer of lead and cadmium in the soil to predict potential sources of contamination [246, 247]. As a microanalytic technique using laser ionization sources, LIMS has the characteristics of large energy dispersion, which can lead to interference between multi-charged ions and multi-atom ions. To solve this problem, Hang and co-workers recently established a buffer-gas-assisted LIMS instrumentation enabling direct solid sampling and semiquantitative analysis of more than 60 elements ranging from 0.3 mg/kg (Be) to 18.6% (Al) levels in soil samples [248]. TIMS can measure precise isotope ratios and has better sensitivity and precision for small sample sizes than ICP-MS. TIMS is currently reported as an isotope ratio analysis method for U, Fe, and Pb in soil or water samples to trace the source of heavy metal contaminants and protect food safety [249, 250]. GD-MS has received increasing attention in various research areas over the past few decades. Because it is often used for the analysis of conductive materials, most agricultural media cannot be used as conductive materials. Therefore, the relevant analytical applications of GD-MS in agrometallomics are still few. For example, after sample preparation using a conducting host matrix, dc-GD-MS is in use for the isotopes in soil and vegetation (nonconducting) samples [251]. SIMS enables sensitive surface and depth analysis, providing detailed elemental and isotope information in parts per million. Nano-SIMS and TOF-SIMS are more suitable for microanalysis and compositional analysis. For microanalysis, nano-SIMS was reported to analyze 15N-enrichment to unveil active diazotrophs in a flooded rice soil where 15N was incorporated into soil DNA in the stable isotope probing (SIP) gradient fractions with 15N2-labeling [252]; to investigate in situ Zn/Cd distribution in the apical root tissues of radish exposed to metal contamination [253]; and to couple with immunolocalization to reveal that iron-loaded ferritin was located at the starch-containing plastids' surface [254]. Inductively Coupled Plasma Optical Emission Spectroscopy (ICP-OES) is a versatile analysis tool for polymetallic and certain non-metallic elements. It can also be used in conjunction with LA, ETV, and LC for solids analysis, trace analysis, or elemental speciation analysis. The wide linear dynamic range makes it suitable for micro- or macro-elements as well as non-metallic elements. MP-OES shows difficulty in emitting more atomic spectra than ICP-OES due to its lower excitation source temperature. However, it has superior performance than FAAS and has a good performance in monitoring environmental problems such as soil and water [30]. Besides, Duan et al. fabricated a novel MP-OES in which microwave plasma was employed for surface sampling elemental analysis. This MP-OES was able to quantitatively or semi-quantitatively analyze metallic elements such as Ag, Au, Ba, Cd, Cr, Cu, Eu, La, Mn, Ni, Pb, Sr, and Y in environmental samples [255]. LTP has the advantages of normal temperature operation, miniaturization, simplicity, and low-energy consumption compared to ICP and MP. Common microplasms are DBD [41–43], GD, corona discharge, PD [38], solution anode/cathode glow discharge [35], and microwave or radio frequency discharge [44], and some studies have applied them to the monitoring of As, Sn, and other elements in environmental samples such as surface water and soil. INAA's non-destructive testing of solid, liquid, and gas samples, no preparation, and other advantages make it have certain advantages in rapid screening, absorption of different elements, and bioaccumulation. IT ISAS has been reported to have been used for the analysis of samples such as soil and atmospheric particulate matter [62]. XRF is a non-destructive method for solid sampling analysis of multiple elements without digestion, and is widely used in multi-element analysis in various agricultural media. For example, XRF is designed to measure essential and toxic heavy metals such as Cd, Pb, As, Mn, Fe, Ni, Cu, and Zn in soil samples to investigate elemental information about existence, absorption, changes, and accumulation for the purpose of environmental quality and food safety [256].

Elemental speciation studies are of great significance in agri-food nutrition, animal and plant growth, and agricultural water and soil studies. At present, the commonly used analytical methods mainly include chromatography coupled with atomic spectroscopy, synchrotron radiation, and X-ray energy spectroscopy. Subclasses of LCs such as IC and HPLC have been widely used in water and soil samples, such as the use of IC-HG-AFS to complete the separation of arsenic species [67], the use of HPLC-CV-AFS for inactive mercury, and active mercury high sensitivity analysis [69]. GC is an effective separation method in the analysis of volatile inorganic and organic components, and after GC separation, it is then introduced into the detector for analysis. There are reports of Hg speciation including inorganic or organic Hg in water, seafood, food packaging, and the agricultural environment can be measured by GC-MS/MS [78]. In the past ten years, capillary electrophoresis has been widely used in speciation analysis, and the speciation of Se, Hg, As and other elements in water, soil, and other samples has been studied [83]. Besides, Dominguez-Alvarez attempted to establish a CE-ESI(−)-MS method as a sensitive and suitable alternative for inorganic and organic As species in water samples [257]. SR technology also requires little to no sample pre-treatment and is suitable for in situ analysis of agricultural samples. FRXRF has been reported for in situ analysis of metals in soils and sediments, thin films/particulates, and lead in paint [94]. Although chromatographic separation is a powerful means of elemental speciation, a series of complex sample preparations are required. In contrast, non-destructive analyses such as XAS are more suitable for in situ analysis capabilities of agriculture-related element speciation. Additionally, the SR technique can also be used to represent the elemental state and compound structures in special material used for heavy metal pollution abatement in the agricultural environment. Xu et al. [258] designed and synthesized a novel calcium-based-magnetic biochar (Ca-MBC) to adsorb As and Cd and employed micro-SRXRF and XANES to confirm the formation of bidentate chelate and ternary surface complex on the surface of magnetic iron oxide during the absorption process of As and Cd [259]. X-ray photoelectron spectroscopy (XPS) is well known as electron spectroscopy for chemical analysis (ESCA). In terms of heavy metal pollution abatement in soil, XPS was utilized to measure Cr composition to prove that prepared green-tea-impregnated attapulgite promoted the conversion of weak acid extractable Cr to the residual state such as stable Cr(III) and Cr–Si oxide in soil in order to reduce the Cr contamination in planted crops [189]. XPS also testified that the thiol group was easily oxidized and there were many S forms in thiol-modified humic acids; as a result, effects of humus on the mobility of As in tailing soil were significantly improved [260]. For the fertilizer-related study, XPS analysis was employed to verify the mechanism underlying P, F, and heavy metals in phosphor gypsum (PG) leachates adsorbed onto biochar as well as prove that heavy metal precipitation and complexation was caused by surface ions and different functional groups, and the combination application to loam or sand soil significantly increased crop yield [261]. As an analytical technique that can measure the structural characteristics of components, XRD is mostly oriented to inorganic materials and conforming material samples, and there are also many soils and fertilizers for testing. The formation of struvite in a calcareous soil was evaluated in three incubation periods by XRD [106].

Spatial distribution and microscopic analysis techniques in agrometallomics mainly include laser ablation inductively coupled plasma mass spectrometry (LA-ICP-MS), electrothermal vaporization (ETV), laser-induced breakdown spectroscopy (LIBS), and single-cell and microparticle analysis. As a direct solids injection technology, LA-ICP-MS does not require digestion, reducing the risk of contamination and elemental loss caused by sample pretreatment. Studies have shown that 266nm LA systems with larger spot sizes are more suitable for rapid solids sampling quantification that do not require high resolution, such as heavy metals in soil samples [196]. ETV is mostly made of graphite, quartz, and other high-melting-point metals and is used as a heavy metal analyzer in soil, water, and fertilizer [132]. Rapid detection of toxic heavy metals such as Cd, Pb, As, Cr, Cu, and Zn in agricultural water and soil is very helpful to control heavy metal contamination in advance [34, 41, 43, 134, 135, 137]. In order to overcome matrix interference, there are studies using quartz traps and a new HG-DBD-AFS system to detect ultra-trace arsenic in surface water. [14a, 41] LIBS is an efficient in situ or in-line solids sampling tool that can be used to assess the quality and composition of agricultural samples. It is precisely because of the Shinichi properties that LIBS has always been committed to the rapid analysis of agricultural samples such as soil, water, and fertilizer. For example, Jantzi and Almirall employed a LIBS with 266 nm laser for soil forensic analysis, in which Ba, Ca, Cr, Cu, Fe, Li, Mg, Mn, Pb, Sr, Ti, and V were determined with a precision lower than 10% and LODs lower than 33 mg/kg [146]. Niu et al. attempted to use aluminum oxide nanoparticles to enrich Cd, Ni, and Cr in water samples to improve the LIBS detection limit, whereas it is not suitable for direct solid sample analysis [262]. Therefore, weaknesses such as insufficient analytical sensitivity and LOD still impose restrictions on trace elemental analysis application, such as Cd, Hg in agri-food or soil samples. At present, researchers have found that cells or particles have individual differences, which is also important in agriculture. SP-ICP-MS has been widely used in the particle size and quantification of particles in various substrates, such as environmental samples. For example, SP-ICP-MS was employed to characterize and quantify engineered nanoparticles in water [168] and at a water treatment facility [244] to alleviate nanoparticle pollution.

List of Abbreviations

AAS	atomic absorption spectroscopy
APGD	flowing afterglow atmospheric pressure glow discharge
CE	capillary electrophoresis
Cryo-TOF-MS	electrospray ionization-TOF-MS
CV	cold vapor generation
DBD	dielectric barrier discharge
DBDI	dielectric barrier discharge ion
d-CPE	dual-cloud point extraction
DMA	direct mercury analyzer
EXAFS	extended X-ray absorption fine structure
FPXRF	field portable X-ray fluorescence
GC/QqQ-MS	gas chromatography–triple quadrupole mass spectrometry
GD	glow discharge
HG	hydride generation
HILIC	hydrophilic interaction liquid chromatography
HS-SPME	headspace solid-phase microextraction
IC	ion chromatography
ICP-oa-TOF-MS	inductively coupled plasma orthogonal acceleration time-of-flight mass spectrometry
IEC	ion-exchange column
INAA	instrumental neutron activation analysis
LA-ICP-MS	laser ablation – inductively coupled plasma-mass spectrometry
LIBS	laser-induced breakdown spectroscopy
LIMS	laser ionization mass spectrometry
MALDI-FT-ICR-MS	matrix-assisted laser desorption ionization Fourier transform ion cyclotron resonance mass spectrometry
MC	multi-collector
MIPDI	microwave-induced plasma desorption/ionization source
MP	microwave plasma
MWP-PTR-MS	microwave plasma-proton transfer reaction-MS
Nano SIMS	nanometer-scale secondary ion mass spectrometry
oa	orthogonal acceleration
P-CT	purge-cryogenic trapping
PD	point discharge

PEVG	plasma electrochemical vapor generation
PGD	pulsed glow discharge
QQQ	triple quadrupole
QT	quartz tube
QTAT	quartz tube atom trap
SC	single cell
SCGD	solution cathode glow discharge
SF	sector field
SP	single particle
SPE	solid-phase extraction
SPME	solid-phase microextraction
SRXRF	synchrotron radiation X-ray fluorescence
TIMS	thermal ionization mass spectrometry
TOF	time-of-flight
TQ	triple-quadrupole
UVG	ultraviolet vapor generation
VSG	volatile species generation
XANES	X-ray absorption near edge structure
XPS	X-ray photoelectron spectroscopy
XRD	X-ray fluorescence spectrometry
XRF	X-ray fluorescence

References

1. Guernsey, J.R. (1999). *J. Agric. Saf. Health* 5: 191.
2. Jones, J.W., Antle, J.M., Basso, B. et al. (2017). *Agric. Syst.* 155: 269–288.
3. Tang, L., Hamid, Y., Liu, D. et al. (2020). *Environ. Int.* 145: 106122.
4. Parks, J.M., Johs, A., Podar, M. et al. (2013). *Science* 339: 1332–1335.
5. Williams, R.J.P. (2001). *Coord. Chem. Rev.* 216: 583–595.
6. Haraguchi, H. (2004). *J. Anal. At. Spectrom.* 19: 5–14.
7. Mounicou, S., Szpunar, J., and Lobinski, R. (2009). *Chem. Soc. Rev.* 38: 1119–1138.
8. Ge, R. and Sun, H. (2009). *Sci. China, Ser. B Chem.* 52: 2055–2070.
9. Chen, B., Hu, L., He, B. et al. (2020). *Trends Anal. Chem.* 126: 115875.
10. Lores-Padin, A., Fernandez, B., Alvarez, L. et al. (2021). *Talanta* 221: 121489.
11. Wu, B. and Becker, J.S. (2012). *Metallomics* 4: 403–416.
12. Arnold, T., Schoenbaechler, M., Rehkaemper, M. et al. (2010). *Anal. Bioanal.Chem.* 398: 3115–3125.
13. Nowell, G.M., Pearson, D.G., Parman, S.W. et al. (2008). *Chem. Geol.* 248: 394–426.
14. Tao, G.H., Yamada, R., Fujikawa, Y. et al. (2001). *Talanta* 55: 765–772.
15. Li, C.-F., Feng, L.-J., Wang, X.-C. et al. (2016). *J. Anal. At. Spectrom.* 31: 2375–2383.
16. Moynier, F., Yin, Q.-Z., and Schauble, E. (2011). *Science* 331: 1417–1420.
17. D. M. Lubman, Lasers and Mass Spectrometry, Oxford University Press, 1990.

18 Odom, R.W. and Radicati di Brozolo, F. (1994). Laser Ionization Mass Spectrometry. In: Yacobi, B.G., Holt, D.B., Kazmerski, L.L. (eds) Microanalysis of Solids. Springer, Boston, MA. https://doi.org/10.1007/978-1-4899-1492-7_10.
19 Wei, H.-Z., Jiang, S.-Y., Hemming, N.G. et al. (2014). *Talanta* 123: 151–160.
20 Li, C.-F., Chu, Z.-Y., Wang, X.-C. et al. (2020). *At. Spectrosc.* 41: 249–255.
21 Robertson-Honecker, J.N., Zhang, N., Pavkovich, A., and King, F.L. (2008). *J. Anal. At. Spectrom.* 23: 1508–1517.
22 Gu, G., DeJesus, M., and King, F.L. (2011). *J. Anal. At. Spectrom.* 26: 816–821.
23 Jecklin, M.C., Gamez, G., Touboul, D., and Zenobi, R. (2008). *Rapid Commun. Mass Spectrom.* 22: 2791–2798.
24 Aggarwal, S.K. and You, C.-F. (2017). *Mass Spectrom. Rev.* 36: 499–519.
25 Gonzalvez, A., Armenta, S., Cervera, M.L., and de la Guardia, M. (2008). *Talanta* 74: 1085–1095.
26 Choi, J.Y., Habte, G., Khan, N. et al. (2014). *Food Addit. Contam. Part B.* 7: 295–301.
27 Sharma, N., Singh, V.K., Lee, Y. et al. (2020). *At. Spectrosc.* 41: 234–241.
28 Mavrodineanu, R. and Hughes, R.C. (1963). *Spectrochim. Acta* 19: 1309–1317.
29 Balaram, V. (2020). *Microchem. J.* 159: 105483.
30 Kamala, C.T., Balaram, V., Dharmendra, V. et al. (2014). *Environ. Monit. Assess.* 186: 7097–7113.
31 Akogwu, R.D., Aguoru, C.U., Ikpa, F., and Ogbonna, I. (2018). *Int. J. Environ. Agri. Biotechnol.* 3: 1546–1550.
32 Gallego Rios, S.E., Penuela, G.A., and Ramirez Botero, C.M. (2017). *Food Anal. Methods* 10: 3407–3414.
33 Carter, J.A., Long, C.S., Smith, B.R. et al. (2019). *Expert Syst. Appl.* 115: 245–255.
34 OZBEK, N. (2018). *J. Turk. Chem. Soc. Sect. A Chem.* 5 (3): 857–868.
35 Liu, X., Liu, Z., Zhu, Z. et al. (2017). *Anal. Chem.* 89: 3739–3746.
36 Guo, X., Peng, X., Li, Q. et al. (2017). *J. Anal. At. Spectrom.* 32: 2416–2422.
37 Dong, J., Yang, C., He, D. et al. (2020). *At. Spectrosc.* 41: 57–63.
38 Li, M., Deng, Y., Zheng, C. et al. (2016). *J. Anal. At. Spectrom.* 31: 2427–2433.
39 Chuan, L., Zhi, L., Pengyu, W. et al. (2020). *Plasma Sources Sci. Technol.* 29: 045011.
40 Li, J., Wang, J., Lei, B. et al. (2020). *Adv. Sci.* 7: 1902616.
41 Mao, X., Qi, Y., Huang, J. et al. (2016). *Anal. Chem.* 88: 4147–4152.
42 Liu, M., Liu, J., Mao, X. et al. (2020). *Anal. Chem.* 92: 7257–7264.
43 Qi, Y., Mao, X., Liu, J. et al. (2018). *Anal. Chem.* 90: 6332–6338.
44 Zhao, Z., Dai, J., Wang, T. et al. (2020). *Talanta* 208: 120468.
45 Zhan, X., Zhao, Z., Yuan, X. et al. (2013). *Anal. Chem.* 85: 4512–4519.
46 Mao, X., Qi, Y., Wang, S. et al. (2016). *Trans. Chin. Soc. Agri. Mach.* 47: 216–227.
47 Gibalov, V.I. and Pietsch, G.J. (2012). *Plasma Sources Sci. Technol.* 21: 024010.
48 Horvatic, V., Michels, A., Ahlmann, N. et al. (2015). *Anal. Bioanal.Chem.* 407: 7973–7981.
49 Miclea, M., Kunze, K., Musa, G. et al. (2001). *Spectrochim. Acta, Part B.* 56: 37–43.

50 Zhu, Z., Zhang, S., Xue, J., and Zhang, X. (2006). *Spectrochim. Acta, Part B.* 61: 916–921.
51 Zhu, Z., Liu, J., Zhang, S. et al. (2008). *Anal. Chim. Acta.* 607: 2-136-141.
52 Zhu, Z., Zhang, S., Lv, Y., and Zhang, X. (2006). *Anal. Chem.* 78: 865–872.
53 Na, N., Zhang, C., Zhao, M. et al. (2007). *J. Mass Spectrom.* 42: 1079–1085.
54 Na, N., Zhao, M., Zhang, S. et al. (2007). *J. Am. Soc. Mass Spectrom.* 18: 1859–1862.
55 Liu, X., Zhu, Z., Xing, P. et al. (2020). *Spectrochim. Acta, Part B.* 167: 105822.
56 Liu, T., Liu, M., Liu, J. et al. (2020). *Anal. Chim. Acta.* 1121: 42–49.
57 Liu, X., Yu, K., Zhang, H. et al. (2020). *Talanta.* 219: 121348.
58 Yang, M., Xue, J., Li, M. et al. (2012). *Chin. J. Anal. Chem.* 40: 1164–1168.
59 Yang, M., Li, M., Xue, J. et al. (2015). *Chin. J. Anal. Chem.* 43: 709–713.
60 YuFeng, L.I., ChunYing, C., YuXi, G.A.O. et al. (2009). *Sci. China, Ser. B Chem.* 39: 580–589.
61 J. Feldmann, K. Bluemlein, E. M. Krupp, M. Mueller, B. A. Wood., (2018). *Metallomics: The Science of Biometals.* 1055: 67–100.
62 Moon, J.-H., Kim, S.-H., Chung, Y.-S. et al. (2008). *Anal. Chim. Acta* 619: 137–142.
63 Begaa, S., Messaoudi, M., and Benarfa, A. (2021). *Biol. Trace Elem. Res.* 199: 2399–2405.
64 Yamamoto, Y., Katoh, Y., and Sato, T. (2009). *Legal Med.* 11: S440–S442.
65 Romolo, F.S., Sarilar, M., Antoine, J. et al. (2021). *Talanta* 224: 121829.
66 Margui, E., Queralt, I., and Hidalgo, M. (2009). *Trends Anal. Chem.* 28: 362–372.
67 Wei, C. and Liu, J. (2007). *Talanta.* 73 (3): 540–545.
68 Popp, M., Hann, S., and Koellensperger, G. (2010). *Anal. Chim. Acta.* 668: 114–129.
69 Ai, X., Wang, Y., Hou, X. et al. (2013). *Analyst.* 138: 3494–3501.
70 Chen, Z., He, W., Beer, M. et al. (2009). *Talanta.* 78: 852-856.
71 Amereih, S., Meisel, T., Kahr, E., and Wegscheider, W. (2005). *Anal. Bioanal.Chem.* 383: 1052–1059.
72 Falta, T., Koellensperger, G., Standler, A. et al. (2009). *J. Anal. At. Spectrom.* 24: 1336–1342.
73 Liwei, L., Yong, C., Jinhong, Y., and Yang, S. *At. Spectrosc.* 41: 242–248.
74 Gilfedder, B.S., Lai, S.C., Petri, M. et al. (2008). *Atmos. Chem. Phys.* 8: 6069–6084.
75 Ruzik, L. (2012). *Talanta.* 93: 18–31.
76 García-Bellido, J., Freije-Carrelo, L., Moldovan, M., and Encinar, J.R. (2020). *Trends Anal. Chem.* 130: 115963.
77 Li, Y., Liu, S.-J., Jiang, D.-Q. et al. (2008). *Chin. J. Anal. Chem.* 36: 793–798.
78 Li, J., He, Q., Wu, L. et al. (2020). *Microchem. J.* 153: 104459.
79 Nevado, J.J.B., Martín-Doimeadios, R.C.R., Bernardo, F.J.G., and Moreno, M.J. (2005). *J. Chromatogr. A.* 1093: 21–28.
80 Chenghui, L., Zhou, L., Xiaoming, J. et al. (2016). *Trends Anal. Chem.* 77: 139–155.

81 Amouroux, D., Tessier, E., Pécheyran, C., and Donard, O.F.X. (1998). *Anal. Chim. Acta.* 377: 241–254.
82 Timerbaev, A.R., Hartinger, C.G., and Keppler, B.K. (2006). *TrAC, Trends Anal. Chem.* 25: 868–875.
83 X.-P. Yan, X.-B. Yin, D.-Q. Jiang, X.-W. He, (2003). *Analytical Chemistry* 75: 1726–1732.
84 Silva da Rocha, M., Soldado, A.B., Blanco, E., and Sanz-Medel, A. (2001). *J. Anal. At. Spectrom.* 16: 951–956.
85 Timerbaev, A.R. (2001). *Anal. Chim. Acta.* 433: 165–180.
86 Álvarez-Llamas, G., Fernández de laCampa, M.A.D.R., and Sanz-Medel, A. (2005). *TrAC, Trends Anal. Chem.* 24: 28–36.
87 Ruzik, L. and Kwiatkowski, P. (2018). *Talanta.* 183: 102–107.
88 Michalke, B. and Vinković-Vrček, I. (2018). *J. Chromatogr. A.* 1572: 162–171.
89 Michalke, B. (2004). *J. Chromatogr. A.* 1050: 69–76.
90 Wang, H.-J., Wang, M., Wang, B. et al. (2009). *J. Anal. At. Spectrom.* 25: 328–333.
91 Wang, B., Feng, W., Chai, Z., and Zhao, Y. (2015). *Sci. China Chem.* 58: 768–779.
92 Yu, P. (2004). *Br. J. Nutr.* 92: 869–885.
93 Peth, S., Horn, R., Beckmann, F. et al. (2008). *Soil Sci. Soc. Am. J.* 72: 897–907.
94 Kalnicky, D.J. and Singhvi, R. (2001). *J. Hazard. Mater.* 83: 93–122.
95 Twining, B.S., Baines, S.B., Fisher, N.S. et al. (2003). *Anal. Chem.* 75: 3806–3816.
96 Beckhoff, B., Fliegauf, R., Kolbe, M. et al. (2007). *Anal. Chem.* 79 (20): 7873–7882.
97 Lei, M., Chen, T.-B., Huang, Z.-C. et al. (2008). *Chemosphere.* 72 (10): 1491–1496.
98 Zhang, H., He, X., Bai, W. et al. (2010). *Metallomics.* 2 (12): 806–810.
99 Shi, J., Yuan, X., Chen, X. et al. (2010). *Biol. Trace Elem. Res.* 141 (1): 294–304.
100 Trunova, V.A., Zvereva, V.V., Churin, B.V. et al. (2010). *X-Ray Spectrom.* 39 (1): 57–62.
101 Yano, J. and Yachandra, V.K. (2009). *Photosynth. Res.* 102: 241–254.
102 Bertsch, P.M. and Hunter, D.B. (2001). *Chem. Rev.* 101: 1809–1842.
103 Cotte, M., Susini, J., Dik, J., and Janssens, K. (2010). *Acc. Chem. Res.* 43 (6): 705–714.
104 Hummer, A.A. and Rompel, A. (2013). *Metallomics.* 5 (6): 597–614.
105 Patil, S.A., Suryawanshi, U.P., Harale, N.S. et al. (2022). *Int. J. Environ. Anal. Chem.* 102: 8270–8286.
106 Mohammadi, N. and Shariatmadari, H. (2020). *Commun. Soil Sci. Plant Anal.* 51 (20): 2581–2591.
107 Vivek, K.C., Navneet, Y., Neeraj, K.R. et al. (2020). *Environ. Sci. Pollut. Res.* 28: 13761–13775.
108 Sohail, Kamran, K., Kemmerling, B. et al. (2020). *PLoS One.* 15 (11): No. e0241568.

109 Ondrasek, G., Kranjčec, F., Filipović, L. et al. (2020). *Sci. Total Environ.* 171: 571–578.
110 Russo, R.E., Mao, X., Liu, H. et al. (2002). *Talanta* 57: 425–451.
111 Lin, X., Guo, W.J., Jin, L., and Hu, S. (2020). *At. Spectrosc.* 41 (1): 1–10.
112 Jackson, S.E., Pearson, N.J., Griffin, W.L., and Belousova, E.A. (2004). *Chem. Geol.* 211: (1–2), 47–69.
113 Gonzalez, J., Mao, X.L., Roy, J. et al. (2002). *J. Anal. At. Spectrom.* 17 (9): 1108–1113.
114 Sabine Becker, J., Matusch, A., Palm, C. et al. (2010). *Metallomics.* 2: 104–111.
115 Fiket, Ž., Mikac, N., and Kniewald, G. (2016). *Geostand. Geoanal. Res.* 41 (1): 123–135.
116 Bayon, G., Barrat, J.A., Etoubleau, J. et al. (2009). *Geostand. Geoanal. Res.* 33 (1): 51–62.
117 Guillong, M., Horn, I., and Günther, D. (2003). *J. Anal. At. Spectrom.* 18 (10): 1224–1230.
118 Delgadillo, M.A., Garrostas, L., Pozo, Ó.J. et al. (2012). *J. Chromatogr. B.* 897: 85–89.
119 Zoriy, M.V., Kayser, M., and Becker, J.S. (2008). *Int. J. Mass Spectrom.* 273 (3): 151–155.
120 Yang, L., Tong, S., Zhou, L. et al. (2018). *J. Anal. At. Spectrom.* 33 (11): 1849–1861.
121 McCulloch, M., Cappo, M., Aumend, J., and Müller, W. (2005). *Mar. Freshwater Res.* 56 (5): 637–644.
122 Chen, L., Yuan, H., Chen, K. et al. (2019). *J. Asian Earth Sci.* 176: 325–336.
123 Jurowski, K., Walas, S., and Piekoszewski, W. (2013). *Talanta.* 115: 195–199.
124 Bauer, O.B., Hachmöller, O., Borovinskaya, O. et al. (2019). *J. Anal. At. Spectrom.* 34 (4): 694–701.
125 Hare, D., Austin, C., and Doble, P. (2012). *Analyst.* 137 (7): 1527–1537.
126 Vassilev, S.V., Vassileva, C.G., and Baxter, D. (2014). *Fuel.* 129: 292–313.
127 Feng, L., Zhang, D., Wang, J., and Li, H. (2014). *Anal. Methods.* 6 (19): 7655–7662.
128 Konz, I., Fernández, B., Fernández, M.L. et al. (2011). *Anal. Chem.* 83 (13): 5353–5360.
129 Feng, L., Wang, J., Li, H. et al. (2017). *Anal. Chim. Acta.* 984: 66–75.
130 Boulyga, S.F., Desideri, D., Meli, M.A. et al. (2003). *Int. J. Mass Spectrom.* 226 (3): 329–339.
131 Zheng, L.-N., Feng, L.-X., Shi, J.-W. et al. (2020). *Anal. Chem.* 92 (21): 14339–14345.
132 Wen, X., Wu, P., Chen, L., and Hou, X. (2009). *Anal. Chim. Acta.* 650 (1): 33–38.
133 Durduran, E., Altundag, H., Imamoglu, M. et al. (2015). *J. Ind. Eng. Chem.* 27: 245–250.
134 Costley, C.T., Mossop, K.F., Dean, J.R. et al. (2000). *Anal. Chim. Acta.* 405 (1–2): 179–183.
135 Badiei, H.R., Liu, C., and Karanassios, V. (2013). *Microchem. J.* 108: 131–136.

136 Feng, L., Liu, J., Mao, X. et al. (2016). *J. Anal. At. Spectrom.* 31 (11): 2253–2260.

137 Bruhn, C.G., Ambiado, F.E., Cid, H.J. et al. (1995). *Anal. Chim. Acta.* 306 (2–3): 183–192.

138 Cui, Y., Jin, L., Li, H. et al. (2020). *At. Spectrosc.* 41: 87–92.

139 Huang, R.-J., Zhuang, Z.-X., Tai, Y. et al. (2008). *Talanta.* 68 (3): 728–734.

140 Zhang, Y., Mao, X., Liu, J. et al. (2016). *Spectrochim. Acta, Part B.* 118: 119–126.

141 Thongsaw, A., Sananmuang, R., Udnan, Y. et al. (2019). *Spectrochim. Acta, Part B.* 160: 105685.

142 Mao, X., Zhang, Y., Liu, J. et al. (2016). *RSC Adv.* 6 (54): 48699–48707.

143 Mao, X., Liu, J., Huang, Y. et al. (2013). *J. Agric. Food. Chem.* 61 (4): 848–853.

144 Senesi, G.S., Cabral, J., Menegatti, C.R. et al. (2019). *Trends Anal. Chem.* 118: 453–469.

145 Kaiser, J., Novotný, K., Martin, M.Z. et al. (2012). *Surf. Sci. Rep.* 67 (11–12): 233–243.

146 Jantzi, S.C. and Almirall, J.R. (2011). *Anal. Bioanal.Chem.* 400 (10): 3341–3351.

147 Hosic, S., Murthy, S.K., and Koppes, A.N. (2015). *Anal. Chem.* 88 (1): 354–380.

148 Nomizu, T., Kaneco, S., Tanaka, T. et al. (1994). *Anal. Chem.* 66: 3000–3004.

149 Miyashita, S.I., Groombridge, A.S., Fujii, S.I. et al. (2014). *J. Anal. At. Spectrom.* 29 (9): 1598–1606.

150 Ho, K.-S. and Chan, W.-T. (2010). *J. Anal. At. Spectrom.* 57 (4): 659–688.

151 Hadioui, M., Peyrot, C., and Wilkinson, K.J. (2014). *Anal. Chem.* 86: 4668–4674.

152 Yu, X., He, M., Chen, B., and Hu, B. (2020). *Anal. Chim. Acta.* 1137: 191–207.

153 Yin, L., Zhang, Z., Liu, Y. et al. (2018). *Analyst.* 144: 824–845.

154 Tanaka, Y.K., Iida, R., Takada, S. et al. (2020). *ChemBioChem.* 21: 3266–3272.

155 Bendall, S.C., Simonds, E.F., Qiu, P. et al. (2011). *Science.* 332: 687–696.

156 Daferera, D.J., Ziogas, B.N., and Polissiou, M.G. (2000). *J. Agric. Food. Chem.* 48: 2576–2581.

157 Peng, J., Elias, J.E., Thoreen, C.C. et al. (2003). *J. Proteom. Res.* 2: 43–50.

158 Corte Rodríguez, M., Álvarez-Fernández García, R., Blanco, E. et al. (2017). *Anal. Chem.* 89: 11491–11497.

159 Wang, H., Wang, M., Wang, B. et al. (2016). *Anal. Bioanal.Chem.* 409: 1415–1423.

160 Shigeta, K., Koellensperger, G., Rampler, E. et al. (2013). *J. Anal. At. Spectrom.* 28: 637–645.

161 Meyer, S., López-Serrano, A., Mitze, H. et al. (2017). *Metallomics.* 10: 73–76.

162 Merrifield, R.C., Stephan, C., and Lead, J.R. (2018). *Environ. Sci. Technol.* 52: 2271–2277.

163 Álvarez-Fernández García, R., Corte-Rodríguez, M., Macke, M. et al. (2019). *Analyst.* 145: 1457–1465.

164 Gomez-Gomez, B., Corte-Rodríguez, M., Perez-Corona, M.T. et al. (2020). *Anal. Chim. Acta.* 1128: 116–128.

165 Laborda, F., Trujillo, C., and Lobinski, R. (2020). *Talanta.* 221: 121486.

166 Loeschner, K., Brabrand, M.S.J., Sloth, J.J., and Larsen, E.H. (2013). *Anal. Bioanal.Chem.* 406: 3845–3851.

167 Dan, Y., Zhang, W., Xue, R. et al. (2015). *Environ. Sci. Technol.* 49: 3007–3014.

168 Laborda, F., Bolea, E., and Jiménez-Lamana, J. (2016). *Trends Environ. Anal. Chem.* 9: 15–23.
169 Slejkovec, Z., van Elteren, J.T., and Byrne, A.R. (2008). *Talanta.* 49: 619–627.
170 Mleczek, M., Niedzielski, P., Rzymski, P. et al. (2016). *J. Environ. Sci. Health Part B.* 51: 469–476.
171 Yao, Z., Liu, M., Liu, J. et al. (2020). *J. Anal. At. Spectrom.* 35: 1654–1663.
172 Marcinkowska, M. and Barałkiewicz, D. (2016). *Talanta.* 161: 177–204.
173 Romarís-Hortas, V., Bermejo-Barrera, P., Moreda-Piñeiro, J., and Moreda-Piñeiro, A. (2012). *Anal. Chim. Acta.* 745: 24–32.
174 Zhang, D., Yang, S., Ma, Q. et al. (2019). *Food Chem.* 313: 126119.
175 Zhou, L., Wang, S., Hao, Q. et al. (2018). *Chin. Med.* 13: 1–8.
176 Olesik, J.W., Kinzer, J.A., and Olesik, S.V. (1995). *Anal. Chem.* 67: 1–12.
177 Zhao, Y., Zheng, J., Yang, M. et al. (2011). *Talanta.* 84: 983–988.
178 Deng, B., Zhu, P., Wang, Y. et al. (2008). *Anal. Chem.* 80: 5721–5726.
179 Yu, L.P. (2007). *Anal. Lett.* 40: 1879–1892.
180 Li, P., Zhang, X., and Hu, B. (2011). *J. Chromatogr. A.* 1218: 9414–9421.
181 Bendahl, L. and Gammelgaard, B. (2004). *J. Anal. At. Spectrom.* 19: 143–148.
182 Liu, L., He, B., Yun, Z. et al. (2013). *J. Chromatogr. A* 1304: 227–233.
183 Xu, C., Chen, X., Duan, D. et al. (2015). *Environ. Sci. Pollut. Res.* 22: 5070–5081.
184 Shen, Y.-T. (2014). *Guang pu xue yu guang pu fen xi = Guang pu* 34: 818–822.
185 Wu, Z., McGrouther, K., Chen, D. et al. (2013). *J. Agric. Food. Chem.* 61: 4715–4722.
186 Li, Y., Li, H., Li, Y.-F. et al. (2017). *J. Trace Elem. Med. Biol.* 50: 435–440.
187 Verbi Pereira, F.M. and Bastos Pereira Milori, D.M. (2009). *J. Anal. At. Spectrom.* 25: 351–355.
188 Shi, J.Y., Chen, Y.X., Huang, Y.Y., and He, W. (2004). *Micron.* 35: 557–564.
189 Wang, Q., Wen, J., Hu, X. et al. (2020). *J. Cleaner Prod.* 278: 123967.
190 Tan, G. and Xiao, D. (2008). *Sep. Sci. Technol.* 43: 2196–2207.
191 Liu, J., Ma, J., He, C. et al. (2013). *New Phytol.* 200: 691–699.
192 Kavčič, A., Mikuš, K., Debeljak, M. et al. (2019). *Ecotoxicol. Environ. Saf.* 184: 109623.
193 Alirzayeva, E., Neumann, G., Horst, W. et al. (2016). *Environ. Pollut.* 220: 1024–1035.
194 Moustakas, M., Hanć, A., Dobrikova, A. et al. (2019). *Materials.* 12: 2953.
195 Hanć, A., Barałkiewicz, D., Piechalak, A. et al. (2009). *Int. J. Environ. Anal. Chem.* 89: 651–659.
196 Neves, V.M., Heidrich, G.M., Hanzel, F.B. et al. (2018). *Chemosphere.* 198: 409–416.
197 Galiová, M., Kaiser, J., Novotný, K. et al. (2011). *Microsc. Res. Tech.* 74: 845–852.
198 Becker, J.S., Zoriy, M., Krause-Buchholz, U. et al. (2004). *J. Anal. At. Spectrom.* 19: 1236–1243.
199 Yu, X., Wei, S., Yang, Y. et al. (2018). *Int. J. Biol. Macromol.* 119: 597.
200 Wang, Y.X., Specht, A., and Horst, W.J. (2010). *New Phytol.* 189: 428–437.
201 Xing, P., Li, X., Feng, L., and Mao, X. (2021). *J. Anal. At. Spectrom.* 36: 285–293.
202 Šerá, L., Loula, M., Matějková, S., and Mestek, O. (2019). *Chem. Pap.* 73: 3005–3017.

203 Ercioglu, E., Velioglu, H.M., and Boyaci, I.H. (2018). *Food Anal. Methods* 11: 1656–1667.
204 Arantes de Carvalho, G.G., Bueno Guerra, M.B., Adame, A. et al. (2018). *J. Anal. At. Spectrom.* 33: 919–944.
205 Sidbury, R. and Kodama, S. (2018). *Clin. Dermatol.* 36: 648–652.
206 Krajcarová, L., Novotný, K., Kummerová, M. et al. (2017). *Talanta* 173: 28–35.
207 Zhao, X., Zhao, C., Du, X., and Dong, D. (2019). *Sci. Rep.* 9: 906.
208 da Silva Gomes, M., de Carvalho, G.G.A., Santos, D., and Krug, F.J. (2013). *Spectrochim. Acta, Part B.* 86: 137–141.
209 Chen, C., Wang, T., Khim, J.S. et al. (2011). *Chem. Ecol.* 27: 165–176.
210 Li, X., Liu, T., Chang, C. et al. (2021). *J. Agric. Food. Chem.* 69: 6100–6118.
211 Jamila, N., Khan, N., Hwang, I.M. et al. (2021). *Anal. Lett.* 55: 159–173.
212 Mielcarek, K., Nowakowski, P., Puścion-Jakubik, A. et al. (2022). *Food Chem.* 379: 132167.
213 Ogbomida, E.T., Nakayama, S.M.M., Bortey-Sam, N. et al. (2018). *Ecotoxicol. Environ. Saf.* 151: 98–108.
214 Chijioke, N.O., Uddin Khandaker, M., Tikpangi, K.M., and Bradley, D.A. (2020). *J. Food Compos. Anal.* 85: 103332.
215 Hejna, M., Moscatelli, A., Onelli, E. et al. (2019). *Ital. J. Anim. Sci.* 18: 1372–1384.
216 Bandoniene, D., Walkner, C., Ringdorfer, F., and Meisel, T. (2020). *Food Res. Int.* 132: 109106.
217 Bozorgzadeh, E., Pasdaran, A., and Ebrahimi-Najafabadi, H. (2021). *Food Chem.* 346: 128916.
218 Franco-Fuentes, E., Moity, N., Ramirez-Gonzalez, J. et al. (2021). *J. Environ. Manage.* 286: 112188.
219 Alberto, A., Francesco, C., Atzei, A. et al. (2021). *Environ. Monit. Assess.* 193: 448.
220 Lozano-Bilbao, E., Jurado-Ruzafa, A., Lozano, G. et al. (2020). *Chemosphere.* 261: 127692.
221 Milenkovic, B., Stajic, J.M., Stojic, N. et al. (2019). *Chemosphere* 229: 324–331.
222 Rajib, A., Tariqur Rahman, M., and Ismail, A.B.M.D. (2021). *J. Phys. Conf. Ser.* 1718: 012015.
223 Baldassini, W.A., Braga, C.P., Chardulo, L.A. et al. (2015). *Food Chem.* 169: 65–72.
224 Veado, M.A.R.V., Heeren, A.O., Severo, M.I. et al. (2007). *J. Radioanal. Nucl. Chem.* 272: 511–514.
225 Jayawickreme, C.K. and Chatt, A. (1990). *Nuclear Analytical Methods in the Life Sciences.* 503–512.
226 Ogiyama, S., Sakamoto, K., Suzuki, H. et al. (2006). *Soil Sci. Plant Nutr.* 52: 114–121.
227 Oliveira, A.F., Landero, J., Kubachka, K. et al. (2016). *J. Anal. At. Spectrom.* 31: 1034–1040.
228 Gawor, A., Ruszczynska, A., Czauderna, M., and Bulska, E. (2020). *Animals (Basel)* 10: 808.

229 Moreda-Pineiro, J., Moreda-Pineiro, A., Romaris-Hortas, V. et al. (2013). *Food Chem.* 139: 872–877.
230 Zhang, X., Liu, Y., Zhang, Z. et al. (2018). *J. Anal. At. Spectrom.* 34: 292–300.
231 Deng, D., Zhang, S., Chen, H. et al. (2015). *J. Anal. At. Spectrom.* 30: 882–887.
232 Mao, X., Chen, B., Huang, C. et al. (2011). *J. Chromatogr. A* 1218: 1–9.
233 Peng, H., Hu, B., Liu, Q. et al. (2014). *J. Chromatogr. A* 1370: 40–49.
234 Becker, J.S., Matusch, A., and Wu, B. (2014). *Anal. Chim. Acta* 835: 1–18.
235 Ackerman, C.M., Weber, P.K., Xiao, T. et al. (2018). *Metallomics* 10: 474–485.
236 Matusch, A., Depboylu, C., Palm, C. et al. (2010). *J. Am. Soc. Mass Spectrom.* 21: 161–171.
237 Tzadik, O.E., Jones, D.L., Peebles, E.B. et al. (2017). *Estuaries Coasts* 40: 1785–1794.
238 Lum, T.-S., Ho, C.-L., Tsoi, Y.-K. et al. (2016). *Int. J. Mass Spectrom.* 404: 40–47.
239 Liaw, M.-J., Jiang, S.-J., and Li, Y.-C. (1997). *Spectrochim. Acta, Part B.* 52: 779–785.
240 Tormen, L., Gil, R.A., Frescura, V.L. et al. (2012). *Anal. Chim. Acta* 717: 21–27.
241 Kumar, A., Yueh, F.-Y., Singh, J.P., and Burgess, S. (2004). *Appl. Opt.* 43: 5399–5403.
242 Wan, X. and Wang, P. (2015). *Optik - Int. J. Light Electron Opt.* 126: 1930–1934.
243 Chung, C.-H., Brenner, I., and You, C.-F. (2009). *Spectrochim. Acta, Part B.* 64: 849–856.
244 Mahbub, K.R., Krishnan, K., Megharaj, M., and Naidu, R. (2015). *Chemosphere.* 144: 330–337.
245 Wysocka, I. (2020). *Talanta.* 221: 121636.
246 Arnold, T., Schönbächler, M., Rehkämper, M. et al. (2010). *Anal. Bioanal.Chem.* 398: 3115–3125.
247 Huang, Y., Zhang, S., Chen, Y. et al. (2019). *J. Hazard. Mater.* 385: 121528.
248 Uchimura, T., Hironaka, Y., and Mori, M. (2013). *Anal. Sci.* 25: 85–88.
249 Smith, S.R., Foster, G.L., Romer, R.L. et al. (2004). *Contrib. Mineral. Petrol.* 147: 549–564.
250 Götz, A. and Heumann, K.G. (1988). *Int. J. Mass Spectrom. Ion Processes* 83: 319–330.
251 Betti, M., Giannarelli, S., Hiernaut, T. et al. (1996). *Anal. Bioanal.Chem.* 355: 642–646.
252 Wang, X., Bei, Q., Yang, W. et al. (2020). *Biol. Fertil. Soils.* 56: 1189–1199.
253 Ondrasek, G., Rengel, Z., Clode, P.L. et al. (2019). *Ecotoxicol. Environ. Saf.* 171: 571–578.
254 Moore, K.L., Rodríguez-Ramiro, I., Jones, E.R. et al. (2018). *Sci. Rep.* 8: 6865.
255 Yuan, X. and Duan, Y.-X. (2015). *Chin. J. Anal. Chem.* 43: 1306–1312.
256 Wan, M., Hu, W., Qu, M. et al. (2019). *Ecol. Indic.* 101: 583–594.
257 Langan, T.J. and Holland, L.A. (2012). *Electrophoresis.* 33: 607–613.
258 Wu, J., Huang, D., Liu, X. et al. (2018). *J. Hazard. Mater.* 348: 10–19.
259 Wu, J., Li, Z., Huang, D. et al. (2020). *J. Hazard. Mater.* 387: 122010.
260 Xu, Y., Wang, K., Zhou, Q. et al. (2020). *Chemosphere.* 259: 127403.
261 Peng, X., Deng, Y., Liu, L. et al. (2020). *J. Cleaner Prod.* 277: 124052.
262 Niu, S., Zheng, L., Qayyum Khan, A., and Zeng, H. (2019). *Appl. Spectrosc.* 73: 380–386.

5

Metrometallomics*

Liuxing Feng

National Institute of Metrology, Division of Chemical Metrology and Analytical Science, No. 18 Bei San Huan Dong Lu, Chaoyang Dist, Beijing 100013, P.R. China

5.1 The Concept of Metrometallomics

With the rapid development of biotechnology, protein quantification, *in situ* analysis, and bioimaging have been playing important roles in life sciences and clinical research. For example, absolute quantification is critical to access the capacity of biomarkers in clinical applications, which involves the determination of amount of substance of the target analyte in given sample. However, since there is a great risk of exchange, acquisition or loss of the metal during separation of proteins, the analysis strategies of absolute quantification of metalloproteins are essential and technically challenging. Another application area is *in situ* analysis of biomarkers, including elements and bio-molecules. Imaging analytical techniques with spatial resolution in the low μm range are today of crucial interest in biological studies to achieve a comprehensive understanding of the role of metals and macromolecules in biological systems. But now, due to the matrix effect and elemental fractionation, achieving quantitative analysis of elements or proteins in biological samples is still challenging.

The "concept of metrometallomics" is defined as measurement activities for metallic analytes with metrological strategies, which includes establishment of reference methods with uncertainty evaluation, certified reference material (CRM) development, and their applications in life science and environmental protection [1]. The emerging field of metrometallomics refers to all research activities aimed at qualitative and quantitative measurement of metallic analytes with metrological strategy, which includes absolute quantitative strategies for the determination of metal-transport protein and metalloenzymes, *in situ* quantitative analysis of

---

* This chapter has been modified to feature as Reviews: (i) Pan, M.Y., Zang, Y., Zhou, X.R., et al. (2021). Inductively coupled plasma mass spectrometry for metrometallomics: the study of quantitative metalloproteins. *At. Spectrosc. 42*: 262–270; (ii) Pan, H.J., Feng, L.X., Lu, Y.L., et al. (2022). Calibration strategies for laser ablation ICP-MS in biological studies: a review. *Trends Anal. Chem. 156*: 116710.

Applied Metallomics: From Life Sciences to Environmental Sciences, First Edition.
Edited by Yu-Feng Li and Hongzhe Sun.
© 2024 WILEY-VCH GmbH. Published 2024 by WILEY-VCH GmbH.

elements or proteins in biological samples [2]. In the metrometallomics studies, metrological properties including accuracy, stability, method validation, and uncertainty evaluation are of vital importance.

5.2 The Analytical Techniques in Metrometallomics

5.2.1 Analytical Techniques of Protein Quantification in Metrometallomics

Mass-spectrometry (MS)-based quantitative strategies mainly include molecular MS (e.g. ESI-MS) and elemental MS (e.g. ICP-MS). Molecular MS (such as ESI-MS, MALDI-MS) using multiple-reaction monitoring (MRM) for peptide or protein quantification has been considered to be a promising high-throughput targeting protein quantification technology (common strategies are based on bottom-up and top-down) in many occasions [3]. MRM technology selects the specific parent ion and product ion pair of the target protein to perform mass spectrometry analysis to eliminate the influence of interfering ions to the greatest extent, and significantly improves the signal-to-noise ratio of the target peptide. This technology has the advantages of high sensitivity, good accuracy, and strong specificity. It is known as the "gold standard" for mass spectrometry quantification and is especially suitable for high-throughput verification of labeled proteins. MRM technology can be combined with a variety of quantitative strategies. Proteomics experiments usually include a large number of separation and enrichment steps. Therefore, the earlier the internal standard is added, the better the experimental error can be reduced. In addition, MRM is suitable for a variety of mass spectrometers, such as high-resolution mass spectrometers (Q-TOF, orbitrap, FT-ICR, which can distinguish mass differences in ppm) and low-resolution mass spectrometers (such as triple quadrupole mass spectrometers) [4]. However, some limitations of the MRM technology must be taken into account. Because the signal intensity they provide is strongly affected by the sample matrix and the solvent used, the ion sources used in MS are not inherently quantitative. On the other hand, it is well known that there is no linear relationship between the target peptide/protein concentration and the measured soft ion source signal intensity. Therefore, a specific standard (usually manual synthesized) of target peptide/protein is required in order to obtain absolute quantification results [4].

ICP-MS, as an elemental analysis tool, uses a hard ionization source that dissociates proteins into elemental ions or molecular ions under ~10 000 K in inductively coupled plasma (ICP). These elemental and molecular ions are introduced and filtered by the mass analyzer based on their mass-to-charge ratio (m/z), and finally measured without significant spectral interference. That is particularly advantageous for quantitative purpose because of its excellent analytical features, including low detection limit, wide linear dynamic range, high sensitivity, matrix-independent ionization, and multi-element analysis [5, 6]. Because of its hard ionization source, ICP-MS is a powerful technique which is able to provide absolute amounts of biomolecules in complex samples without the need of specific targeted standards [7, 8]. Moreover, the use of stable isotopes, as non-radioactive tags for biomarker and for accurate biomolecule quantification with isotope-dilution analysis (IDA), is a great bonus of ICP-MS in tackling the long-standing problem of absolute determinations of proteins [9]. IDA is an analytical technique based on the measurement of isotope ratios after the mixture of the spike (i.e. a known amount of an enriched isotope) [10]. For an IDA study, the equilibration of the spike with the analyte is a prerequisite. Isotope dilution mass spectrometry (IDMS) has many unique advantages. Firstly, the final results obtained by IDMS are independent of the drift of instrumental signals and the matrix effects of samples. Secondly, the results can provide definable uncertainty values, and can achieve very high accuracy and precision with a mass spectrometer. Last but not least, once the complete isotope equilibration is achieved between sample and spike (e.g. full equilibration for solid samples can be reached after an appropriate digestion), possible losses of the isotope-diluted analyte have no influence on the final results, because all fractions in the isotope-diluted sample contain the same isotope ratio. Therefore, IDMS is regarded as one of the highly qualified primary methods [11]. The detected heteroatoms may be naturally occurring, such as S, Se, P and metals in proteins, selenoproteins, phosphoproteins, and metalloproteins, respectively, or be intentionally bioconjugated as an elemental tag. However, those assets of ICP-MS as reference technique also have its limitations for quantitative proteomic analysis. Due to the loss of structural information or identification from an atomized peptide or protein taking place in ICP, reliable analysis of target proteins depends on the separation process in advance such as liquid chromatography (LC) and gel electrophoresis (GE) [12, 13] in order to isolate the different compounds containing detectable heteroatom before final measurement.

Protein quantification techniques based on ICP-MS are shown in Figure 5.1, which focuses on several techniques and methods for quantitative protein research using ICP-MS detectable heteroatom, particularly metal-coded affinity tags (MECAT) and nanoparticles (NPs) immunoassay. What's more, because of the growing potential of direct analysis via laser ablation (LA) ICP-MS for metalloprotein quantification, the progress of this methodology is also included.

5.2.2 Analytical Techniques of Quantitative *In Situ* Analysis in Metrometallomics

Laser ablation inductively coupled plasma mass spectrometry (LA-ICP-MS), a combination of laser ablation (LA) system and ICP-MS, can directly measure the content of elements in solid samples without laborious sample preparation. The laser ablation system focuses the laser beam to ablate the sample into aerosol state, and then with the carrier gas, the sample aerosols are directly and quickly introduced to ICP-MS for element detection. The advantages of LA-ICP-MS are ease of sample preparation, sensitivity in the µg/g and even ng/g level, dynamic range which is up to 12 orders of magnitude, quantitative capabilities, spatial resolution in the low mm range, and the fact that it is quasi-nondestructive to samples [14–17].

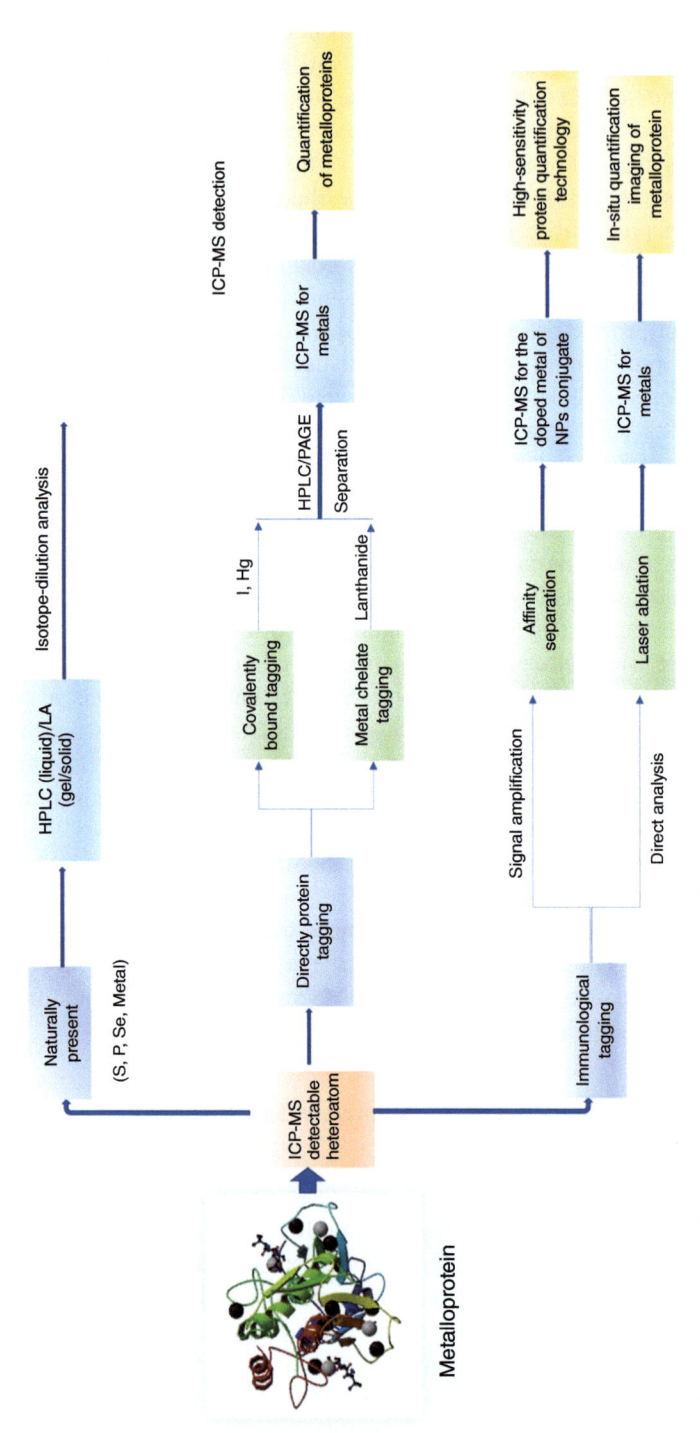

Figure 5.1 Protein quantification strategies based on ICP-MS. Source: Liuxing Feng (author).

Compared with solution-based ICP-MS, LA-ICP-MS simplifies the pretreatment procedures, reduces the interference of atoms, and can more intuitively show the element distribution on the sample surface and quantitative imaging by calibrating each abated pixels.

Numerous imaging studies [18, 19] by LA-ICP-MS have been published after the first reported example by Wang et al [20] in 1994. However, when it comes to quantitative imaging, matrix effect [21] and elemental fractionation [22] have restricted the accurate and precise quantitative applications of this technique. The matrix effect of LA-ICP-MS is the different performance in the interaction between the laser beam and the sample surface of various matrices, causing changes in the mass of analyte ablated per pulse due to differences in the properties of the matrices investigated (e.g. absorptivity, reflectivity, and thermal conductivity). The aerosol particles produced during ablation with different matrices may vary in size and geometry, thus having an effect on the aerosol transport efficiency from the ablation cell to the plasma [23]. Because the vaporization, atomization, and ionization efficiencies of the analytes introduced into the plasma depend on the mass load, both the size and geometry variances could contribute to differences in the mass load of the plasma and result in matrix effects. Sample-related matrix effects therefore hamper the accuracy of quantification by LA-ICP-MS. [24, 25] The other major drawback of LA-ICP-MS is the elemental fractionation which means the abundances of the ions detected after m/z separation are often not entirely representative of the composition of the original sample.

Besides the ablation process, the transport of the aerosols from the ablation chamber into the ICP (e.g. differences in gravitational settling between smaller and larger particles) and vaporization, atomization, and ionization in the ICP are also important sources to fractionation effects [26–28]. In many cases, matrix effects and elemental fractionation occur simultaneously, leading to LA-ICP-MS signals that are not representative of the elemental composition of the real sample. The sensitivity or signal intensity can vary significantly for samples with the same analyte concentrations, but different matrix compositions and/or physical properties. As a result, most studies seem to be the qualitative analysis, and it is still necessary to improve the stability and accuracy of the quantitative imaging applications. To overcoming the aforementioned drawbacks, attempts were made to address the limitations of LA-ICP-MS by improving the instrumental parameters relevant to aerosol formation. However, complete elimination of matrix effects and elemental fractionation by optimizing instrumentation is still not possible. Many studies are therefore dedicated to calibration strategy developments that realize quantification with the currently available instrumentation for LA-ICP-MS analysis.

To accomplish accurate quantification by LA-ICP-MS, preparation and calibration procedure must be taken carefully and match to the nature, structure, and composition of the analytes [5]. The term "calibration" comes from the field of metrological and analytical areas, which is an essential procedure for quantitative analysis. According to the IUPAC, the calibration is the operation that determines the functional relationship between measured values (signal intensities) as y-variable and analytical quantities charactering the types and amount (content,

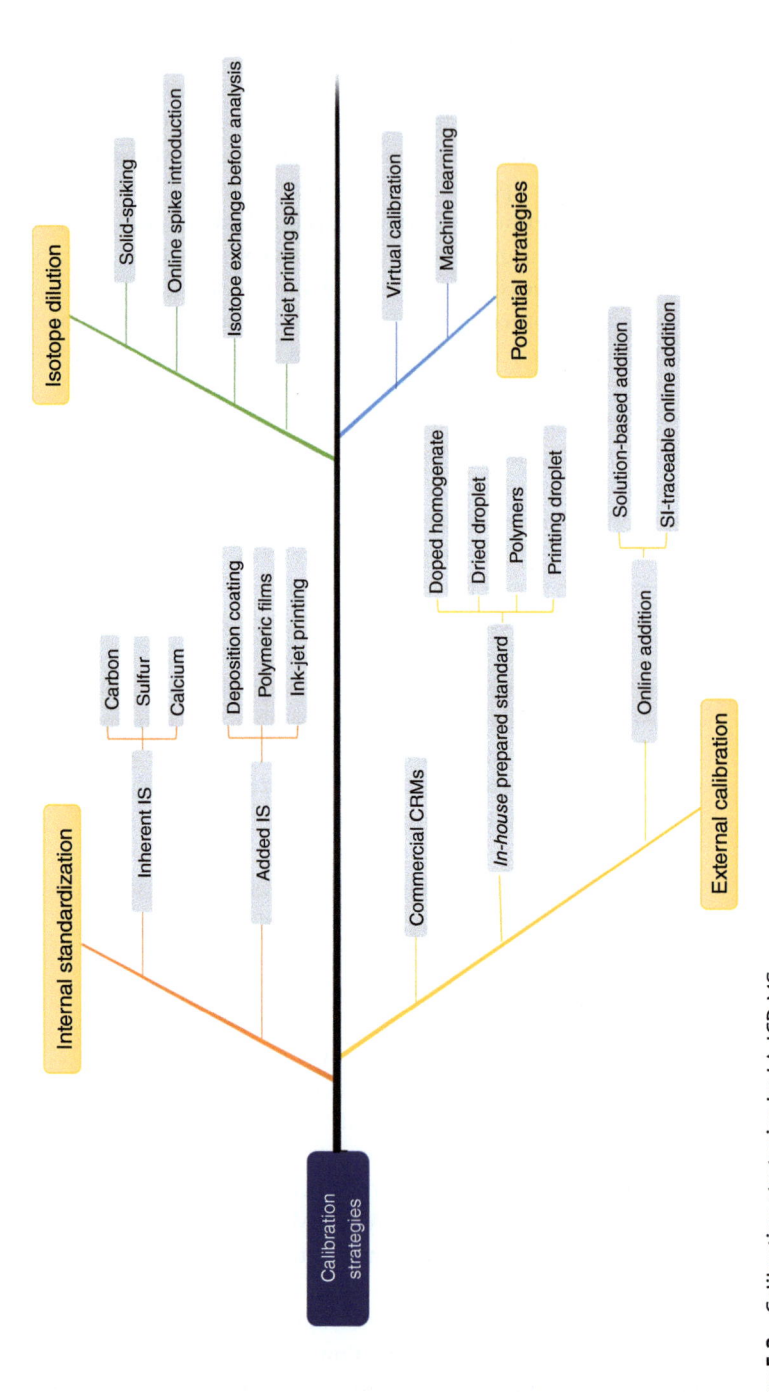

Figure 5.2 Calibration strategies by LA-ICP-MS.

concentration) of analytes as *x*-variable [29]. Quantitative analysis by LA-ICP-MS particularly relies upon suitable calibration strategies. In an ideal situation, matrix-matched certified reference materials (CRMs) are available to establish standard curves for LA-ICP-MS quantitative analysis. For example, SRM 610 serials CRMs with glass matrix are usually used in the quantitative analysis by LA-ICP-MS in geology analysis [30–33]. However, because of variations in composition, the range of available sample states in biological studies are highly diverse. Therefore, suitable commercially CRMs are usually unavailable for LA-ICP-MS calibration, and internal standard (IS), *in-house* prepared standards and solution-based online addition are most commonly used calibration approaches. Although the calibration approaches do not have metrological property and the measurement results cannot be traceable to SI unit, the adoption of these calibration layouts can improve the quantitative results obviously. Isotope dilution mass spectrometry (IDMS), directly traceable to SI unit, is one of the most accepted methods capable of achieving analytical results that are less affected by signal drifts, matrix effects, or analyte losses. As a result, the incorporation of isotope dilution with LA-ICP-MS can eliminate some common fractionation and matrix effects that cannot be addressed using other calibration procedures [34].

Figure 5.2 summarizes state-of-the-art procedures and recent developments in quantitative LA-ICP-MS analysis in biological applications. It concerns the quantification of elements and constituents by LA-ICP-MS, several calibration strategies including internal standardization, external calibration (including commercial CRMs, *in-house* prepared standards, online addition calibration), isotope dilution, and calibration based on machine learning strategies.

5.3 The Application of Metrometallomics in Life Science and the Perspectives

5.3.1 Absolute Quantification of Metalloproteins in Metrometallomics

5.3.1.1 Naturally Present Elements (P, S, Se, Metals)

So far, protein quantification by molecular MS is greatly limited by the need of specific standards. Such standards must be synthesized first, then they have to be characterized and finally accurately certified, which lead to severe quantitative limitations. However, ICP-MS may achieve absolute protein quantification without the need of such specific standards due to its robust species-independent signal. There are some ICP-MS detectable heteroatoms (S, P, Se, and metals) naturally present in metalloproteins. All species in sample need to be separated first due to ICP-MS, which cannot distinguish the molecules of origin of detectable element. Among the methods, the coupling of HPLC to ICP-MS appears to be one of the most common methods for protein analysis because of its ease of sample preparation and the simplicity of the interface [35]. However, the hyphenated techniques possess some problems when liquid chromatography column is coupled to ICP-MS by a nebulizer. GE-LA-ICP-MS has been increasingly employed for the quantification and characterization of metal-containing (e.g. Zn and Fe) proteins, for example in brain or human serum [36–38]. As there is lack of suitable matrix-matched standard reference materials in GE-LA-ICP-MS, accurate quantification of metal-containing proteins still has an important challenge. Some relevant applications of macrobiomolecules analysis by using ICP-MS as detector are given in Table 5.1.

In the determination of heteroatom naturally occurring in proteins and peptides by ICP-MS for protein quantification, the quality and reliability of quantification results are less affected by isotope or chemical labeling of proteins and peptides. However, there are still many limitations in using heteroatom to quantify proteins [39, 40]. Metals are not covalently bonded to protein's moiety. It does not maintain the integrity of metalloproteins during separation which may lead to loss of metal. Buffers and solutions for gel staining may be sources of metal contamination, and metal loss also occurs during the decolorization process. Such loss will result in crucial errors in protein quantification. Due to the low ionization efficiency of heteroatom (such as phosphorus and sulfur which are covalently bonded to the protein primary structure) and its susceptibility to severe spectral line interference [41, 42], it is difficult to measure heteroatom by ICP-MS in directly, and the measurement of heteroatom is also interfered by polyatomic ions [43, 44]. In fact, the use of collision reaction cells and high-resolution ICP-MS instrumentation to remove polyatomic interferences may hinder the applicability of ICP-MS quantification in biological applications, usually requiring low limit of detection (LOD) [45, 46]. In this sense, the introduction of tandem configurations into ICP-MS instrumentation (ICP-QQQ) provided an efficient way for elimination of polyatomic interferences and so for highly sensitive detection of non-metallic heteroatoms [47]. As a matter of fact, tandem ICP-MS enables today highly sensitive quantification of S- and P-containing peptides and proteins [48].

5.3.1.2 Elemental Labeling

The sensitivity of ICP-MS for the detection of heteroatoms naturally present in proteins, even using tandem MS configuration, might be not sufficient for certain quantitative applications in biomedical research. In this sense, elemental labeling is becoming a trend to achieve higher selectivity and sensitivity in spite of the laborious and time-consuming labeling process. Usually, the label is easily ionized and has low background, which makes better sensitivity than that of S, P, and Se [49]. By element labeling, peptides and proteins without heteroatoms can be detected and quantified by ICP-MS, which expands the analysis of proteins [50]. Protein labeling can be done directly on the target protein, or indirectly through its corresponding antibody (immunological strategy). The main limitations of the two strategies are dealing with excessive reagents, stoichiometric determination, and the possibility of generating undesirable products or fragments during the labeling process. Therefore, the labeling procedure must be properly characterized, and separation techniques are often employed to resolve potential interfering compounds in most proteomics applications [49–51].

Table 5.1 Applications of biomolecules analysis by using ICP-MS.

Elements	Sample/analyte	Technique	Analytical performance	References
Mn, Co, Cu, Se,	human milk	SEC-ICP-MS	Mn: LOD: 2.81 g/l Linear range: 8.51~100 g/l Co: LOD: 3.39 g/l Linear range: 10.27~100 g/l Cu: LOD: 1.46 g/l Linear range: 4.43~100 g/l Se: LOD: 5.86 g/l Linear range: 17.75~100 g/l	[16]
Fe, Cu, Zn, I	human milk	RP-HPLC-ICP-QQQ MS	mothers of pre-term (Fe: 0.997, Cu: 0.506, Zn: 4.15 and I: 0.458 mg/l) mothers of full-term (Fe: 0.733, Cu: 0.234, Zn: 2.91 and I: 0.255 mg/l)	[17]
S	βamyloid peptide ($A\beta_{42}$)	SEC-ID-ICP-MS	Aβ: 0.763 ± 0.0044 g/g LOD: 140 pg/g Linear range: 5~60 μg/g	[35]
S	proteomics in snake venoms	RP-HPLC-ID-ICP-QQQ MS		[39]
S	hGh	SEC-ID-ICP-MS	hGH: 18.86 ± 0.77 mg k/g.	[40]
S	transferrin and albumin in human serum	GE-LA-ID-ICP-MS	Linear range: 1~400 μg/g Albumin: 36.9 ± 1.55 mg/ml Transferrin: 2.45 ± 0.06 mg/ml	[36–38]
Ca	serum	ID-ICP-MS		[41]
S	SOD, BSA, MT-II	SEC-ICP-CC-MS	LOD for BSA, SOD, and MT-II are 8, 31, and 15 pmol/g	[42]
Fe	Haemoglobin (HGB)	ID-ICP-MS	HGB: 115.3 ± 2.4 mg/g LOD: 1.0×10^{-7} mg/g RSD < 3%	[43]
Fe	Haemoglobin (HGB)		HGB: 122.1 ± 1.8 mg/g	[44]

5.3.1.3 Directly Protein Tagging (I, Hg, Chelate Complexes)

As one of the most active functional groups in protein, sulfhydryl group (–SH) has unique physiological characteristics and plays an important role in protein stabilization, enzyme catalysis, heavy metal detoxification, and radicals reactivity [52]. Because of the strong affinity of mercury and sulfur, organic mercury compounds are the most specific and sensitive reagents reacting with sulfhydryl groups in peptides and proteins. The advantage of using monofunctional organic mercury ion (R–Hg) as a label is that it can react with a sulfhydryl group to provide a definite mass shift, especially a stable and characteristic non-radioactive isotope distribution. Y. Guo et al. [53] studied the interaction between methylmercury and the -SH of peptides or proteins at room temperature, and provided an alternative strategy to calculate the number of –SH and –S–S– involved. Compared with non-metallic elements, the method using mercury labeling has no spectral interference. However, since mercury has a high ionization energy, only about 20% of the mercury is ionized in the plasma. In addition, mercury is highly toxic, which limits the practical application of mercury in quantitative protein analysis in ICP-MS. Although 96.6% of the proteins contain cysteine residues or disulfide bonds with –SH groups, the reactivity of -SH groups may vary depending on their location and surrounding microstructure [54].

Halogen, as a label for protein quantification, can be conjugated with the organic part of protein. But currently, due to the high ionization energy, only iodine has been developed in the quantitative analysis of proteins. The principle is that a simple reagent (NaI) can be oxidized to I^+ and react with the aromatic ring of amino acid residues. Each tyrosine residue introduces two iodine atoms and each histidine residue introduces one iodine atom. Since iodination may be specific to more than one functional group of a protein, the non-specific iodination of tyrosine and histidine residues must be considered [55]. More complex iodination reagents (such as pyridine) were used to iodide tetrafluoroborate, and the result showed that only tyrosine residues in standard peptides are completely and specifically derivatized [56]. The general disadvantage of iodine labeling is that the sensitivity of iodine in ICP-MS is not as good as that of metals (3-4 orders of magnitude lower than lanthanides) but more sensitive than non-metals naturally present in proteins (such as P or S). In addition, biological samples may contain a significant amount of natural iodine background. Waentig et al. [57] examined the iodination of proteins, the entire proteome, and antibodies to analyze Western Blot membranes by LA-ICP-MS. Iodinated antibodies have been applied for the sensitive determination of transferrin in breast cancer cell lines using a novel immunoassay coupled to ICP-MS detection of iodine. Alonso-García revealed an iodine on the antibody iodination efficiency, the transferrin molar ratio was of 27:1 which corresponds to the iodination of all the tyrosine residues present in the antibody [58].

So far, many commercially available reagents have been widely used in real samples. However, these methods are limited to the derivatization of one or two functional groups with a single element (Hg or I). In contrast, metal-chelating tags show greater versatility because they potentially provide a combination of different metals and functional groups. By using coordination ligands or dual-functional ligands containing detectable heteroatoms, metals detectable by

ICP-MS can be introduced into proteins. Tetraazacyclododecane tetraacetic acid (DOTA) are commonly used ligands. Lanthanide-chelating tags containing reactive groups that allow binding to functional groups on biomolecules can be used to achieve specific lanthanide labeling (bio-binding) of biomolecules, which forms a covalent bond between the biomolecule and the chelating tag. In addition, the lanthanide-chelate label is with very high thermodynamic stability. The use of lanthanide tags for protein labeling has many advantages. Compared with stable isotope reagents, lanthanide tags are cheaper, and there are more lanthanide elements available, which expands the selection range of protein metal labels. And the lanthanide elements have high detection sensitivity, higher ionization efficiency, low background, and large quantitative dynamic range, which make the lanthanide ideal tags for quantitative analysis of protein by ICP-MS. Despite excellent features of lanthanides as labels for ICP-MS detection, this method also has some disadvantages. Firstly, the sensitivity is still limited by the number of metal ions that can be loaded in such chelates. Moreover, relatively high polarity of these derivatives makes their separation in reversed-phase columns difficult in peptide/protein separations. In addition, metal atoms may be exchanged during the pretreatment process. The most important thing is that the quantified peptide must be able to achieve baseline separation. At the same time, other MS methods such as ESI-MS and MALDI-TOF must be used to identify their sequences, and the structure of proteins and peptides must be determined to achieve quantification. Whetstone et al. proposed a lanthanide-DOTA-based element-coded affinity tags (ECAT) method by avoiding the use of stable isotopes as mass labels, and a specific peptide containing a thiol group was labeled through specific reaction with the thiol group [59]. Jakubowski et al. [60] used the bifunctional reagent 2-(4-isothiocyanophenyl)-1,4,7,10-tetraazacyclododecane-1,4,7,10-tetraacetic acid (p-SCN-Bn-DOTA) as label and measured two different proteins (bovine serum albumin and chicken egg white lysozyme). The proteins were labeled with elements such as Eu, Tb, and Ho, and the proteins were separated by sodium dodecyl sulfonate (SDS)-PAGE. After the target protein was transferred to nitrocellulose, LA-ICP-MS was used for protein quantification. The obtained linear range of BSA is 0.015–15 pmol/g, and the detection limit can be as low as 15 fmol/g. Yan et al. [61] established a specific and efficient method for labeling intact proteins, and achieved absolute quantification of proteins with isotope dilution ICP-MS. 1,4,7,10-tetraazacyclododecane-1,4,7-trisaceticacid-10-maleimidoethylacetamide-europium (MMA-DOTA-Eu) was used as the marker to label lysozyme (lysozyme), insulin (insulin) and ribonuclease A (Rnase A). After the labeling conditions were optimized by ESI-MS, it was quantified by isotope dilution ICP-MS. The LODs of the three proteins obtained were 0.819, 1.638, and 0.819 fmol/g, and the recovery rate of this method could reach 97.9%. This method further verifies that ICP-MS binding element labeling can be well applied to the absolute quantitative analysis of proteins. In order to improve the efficiency and specificity of protein metal labeling, Rappel and Schaumloffel [54] optimized the acid anhydride bifunctional reagent (diethylenetriamine pentaacetic anhydride, Lu-DTPA) to label proteins, and peptide reaction steps and reaction buffers were investigated to improve the labeling efficiency. The specificity of the labeling reaction was verified by ESI-MS detection, and the recovery of the labeled peptide was 100% with the precision of 4.9%. The LOD of Lu-DTPA-labeled peptides can reach 179 pmol/g, which are 4 orders of magnitude more sensitive than ICP-MS quantification by detecting sulfur and phosphorous atoms of proteins.

5.3.1.4 Immunological Tagging

The antibody-based immunoassay is another main tool for quantifying protein due to its good sensitivity, selectivity, and throughput, especially the method based on enzyme-linked immunosorbent assay (ELISA). In essence, suitable heteroatom-labeled antibodies can specifically recognize and bind to the target antigen protein or analyte even in complex matrices. However, traditional tags such as colorimetric tags and fluorescent tags are greatly affected by complex biological matrix and poor dynamic range. Because of its good stability and great potential in multiplex analysis, element-tagged immunoassay may be a good choice for immunoassay. In 2001, Zhang et al. [62] firstly published an ICP-MS-based method for detecting proteins with lanthanide-labeled antibodies. The antigen is immunoreacted with a biotin-labeled antibody, which then binds to Eu-labeled streptavidin. Research in the next few years further demonstrated the potential of lanthanide-labeled antibodies for multiple protein analysis using ICP-MS [63]. Due to the success of these experiments, various bioconjugation technologies and lanthanide chelation tags have been developed and verified to improve the sensitivity and multiplexing capabilities of these assays. However, element-tagged immunoassay requires specific antibody design, and the sensitivity of the method of labeling antibodies with lanthanide-chelate complexes is affected by the reduction in the number of detectable heteroatoms of labeled antibodies. Tanner et al. [64] solved this problem by using polymer-based metal tags to label antibodies. These polymer tags bind up to 60–120 lanthanides per antibody and so increase the final ICP-MS sensitivity obviously. Perez et al. [65] labeled specific monoclonal antibodies against four cancer biomarkers (CEA, sErbB2, CA 15.3, and CA 125) with different polymer-based lanthanides, and separated by size exclusion chromatography followed by ICP-MS detection of antigen–antibody complex. The lanthanide loading in the polymer tag may be 30 times that of lanthanide-DOTA to improve the sensitivity and detection limit. This labeling method has a recovery rate ranging from 95% to 110% for all biomarkers studied, and the inter-assay and intra-assay accuracy is less than 8%. However, polymer preparation is complicated, and antibody cognitive ability is liable to be affected by a large number of polymer chains introduced. In fact, the concentrations of some target of analyte in real sample are extremely low, requiring the development of new signal amplification techniques for the highly sensitive detection. A recent alternative strategy was developed by using metal-containing nanoparticles (NPs) as ICP-MS tags. In this strategy, if the exact number of metal atoms per nanoparticle and the stoichiometric number of the nanoparticle per antibody are known, the protein/peptide can be quantified without specific standards. However, the surface of nanoparticle is compatible with a large number of surface sealants and cleaning steps to prevent non-specific adsorption. Yang et al. [66] reported a protocol for the detection of tumor cells by using ICP-MS with lead sulfide nanoparticles (PbS NPs) as the elemental tag. Under the optimized conditions, the linear range of 800–40 000 was obtained, and the relative standard deviation was 5.0%. The proposed method has several advantages, including easy sample preparation, high sensitivity and selectivity. More importantly, this methodology could be extended to the detection of other cells based on their cellular biomarkers. Gold nanoparticle (AuNP) is most widely used due to its high sensitivity, good biocompatibility, and low background in biological samples. Xiao et al. [67] proposed a simple, sensitive, and specific assay for DNA by ICP-MS detection with AuNPs amplification and isothermal circular strand-displacement polymerization reaction (ICSDPR). Under the optimized condition, the proposed method could detect target DNA as low as 45 zmol (8.9 fM in 5 µL) in a relative short time (about 4.5 hours) with good specificity, and the linear range of this method is 0.1–10 000 pmol. Li et al [68] developed a highly sensitive detection method for alpha-fetoprotein (AFP) by immunoassay ICP-MS strategy, which was based on tyramide signal amplification (TSA) and AuNPs labeling. Under the optimized conditions, the LOD of the developed method is 1.85 pg/ml, and the linear range is 0.005–2 ng/ml. The relative standard deviation (RSD) of seven replicates was 5.2%. This strategy is highly sensitive and easy to operate, and can be extended to sensitive detection of other biomolecules in human serum. Ko et al. [69] synthesized three different metal/dye-doped silica nanoparticles (SNPs) as probes for multiple detection of several clinical biomarkers. In the quantification, the doped metal of SNP conjugate was measured by ICP-MS, and the LOD is 0.35–77 ng/ml of deferent biomarkers.

5.3.1.5 Direct Quantification of Proteins by LA-ICP-MS

Laser ablation inductively coupled plasma mass spectrometry (LA-ICP-MS) has already been established as a powerful tool for direct analysis of a wide variety of solid samples in life science [14, 70–73]. In LA-ICP-MS, sample aerosol is generated by LA and introduced into the plasma for vaporization, atomization, and ionization. The elements on the surface of solid samples can be analyzed directly without sample pretreatment. Moreover, LA-ICP-MS can offer bioimaging with high spatial resolution in the low micrometer range, excellent LOD, and the possibility of quantitative data. One of the main drawbacks of quantitative analysis by LA-ICP-MS, however, is sample-related "matrix effects" due to the difference in the interactions (e.g. absorptivity, reflectivity, and thermal conductivity) between the laser beam and different sample matrices. This brings noticeable changes in the mass of analyte ablated per laser pulse in each case. An extensive overview about overcoming such limitations via internal standardization for LA-ICP-MS quantifications has been published in several recent reviews [14, 73, 74].

In separation process of metalloproteins such as chromatography and electrophoresis, the metal may be lost. In order to maintain the metal-protein binding, a non-denaturing gel separation method (natural PAGE) is recommended. The PAGE and LA-ICP-MS detection can be combined with elemental labeling after the tissue is incubated with an antibody that holds an elemental label. Among the external calibration strategies, the most commonly used one is protein standard for electrophoretic separation or hydration of gels and standard solutions for external calibration methods for analysis of selenoproteins, phosphoprotein, and metalloprotein [75]. In order to overcome problems caused by incomplete or changeable ablation and/or ionization behavior of the analytes, the ^{13}C signal is generally used as internal standard [76]. It has also been reported to use isotope dilution method to absolutely quantify protein by GE-LA-ICP-MS. Konz et al. [36] reported for the first time the absolute quantification of natural transferrin (Tf) in human serum samples separated by GE-LA-ICP-MS in combination with species-specific isotope dilution mass spectrometry (SS-IDMS). However, unavailable isotopically enriched spike and metal losses in GE separation are main limitations for SS-isotope dilution GE-LA-ICP-MS. To overcome this problem, Feng et al. [37] reported for the first time to determine Fe transferrin and albumin in human serum by GE-LA-ICP-MS combined with species-unspecific isotope dilution. Compared with alternative quantification methodologies, no calibration curves and species-specific spike are necessary. The use of LA-ICP-MS analysis approach based on such species-unspecific isotope dilution offers important potential to achieve reliable, direct, and simultaneous quantification of proteins after conventional 1D and 2D gel electrophoretic separations. The combination of immunoassay with LA-ICP-MS has also been reported. Gao et al. [2] established a method using antibody-conjugated gold nanoparticles (AuNPs) to quantitatively image β-amyloid peptide (Aβ) in the brains of AD mice by LA-ICP-MS. The Aβ antibody (anti-Aβ) was labeled with AuNPs to form a conjugate AuNPs-Anti-Aβ that immunoreacted with Aβ in mouse brain slices. Homogeneous brain slice matrix matching standards are used as external calibrators for quantitative imaging of Au. In recent years, there has been a lot of rising interest in studying the combined elemental and molecular distribution of elements in biological tissues, while LA-ICP-MS has emerged as one of popular and powerful elemental imaging techniques for heteroatom-containing proteins [77, 78]. However, because the use of LA-ICP-MS to detect biological tissues may have matrix effects during ablation, appropriate calibrations must be designed to compensate or mitigate this possible difference between samples and standards in order to obtain reliable quantitative data [25, 79]. The use of internal standards can help to correct the changes in the signal intensity of ectopic atoms for imaging and analyzing the matrix composition of the sample. Alternatively, an external appropriate internal standard can be used as a dry thin-layer aerosol, or applied by depositing a uniform gold film on the surface of the tissue [77]. In addition, an attempt was made to perform spatially resolved imaging of actual proteins by using exogenous tags for protein labeling and final LA-ICP-MS measurement. Different labeling reagents have been used for this exogenous immunoimaging, including metals (e.g. lanthanides), conventional dual-functional ligands, isotope-enriched polymer tags, and/or nanoparticles. Hutchinson et al [80] first reported the use of LA-ICP-MS for imaging of Eu and Ni coupled antibodies against amyloid precursor protein and Aβ in histological sections of a transgenic mouse model of Alzheimer's disease.

5.3.1.6 Calibration for Metalloprotein Quantification by ICP-MS

One of the attractions of quantification by elemental labeling is the simple calibration using only universal element standards. Conventional external calibration methods using such elemental standards can provide absolute protein quality by directly measuring target heteroatoms. However, this method will be hindered by the occurrence of contamination, because in addition to the target protein, any substance that contains the detected element would contribute to the final heteroatom signal detected by ICP-MS. This is the primary reason why this method is mainly applied to pure or highly purified samples [81]. In most occasions, two or more proteins need to be quantified in the same sample. In this case, if an ICP-MS-based method is to be used, a reliable platform is required to ensure the pre-MS separation of each protein. It can be quantified by adding standards containing heteroatoms to the sample before chromatographic or electrophoresis separation. This method can determine the element response factor (peak area of the injected element per ng) of the instrument, which can be used to quantify the element content of other chromatographic peaks (i.e. certain element-containing proteins in the sample mixture). However, when using online reversed-phase LC-ICP-MS coupling, the influence of the change in the amount of organic solvent introduced during the gradient elution process on the sensitivity of the considered metal must be considered. At each time point of the gradient, the system should be calibrated. In fact, internal standards that show similar ionization behavior in plasma can be used to balance gradient effects. The most elegant solution is to add stable isotopes of the relevant metal after the column. This post-column isotope dilution technique can not only compensate for the influence of the gradient, but also improve the precision and accuracy of the ICP-MS measurement, because only the isotope ratio is determined instead of the absolute intensity. Therefore, the ability of ICP-MS to measure isotopes can combine ICP-MS element detection with isotope dilution procedures to quantify elements with multiple stable isotopes. The method involves adding the isotope-enriched species of known concentration and isotopic composition to the sample, so that the analyte can be quantified by the measuring isotope ratio. The added standard substance can be an isotope-enriched analogue of the target analyte (i.e. species-specific ID species), which can be added before chromatographic analysis to correct for sample loss or matrix effects. Alternatively, non-specific (universal) compounds (i.e. species-nonspecific ID speciation) containing enriched isotopes of the detected heteroatoms can be added online to the chromatographic eluent. Of course, the species-nonspecific ID method cannot correct the sample loss or incomplete recovery. However, it can quantify all the element species containing the target element using only quantitative analysis, because after adding spikes online and measuring the light/heavy isotope ratio, any possible signal changes during the chromatographic gradient can be corrected. Another latest strategy is to combine the internal standard (IS) containing common elements with species non-specific ID quantification. As a result, the complexity of the isotope dilution calculation is reduced and the injection error is compensated. On the one hand, immunoassay quantification of element labeling is usually performed according to a typical immunoassay calibration curve. If the stoichiometric element is labeled and the antibody is known and controlled, no specific standards are needed to achieve quantification of biomolecules.

Certified reference material (CRM) is a very important calibration strategy for protein analysis. In the protein CRM development, although the ICP-MS approach was successfully demonstrated in method development of protein quantification, its application on CRM certification has just been reported recently. In Feng et al. [35] for the first time, isotope dilution high-performance liquid chromatography mass spectrometry (ID-LC-MS) and high-performance liquid chromatography isotope dilution inductively coupled plasma mass spectrometry (HPLC-ID-ICP-MS) strategies were employed to certify the candidate Aβ solution CRMs. These CRMs are primarily intended to be used for value assignment to secondary calibrators or CRMs with a clinical matrix, which will help in early diagnosis of AD.

5.3.1.7 Perspectives of Absolute Quantification of Metalloproteins

We can imagine that the application of ICP-MS in the quantification of metalloproteins will be highly valued in order to obtain new insights into the key role of metalloproteins. The multiplicity of protein determination by ICP-MS is predictable and may constitute a valuable supplementary technique in the near future. In addition, isotope dilution strategy can expand the multiplexing capabilities of ICP-MS with high accuracy and small uncertainty. At present, basic research on the characterization of metal proteome in cells, tissues, and organisms (such as microbial proteome) is also a clear field for ICP-MS-guided proteomics. Therefore, we can imagine multiple quantitative analyses of proteins in the microarray, resorting to bioassays and finally detection by LA-ICP-MS (and quantitative imaging work). By using ICP-MS-certified peptides for absolute quantification of the target protein, perhaps the application of ICP-MS in a wide range of molecular MS proteomics may be closer. From an instrumental point of view, the recently introduced flow cytometer based on ICP-MS (which may allow the analysis of more than 100 different antibodies at the single-cell level) may also provide a direct, innovative solution to the current situation.

5.3.2 Calibration Strategies of Quantitative *In Situ* Analysis in Metrometallomics

5.3.2.1 Internal Standardization

In the quantitative analysis of LA-ICP-MS, the internal standard (IS) could partly correct the measurement deviation caused by matrix effect and elemental fractionation. Variations in sample ablation and transport as well as ICP-related alterations in signal intensity can be partly corrected by using an internal standard. A precondition for the successful application of this approach is that the internal standard element is homogeneously distributed within the sample, and that the behavior of IS and analyte during ablation, transport, and ionization is similar. Optimally, the concentration of the element used as internal standard in the sample is known. In the biological samples analysis, three elements are commonly used as internal standards in LA-ICP-MS quantification. The representative applications by internal standardization are summarized in Table 5.2.

Table 5.2 Representative applications of LA-ICP-MS by internal standardization.

IS element	Sample	Analyte	Analytical performance or comments	References
Carbon	Sweet basil	Cs, S, Ca, K	The R^2 are of 0.991 for K and 0.999 for Cs	[82]
	Dried serum	Al, Be, Ca, Cu, Fe, K, Li, Mg, Mn, Mo, Na, P, Rb, Se, V, Zn	The LODs range from 21 µg/l to 221 mg/l, and the RSDs are below 12% for all analytes	[83]
	Polymers with carbon concentration ranging from 45.620 to 85.543%	Cr, Br, Cd, Hg, Pb	By using fsLA-ICP-MS, R^2 of calibration curves are improved from 0.8067–0.9880 to 0.9556–0.9993 with IS. The determined concentrations agree with the certified value with deviation within 30%	[84]
	Samples of carbon matrix includes organic and inorganic materials	Carbon-containing gaseous species (CCGS) and carbon-containing particle (CCP)	A matrix-dependent partitioning of carbon into CCGS and CCP is observed. And the production of CCGS critically depends on the presence of oxygen	[84]
	PMMA thin films	Mg, Fe, Cu, Zn	^{13}C is generally less effective at compensating for changes in mass ablated, and the sample must generate a ^{13}C signal at least 6% of the total gross signal	[84]
	Leaves	K, Mg, Mn, Cu, P, S, B	The concentrations of measured elements are 20 µg/g for Cu and 14 000 µg/g for K	[84]
	Brain tissue	Fe, Cu, Zn	RSDs of the *in-house* standards are of 6.1%, 7.4%, and 8.2% for ^{56}Fe/^{13}C, ^{63}Cu/^{13}C, and ^{64}Zn/^{13}C, respectively	[84]
Sulfur	Human hair	Hg	With the LOD of 0.2 µg/g	[84]

(Continued)

Table 5.2 (Continued)

Elements	Sample/analyte	Technique	Analytical performance	References
	Human hair	Hg, Pt	Dynamic reaction cell is employed with the LODs of 0.3 and 0.5 µg/g. Standard uncertainties are 10% (Hg) and 15% (Pt);	[85]
Calcium	Human tooth enamel	Cu, Fe, Mg, Sr, Pb, and Zn	Elemental intensities show the following order: Sr > Mg > Zn > Pb > Fe > Cu	[86]
	Tooth and bone	Mn	^{55}Mn/^{43}Ca is a better approach for reporting the content of Mn	[86]
Added IS	Histological tissue	Mg, Fe, Cu	Based on the deposition of a homogeneous thin gold film on the tissue surface and the use of the ^{197}Au signal as IS	[86]
	Biological soft tissues	Cu, Zn	By Rh or Yb in the underlying PMMA film as IS, R^2 is better than 0.999 for Cu and Zn	[86]
	Biological tissues	Pt	By printing an Ir-spiked ink onto the surface, reproducible and homogeneous deposition of the IS is realized	[86]
	Biological tissues	Immunoassays for macromolecules	Using Ir as the IS can correct both for matrix effects and for ablated mass. The RSD is as 5% and LOD is as 1–4 fmol	[86]

Carbon Due to the availability, abundant and apparent homogeneity of carbon in living organisms, carbon is frequently used as an internal standard in many biological studies. Since the signal at m/z 12 is deflected due to the high amounts of carbon in the tissue samples, carbon 13 (^{13}C) is preferentially monitored as IS in the LA-ICP-MS analysis [87–92]. Signal intensities can be normalized by ^{13}C prior to linear regression analysis, and normalization to ^{13}C is an essential procedure to correct for variation in the ablation process. However in some situations, as its ionization potential is significantly higher than those of targeted elements, ^{13}C as an internal standard does not fully reflect the behavior of sample transmission and elemental fractionation. Furthermore, the transport of carbon into the ICP can partly occur in the form of carbon dioxide, which will lead to transport property and efficiency that can obviously differ from the targeted elements that are transported as particulate matter [93]. In the research by Todolí and Mermet [94], there was up to 80% carbon released as gaseous or ultra-fine particulate species during laser ablation, which lead to the fractionation of carbon from different ablation and transport conditions. Frick et al. [93] analyzed twelve common carbon matrices using a gas-exchange device (GED) to investigate the formation of carbon-containing gaseous species (CCGS) during laser ablation. The results showed that a matrix-dependent partitioning of carbon into CCGS and carbon containing particles (CCP) was observed while trace element analytes were exclusively transported as the particulate phase. The production of CCGS was also found to critically depend on the presence or absence of oxygen (matrix or gas impurities) and the affinity of matrix constituents toward oxygen. Also, Austin et al. [95] found out that only if the carbon signal accounts for at least 6% in the total signal can it be used as an internal standard. Another concern is that the signal of carbon in biological matrices appears to decrease with increasing water content [96]. That is to say, carbon may not be a perfect element choice for internal standard for its diffusion loss during the transport procedure [93, 97].

In many cases, both internal standardization by ^{13}C and external calibration based on *in house* prepared matrix-matched standards are employed for calibration. Using ^{13}C for internal normalization, tissue density of each point on tissue section could be corrected. However, since the ^{13}C intensity is closely correlated with tissue density, in some areas with low tissue density or even no tissue at all, the normalized intensities of the targeted analytes could be erroneous. Therefore, although the ^{13}C normalization could partly correct the tissue density and ablated mass in LA-ICP-MS analysis, attention should be specially paid to the inconsistent areas which sometimes make the displayed intensities erroneous [98].

Sulfur Similar to ^{13}C, sulfur is another abundant element in biological samples. As a main element component of amino acid, sulfur is present in most proteins, which makes it possible to be as an internal standard. Melissa [2] chose ^{34}S as internal standard to analyze mercury in single human hair strand by LA-ICP-MS. Since sulfur can be found in hair due to its inclusion in several amino acids that made the signal of sulfur larger than that of ^{13}C in hair sample, it was suitable to choose ^{34}S rather than ^{13}C as the internal standard. The ratio of ^{202}Hg to ^{34}S changed along the sites of the hair strand, and the relative standard deviation (RSD) of ^{34}S signal was found to be 6%. Another work by Stadlbauer [99] investigated the heavy-metal intoxication in single hair by LA coupled to quadruple ICP-MS with dynamic reaction cell (DRC) system. In this research, sulfur was chosen to be the internal standard, and after being normalized by S, the detection limits for Hg and Pt in human hair were found to be 0.3 µg/g and 0.5 ng/g. The results suggested that the accumulation of metals from the bloodstream during growth of the hair was more noticeable rather than uptake of elements secreted by the sweat and/or sebaceous glands.

However, since the first ionization potential of sulfur is 10.36 eV, which make the ionization efficiency in ICP very low. Moreover, the interference of sulfur is another limitation for sulfur detection. As the most abundant isotopes of S are 32 and 34, there can be a lot of interference such as $^{16}O^{16}O$, $^{16}O^{18}O$, ^{31}PH, $^{14}N^{18}O$, and $^{15}N^{16}OH$. To eliminate the interference caused by polyatomic ions, collision gas and high-resolution instrument are needed. The use of reaction gas could shift the monitored mass to an interference free m/z. Stadlbauer [100] used O_2 as reaction gas and $^{32}S^{18}O^+$ as internal standard to correct the variation of the volume ablated. Another approach to avoid the spectral interference is to employ high-resolution ICP-MS instrument, which improves the resolution by distinguishing the tiny m/z difference between targeted isotope and the polyatomic interference ions [101].

Calcium For the hard tissues like bones and teeth, which contains calcium-rich compounds, researchers usually choose Ca as internal standard element for normalization. All the Ca isotopes, apart from the most abundant (^{40}Ca) and the least abundant (^{46}Ca) isotopes, have been reported as an internal standard in LA-ICP-MS analysis. However, due to the interferences of ArH, ^{43}Ca and ^{44}Ca are the mostly used internal standard isotopes.

To investigate spatial distributions of the trace metals in human tooth, Daniel et al. [102] used ^{43}Ca as internal standard to normalize six trace metals (Cu, Fe, Mg, Sr, Pb, and Zn) intensities obtained by LA-ICP-MS. The elemental intensities after normalization in the tooth tissue regions followed the general pattern: Sr > Mg > Zn > Pb > Fe > Cu. Meredith [103] used Ca as an internal standard to investigate four calibration strategies for quantifying Mn content in tooth and bone by LA-ICP-MS. Quantitative results were acquired based on the intensities of ^{55}Mn and the ratio of $^{55}Mn/^{43}Ca$. The results showed that the ratio of $^{55}Mn/^{43}Ca$ represented a better strategy for compensating for the matrix effect and elemental fractionation. However, Manish [104] also claimed that the distribution of Ca varied obviously within the tooth, which should be taken into account when using Ca as internal standard.

Added Internal Standard When the internal standard elements contained in the biological sample itself are difficult to obtain, the internal standard correction of the biological sample can be achieved by adding self-made internal standard elements. Deposition of elemental coatings [79], thin polymeric films containing an internal standard on the sample surface [105], and ink-jet printed elements as internal standard [106] have now interested scientists.

A novel internal standardization procedure for immunoassays was investigated by Frick et al. [108] in which mouse tissues were properly treated, labeled with antibodies, and immersed in a solution of an Ir compound. Standards of cellulose acetate doped with rare-earth elements (REEs), Rh, Ir, and Pt were prepared and the respective signal intensities were normalized to Ir (as IS). The authors cited that the method was fast and readily applicable, and standardization with iridium was an alternative to traditional iodine staining. Christine [105] proposed a method for quantitative analysis of biological soft tissues by LA-ICP-MS using metal-spiked polymethylmethacrylate (PMMA) films. Polymer film standards were produced by spin coating spiked solutions of PMMA onto quartz substrates, and calibration curves throughout the range of 0–400 μg/g yielded correlation coefficients better than 0.999 for ^{66}Zn and ^{63}Cu. Thin gold layers as IS and patterns printed with commercially available ink-jet printers as standards were proposed in recent years [109]. In this ink-jet printed approach, ink was spiked with different elements at known concentration and was sprayed onto the top of the analytical sample uniformly by commercial printer. By normalization of the ratio between spiked elements and the targeted analytes, the deviation caused by laser ablation efficiency and instrument drift could be partly corrected. By printing pattern on paper with gold layer as internal standard, the concentration of platinum on malignant mesothelioma samples was obtained which improved the reliability of LA-ICP-MS quantification result. In this way, it is possible to spike any element according to the targeted analytes with a simple spiking procedure. Hoesl [110] used ink spiked with different elements to print onto the top of sample as internal standard and calibrants in LA-ICP-MS quantification. Spiked indium in the ink was selected as an internal standard and multiple lanthanides in the ink were used for calibration series. With this approach, the lanthanides tagged with the proteins in the electroblotting sample can be quantified by LA-ICP-MS. Moreover, the quantitative results were validated by conventional-solution-based ICP-MS after sample digestion. The RSD obtained by this method improved obviously, and the homogeneity was as low as 5%. The limit of detection (LOD) is approximately 1–4 fmol, which showed great potential for protein quantification by LA-ICP-MS. By employing the same printing approach and in combination with IHC staining, quantitative multiplex assay of proteins in archived FFPE sample was also realized [97], which broadened the quantification application of LA-ICP-MS (Figure 5.3). By printing an indium-doped ink as internal standard on gelatin, Boris et al. [111] applied a new approach to realize quantitative imaging of brain tissue of parkinsonian mouse. Different ablation spot sizes with respect to resolution and signal-to-noise ratio were evaluated, and the results proved that the ink-jet printing

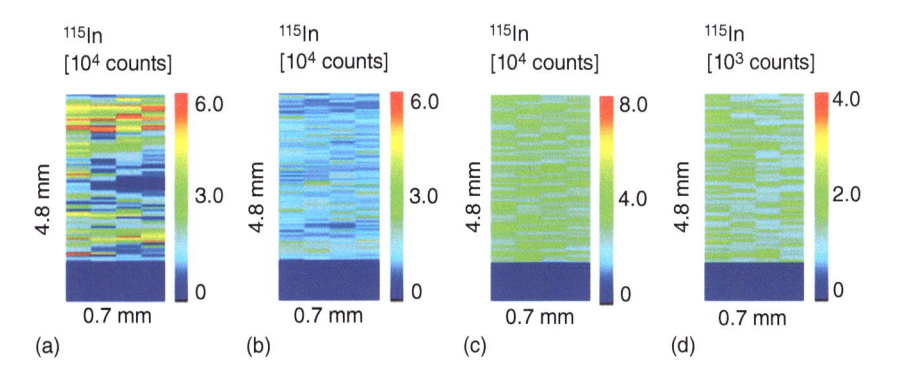

Figure 5.3 The effect of gelatin-coating on ablation of the internal standard ^{115}In. Representative intensity images zoomed for illustration of ^{115}In-spiked glass slides non-coated (a) and coated with 2% gelatin (b) or 5% gelatin (c) and ablated with a laser spot size of 130 μm (a–c) or 35 μm (d). The internal standard ^{115}In was applied using a commercial inkjet printer. Source: From Ref. [107], with permission of the Royal Society of Chemistry.

approach can be an efficient way to overcome the instrumental drifts. In all, the ink-jet printed internal standardization strategy offers a broad range of applications of LA-ICP-MS quantitative analysis for biological samples [111].

5.3.2.2 External Calibration

External calibration, where a set of known standards are used to construct multi-point calibration curves from which linear regression analysis can be performed, is the preferable approach when quantitative analysis is required [112]. External standard method based on CRMs is the most commonly used calibration strategy in chemical analysis. CRMs in chemical analysis could be used for establishing standard curves and to validate the accuracy and trueness of method. The external standardization is based on the principle of matrix matching, which calls for the similar composition between the calibration standard and the sample [82, 83]. In LA-ICP-MS analysis, since the presence of matrix effect is throughout the procedure of sample ablation, transportation, atomization, and ionization, the standards used as external calibrants must match the sample matrix and not be affected by element fraction. Therefore, matrix-matched CRMs are the most ideal external calibrants for quantitative analysis by LA-ICP-MS. However, in nowadays, since the complex and heterogeneous component of biological samples is available, the CRMs applied in LA-ICP-MS are mostly geological materials and there is a large lack of commercial biological CRMs dedicated to LA-ICP-MS analysis. As a result, the most frequently external calibration strategy still relies on *in-house* calibration standards prepared by the technique of pressing pellet or spiking similar matrix material with known amounts of the analytes for biological samples [88, 113].

In the preparation of matrix-matching *in-house* calibrants, similar composition and physicochemical properties materials are needed to match the analyte concentration and sample matrix. Various standards are prepared according to the sample type and analyte concentration [84, 107, 114]. To imitate the composition of soft tissue, polymer and gelatin are the most commonly used standard materials [84, 115–117]. The concentrations of the spiked elements are in accordance with the analyte content in the sample. More importantly, the distribution of the spiked analytes of the *in-house* calibrants should be homogeneous, and the calibrants should have good stability. To determine the analyte concentration in the standards, replicates of the prepared standards are submitted to bulk analysis with acid digestion and for further analyte determination using solution-based ICP-MS [118]. It is worth noting that the procedure is not straightforward if the starting material is a native tissue, which needs to be properly preserved and doped with targeted analytes before grinding. Therefore, despite the significant number of applications and some detailed protocols for the *in-house* standards, the complexity and skills needed to produce them make such a calibration approach not well established yet. The representative applications by external calibration are summarized in Table 5.3.

Calibration by Commercial CRMs Signal quantification of solid samples could be accomplished by using CRMs with biological matrix. Presently, few commercial CRMs used for biological samples analysis by LA-ICP-MS exist. Most of the CRMs

Table 5.3 Representative applications of LA-ICP-MS by external calibration.

Calibration approach	Sample	Analyte	Analytical performance or comments	References
Commercial CRMs	Wine	Li, B, Mg, Al, V, Mn, Co, Cu, Zn, Rb, Sr, Y, Cs, Ba, La, Ce, Pr, Nd, Pb, U	SRM 610 glass or aqueous standard solutions are employed, and the LODs range from 0.01 to 1 ng/ml	[83]
	Vanilla	Ca, Mg, P, Fe, Zn, Mn, Sr, Br, Rb, Ba, Cu	CRMs of NIST SRM 1549, SRM 1575a, SRM 1515, SRM 1547, SRM 1570a are employed	[84]
	Rat brain	Cu, Zn, Fe	CRMs of TORT-2, DOLT-2, and DORM-2 are employed, and the R^2 are 0.997 (Zn), 0.997 (Cu), and 0.982 (Fe), respectively	[84]
	Coal fly ash	18 elements	CRM of NIST SRM 612 is employed	[84]
In-house prepared standard	Oral mucosa	Ti, Al, V	Two calibration approaches based on matrix matched solid standards with analytes addition are employed with the LOD of 4.8, 0.84, 0.58 µg/g for Ti, Al, and V	[84]
	Mouse tissue	Tm	Matrix-matched standards based on egg yolk are employed. The LOD and LOQ are 2.2 and 7.4 ng/g. Linearity over the concentration ranges from 0.01 to 46 µg/g	[84]
	Rat brain	Zn, Mg	Standard addition method is used for *in-house* calibrant preparation. The R^2 for Zn and Mg are 0.944 and 0.989	[84]
	Brain tissue	Co, Cu, Fe, Mg, Mn, Sr, Se, Zn	Spiked sheep brain is homogenized for calibration. The R^2 ranges from 0.9874 (Mg) to 0.9991 (Sr)	[86]

(Continued)

Table 5.3 (Continued)

Calibration approach	Sample	Analyte	Analytical performance or comments	References
	Biological tissue	Mn, Ni, Cu, Zn, etc	Standard addition via dried pL-droplet is employed, which allows at least 6 orders of magnitude with LOD of a few fg;	[84]
	Liver tissue	Cu, Zn	Both matrix-matched with spiked homogenates and gelatin droplet are employed, and no differences are observed	[84]
	Histological tissue	U	Element solution is spiked into the polymer that generates a soft matrix-matched standard	[84]
	Eye tissue	MT1/2 protein via AuNCs	5 µm Au-spiked gelatin ranges from 0 to 60 µg Au/ml as the external standard;	[84]
	Brain tissue	FPN and Fe	*In-house* prepared gelatin standards containing Fe and Au are used for calibration;	[84]
	Cell	AuNPs	Printed droplets as a calibration method. The LOQ is 1.7 fg for Au;	[84]
Online addition standard	Mouse brain	Li, Mn, Fe, Cu, Zn, Rb, etc	Aqueous and synthetic laboratory standards are employed and compared. And sensitivity differences are investigated;	[84]
	Hair	26 elements	The LODs of the elements were from µg/g to ng/g; Differences in aerosol generation and transport efficiencies between solution nebulization and LA are assessed	[84]
	NIST SRM 610 and 612	Pb, Rb	This method enables SI-traceability without matrix-matched CRMs	[84]

come as a powder or lyophilized powder, which can be seen as inconvenient in many cases for LA-ICP-MS. Through calibration by pressed pellets of CRMS, the quantitative results of LA-ICP-MS could be directly traceable to SI unit. The approaches most frequently applied for the preparation of compact samples from powders include milling/grinding/sieving for sample homogenization, combined with pelletization [118–124], fusion to sample disks [125–128], or mounting/embedding of the sample in a polymeric resin [129, 130].

The first application of CRMs pellets as external calibrant was conducted by Jackson et al. [131] In this study, CRMs of TORT-2, DOLT-2, and DORM-2 were pressed into pellets for calibration. The representative calibration curves reached a linear correlation coefficient to 0.997 (^{66}Zn), 0.997 (^{63}Cu), and 0.982 (^{56}Fe), respectively. This application was one of the earliest approaches to quantify metals in thin sections of rat brain by using pressed pellets of lyophilized marine organisms CRMs, which correlated well with the results of solution-based bulk analysis method. Considering the similar construction and physical properties, this calibration strategy can be useful especially for hard samples like bones and teeth rather than biological soft tissue. However, the complex sample pretreatment like milling and grinding may cause the analyte dilution and decrease the detection power of the analysis approach [132].

Hondrogiannis et al. [124] used LA-ICP-TOF-MS to successfully classify 25 vanilla samples according to their origin. Vanilla powder was directly pressed into a sample pellet, and external calibrants were prepared by NIST SRM 1549 (non-fat milk powder), NIST SRM 1575a (trace elements in pine needles), NIST SRM 1515 (apple leaves), NIST SRM 1547 (peach leaves), and NIST SRM 1570a (trace elements in spinach leaves). And this method was validated by using NIST SRM 1573a (tomato leaves). Eze et al. [133] used LA-ICP-MS to investigate the composition of coal fly ash. Fusion disks of each sample were prepared according to an automatic gas fusion procedure. Quantification of 18 elements was achieved via external calibration by NIST SRM 612 and using ^{29}Si as an internal standard, and USGS BCR-2 or BHVO 2G CRMs were used for method validation. Further applications include the analysis of Sahara dust samples, desert varnish [134], soil samples [135], biomass ashes [125], fly ash samples [126], ash-related deposits [136], coral skeletons [137], and forensic investigations [121].

Calibration by In-house Prepared Standard If no suitable commercial CRMs are available, or the range of analytes cannot be covered, the preparation of *in-house* standards is another possibility for LA-ICP-MS quantification. In this approach, the ablation behavior of the *in-house* prepared standard and sample are assumed as identical for their similar constitution. The procedures of this approach are usually as follows: First, the biological materials (such as brain tissue, liver, heart) are homogenized. Then the homogenate portions are transferred to the vial, and elemental standard solutions with accurate concentration are spiked into the biological sample homogenates. After being frozen and cryo-cut, the homogenate is coated to the microscope slides and stored at $-20\,°C$ before analysis. Concentrations of the analytes of the *in-house* standard are certified by other solution-based bulk analysis methods, and homogeneity of the standard should be guaranteed.

Reifschneider et al. [138] proposed egg yolk as matrix matching for Tm imaging in mouse tissues. Egg yolk spiked with Tm was heated up to 90 °C for 10 minutes to form a solid, which was sectioned by cryo-cutting, fixed on a glass slide, and then ablated. By calibration with such standards, quantitative and spatially resolved data were obtained for tumor cells and macrophages, which complemented the information given by magnetic resonance imaging (MRI). In another study, matrix-matched standards were developed for calibration and quantitative imaging of Zn and Mg in brain tissue analyzed by LA-ICP-TOF-MS [139]. The brain tissue homogenate was mixed with nonacid standard solution, followed by mixing and immediately freezing with liquid nitrogen, which guaranteed the homogeneous distribution of analytes and the linearity of calibration curves. Internal standardization with ^{13}C was also carried out to compensate differences of ablated mass. Hare et al. [140] published a general guide for producing matrix-match standards for the analysis of trace metals in brain tissues. In this method, they spiked the sheep brains with 8 elements and measured the analyte amount by conventional nebulization ICP-MS. Jennifer [141] reported a straightforward standard procedure for matrix-matched calibration to quantify the Fe concentration in brain tissue. In this method, homogenized sheep brain was merged into the corresponding solution of Fe (0.5–20 mg/kg) in methanol. The Fe concentration of the standards were determined using microwave digestion ICP-MS. Assessment of the accuracy of the method for the quantitative imaging of Fe in tissues was undertaken by comparison of the LA-ICP-MS data with that obtained by micro-X-Ray Fluorescence (µ-XRF) approach.

Because of the time-consuming and complicated preparation procedures of the homogenate calibration approach, dried-droplet is a calibration approach with increasing interest, which represents an easy-to-handle alternative to typically *in-house* matrix-matched tissue standards [142]. This method has the advantage of simplicity, easy handling, and a wide analyte concentration range. This calibration strategy was applied for Mn, Ni, Cu, and Zn determination in human malignant mesothelioma biopsy [143], Pt imaging in human malignant pleural mesothelioma [144], several elements in mouse kidney [145], and Cd, Co, Pb, and Cu determination in blood [146]. Marta et al. [147] compared two calibration approaches (matrix-matched with spiked homogenates and spiked gelatin droplet standard) for the quantitative Cu bioimaging of liver cryo-sections, which showed no statistical differences between these two calibration strategies. Thus, considering the simplicity and availability of the material, spiked gelatin droplet might be a preferred choice in some occasions. However, the concurrent problem of non-uniform distribution of analyte onto the support surface, also called the "chromatographic" effect, does exist and the liquid volume deposited on the surface must be evaluated.

In another *in-house* calibration strategy, especially for quantitative bioimaging, standards are prepared by the application of polymers, such as gelatin and resin Technovit [148]. The main idea of this concept is to spike element solution standards into the polymer material that generate a soft matrix-matched standard. Homogeneous thickness and element distribution are of vital importance in standard preparation, and the element concentrations of spiked standard should be accurately measured by conventional bulk analysis [149]. To quantify the Au concentration in human retina tissue, María et al [150] prepared 5 μm (the same thickness to the eye tissue section) Au-spiked gelatin range from 0 to 60 μg Au/ml as the external standard. The accurate concentration of the standard was determined by conventional nebulization after digestion. By this calibration method, the calibration correlation coefficient was obtained to be 0.999, and the MT1/2 concentration of different donors can be quantified, respectively. Recently, Mika et al. [116] presented a novel approach for the preparation of gelatin standards using both commercial and laboratory-made molds. The mold-prepared standard showed an excellent thickness consistency and signal precision for robust quantification. And the comparison of mold-prepared standard and normal animal sources demonstrated a satisfactory LOD. The dynamic calibration range was further improved for gelatin standards by employing an additional metal extraction step during standard preparation with various resins. Cruz-Alonso et al. [85] realized simultaneous quantitative imaging of iron and ferroportin in hippocampus of human brain tissues with Alzheimer's disease, and the *in-house* prepared gelatin standards containing Fe and Au were used for LA-ICP-MS calibration (Figure 5.4).

In light of the common use of gelatin and Technovit standards for elemental bioimaging by LA-ICP-MS, the matrix and fractionation effects mainly happened in the duration of ablation; transportation, vaporization, atomization, and ionization were also investigated. To investigate the aerosol characteristics, Rebecca et al. [151] used an optical particle counter inserted in-line between the LA system and ICP-MS to evaluate the aerosol characteristics (such as particle counts and size) after ablation. In the process of transportation, they varied the transportation tube length to study the differences between gelatin aerosol and wet aerosol particles. Two standard materials, including gelatin standard and Technovit standard, were used in this experiment. The results can be concluded as: (i) The size of the particles produced by the gelatin standard was smaller than Technovit standard which caused higher signal intensity in ICP-MS; and the two standard materials produced a large number of μm-level particles with the increase of laser energy; (ii) Compared to wet aerosols, the transportation and ionization efficiency of Technovit standards was low; however, the ionization efficiency of gelatin aerosols was similar to that of the wet aerosols, which indicated a higher vaporization efficiency of the gelatin particles; (iii) In the process of ablation, due to the preferential evaporation of elements with higher temperature, the isotope ratio of targeted analyte in gelatin standard was different with that in its corresponding solution sample.

Arakawa et al. [152] proposed a quantitative imaging approach of silver nanoparticles and essential elements in thin sections of fibroblast multicellular spheroids. They designed matrix-matched calibration standards for this purpose and printed them using a noncontact piezo-driven array spotter with a AgNP suspension and multielement standards. The LODs for Ag, Mg, P, K, Mn, Fe, Co, Cu, and Zn were at the femtogram level, which is sufficient to investigate intrinsic minerals in thin sections. In the application on single-cell analysis by LA-ICP-MS, one of the

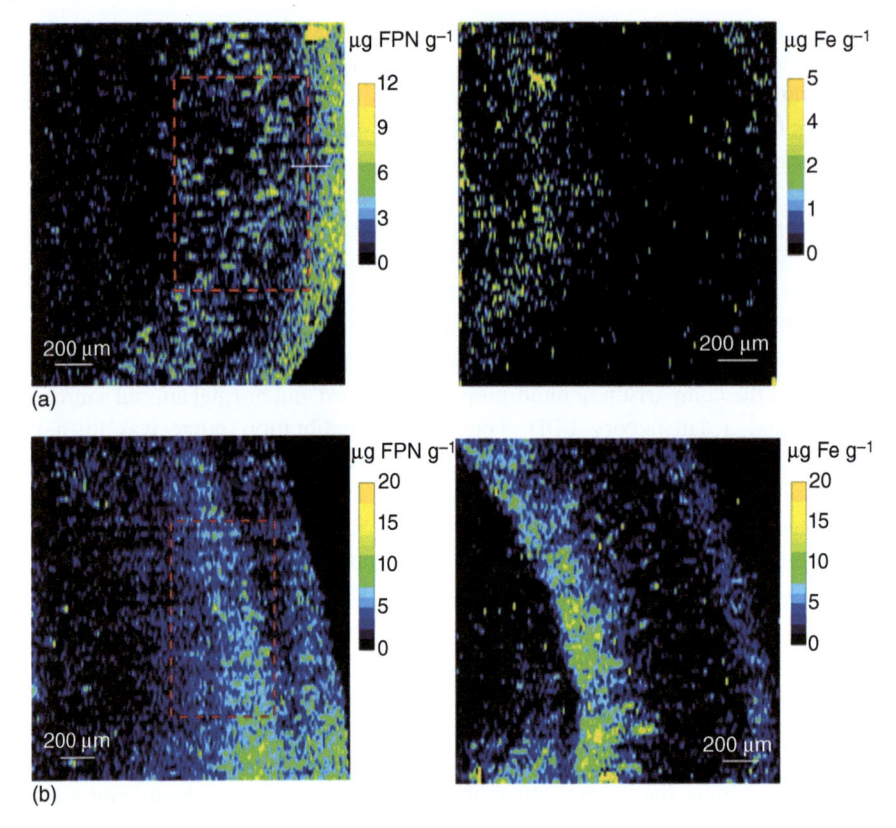

Figure 5.4 Quantitative images for FPN and Fe distribution (expressed as µg/g) obtained by LA-ICP-MS from the stratum pyramidale of hippocampus CA1 region of human brain tissues after IHC with AuNC bioconjugate for FPN. (a) HC human brain, and (b) AD human brain. Ref. [85], with permission of the ELSEVIER.

challenges for LA-ICP-MS is the lack of matrix-matched standards whose sizes are similar to single cells. Since inkjet printing can accurately and reproducibly eject picoliter volumes of liquid from micro-sized apertures onto a substrate in a defined pattern, Wang et al. [153] described the use of individual printed droplets as a calibration method for the quantification of AuNPs in single cells. Experimental results showed that the ink-jet printing standards simulated the ideal matrix-matched standards and offer a viable route for calibration.

In general, using the *in-house* prepared standard is a practical strategy to realize quantitative analysis by LA-ICP-MS. But there are still some key points to be addressed. Firstly, how to assure satisfactory homogeneity of the gelatin films or sections? Additionally, the similarity between *in-house* calibrants and samples should be good enough to compensate the fractionation effects, which can lead to essentially equivalent and comparable ion intensity signals. Therefore, except for the instrument parameters, researchers should consider these questions before employing the *in-house* standard calibration strategy.

Calibration by Online Addition Standard Limited by matrix-matched CRMs and time-consuming sample preparation procedures, another easy and rapid quantification procedure is solution-based calibration, whereby dual-gas and single-gas sample introduction systems have been proposed. The calibration strategy of online addition was firstly established in 1989 [154]. In this method, the laser ablation and spray chamber were combined with a "Y" or "T" connector, and the wet aerosols were mixed on-line with the dry aerosols coming from the ablation chamber in the injector tube of the ICP torch.

Pozebon et al. [155] developed a solution-based calibration strategy for imaging the elements in mouse brain. Calibration curves of elements (Li, Na, Al, K, Ca, Ti, V, Mn, Ni, Co, Cr, Cu, Zn, As, Se, Rb, Sr, Y, Cd, Ba, La, Ce, Nd, Gd, Hg, Pb, Bi, and U) were obtained using (i) aqueous standards or (ii) the set of synthetic laboratory standards prepared from a mouse brain homogenate doped with elements with known concentrations. The ratio of the slope of the calibration curves (obtained by using aqueous standards and solid standards) was applied to correct the differences of sensitivity among ICP-MS and LA-ICP-MS. Quantitative imaging of Li, Mn, Fe, Cu, Zn, and Rb in mouse brain was obtained under wet plasma condition (nebulization of diluted HNO_3 solution in parallel with ablation of solid brain sample). This calibration, by using pneumatic nebulization of standard solution and aerosol desolvation combined with laser ablation of brain tissue, demonstrated possibility for trace and major elements imaging in brain. Moreover, once the difference in sensitivity between ICP-MS and LA-ICP-MS was known for each investigated element in brain tissue, solution-based calibration can be used for elements imaging in tissue sections. Dressler et al [156] pursued calibration of LA-ICP-MS measurements of mouse and human hair by simultaneous aspiration of multi-element solutions (at several concentration levels) via a conventional nebulizer. The LODs of the elements were from µg/g to ng/g, and even 26 elements can be detected at the same time which extended the use of hair analysis. Differences in aerosol generation and transport efficiencies between solution nebulization and LA were assessed by ablating *in-house* hair material with known analyte element concentrations. The *in-house* hair standard was prepared by immersion of hair strands in a multi-solution, subsequent drying, and digestion thus obtained for the determination of reference concentrations via conventional solution nebulized ICP-MS. In combination with standard addition, Lena et al. [157] established a new SI-traceable method for LA-ICP-MS quantification. Based on simultaneous introduction of ablated solid sample and aerosols of standard solution with the principle of standard addition, the volume or mass flow need not be determined in this method. After mathematical derivation, the role of the unknown mass flows can be represented by the regression parameters and the mass fraction of the reference element which meant that the sample can act as a perfect matrix-matching standard. It's the first time that solid sample acts as the reference material, which meant there was no need to use CRMs or prepare matrix-matching standard. But there were still some drawbacks: (i) mass fraction should be known; (ii) the procedure of references and preparation of analyte element solutions was time consuming.

In summary, although the online addition calibration strategy has several advantages, especially if no suitable reference materials are available or the preparation of synthetic laboratory standards is difficult or impossible, it cannot totally compensate the matrix effect and elemental fractionation caused from sample ablation, aerosols transportation, and ionization process. And moreover, for this technique, the ablated sample mass should always be quantified accurately in calculating concentration of the analyte, which is usually difficult in practice.

5.3.2.3 Calibration by Isotope Dilution

Isotope dilution mass spectrometry (IDMS) is one of the accepted approaches capable of achieving analytical results that are less affected by signal drifts, matrix effects, or analyte losses, and the measuring results could be directly traceable to SI unit [86, 158]. The combination of isotope dilution with LA-ICP-MS can not only realize quantitative imaging but also eliminate some common fractionation and matrix effects that cannot be addressed using other calibration procedures [34, 159]. Up to now, for the quantification by using ID-LA-ICP-MS method, four calibration strategies have been proposed. The representative applications by isotope dilution are summarized in Table 5.4.

The first strategy is the direct analysis after solid spiking [34, 159]. In this arrangement, the isotope-diluted powder is prepared by using time-consuming procedures in which they are spiked with enriched isotope standards, dried, homogenized, and pressed into form of a pellet. Although the experimental procedures are laborious, the IDMS arrangement do obviously improve the results accuracy and uncertainty. However, because no mixing, pressing, and homogenizing process could be introduced in sample preparation, the merits of this ID-LA-ICP-MS approach are not applicable to native biological section quantitative imaging.

The second strategy is the on-line isotope dilution LA-ICP-MS equipped with gas sample introduction system. Feng et al. [160] determined the Pb concentration in NIST SRM 610 sample by on-line ID-ICP-MS layout, and the results were in good agreement with the certified value. Moreover, to improve the mixing efficiency and achieve a good isotopic equilibrium, four different mixing devices were designed and evaluated. The signal sensitivity, trueness, and precision of the blended ^{208}Pb/^{207}Pb ratio of each device in different connection modes (pre- or post-cell introduction of a ^{207}Pb spike) were also investigated. Another on-line ID-LA-ICP-MS was applied for quantitative Fe mapping in sheep brain tissues [161]. A solution of enriched ^{57}Fe was added to the dry aerosol before entering the ICP, and high-resolution ICP-MS instrument operated at medium mass resolution was used to resolve polyatomic interference on ^{56}Fe and ^{57}Fe. Recoveries of 80–109% of the expected Fe concentration in the model tissue were obtained, and the overall combined expanded uncertainty was 15~27%. For this on-line isotope strategy, still two key points need to be further investigated. The first point is whether the isotopic equilibrium is reached after uncompleted mixing process of the aerosols. The other is how to quantify the mass of spike solution and ablated sample in the IDMS formula. For the mass calculation of sample and spike, the calculation of the actual mass flow rate of the ablated sample and a total consumption of spike nebulizer are mainly employed for mass

Table 5.4 Representative applications of LA-ICP-MS by isotope dilution.

Calibration approach	Sample	Analyte	Analytical performance or comments	References
Solid spiking	Soils, sediment	Cu, Zn, Sn, Pb	Precisions are lower than 10% RSD;	[83]
	Coal	Cl, S, Hg, Pb, Cd, U, Br, Cr, Cu, Fe, Zn	LODs are 450 ng/g, 18 ng/g, 9.5 pg/g and 0.3 pg/g for Cl, S, Hg, and U, respectively;	[84]
Online spike introduction	Glass	Pb	Good precision (1.5–2.5%) of ^{208}Pb/^{207}Pb ratio is achieved;	[84]
	Sheep brain	Fe	Recoveries of 80% to 109% are obtained, and the combined expanded uncertainty ($k=2$) is of 15–27%	[84]
Isotope exchange before analysis	Mouse brain	Fe, Cu, Zn	A "border" is constructed to make spike droplet stay on the tissue for isotope exchange and equilibrium	[84]
	Human serum	Transferrin	Species-specific GE-LA-ICP-IDMS method is developed for protein analysis, and the RSD is in the range of 0.9–2.7%	[84]
	Human serum	Transferrin	Species-unspecific GE-LA-ICP-IDMS method is developed for protein analysis with small uncertainty (1.5–3%)	[84]
Ink-jet printing spike	Kidney	Pt	The LOD is 50 pg with recovery higher than 90%	[84]
	Single cell	AgNP	The LOD is 0.2 fg Ag per cell	[84]
	Polymer	Pb	The expanded uncertainties for single- and double-isotope dilution are 11% and 8%	[84]

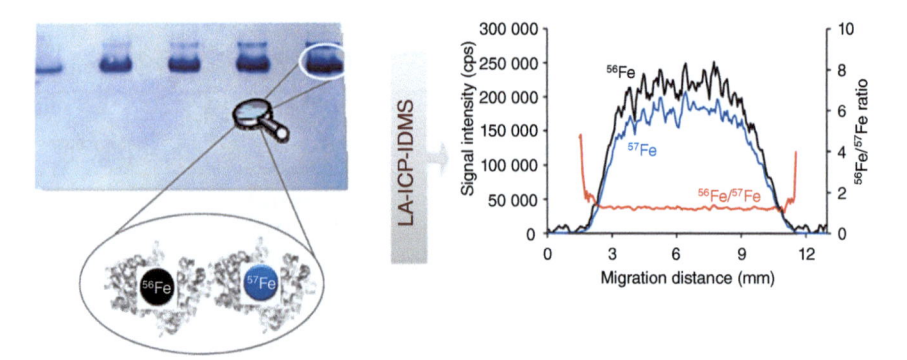

Figure 5.5 Absolute quantification of Tf separated by nondenaturing GE using LA-ICP-MS in combination with species-specific IDMS strategy. Profiles obtained by LA-ICP-MS for ^{56}Fe, ^{57}Fe, and their isotopic ratio using a mixture of the CRM and the isotopically enriched ^{57}Fe-Tf. Source: Konz et al. [36]/Reproduced with permission from Springer Nature.

confirmation. However, the transport efficiency of nebulization needs to be clearly confirmed and sometimes unstable in practice.

The third strategy is to achieve isotope-exchange treatment first and then isotope ratio measurement. Ioana [36] reported a species-specific quantification of metalloprotein separated by nondenaturing gel electrophoresis (GE) using LA-ICP-MS combined with IDMS for the first time. ^{57}Fe was selected as analytical element to quantify the transferrin (Figure 5.5). The natural and isotope-enriched process were conducted by using freshly synthesized Fe-citrate solution. To demonstrate the accuracy, ERM-DA470k was used for method validation and the results showed good arrangement with the certified value. Another species-unspecific ID-LA-ICP-MS approach was proposed as an absolute quantitation strategy for transferrin and albumin in human serum [37]. In order to achieve homogeneous distribution of both protein and isotope-enriched spike, immersing the protein strips with ^{34}S spike solution after gel electrophoresis was demonstrated to be an effective way of spike addition. Furthermore, effects of immersion time and ^{34}S spike concentration were fully investigated to obtain optimal conditions of the post-electrophoresis isotope dilution method. The relative mass of spike and ablated sample (m_{sp}/m_{sam}) in IDMS equation was calculated by standard Tf and Alb proteins. The results were in agreement with the certified value with good precision and small uncertainty (1.5–3%). In this method, species-specific spike protein is not necessary and the integrity of the heteroatom-protein could be maintained in sample preparation process. To realize quantitative imaging of biological section sample by isotope dilution LA-ICP-MS, Feng et al. [37] published a method that constructed a "border" to make spike droplet (containing ^{54}Fe ^{65}Cu ^{67}Zn) stay on the tissue for isotope exchange and equilibrium thoroughly. The prepared homogeneous *in-house* standard was used to validate the approach and good agreement with the bulk analysis was achieved. On this basis, quantitative imaging of Fe, Cu, and Zn in real mouse brain of Alzheimer's disease (AD) was first realized by the improved methodology. Assessment of the method for real sample was undertaken by comparison of the LA-ICP-MS image with that obtained by μ-XRF and immunohistochemistry approaches.

The fourth strategy is by printing the enriched spike with ink-jet printer [162]. Moraleja [162] printed the isotope-enriched inks onto kidney slices treated with Pt-based drugs by a commercial ink-jet device. ^{194}Pt was selected as enriched isotope and this approach was validated by the deposition of natural Pt standard droplets with known amount of Pt. Images from different scanning area also showed the suitability of this method. With a new approach termed "single-cell isotope dilution analysis" (SCIDA), Zheng et al [163] obtained an accurate quantification of AgNP in single cells and the detection limit reached to 0.2 fg Ag per cell. In this approach, the single cell was placed in an array and transferred to a substrate by a microfluidic technique; each cell in the array was precisely dispensed with a known picoliter droplet of an enriched spike solution with a commercial inkjet printer; the single cells and printing droplets were simultaneously ablated and determined by ID-LA–ICP–MS.

Compared to other calibration strategies, the quantitative results by using isotope dilution LA-ICP-MS could be directly traceable to SI unit, with small uncertainty and sometimes no laborious sample pretreatment. However, how to homogeneously distribute enriched isotope spike on tissue section and how to confirm isotope equilibration between sample and spike are two important challenges. Since complete equilibration between isotopes of elements in biological molecule and spiked isotopes with different state is not easy to be achieved, the discrimination of aerosols with different state should also be taken into account in many situations.

5.3.2.4 Perspectives of Quantitative In Situ Analysis in Metrometallomics

In this part, we review the calibration strategies of quantitative analysis by LA-ICP-MS in biological applications. Although the calibration strategies for LA-ICP-MS have made remarkable achievements in recent years, hesitation still exists for its universal applicability. Because of the complicated components of biological samples, it is important to note that there is no perfect element for internal standard calibration and even external calibration. Moreover, due to the lack of commercial CRMs, fully matrix-matching sample preparation procedures, and professional data process software, the quantitative applications of LA-ICP-MS in biological analysis are still limited. Considering the defects of calibration methods, developing new calibration strategies is still highly challenging for the research of LA-ICP-MS.

Recently, another new approach has been employed for mass spectrometry imaging. To help with the statistics and deal with the large data generated by desorption electrospray ionization (DESI) mass spectrometry, Zeper Abliz [164] reported a virtual calibration (VC) strategy for quantitative mass imaging based on machine learning algorithms. In this strategy, the endogenous metabolite ions acquired in the mass spectrum of each organ served as potential natural IS candidates. Among these, analyte-response-related metabolite ions were screened out and utilized as input features for the regression model of inter-region analyte response variation. Then, the analyte ions were corrected with virtual calibration factors predicted by this regression model with the aid of machine learning technique. Although this VC strategy was proposed based on DESI-MS analysis and its traceability was still ambiguous, this computational approach did realize quantitative imaging of many analytes with a small number of standards, which may have the potential for quantitative application of LA-ICP-MS calibration.

Acknowledgments

This work was financially supported by the National Natural Science Foundation (Nos. 12075230; 11475163) and the National Key Research and Development Program (No. 2017YFF0205402).

References

1 Pan, M., Zang, Y., Zhou, X. et al. (2021). *At. Spectrosc.* 42: 262.
2 Gao, X., Pan, H., Han, Y. et al. (2021). *Anal. Chim. Acta* 1148: 238197.
3 Aebersold, R. and Mann, M. (2003). *Nature* 422: 198.
4 Lambert, J.P., Ethier, M., Smith, J.C., and Figeys, D. (2005). *Anal. Chem.* 77: 3771.
5 Calderon-Cells, F. and Encinar, J.R. (2019). *J. Proteomics* 198: 11.
6 Tholey, A. and Schaumloffel, D. (2010). *Trends Anal. Chem.* 29: 399.
7 Cid-Barrio, L., Calderon-Celis, F., Abasolo-Linares, P. et al. (2018). *Trends Anal. Chem.* 104: 148.
8 Amais, R.S., Donati, G.L., and Arruda, M.A.Z. (2020). *Trends Anal. Chem.* 133: 116094.
9 Liu, Z., Li, X., Xiao, G. et al. (2017). *Trends Anal. Chem.* 93: 78.
10 Bettmer, J., Bayon, M.M., Encinar, J.R. et al. (2009). *J. Proteomics* 72: 989.
11 Feng, L., Zhang, D., Wang, J. et al. (2014). 6: 7655.
12 Acosta, M., Torres, S., Marino-Repizo, L. et al. (2018). *J. Pharm. Biomed. Anal.* 158: 209.
13 Trinta, V.D., Padilha, P.D., Petronilho, S. et al. (2020). *Food Chem.* 326: 126978.
14 Pozebon, D., Scheffler, G.L., and Dressler, V.L. (2017). *J. Anal. At. Spectrom.* 32: 890.
15 Konz, I., Fernandez, B., Luisa Fernandez, M. et al. (2012). *Anal. Bioanal.Chem.* 403: 2113.
16 Sussulini, A., Becker, J.S., and Becker, J.S. (2017). *Mass Spectrom. Rev.* 36: 47.
17 Martinez, M. and Baudelet, M. (2020). *Anal. Bioanal.Chem.* 412: 27.
18 Doble, P.A., de Vega, R.G., Bishop, D.P. et al. (2021). *Chem. Rev.* 121: 11769.
19 Zhou, J., Ni, X., Fu, J. et al. (2021). *At. Spectrosc.* 42: 210.
20 Wang, S., Brown, R., and Gray, D.J. (1994). *Appl. Spectrosc.* 48: 1321.
21 Greenhalgh, C.J., Karekla, E., Miles, G.J. et al. (2020). *Anal. Chem.* 92: 9847.
22 Kroslakova, I. and Guenther, D. (2007). *J. Anal. At. Spectrom.* 22: 51.
23 Günther, D. and Heinrich, C.A. (1999). *J. Anal. At. Spectrom.* 14: 1369.
24 Resano, M., Aramendia, M., Rello, L. et al. (2013). *J. Anal. At. Spectrom.* 28: 98.

25 Hare, D., Austin, C., and Doble, P. (2012). *Analyst* 137: 1527.
26 Fernandez, B., Claverie, F., Pecheyran, C., and Donard, O.F.X. (2007). *Trends Anal. Chem.* 26: 951.
27 Pisonero, J. and Guenther, D. (2008). *Mass Spectrom. Rev.* 27: 609.
28 Resano, M., Garcia-Ruiz, E., and Vanhaecke, F. (2010). *Mass Spectrom. Rev.* 29: 55.
29 International Union of Pure and Applied Chemistry, *Compendium of chemical terminology Gold Book*, 2006.
30 Liao, X., Hu, Z., Luo, T. et al. (2019). *J. Anal. At. Spectrom.* 34: 1126.
31 Lv, N., Chen, K., Bao, Z. et al. (2021). *At. Spectrosc.* 42: 51.
32 Iwano, H., Danhara, T., Danhara, Y. et al. (2020). *Isl. Arc* 29: e12348.
33 Luo, T., Zhao, H., Zhang, W. et al. (2021). *Sci. China Earth Sci.* 64: 667.
34 Fernandez, B., Claverie, F., Pecheyran, C., and Donard, O.F.X. (2008). *J. Anal. At. Spectrom.* 23: 367.
35 Feng, L.X., Huo, Z.Z., Xiong, J.P., and Li, H.M. (2020). *Anal. Chem.* 92: 13229.
36 Konz, I., Fernandez, B., Fernandez, M.L. et al. (2011). *Anal. Chem.* 83: 5353.
37 Feng, L.X., Zhang, D., Wang, J. et al. (2015). *Anal. Chim. Acta* 884: 19.
38 Zhang, D., Feng, L.X., Wang, J. et al. (2014). *Chem. J. Chinese Universities* 35: 1889.
39 Calderon-Celis, F., Cid-Barrio, L., Encinar, J.R. et al. (2017). *J. Proteomics* 164: 33.
40 Lee, H.-S., Kim, S.H., Jeong, J.-S. et al. (2015). *Metrologia* 52: 619.
41 Han, B., Ge, M., Zhao, H. et al. (2018). *J. Anal. At. Spectrom.* 56: 51.
42 Wang, M., Feng, W.Y., Lu, W.W. et al. (2007). *Anal. Chem.* 79: 9128.
43 Pan, M.Y., Feng, L.X., and Li, H.M. (2020). *Chem. J. Chinese Universities* 41: 1983.
44 Brauckmann, C., Frank, C., Schulze, D. et al. (2016). *J. Anal. At. Spectrom.* 31: 1846.
45 Bolea-Fernandez, E., Balcaen, L., Resano, M., and Vanhaecke, F. (2017). *J. Anal. At. Spectrom.* 32: 1660.
46 Palacios, O., Encinar, J.R., Schaumloffel, D., and Lobinski, R. (2006). *Anal. Bioanal.Chem.* 384: 1276.
47 Fernandez, S.D., Sugishama, N., Encinar, J.R., and Sanz-Medel, A. (2012). *Anal. Chem.* 84: 5851.
48 Calderon-Celis, F., Diez-Fernandez, S., Costa-Fernandez, J.M. et al. (2016). *Anal. Chem.* 88: 9699.
49 Chahrour, O. and Malone, J. (2017). *Protein Pept. Lett.* 24: 253.
50 Kretschy, D., Koellensperger, G., and Hann, S. (2012). *Anal. Chim. Acta* 750: 98.
51 Sanz-Medel, A., Montes-Bayon, M., Bettmer, J. et al. (2012). *Trends Anal. Chem.* 40: 52.
52 Xu, M., Yang, L.M., and Wang, Q.Q. (2012). *Chem. Eur. J.* 18: 13989.
53 Guo, Y.F., Chen, L.Q., Yang, L.M., and Wang, Q.Q. (2008). *J. Am. Soc. Mass Spectrom.* 19: 1108.

54 Rappel, C. and Schaumloffel, D. (2009). *Anal. Chem.* 81: 385.
55 Jakubowski, N., Messerschmidt, J., Anorbe, M.G. et al. (2008). *J. Anal. At. Spectrom.* 23: 1487.
56 Navaza, A.P., Encinar, J.R., Ballesteros, A. et al. (2009). *Anal. Chem.* 81: 5390.
57 Waentig, L., Jakubowski, N., Hayen, H., and Roos, P.H. (2011). *J. Anal. At. Spectrom.* 26: 1610.
58 Alonso-Garcia, F.J., Blanco-Gonzalez, E., and Montes-Bayon, M. (2019). *Talanta* 194: 336.
59 Whetstone, P.A., Butlin, N.G., Corneillie, T.M., and Meares, C.F. (2004). *Bioconjugate Chem.* 15: 3.
60 Jakubowski, N., Waentig, L., Hayen, H. et al. (2008). *J. Anal. At. Spectrom.* 23: 1497.
61 Yan, X.W., Xu, M., Yang, L.M., and Wang, Q.Q. (2010). *Anal. Chem.* 82: 1261.
62 Zhang, C., Wu, F.B., Zhang, Y.Y. et al. (2001). *J. Anal. At. Spectrom.* 16: 1393.
63 Giesen, C., Waentig, L., Panne, U., and Jakubowski, N. (2012). *Spectrochim. Acta Part B: Atom. Spectrosc.* 76: 27.
64 Tanner, S.D., Bandura, D.R., Ornatsky, O. et al. (2008). *Pure Appl. Chem.* 80: 2627.
65 Perez, E., Bierla, K., Grindlay, G. et al. (2018). *Anal. Chim. Acta* 1018: 7.
66 Yang, B., Zhang, Y., Chen, B.B. et al. (2017). *Talanta* 167: 499.
67 Xiao, G.Y., Chen, B.B., He, M. et al. (2019). *Talanta* 202: 207.
68 Li, X.T., Chen, B.B., He, M. et al. (2018). *Talanta* 176: 40.
69 Ko, J.A. and Lim, H.B. (2016). *Anal. Chim. Acta* 938: 1.
70 Mokgalaka, N.S. and Gardea-Torresdey, J.L. (2006). *Appl. Spectrosc. Rev.* 41: 131.
71 Maloof, K.A., Reinders, A.N., and Tucker, K.R. (2020). *Curr. Opin. Environ. Sci. Health* 18: 54.
72 Chew, D., Drost, K., Marsh, J.H., and Petrus, J.A. (2021). *Chem. Geol.* 559: 119917.
73 Pozebon, D., Scheffler, G.L., Dressler, V.L., and Nunes, M.A.G. (2014). *J. Anal. At. Spectrom.* 29: 2204.
74 Miliszkiewicz, N., Walas, S., and Tobiasz, A. (2015). *J. Anal. At. Spectrom.* 30: 327.
75 Ballihaut, G., Claverie, F., Pecheyran, C. et al. (2007). *Anal. Chem.* 79: 6874.
76 Nunes, M.A.G., Voss, M., Corazza, G. et al. (2016). *Anal. Chim. Acta* 905: 51.
77 Becker, J.S. (2013). *J. Mass Spectrom.* 48: i.
78 Becker, J.S., Matusch, A., and Wu, B. (2014). *Anal. Chim. Acta* 835: 1.
79 Konz, I., Fernandez, B., Luisa Fernandez, M. et al. (2013). *Anal. Bioanal.Chem.* 405: 3091.
80 Hutchinson, R.W., Cox, A.G., McLeod, C.W. et al. (2005). *Anal. Biochem.* 346: 225.
81 Wang, M., Feng, W.Y., Zhao, Y.L., and Chai, Z.F. (2010). *Mass Spectrom. Rev.* 29: 326.
82 VanderSchee, C.R., Frier, D., Kuter, D. et al. (2021). *J. Anal. At. Spectrom.* 36: 2431.

83 Liao, X., Luo, T., Zhang, S. et al. (2020). *J. Anal. At. Spectrom.* 35: 1071.
84 Fingerhut, S., Niehoff, A.C., Sperling, M. et al. (2018). *J. Trace Elem. Med. Biol.* 45: 125.
85 Cruz-Alonso, M., Fernandez, B., Navarro, A. et al. (2019). *Talanta* 197: 413.
86 Clases, D., Gonzalez de Vega, R., Adlard, P.A., and Doble, P.A. (2019). *J. Anal. At. Spectrom.* 34: 407.
87 Ko, J.A., Furuta, N., and Lim, H.B. (2018). *Chemosphere* 190: 368.
88 Chantada-Vázquez, M.P., Moreda-Piñeiro, J., Cantarero-Roldán, A. et al. (2018). *Talanta* 186: 169.
89 Kim, J.Y., Park, J., Choi, J., and Kim, J. (2020). *Minerals* 10: 1.
90 Makino, Y. and Nakazato, T. (2021). *J. Anal. At. Spectrom.* 36: 1895.
91 Feldmann, J.r., Kindness, A., and Ek, P. (2002). *J. Anal. At. Spectrom.* 17: 813.
92 Augusto, A., Castro, J., Sperança, M., and Pereira, E. (2018). *J. Braz. Chem. Soc.*
93 Frick, D.A. and Günther, D. (2012). *J. Anal. At. Spectrom.* 27: 1294.
94 Todolí, J.L. and Mermet, J.M. (1998). *Spectrochim. Acta. Part B* 53: 1645.
95 Austin, C., Fryer, F., Lear, J. et al. (2011). *J. Anal. At. Spectrom.* 26: 1494.
96 Wu, B., Zoriy, M., Chen, Y., and Becker, J.S. (2009). *Talanta* 78: 132.
97 Hoesl, S., Neumann, B., Techritz, S. et al. (2016). *J. Anal. At. Spectrom.* 31: 801.
98 Feng, L., Wang, J., Li, H. et al. (2017). *Anal. Chim. Acta* 984: 66.
99 Legrand, M., Lam, R., Jensen-Fontaine, M. et al. (2004). *J. Anal. At. Spectrom.* 19.
100 Stadlbauer, C., Prohaska, T., Reiter, C. et al. (2005). *Anal. Bioanal.Chem.* 383: 500.
101 Rodushkin, I. and Axelsson, M.D. (2003). *Sci. Total Environ.* 305: 23.
102 Kang, D., Amarasiriwardena, D., and Goodman, A.H. (2004). *Anal. Bioanal.Chem.* 378: 1608.
103 Praamsma, M.L. and Parsons, P.J. (2016). *Accredit. Qual. Assur.* 21: 385.
104 Arora, M., Hare, D., Austin, C. et al. (2011). *Sci. Total Environ.* 409: 1315.
105 Austin, C., Hare, D., Rawling, T. et al. (2010). *J. Anal. At. Spectrom.* 25.
106 Moraleja, I., Esteban-Fernández, D., Lázaro, A. et al. (2016). *Anal. Bioanal.Chem.* 408: 2309.
107 Turková, S., Vašinová Galiová, M., Štůlová, K. et al. (2017). *Microchem. J.* 133: 380.
108 Frick, D.A., Giesen, C., Hemmerle, T. et al. (2015). *J. Anal. At. Spectrom.* 30: 254.
109 Bonta, M., Lohninger, H., Marchetti-Deschmann, M., and Limbeck, A. (2014). *Analyst* 139: 1521.
110 Hoesl, S., Neumann, B., Techritz, S. et al. (2014). *J. Anal. At. Spectrom.* 29: 1282.
111 Neumann, B., Hösl, S., Schwab, K. et al. (2020). *J. Neurosci. Methods* 334: 108591.
112 Hare, D.J., Kysenius, K., Paul, B. et al. (2017). *J. Visualized Exp.* e55042.
113 Löhr, K., Traub, H., Wanka, A. et al. (2018). *J. Anal. At. Spectrom.* 33: 1579.
114 Sajnóg, A., Hanć, A., Makuch, K. et al. (2016). *Spectrochim. Acta. Part B* 125: 1.
115 Li, Y., Guo, W., Hu, Z. et al. (2019). *J. Agric. Food. Chem.* 67: 935.

116 Westerhausen, M.T., Lockwood, T.E., Gonzalez de Vega, R. et al. (2019). *Analyst* 144: 6881.
117 Marković, S., Uršič, K., Cemazar, M. et al. (2021). *Anal. Chim. Acta* 1162: 338424.
118 Wagner, B., Syta, O., Kepa, L. et al. (2018). *J. Mex. Chem. Soc.* 62: 323.
119 Fernandez, B., Rodriguez-Gonzalez, P., Garcia Alonso, J.I. et al. (2014). *Anal. Chim. Acta* 851: 64.
120 Jimenez, M.S., Gomez, M.T., and Castillo, J.R. (2007). *Talanta* 72: 1141.
121 Jantzi, S.C. and Almirall, J.R. (2014). *Appl. Spectrosc.* 68: 963.
122 Coedo, A.G., Padilla, I., and Dorado, M.T. (2005). *Talanta* 67: 136.
123 Su, P., Ek, P., and Ivaska, A. (2012). *Holzforschung* 66: 833.
124 Hondrogiannis, E., Ehrlinger, E., Poplaski, A., and Lisle, M. (2013). *J. Agric. Food. Chem.* 61: 11332.
125 Vassilev, S.V., Vassileva, C.G., and Baxter, D. (2014). *Fuel* 129: 292.
126 Piispanen, M.H., Arvilommi, S.A., Broeck, B.V.d. et al. (2009). *Energy Fuels* 23: 3451.
127 Malherbe, J., Claverie, F., Alvarez, A. et al. (2013). *Anal. Chim. Acta* 793: 72.
128 Claverie, F., Malherbe, J., Bier, N. et al. (2013). *Anal. Chem.* 85: 3584.
129 Macholdt, D., Jochum, K., Stoll, B. et al. (2014). *Chem. Geol.* 383: 123.
130 Lloyd, N., Parrish, R., Horstwood, M., and Chenery, S. (2009). *J. Anal. At. Spectrom.* 24: 752.
131 Jackson, B., Harper, S., Smith, L., and Flinn, J. (2006). *Anal. Bioanal.Chem.* 384: 951.
132 Limbeck, A., Galler, P., Bonta, M. et al. (2015). *Anal. Bioanal.Chem.* 407: 6593.
133 Chuks, O.F., Eze, P., Madzivire, G. et al. (2013). *Chem. Didac. Ecol. Metrol.* 18: 19.
134 Nowinski, P., Hodge, V., Lindley, K., and Cizdziel, J. (2010). *Open Chem. Biomed. Methods J.* 3.
135 Piispanen, M., Niemelä, M., Tiainen, M., and Laitinen, R. (2012). *Energy Fuels* 26: 2427.
136 Piispanen, M., Tiainen, M., and Laitinen, R. (2009). *Energy Fuels* 23: 3446.
137 Mertz-Kraus, R., Brachert, T., Jochum, K. et al. (2009). *Palaeogeography* 273: 25.
138 Reifschneider, O., Wentker, K., Strobel, K. et al. (2015). *Anal. Chem.* 87: 4225.
139 Jurowski, K., Szewczyk, M., Piekoszewski, W. et al. (2014). *J. Anal. At. Spectrom.* 29: 1425.
140 Hare, D., Lear, J., Bishop, D. et al. (2013). *Anal. Methods* 5: 1915.
141 O'Reilly, J., Douglas, D., Braybrook, J. et al. (2014). *J. Anal. At. Spectrom.* 29: 1378.
142 Kuczelinis, F., Petersen, J., Weis, P., and Bings, N. (2020). *J. Anal. At. Spectrom.* 35: 1922.
143 Bonta, M., Hegedus, B., and Limbeck, A. (2016). *Anal. Chim. Acta* 908: 54.
144 Bonta, M., Lohninger, H., Laszlo, V. et al. (2014). *J. Anal. At. Spectrom.* 29: 2159.
145 Shariatgorji, M., Nilsson, A., Bonta, M. et al. (2016). *Methods* 104: 86.

146 Aramendia, M., Rello, L., Berail, S. et al. (2015). *J. Anal. At. Spectrom.* 30: 296.
147 Costas-Rodriguez, M., Van Acker, T., Hastuti, A.A.M.B. et al. (2017). *J. Anal. At. Spectrom.* 32: 1805.
148 Grijalba, N., Legrand, A., Holler, V., and Bouvier-Capely, C. (2020). *Anal. Bioanal.Chem.* 412: 3113.
149 Qiao, L., Zhang, R., Qiao, J. et al. (2021). *RSC Adv.* 11: 6644.
150 Cruz-Alonso, M., Fernandez, B., García, M. et al. (2018). *Anal. Chem.* 90: 12145.
151 Niehaus, R., Sperling, M., and Karst, U. (2015). *J. Anal. At. Spectrom.* 30: 2056.
152 Arakawa, A., Jakubowski, N., Koellensperger, G. et al. (2019). *Anal. Chem.* 91: 10197.
153 Wang, M., Zheng, L.N., Wang, B. et al. (2014). *Anal. Chem.* 86: 10252.
154 Thompson, M., Chenery, S., and Brett, L. (1989). *J. Anal. At. Spectrom.* 4: 11.
155 Pozebon, D., Dressler, V., Mesko, M. et al. (2010). *J. Anal. At. Spectrom.* 25: 1739.
156 Dressler, V.L., Pozebon, D., Mesko, M.F. et al. (2010). *Talanta* 82: 1770.
157 Michaliszyn, L., Ren, T., Röthke, A., and Rienitz, O. (2020). *J. Anal. At. Spectrom.* 35: 126.
158 Sargent, M., Harte, R., and Harrington, C. (ed.) (2002). *Guidelines for Achieving High Accuracy in Isotope Dilution Mass Spectrometry (IDMS)*. The Royal Society of Chemistry.
159 Pisonero, J., Fernández, B., and Günther, D. (2009). *J. Anal. At. Spectrom.* 24: 1129.
160 Feng, L. and Wang, J. (2014). *J. Anal. At. Spectrom.* 29: 2183.
161 Douglas, D., O'Reilly, J., O'Connor, C. et al. (2016). *J. Anal. At. Spectrom.* 31: 270.
162 Moraleja, I., Mena, M.L., Lázaro, A. et al. (2018). *Talanta* 178: 166.
163 Zheng, L.N., Feng, L.X., Shi, J.W. et al. (2020). *Anal. Chem.* 92: 14339.
164 Song, X., He, J., Pang, X. et al. (2019). *Anal. Chem.* 91: 2838.

6

Medimetallomics and Clinimetallomics

Guohuan Yin[1,2], Ang Li[1,2], Meiduo Zhao[1,2], Jing Xu[1,2], Jing Ma[3], Bo Zhou[3], Huiling Li[3,*], and Qun Xu[1,2,*]

[1] Chinese Academy of Medical Sciences, Institute of Basic Medical Sciences, School of Basic Medicine Peking Union Medical College, Department of Epidemiology and Biostatistics, No.5 Dongdan Santiao, Dongcheng, Beijing 100005, China
[2] Chinese Academy of Medical Sciences, Center of environmental and Health Sciences, Peking Union Medical College, No.9 Dongdan Santiao, Dongcheng, Beijing 100005, China
[3] Capital Medical University, Beijing Chao-yang Hospital, Department of Occupational Medicine and Clinical Toxicology, No.8 Gongtu South Road, Chaoyang, Beijing 100020, China

Trace metal element persistently and naturally exists in the environment [1]. Some natural events (e.g. volcanic explosions and weathering) [2] or metallic elements degradation and migration [3] can be a type of environmental contamination. Moreover, some anthropogenic factors, such as the development of industry and agriculture, also contribute to the contamination of the environment with these substances [4]. Therefore, trace metal element can be detected in various media such as air, water, and soil. Furthermore, these widespread trace metal elements can be absorbed into human body mainly by three routes (i.e. inhalation, digestion, and dermal absorption). However, these elements are not always beneficial to humans. Some metallic elements, such as Zn, Fe and Cu, play a vital role in certain physiological and biochemical functions, and insufficient amounts may lead to deficiency-related diseases [5], while large doses may lead to toxicity. However, some metals play nonessential roles, and levels higher than a low background level may cause toxicity [1]. Many studies have shown that human exposure is essentially a mixture of multiple metals and that mixed exposure to heavy metals is inextricably linked to human disease outcomes.

6.1 The Concept of Medimetallomics and Clinimetallomics

Due to the extensive distribution and potential adverse effects of trace metal element, a systematic approach called metallomics was proposed. Metallomics is

[*] Corresponding authors

Applied Metallomics: From Life Sciences to Environmental Sciences, First Edition.
Edited by Yu-Feng Li and Hongzhe Sun.
© 2024 WILEY-VCH GmbH. Published 2024 by WILEY-VCH GmbH.

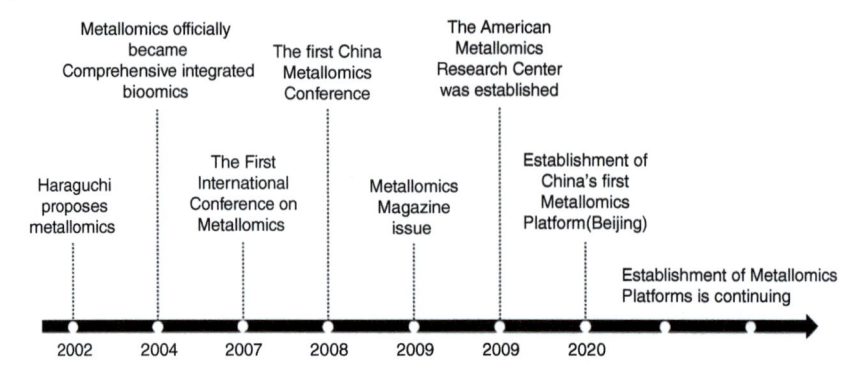

Figure 6.1 The development history of metallomics [7]. Source: Guohuan Yin.

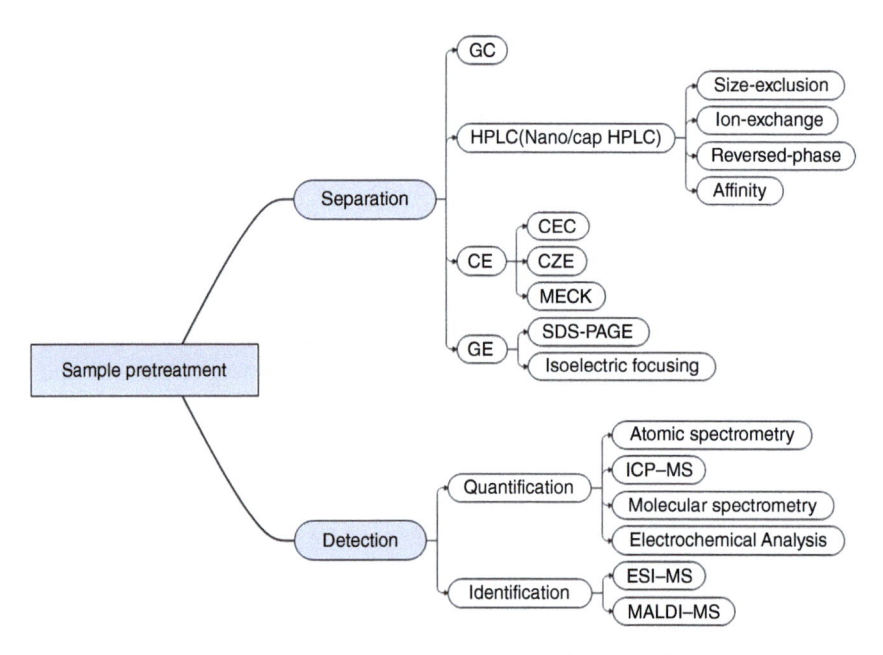

Figure 6.2 Hyphenated techniques for elemental speciation [20a]. Source: Huiling Li.

a comprehensive discipline developed in 2002 to study the distribution, content, chemical states, and functions of all free or complex metal elements in living bodies, following genomics, proteomics, and metabolomics [6]. The development history of metallomics is shown in Figure 6.1.

This approach aims to study the metallome, interactions and functional connections of metal ions and their species with genes, proteins, metabolites, and other biomolecules within organisms and ecosystems [8]. In order to study the effects of trace metal element on the habitat or ecosystem and propose targeted environmental protection policy, Chen et al. further propose a new term called environmentallomics [9]. However, metallomics and its derivative disciplines (e.g. environmentallomics) highlight all the potential implications of metal contamination in the entire living organism and natural environment. Nevertheless, the ultimate aim of metallomics is to reveal the full range of effects of various metallic elements on human, so as to guide policy making and further promote public health. Therefore, more importance should be attached to the adverse health effect of trace metal element at the population level. Therefore, integrating epidemiological study design and ideology into metallomic research can bring novel insights and new chances for the development of metallomic studies.

6.1.1 Medimetallomics

An explicit definition of the full combination between medical and metallomics, called medimetallomics, has been proposed. Generally, medimetallomics adopts the epidemiological distributions (i.e. time, place, and person) as a principle and views standard and rigorous epidemiological design as a guiding ideology, aiming to draw a blueprint for the study of the effects of trace metal element on public health. Medimetallomics studies should underpin the following aspects: (i) draw up a strict and standardized study design; (ii) determine and monitor the markers of trace metal element exposure; (iii) determine biomarkers of related health outcomes; (iv) link exposure markers with outcome markers using rigorous statistical methods; and (v) set limit values for hazardous trace metal element to provide a basis for policy makers to devise effective management strategies.

6.1.2 Clinimetallomics

Clinimetallomics is a branch of metallomics, which systematically studies the content, morphology, distribution, and function of all metal elements in human body. People use related instruments to detect the content of metal elements in biological samples such as blood, urine, tissues, and organs, so as to understand the correlation between metal elements and diseases, and provide evidence for the prevention and treatment of diseases [10].

In this book, metals are generally solid (except Hg) at room temperature, with metallic luster (i.e. strong reflection of visible light). Most of them are excellent conductors of electricity and heat, with ductility, high density, and high melting point. Heavy metals are defined as metals with a density of $4.5\,g/cm^3$ or more [11], such as Cu, Pb, Zn, Fe, Co, Ni, manganese, Cd, Hg, and W.

6.2 The Analytical Techniques in Medimetallomics and Clinimetallomics

There are more than 60 elements that make up the human body, some of them are essential for human body, while others are elements with unknown functions or toxic and harmful. A total of 27 elements is necessary for human growth and development, and these elements maintain a certain concentration in human body fluids and various organs. The main physiological function of these elements is to form a part of the human body, and it is also a part of enzymes and vitamins. It maintains the pH and electrolytic balance of the blood, participates in endocrine, promotes gonadal development, fertility, sexual function, and sugar metabolism. And it also assists human organs and tissues to transport essential substances throughout the whole body for metabolic needs [1, 12].

The proportion of trace elements in the human body is very low, but plays an essential role. Too much or too little trace elements in the body may cause obvious symptoms or irreversible lesions. In addition, toxic heavy metals such as Hg, As, and Cd in the environment will also have a bioaccumulation effect, which could accumulate in the human body and cause diseases. With the improvement of people's living standards and health awareness, more attention has been paid to the content of trace elements and toxic elements in the body.

6.2.1 Total Analysis of Clinical Elements

The detection methods of elements in the human body are also becoming more and more mature, and the detection limits are getting lower and lower. There are many methods that can be used for element detection in clinic, such as mass spectrometry, molecular spectroscopy, neutron activation analysis, atomic emission spectroscopy, atomic absorption spectroscopy, and electrochemical analysis.

Common detection methods for trace elements include atomic absorption spectrometry (AAS), potentiometric dissolution method, and inductively coupled plasma mass spectrometry (ICP-MS) [13].

The following is a brief introduction to the methods currently available for clinical element detection.

6.2.1.1 Atomic Spectroscopy Detection Technology

Atomic spectroscopy is an effective method for the determination of inorganic elements. Metalloproteins are often used as catalysts in biological systems and play a crucial role in signal transmission and gene expression. These trace proteins can be detected from complex biological systems. Although there exist many methods, the direct determination of metal by atomic spectroscopy, which is simple and fast, is an important analytical tool in metallomic research.

Atomic Absorption Spectroscopy (AAS), also known as atomic spectrophotometry, is an instrumental analysis method to determine the content of target element in the sample based on the absorption intensity of resonance radiation by the ground-state atoms in the vapor phase [14]. According to different atomization methods, AAS can be divided into Flame Atomic Absorption Spectrometry (FAAS) and Electrothermal Atomic Absorption Spectrometry (ETAAS, also known as high-temperature furnace atomic absorption spectrometry). AAS is an effective method for the determination of trace and ultra-trace elements. It has a variety of features, including low detection limit, good selectivity, high precision, strong anti-interference ability, simple equipment, and small dosage. The disadvantage of AAS is that only single-element determinations can be achieved. Together with mass spectrometry and neutron activation, ETAAS is recognized as the three main methods for the determination of ultrarace elements [14].

Flame Atomic Absorption Spectroscopy (FAAS) is an atomic absorption spectrometry method that uses chemical flame as the heat source to realize the atomization of compound elements. Non-flammable Electrothermal Atomic Absorption Spectrometry (ETAAS) has higher sensitivity than Flame Atomic Absorption Spectrometry, especially the graphite furnace method has the advantage of requiring less sample and is widely used in sanitary inspection.

Atomic fluorescence spectrometry (Atomic Fluorescence Spectrometry [AFS]), is a spectral analysis technology between atomic emission spectrometry and atomic absorption spectrometry. The characteristics and intensity of atomic fluorescence are used to carry out the qualitative and quantitative methods of elements. Atomic fluorescence spectrometry has simple spectral lines, less interference, high sensitivity, and low detection limit, and is used for the determination of Hg, As, Se, and other elements in clinical.

Volatile species generation (VSG) is a commonly used injection method in atomic fluorescence. VSG is a detection technology by using some reducing agents that can generate nascent hydrogen, such as potassium borohydride, or chemical reactions, to form volatile covalent hydrides or cold vapors with the analyte elements in the sample solution, and then the volatile covalent hydrides or cold vapors can be imported in atomic spectral analysis system through carrier gas stream to achieve analysis. According to the different volatile species generated, VSG can be divided into hydride generation technology (hydride generation [HG]) and cold vapor generation technology (cold vapor [CV]). It is suitable for elements that are easy to generate hydrides such as As, Sb, Bi, Ge, Sn, Pb, Se, and Tc, and Hg (volatile at room temperature) [14]. VSG-AFS technology, using hydride-generating atomizer, is a trace element analysis method with great practical value. At present, China has a leading position in the world.

If you want to achieve multi-element determination, you need to use atomic emission spectroscopy (Atomic Emission Spectrometry [AES]). Atomic emission spectroscopy is used for qualitative and quantitative analysis of elements by using the characteristic spectra of atomic or ion emission of each element under thermal or electric excitation.

Atomic emission spectrometry can analyze about 70 elements (metal elements and non-metal elements such as phosphorus, silicon, As, carbon, and boron). However, for some elements, AES has poor sensitivity and cannot meet the multi-element determination of actual samples.

Inductively Coupled Plasma Atomic Emission Spectroscopy (ICP-AES) is a spectral analysis method using an inductively coupled plasma torch as the excitation light source. This method possesses advantages like simplicity, rapidity, low detection limit, wide measurement dynamic linear range, simultaneous analysis of multiple elements, no chemical separation, no obvious interference of the measured elements, small matrix effect, high precision, and good accuracy [14].

6.2.1.2 Mass Detection Technology

The mass spectrometer mainly consists of three parts: ion source, mass analyzer, and detector. The working principle of mass detection is to use a specific technology to vaporize and charge the target analyte. Through the designed interface device, the ions are extracted into the vacuum chamber, and then the ions are separated according to the mass-to-charge ratio according to electromagnetic principle. Finally, the target ions are selected to enter the detector, and the signal is displayed the recorded in the form of a map in the computer [15].

Inductively Coupled Plasma Mass Spectrometry (ICP-MS) can analyze most of the elements on the periodic table except for a very few elements such as C, H, and O. Compared with inductively coupled plasma optical emission spectrometer (ICP-AES), the detection limit of ICP-MS is about 3 orders of magnitude higher; compared with graphite furnace atomic absorption spectrophotometer (GF-AAS), ICP-MS has an advantage of fast measurement of multiple elements; in addition, ICP-MS has the ability to detect isotopes. At present, ICP-MS has been recognized as one of the most authoritative chemometric methods by the International Commission for Weights and Measures (ICWM) [16].

6.2.1.3 Electrochemical Analysis

Electrochemical analysis is a kind of analysis method according to the principle of electrochemistry and the electrochemical properties and changes of substances in solution.

The electrochemical method can only be applied to the heavy metal detection process under the premise that the detection data of the electrochemical method is sufficiently accurate. The electrochemical detection method can quickly and accurately separate heavy metals and free substances. At the same time, the instruments are relatively simple, the detection cost is relatively low, and repeated detection and data verification can be realized. Electrochemical analysis method can be used for the determination of fluorine in clinical.

The anodic stripping method is widely used as an effective electrochemical detection method. The research principle is based on the chemical properties of heavy metal elements in water. It could record the current and voltage changes in chemical reactions, and according to the relationship between element concentration and current, voltage to design the function expression, the function image can be depicted. The anodic dissolution method is suitable for the redox reaction of voltametric melting. Since the concentration analysis of metal ions needs to be combined with the oxidation reaction, the anodic dissolution method is often selected in the actual detection work. In addition, the anodic dissolution method is not only more automated than other detection methods, but the amount of reagents consumed in the reaction process is also very small, which has the characteristics of high precision and accurate results [17].

6.2.1.4 Neutron Activation Analysis

The neutron activation method, also known as Neutron Activation Analysis (NAA), is the most important method in activation analysis. It uses neutrons generated by reactors, accelerators, or isotopic neutron sources as the activation analysis method to bombard particles. It is a qualitative and quantitative analysis method to determine the composition of material elements.

The neutron activation method is characterized by extremely high sensitivity, which can be used for ultra-trace analysis below the ppt level; its accuracy and precision are also high. Due to the high accuracy and precision of neutron activation method, it is often used as an arbitration analysis method. It can measure a wide range of elements including atomic numbers from 1 to 83, and has the function of simultaneous measurement of multiple components. At the same time, about 30–40 kind elements can be measured in a same specimen. Therefore, it is suitable for the simultaneous analysis of multiple elements in environmental solid samples, such as metal elements measurement in atmospheric particulate matter, industrial dust, and solid waste. Because neutron activation analysis is a multi-element and nuclide analysis technology, it has the advantages of high accuracy and sensitivity and less sample consumption than other analysis methods, and has become a hot spot in bioscience research [18]. The neutron activation method is used to determine the content of various trace elements in organisms, which can study the relationship between trace elements and diseases and the pathogenesis. Meanwhile, neutron activation method is used in nuclear medicine and clinical medicine [19]. Neutron activation analysis is to irradiate the sample with neutrons to make the atoms of the element have radioactivity after nuclear reaction, and then the content of the element can be identified and measured by analyzing the radioactivity of the atom. Neutron analysis and activation technology can be divided into instrumental neutron activation analysis (instrumental NAA) and separation neutron activation analysis (NAA with radiochemical separations). The former can be directly analyzed without any chemical treatment after activation. The benefits of this technology include easy, time saving, and automation, but the disadvantage is poor sensitivity, large interference between elements. The latter are activated, chemically separated, and then analyzed [20].

6.2.2 Clinical Element Morphology and Valence Analysis Technology

While many problems related to biological systems in the past were explained by the total amount of elements, there is now a growing recognition of the importance of chemical forms of elements in living organisms, such as oxidation states, ligand properties, and molecular structures. Different forms of elements have different toxicity, characteristic chemical properties, and physiological functions [12]. It is crucial to develop new speciation analysis techniques and methods for metals and their compounds in living organisms, as well as methodologies for the study of the interactions between the speciation of metal elements and biologically active molecules.

According to the definition of International Union of Pure and Applied Chemistry (IUPAC), chemical species refers to the existence form of an element, including isotopic composition, charge/oxidation state inorganic/small molecule complexes, organic complexes (coordination bonds), organometallic compounds (covalent bonds), and macromolecular compounds or complexes (metalloproteins, etc.). Speciation analysis is the process of qualitative and quantitative analysis of chemical species of elements in a sample. Elements and their morphological analysis in living systems, including exploring the composition, concentration, and distribution of free metal ions and metal biomolecules in cells or tissues, and based on this, exploring the biological functions of elements in living bodies and their relationship with genome, proteome, and metabolism relationship between groups [21].

In 2004, HARAGUCHI proposed the concept of metallomics, which is also known as biometal science. Metallomics focuses on the role and function of metal-bound biomolecules in biological systems, so it is also called metal-assisted functional biochemistry. Metallomics has become another important frontier subject to reveal the laws of living after genomics, proteomes, and metabollomes [22]. And metallomics holds the same important status as genomics and proteomics research.

Different forms of elements have different toxicity characteristics, chemical properties, and physiological functions. Qualitative and quantitative analysis of the different chemical forms of elements is a challenging subject for analytical chemistry in life sciences. Since the 1980s, the study of morphological analysis of elements has made great progress. Modern chromatographic separation techniques consisting of liquid chromatography, gas chromatography, and capillary electrophoresis have become the basis for elemental speciation analysis. The high-efficiency separation technology combined with the high sensitivity and identification technology of mass spectrometers plays an important role in the speciation analysis of elements in complex systems. Researchers have focused on the morphological analysis of As, Pb, Hg, Sn, Cr, and other elements for morphological analysis applications [12].

From the perspective of analytical techniques, the most effective method for morphological analysis of elements is the combination technique, i.e. highly selective separation and highly sensitive detection techniques [21]. As shown in Figure 6.2, the main separation methods on the left are: gas chromatography (GC), high-performance liquid chromatography (HPLC), capillary electrophoresis (CE), and gel electrophoresis (GE). The right side of Figure 6.2 shows the detection method. The detection methods in the combined technology are mainly divided into two categories: quantitative detection and structure identification. Quantitative detection methods mainly include Inductively Coupled Plasma Mass Spectrometry (ICP-MS), Inductively Coupled Plasma Atomic Emission Spectrometry (ICP-OES), Atomic Absorption Spectroscopy (AAS), Atomic Fluorescence Spectroscopy (AFS), Molecular Spectroscopy spectrophotometry (Ultraviolet–visible spectroscopy [UV–Vis]), fluorescence spectroscopy, chemiluminescence analysis, and electrochemical analysis. Among them, the element-specific detection method-atomic spectroscopy/mass spectrometry has significant advantages in the analysis of elements and their forms. For the unknown element species present in the sample, biological mass spectrometry (such as electrospray mass spectrometry [ESI-MS] and matrix-assisted laser desorption mass spectrometry [MALDI-MS]) can be used for structural identification analysis [21].

6.2.2.1 Atomic Spectroscopy Detection Technology

Combining atomic spectroscopy analysis technology with chromatography and capillary electrophoresis (Capillary Electrophoresis [CE]) technology can fully utilize the advantages of the previous high sensitivity and selectivity and the latter high separation performance to achieve complementary advantages, which is an important way to solve the trace element morphology in complex matrixes [23].

CE has become an attractive technique for morphological separation due to its high resolution, rapidity, low reagent consumption, and low perturbation of the equilibrium between different species. CE combined with an element-selective detector not only has good selectivity and high sensitivity, but also only needs to separate the different forms of a specific element. CE is used in conjunction with an atomic spectrometer detector. The selectivity of the detector can eliminate the interference between co-migrating species of different elements, and the high sensitivity of the detector can further reduce the detection limit of the method [14].

Chromatography or CE combined with VSG-AFS technology: when VSG-AFS is used directly, only the total amount of elements can be detected, and speciation analysis is difficult [10]. Chromatography or CE combined with VSG-AFS technique is a powerful means for the speciation analysis of As, Sb, Bi, Ge, Sn, Pb, se, Tc, and Hg [23b]. Yan et al. developed a CE-AFS interface and it is applied to speciation analysis of As, Hg, and selenium [14].

Atomic absorption spectrometry (AAS) coupled with HPLC was the earliest used method for the determination of metalloproteins. It is mainly used in the speciation analysis of elements with high sensitivity, such as Cd, Zn, Cu, or As, selenium and Cd that can generate hydrides [24].

Inductively Coupled Plasma Optical Emission Spectroscopy (ICP-AES) can be used in combination with flow injection and chromatography for online separation, enrichment, and speciation analysis. Xu Qiumei used CE-ICP-AES to analyze the calcium element in rat erythrocytes for the first time and measured free calcium ion concentration [25]. ICP-AES can be used as a multi-element detector for HPLC. The advantage of ICP-AES includes the ability to detect S, P, and C in addition to heavy metals. Therefore, multiple elemental compositions can be determined in a given analysis, which is beneficial to study the biological reactions of multiple elements in organisms [25].

6.2.2.2 Mass Spectrometry Detection Technology

Since the 1980s, mass spectrometry has gradually developed into one of the most important detection techniques in the field of analytical chemistry. In particular, the importance of elemental morphological analysis is unquestionable [15].

Atomic mass spectrometry (ICP-MS) is currently the most sensitive detection method in the field of elemental speciation analysis. It can be used not only for the analysis of trace/ultrarace elements in environmental and biological samples, but also for elemental changes (total or form) monitoring. Some metal elements (such as As, Cd, Pd, Pb, and U) are potentially toxic to living organisms. Although mass spectrometry can provide rich structural and content information, its application to elemental speciation analysis in real samples often needs to be combined with various modern separation techniques. In fact, the combination of various chromatographic separation techniques (including Gas Chromatography [GC], Capillary Electrophoresis [CE], and High-Performance Liquid Chromatography [HPLC]) and mass spectrometry has become the main tool for elemental morphological analysis currently.

GC is suitable for the analysis of self (or after derivatization) thermally stable and volatile substances. It has the advantages of fast separation speed, high resolution, and simple separation system. The generated gas phase does not require special interfaces and can be well separated. Thus it is used in conjunction with various mass spectrometry instruments. Wang Qiuquan's group designed an optional thermal diffusion device as the interface between capillary GC and ICP-MS, and synthesized triopoly Pb chloride as an internal standard to analyze inorganic Pb, inorganic Hg, alkyl Pb, and alkyl Hg [24].

CE has the advantages of high resolution, fast separation, minimal sample consumption, and low cost, making it widely used in elemental speciation separation. Especially in the field of As speciation analysis, it has become one of the important tools [26]. However, the combination with mass spectrometer has disadvantages such as poor reproducibility, susceptibility to sample matrix interference, and complex interface technology, which limits the wide application of CE.

HPLC requires fewer sample properties and offers a wide range of separation modes. By adjusting chromatographic parameters, a strong separation ability can be obtained. The interface used with various instruments is simple, and it has the advantages of high separation efficiency, strong matrix tolerance, and good reproducibility. It is the most widely used separation technique in elemental speciation analysis combined with mass spectrometry [15]. Peng Hanyong [13] established a method of HPLC coupled with ICP-MS and electrospray ionization mass spectrometry (ESI-MS), and it was applied to the speciation analysis of As in chicken liver, which could detect various As species.

Combined technology can effectively realize the complementary advantages of various detection technologies. The different forms of the target element are separated and the sample matrix is purified. The flow rate used in conjunction with HPLC can also be controlled by suitable interfaces, splitters, or novel nebulizers. The retention time of the target analytes is consistent whether using atomic mass spectrometry or molecular mass spectrometry as the detector. Therefore, the combination of HPLC and mass spectrometry has also become a powerful detection method.

Atomic mass spectrometry can provide quantitative information on elemental species, while molecular mass spectrometry can provide structural information. The quantitative and qualitative information provided by the hyphenated technique makes it widely used in elemental speciation analysis, especially the identification of unknown species. On the other hand, after using the heteroatoms (such as S, P, I, and Cu) contained in biological macromolecules to characterize their binding states, sites, and ratios, the biological macromolecules can be analyzed by simple conversion. Quantitative analysis has also emerged in the application of quantitative proteomics [15].

Due to the high temperature of the ICP ion source (6000–8000 K), the original structure and morphology of the target analyte are completely destroyed at high temperature, and the structural information of the original molecule cannot be obtained.

Therefore, ICP-MS usually needs to be combined with techniques such as molecular mass spectrometry (such as electrospray tandem mass spectrometry [ESI-MS/MS] and electrospray time of flight mass spectrometry [ESI-TOF-MS]) to obtain complete structural data information for the identification of confirming/unknown element species [15].

It should be pointed out that although the importance and application of hyphenated techniques in elemental speciation analysis has become increasingly prominent, when the content of target analytes is low or the sample matrix is relatively complex, hyphenated techniques cannot give accurate results. Therefore, the sample pre-treatment technique can be used to effectively remove complex matrices, purify and enrich the target analytes, and bring the sample to a state suitable for mass spectrometry detection [15].

6.2.3 Summary and Outlook

The past three decades have been a period of vigorous development of ICP-MS and its hyphenated technology. With the continuous establishment of new methods of ICP-MS and its combined technology, the research on the analysis of elements and their forms in biological samples has continued to deepen, and scientists are no longer satisfied with only obtaining the content information of trace elements and their forms. Instead, it is expected to further explore the interaction between elements and biomolecules on this basis, and provide more useful information for life science, medicine, pharmacy, nutrition, and other fields.

The complexity of biological samples and the diversity of target elements or forms place increasing demands on analysts. One of the effective solutions is to use chromatographic separation and combination according to the characteristics of each component in the sample to achieve efficient separation of complex samples [21]. Various new separation technologies based on this purpose have also been reported, such as dispersive liquid-phase microextraction technology, which is the application of ionic liquids in elemental speciation analysis.

Nowadays, hyphenated techniques in bio-inorganic analytical chemistry are applied exploratory to find new metal species rather than to determine known compounds [24]. Hybrid techniques are useful tools in the detection and characterization of metal–macromolecular complexes in biological samples. In the future, with the continuous development of ICP-MS and electrospray (tandem) mass spectrometry and the emergence of new and more efficient separation techniques, more selective and more sensitive analytical methods will be developed. The formulation of new standards and reference will promote the further application of combined technology in biochemistry and clinical chemistry [24].

Metal ions perform a highly important role in human activities. The establishment of metallomics will promote the development of biological metal science. In the post-genome era, metallomics research should be valued for common development along with proteomics and metabolomics research [20a]. It is foreseeable that in order to further study the relevant information of metal metabolism, toxicology, and biological activity, the potential of the novel method of ICP-MS-based coupling technology is extremely promising in the field of analysis of elements and their morphology in biological systems [21].

6.3 The Application of Medimetallomics and Clinimetallomics in Medical and Clinical Science and the Perspectives

In recent years, metallomics has been applied to study many disease-related studies, which provides new opportunities to explore the mechanisms of these diseases and find new diagnostic solutions and therapeutic targets [27].

Metallomics is an emerging "omics" subject developed in recent years. It utilizes advanced techniques to study the molecular mechanisms of the composition, distribution, metabolism, homeostasis, and disturbance of all trace elements in organisms, their changes under different physiological and pathological conditions, as well as the study of metalloproteins or metalloenzymes. A large number of metallomic studies have been carried out for a variety of complex diseases, providing theories and enlightenments for their etiology, early diagnosis, prognosis, and treatment.

6.3.1 Medimetallomics

6.3.1.1 Global or National Medimetallomics Research

The human body is exposed to environmental pollutants through dermal absorption, ingestion, and inhalation. Human biomonitoring (HBM) evaluates the level of environmental chemicals or their metabolites in biological matrices with consideration to all these routes of exposure. Medimetallomics assesses the total burden of exposure from any source to the body by measuring the concentrations of trace metal element in human matrices (such as blood, urine, and hair). Under the previous research framework, medimetallomics is an important part of HBM, and various countries have carried out regional, national, and global HBM surveys. The goal of this section is to summarize the basic information regarding the trace metal element of interest and the study design in ongoing HBM projects. By highlighting medimetallomics research and summarizing these projects, we can provide a clearer picture of the current progress in this field, for example, by identifying the trace metal element of greatest interest, the targeted population, and the variation in temporal and spatial trends, which can provide insights and standardized protocols for medimetallomics research.

National Health and Nutrition Examination Survey (NHANES) National Health and Nutrition Examination Survey (NHANES) is a nationwide annual cross-sectional health surveillance program in America that collects exposure and health examination data simultaneously. In NHANES, stratified, clustered four-stage samples are selected to represent the general civilian population of the United States [28]. Approximately 7000 randomly selected residents of the United States participate in NHANES each year. The exposure data are regularly monitored for more than 300 substances and contain two parts: (i) the toxic chemical substances dataset, the National Report on Human Exposure to environmental Chemicals, and (ii) the essential nutrients dataset, the National Report on Biochemicals of Diet and Nutrition. Chemicals that are measured include: metals (metalloids), flame retardants, herbicides, phenols, organochlorine pesticides, fungicides, insect repellents, carbamate or organochlorine pesticides, phthalates, polycyclic aromatic hydrocarbon, and volatile organic compounds. Interestingly, the priority of metals (metalloids) as monitored toxic substances in NHANES has increased over the years, from 8.4% in 2005 to 14.4% in 2013 [29].

Canadian Health Measures Survey (CHMS) In Canada, the Canadian Health Measures Survey (CHMS) is the most comprehensive state-directed health survey. It uses a cross-sectional design and is launched in biennial cycles [30]. The CHMS has completed the collection for cycles 1–6; the fieldwork for cycle 6 started in January 2018 and was finished at the end of 2019. The survey includes two procedures: a personal household interview for collection of sociodemographic information and health history plus a visit to a mobile examination center for anthropometric and biological sampling. To obtain data on exposure to environmental chemicals and evaluate the validity of policies, the CHMS has established a robust research plan, design, and sampling framework, as well as an excellent quality assurance and quality control system relying on Statistics Canada and Health Canada. The CHMS measures a wide range of environmental chemicals in the blood, urine, and hair of survey participants. To date, the CHMS has measured 27 trace metal elements in participants' blood, urine, or both [31].

6.3.1.2 Standardized Protocol for Medimetallomics Research

Study Design Cross-sectional medimetallomics studies assess the internal level of trace metal element in humans at a certain time. Random sampling from the target population is the key to ensuring the extrapolation of conclusions. The selection bias of the participants may confound the association between trace metal element and outcome. Due to the inaccessibility of follow-up studies in large populations, studies such as NHANES and CHMS all adopt cross-sectional designs, which results in some inherent weaknesses. It lacks temporality (cannot determine whether trace metal element exposure occurred prior to its health effects) and may cause response or recall bias during the investigation [31].

Case–control medimetallomics studies are a type of observational design aiming to find out factors associated with outcome of interest. This kind of study usually starts with a group of cases (i.e. individuals with the outcome of interest) and then matches comparable controls (i.e. individuals without the outcome of interest) to identify some exposures that are more commonly found in cases than in controls. Generally, case–control study design is not a common choice for medimetallomics studies due to its limitations, such as the selection bias, recall bias as well as the unclear temporal relationship.

In longitudinal cohort medimetallomics studies, participants are followed-up with over time, and several matrices are usually obtained. Simultaneous measurement of trace metal element exposure and outcomes makes it difficult to clarify the temporal relationship, which is not conducive to causal inference. Prospective studies provide an excellent opportunity to measure the burden of individual trace metal element exposure at baseline and observe the incidence of disease through follow-up, thereby revealing the true relationship between exposure and outcome. In addition, the intervariability of the trace elements in biological matrices varies according to living environment and lifestyle. Procurement of multiple biological samples from participants in the longitudinal study will be helpful for better characterizing the real burden of trace metal element in the human body.

The birth cohort is a kind of longitudinal study design. It first assesses perinatal exposure (e.g. trace metal element) in multiple matrices of pregnant women and follows up with infants over time to explore the potential association between perinatal exposure and infants' related health outcomes.

Study Population/Recruitment Criteria The selection of the study population is an integral part of the medimetallomics study process. An ideal study population should exhibit two features, namely susceptibility and representativeness. For susceptibility, previous studies have found that the health effects of trace metal element vary according to the target population. For example, merely low levels of methyl Hg exposure may result in several cognitive and neurobehavioral effects among children [32]. Moreover, susceptibility can also be reflected in different life course stages. Recent studies indicated that trace metal element exposure may be associated with abnormal pregnancy [33]. Therefore, pathophysiological mechanisms underlying the health effects of trace metal element and different metabolic characteristics among different population should be carefully considered when selecting study populations on the basis of susceptibility. It is beneficial for protecting susceptible population and drafting targeted policies.

At the same time, the representativeness of the study population should also be considered. However, it is impossible to investigate the whole populations due to the limitation of time, resource, and efforts. Standard sampling methods can be an ideal solution to get a representative sample, which can allow you to abstract the collected information to a larger population. Nevertheless, in medimetallomics studies, uncertainties and inherent variability exist in the whole process underlying the entry of trace metal element into human body including absorption, distribution, metabolism, and excretion [34]. Moreover, such uncertainties and variability are both process dependent and chemical dependent. Therefore, standardized sampling methods should be adopted to capture the variability of trace metal element in humans. For example, in NHANES, stratified, clustered four-stage samples are selected to represent the general civilian population of the United States [28].

Furthermore, during study design, recruitment criteria should be proposed according to specific scientific issues. This would prevent the inclusion of some unqualified individuals who may bias the real results. For example, participants who have eaten seafood should be excluded when focusing on the effect of specifically derived As elements, not seafood-derived elements; seafood is rich in inorganic As [35].

6.3.1.3 The Application of Medimetallomics Results

Determine the Susceptible Population The ultimate aim of medimetallomics is to protect humans from the adverse effects of metal elements and promote public health. However, one of the essential goals is to protect the susceptible population. Therefore, we should first find out the susceptible population, such as occupational workers, children, pregnant women, and the elderly.

Workers face substantial health risks due to their exposure to metal elements existing in their working environment. Occupational exposure to metal elements mainly occurs in refining, alloy production, mining, electroplating, and welding [36]. One of the major concerns is the adverse effect on lung function. Studies conducted among miners and smelter workers, who had inhalational As exposure, found increased rates of lung cancers [37]. Moreover, Navarro Silvera et al. found that industrial Ni-exposure increased the incidence of cancers (e.g. larynx, lung, and nose cancers) among workers [38]. Exposure to some trace metal element can also cause kidney issues. For example, Järup et al. found that Cd and/or Pb exposure were associated with kidney stones and tubulointerstitial nephritis [39].

A study has found that the implications of the health effects of some metal elements are more severe in children than adults [40]. The most common adverse health effects in children have been stated to be impaired neurological development. For example, it was found that Hg exposure among children was associated with mental retardation [41]. Meanwhile, Reuben et al. also stated that Pb exposure during childhood is associated with cognitive impairment, decreased intelligence, and lower socioeconomic status in subsequent adulthood [42]. Furthermore, a study conducted in Bangladeshi children illustrated that those with high levels of Pb exposure had lower cognitive scores [43]. Another notable adverse health effect on children is cancer. Sherief et al. launched a study to compare Cd levels between pediatric cancer patients and their counterparts [44]. The results showed a positive association between Cd status and malignancy. Evidence regarding cancer due to Cd exposure in children has also been shown in another study [45]. Other adverse effects of metal exposure can be seen in individuals with behavioral problems [46] and diabetes [47].

Great importance should also be attached to pregnant women. Studies have shown that exposure to some trace metal element may cause abnormal pregnancy. A study conducted among pregnant women found that serum levels of As and Pb were associated with pregnancy complications [48]. Milton et al. [49] showed that women with higher As exposure (>50 µg/l) via contaminated water seemed to have greater spontaneous abortion risk (OR = 2.5, 95%CI: 1.5–4.3) compared to those exposed to low levels (i.e. <50 µg/l). Studies have proven that some metal elements (e.g. As, Hg, and Pb) can cross the placental barrier [50], disturb cell division and spindle formation, and cause maternal reproductive toxicity [33a]. All these factors will eventually lead to abnormal pregnancy.

For elderly people, a recent study has found that chronic environmental metal exposure can have some effects related to oxidative stress, including the depletion of antioxidants and the disorder of mitochondrial lipid metabolism [51]. Moreover, exposure to multiple elements also contributed to the significant decline in kidney function among elderly people [52].

Guide the Policy Drafting and Limit Value Setting Related government departments can draft targeted policies according to the medimetallomics results. For example, spatiotemporal variation of trace metal element obtained from the HBM programs can help to evaluate the effects of the policies that have been implemented and to make targeted new policies. Moreover, limit values for various harmful trace metal element can be set according to medimetallomics results.

6.3.1.4 Next Steps and Opportunities for Medimetallomics

Speciation Analysis of Trace Metal Element Previous studies have mainly investigated the relationship between total metal concentration and health. However, medimetallomics should consider the chemical form (i.e. species) because it affects the toxicokinetics and toxicodynamics of metals. Toxicology studies have confirmed that the metal speciation of As, Hg, Cr, Pb, manganese, aluminum, and Fe influences mammalian toxicity [53]. Exposure to low-dose methyl Hg (MeHg), an organic Hg compound, can cause subclinical neurotoxicity and presents the highest risk during prenatal development [54]. Unlike MeHg, little is known about the developmental neurotoxicity of inorganic Hg compounds or elemental Hg [55]. Even though elemental Hg can cross the placental barrier, its accumulation in the brain is far less in fetuses than in mothers [56].

Longitudinal Study on Medimetallomics Temporal variability attributed to changes in the environment and life course, temporality criteria for causal inference, and the use of different biological matrices may result in variations in the conclusions [57]. Longitudinal investigations that include all kinds of social groups, including susceptible people (e.g. children, pregnant women, and the elderly), offer a good chance to comprehensively monitor spatiotemporal patterns of trace metal element exposure among the whole population. Furthermore, measuring the exposure of interest at multiple time points can accurately capture the actual exposure over time, which is essential for exploring the critical/sensitive periods during the life process [58]. Thus, it is pragmatically worth exploring the various metal profiles at multiple points within the life course.

6.3.2 Clinimetallomics

6.3.2.1 Diseases Associated with Trace Elements

The content of dozens of elements such as Fe, Zn, Cu, manganese (Mg), molybdenum (Mo), Cr, and Co in human body is extremely small, accounting for only 0.01% of the body mass, which is called trace elements. Trace elements are the basic components of the human body, participate in the homeostasis of the body, and play an important role in endocrine, reproductive development, energy metabolism, and other aspects. When its lack or proportion is out of proportion, it will affect the normal metabolism and function of the body, and even cause structural changes and diseases [59].

Neurodegenerative Diseases Globally, neurological disorders are the leading cause of DALYs, with approximately 276 million, both the leading cause of disability and the second leading cause of death, with approximately 9 million. The main neurological disorders affecting DALYs are stroke (42.2%), migraine (16.3%), Alzheimer's and other dementias (10.4%), and meningitis (7.9%) [60].As early as 1989, Emmett proposed that ICP-MS could be used to examine the levels of various trace elements in brain tissue, cerebrospinal fluid (CSF), and serum of Alzheimer's disease (AD) patients and matched control subjects [61]. Neurodegenerative diseases represent a group of chronic progressive diseases or disorders, mainly in old age, characterized by the deterioration and eventual loss of neuronal cells in specific brain regions [62].

Major neurodegenerative diseases include AD, Parkinson disease (PD), Huntington's disease (HD), and amyotrophic lateral sclerosis (ALS). Previous studies have identified a relationship between dysregulation of trace element homeostasis and the onset and progression of neurodegenerative diseases [63]. In recent years, metallomics has been applied to the study of these diseases, which is more helpful to understand the biochemical changes of these diseases and discover new drug targets [64].

Although the pathogenesis of AD is unclear. Gerhardson et al. measured the concentrations of 19 elements in the plasma and cerebrospinal fluid of AD patients, AD patients and minor vascular components (AD + vasc), and healthy controls [65]. Elevated concentrations of Mn and Hg in plasma and decreased concentrations of V, Mn, Pb, and several other trace metals in cerebrospinal fluid were observed in subjects with AD and AD + vasc compared to controls. However, no consistent metallic pattern was detected in the plasma or cerebrospinal fluid of patients with AD. This indicates the specificity of metallomic characteristics in different body fluids. To investigate the progression of AD. González-Domínguez et al. analyzed serum ion group such as Fe, Cu, Zn, and aluminum (Al) in healthy individuals, AD, and patients with mild cognitive impairment (MCI), which appear to occur with the development of neurodegeneration progressive changes, while Mn and V were found to be closely related to early neurodegeneration changes, suggesting that metal abnormalities may be associated with different biological processes involved in cognitive decline [66]. Koseoglu et al. measured the concentrations of various trace elements in the hair and nails of AD patients at different clinical stages and revealed significant tissue-specific differences between patients and healthy controls [67]. Some previous studies have shown that AD mainly affects the structure and function of the brain, and the dysregulation of metal homeostasis in different brain regions is closely related to the development of AD [68]. Ciavardelli et al. analyzed the ionic group changes of mouse brain after long-term dietary Zn supplementation [69]. Zheng et al. performed a multi-time-point metallomic analysis using 3xTg-AD mice receiving high-dose selenium supplementation to study the interaction of 15 elements in the brain, which revealed that Se intake could significantly reduce the concentrations of a variety of elements in the brain, especially Fe (a known risk factor for AD) [70], which revealed that Se intake could significantly reduce the concentrations of a variety of elements in the brain, especially Fe (a known risk factor for AD). Such highly complex and dynamic crosstalk between selenium and other elements may provide new mechanistic clues for the potential use of selenates in ameliorating AD-related symptoms and pathology [71].

Parkinson's disease (PD) is recognized as the second largest neurodegenerative disease in the world. The current number of patients in China has exceeded 2.45 million, and it has become a major social problem affecting the health of the population in China. To study the relationship between PD and trace elements, Ford et al. quantified the levels of multiple metals in urine, serum, blood, and CSF of PD-affected patients and age-matched control patients, indicating the significant metal changes occurred in body fluids, such as a downward trend in Al in all studies PD patients had body fluids and elevated Ca levels in urine, serum, and blood [72]. In recent years, many metallomic studies have been carried out, providing more reliable evidence for the interaction between trace elements and the development of PD. For example, Zhao et al. measured the levels of several trace elements in the plasma of 238 PD patients and 302 controls recruited from eastern China, and found that lower plasma Se and Fe levels may reduce the risk of PD, while lower plasma Zn may be a risk factor for PD. They proposed a new model for predicting PD patients, with an accuracy of more than 80% [73]. Sanyal et al. conducted an Indian population cohort study to analyze changes in element distribution in CSF and serum of PD patients and found significant changes in several elements such as Ca, Mg, and Fe [74].

Besides AD and PD, there are also metallomic studies on several other neurodegenerative diseases such as Huntington's disease (HD) and Amyotrophic Lateral Sclerosis (ALS). HD is an autosomal dominant neurodegenerative disorder that causes progressive motor, mental, and cognitive impairment, and dysregulation of the homeostasis of certain metals in the brain may contribute to the neuropathogenesis of HD [75, 76]. Squadrone et al. reported abnormal concentrations of several trace metals in the blood of HD patients, suggesting that the use of blood metal profiling may be useful in identifying potential metal targets for new treatments targeting HD [77]. ALS is an adult-onset progressive and fatal motor neuron degenerative disease of unknown etiology. Over the past decade, metal dysregulation has been proposed to be involved in ALS pathophysiology [78]. Another study analyzed serum and whole blood concentrations of broad-spectrum metals in patients with sporadic ALS and found that higher concentrations of selenium, manganese, and aluminum, as well as lower serum As concentrations, were associated with the disease [79]. Oggiano et al. analyzed the blood, urine, and hair levels of various essential metals and heavy metals in patients with different stages of ALS, which indicated a protective effect of selenium and a risk factor for the presence of Pb in the blood [80].

Metals can cause serious damage to the nervous system. Because of the damage to the peripheral nervous system, Hg can also cause neurological damage and brain damage, and people exposed to too much Hg may develop cognitive and learning disabilities; damage to the human nervous system can also lead to metabolic disorders. We should be aware of the neurological effects of heavy metal mixtures. Most of the current studies focus only on the effects of single heavy metals. However, Hg alone does not lead to loss of enzyme function; it is only when it interacts with other metals and occupies the active target site of the enzyme that it truly has an effect. Compared to single metals, polymetallic exposures can be more toxic to cells and increase the risk of disease. In the future, the effects of metal mixtures on the nervous system should be explored in depth, not just single metal.

Metabolic Diseases Diabetes mellitus is a group of metabolic diseases characterized by chronic hyperglycemia, which may lead to long-term damage, dysfunction, and failure of various organs [81]. Approximately 463 million adults aged 20–79 years had diabetes worldwide in 2019 (1 in 11 people had diabetes), projected to reach 578.4 million by 2030 and projected to reach 700.2 million by 2045 [82]. Studies have shown that ion homeostasis may play an important role in T2D and some other associated metabolic abnormalities [81].

Sun et al. measured fasting plasma element concentrations in approximately 1000 middle-aged Chinese men and women to study the association of ionic modules/networks with obesity, metabolic syndrome, and T2D [83], and showed that among the elements associated with metabolic abnormalities Cu and P were consistently ranked as the top two elements that could potentially be associated with T2D and other metabolic disorders. Liu et al. analyzed the urine ionomer of more than 2100 Chinese aged 55–76 years and showed that increased urinary Ni concentrations were associated with increased prevalence of T2D, suggesting the potential harm of Ni exposure in the pathogenesis of T2D [84]. Another case–control study based on a Chinese population investigated the relationship between the risk of T2D and plasma levels of 20 trace elements and heavy metals, and found that several heavy metals (e.g. Mn, Cu, Zn, As, and Cd) were associated with the onset of diabetes, rates were positively correlated, implying that environmental metal levels had a significant impact on the risk of T2D [85]. Badran et al. assessed the serum levels of 24 trace elements in Egyptian T2D patients and showed that Mg, Fe, Cu, and Zn appear to be the most critical factors associated with T2D [86]. Recently, Marín-Martínez et al. analyzed the levels of 19 trace elements in saliva and plasma of patients with type 2 diabetes and their association with metabolic control and chronic complications [87]. Decreased levels of Co in saliva and elevated levels of strontium (Sr) in plasma have been observed in diabetic patients with chronic complications, providing complementary information for predicting diabetic complications. Overall, it appears that populations of diabetic patients from different regions/ethnics and different body fluids/tissues from the same patient have specific elemental profiles, suggesting a high degree of complexity and tissue specificity of the ionome.

Metallomic studies of T1D diabetes and gestational diabetes mellitus (GD) have also been reported. Ford et al. analyzed blood levels of various trace elements in subjects with T1D and T2D living in Sardinia (an Italian island region with a high incidence of T1D) and showed that T1D was associated with Cr, Mn, Ni, Pb, and Zn deficiencies, while T2D was higher in Cr, Mn, and Ni deficient in Cr, Mn, and Ni, and Pb, the only metal that was significantly different between the two diabetes types, was higher in T2D compared with controls [88]. Peruzu et al. further evaluated the correlation of these elements with blood lipids and glycemic control in T1D patients in the same region, and found that Zn, Fe, Se was associated with different types of blood lipids, while Cr and Cu were significantly correlated with fasting blood glucose and glycated hemoglobin levels in males, respectively [89]. Elevated serum Cu and molybdenum levels and decreased manganese, Zn, and selenium levels have also been reported in T1D subjects in the United States, suggesting that altered balance of these essential elements may be an important hallmark of T1D [90]. Roverso et al. analyzed the placental ionomer of GD-affected and healthy women and showed that Cd levels were reduced and Se levels were elevated in the GD placenta, implying that they are two key elements for understanding the molecular pathway of GD [91]. More recently, they examined the ionomer of placenta, maternal whole blood, and cord blood samples from more GD and control pregnant subjects [92]. Results indicated that many of the elements detected in fetal cord blood were found to be significantly associated with GD, making it possible to use cord blood to help understand the biochemical processes that occur during GD and to delineate more accurate nutritional guidelines for pregnant women [93].

Thyroid dysfunction mainly includes hyperthyroidism and hypothyroidism. Studies in recent years have shown that thyroid dysfunction is not only related to iodine, but may also be related to other trace elements. Mu Xiaodong et al. studied the correlation between serum trace elements and thyroid diseases, collected and sorted 375 cases of inpatients with thyroid diseases, and analyzed the relationship between different genders, thyroid disease types, and trace elements. It was found that different thyroid diseases have different abnormal rates of trace elements. The trace elements in patients with thyroid disease are mainly reduced, especially potassium ion, calcium ion, and sodium ion, which are closely related to the pathogenesis of thyroid disease. Wu Zhaoyu conducted a study on the correlation between trace elements iodine, selenium, magnesium, fluorine, and Fe, and thyroid diseases, and found that the content of trace elements in thyroid tissue of normal people was significantly different from that of patients with thyroid diseases. There are differences in the content of human body, fluorine cannot be effectively measured, and the content of Fe in nodular goiter and thyroid cancer is much different from that of normal people, and the content of trace elements iodine and Fe in three thyroid diseases is also different. Deficiency or excess of iodine, selenium, magnesium, Fe, etc. may become risk factors for thyroid disease, that is, it may cause the occurrence of thyroid disease.

Hyperthyroidism is a disease characterized by hypermetabolism and increased neural excitability caused by excessive thyroid hormone secretion. Li Bingzheng et al. analyzed 5 kinds of trace elements in the serum of patients with hyperthyroidism to explore the relationship between the trace elements in the serum of patients with hyperthyroidism and the changes of the disease. It was found that the serum Zn^{2+} and K^+ of the patients with hyperthyroidism was lower than the normal group, Cu^{2+} was higher than the normal group, Ca^{2+}, Fe^{3+}, no significant difference. The results indicate that supplementing an appropriate amount to patients with hyperthyroidism is helpful for the recovery of the disease [94].

Cancer Cancer has become a major threat to human health and has become the number one cause of death. A total of 14.1 million new cancer cases and 8.2 million deaths worldwide were reported in 2012 [95]. However, by 2017, there were 24.5 million new cancer cases and 9.6 million cancer deaths worldwide [96]. On January 8, 2020, the 2020 U.S. Cancer Statistics report was released in the top international publication *CA: A Cancer Journal for Clinicians*, with cancer incidence data included through 2016 and mortality data included through 2017. It is expected that 1 806 590 new cancer cases will occur in the United States in 2020, and 606 520 people will die from cancer, equivalent to more than 1600 deaths per day [97].

Cancer development is a very complex process, and numerous studies have been conducted to analyze the relationship between one or several metals and different cancers, such as the relationship between dietary intake of metals and cancer risk, possible carcinogenic mechanisms of metals, and long-term heavy metal exposure and cancer occurrence [98]. Recently, metallomic methods have been used to examine changes in the distribution of elements in different body fluids and tissues in several cancers (such as prostate, lung, pancreatic, and gastrointestinal). The complex interactions between them provide valuable information and also offer considerable potential for metallomic studies for cancer biomedical purposes.

Prostate cancer (PC) is the most common cancer in men with a high mortality rate, and exposure to toxic metals is one of the most important factors in the etiology of prostate cancer. The study analyzed elemental profiles in the blood, scalp hair, and nails of PC patients and found increased concentrations of Fe, Cd, Mn, Ni, and Pb, and decreased concentrations of Zn, in all or nearly all tissues examined of patients compared with healthy controls [99]. Saleh et al. study suggests that decreased serum levels of selenium, Zn, and manganese and elevated levels of Cu and Fe may play an important role in the initiation of PC [100].

Lung cancer (LC) is a cancer with a high mortality rate in the world. Many studies have shown that trace elements play an important role in the carcinogenesis of lung cancer. Lee et al. examined the concentrations of 14 elements in the pleural effusion of LC patients and found that only Zn concentrations were significantly lower in all smokers than in non-smokers, suggesting that cigarette smoke may contribute to Zn deficiency in the pleural effusions of smokers [101]. Callejón-Leblic et al. conducted a cross-sectional study analyzing trace elements in serum, urine, and for the first time in bronchoalveolar lavage fluid (BALF) samples from patients with LC, several metals (such as V, Cr, and Cu in serum, Cd in urine, and Mn in BALF) and metal ratio/correlation can be used as important markers of LC in different body fluids [102]. A recent review discussed the key role of metals in LC, especially the high metal exposure, metal homeostasis and interaction, and the importance of metal morphology in the pathogenesis and progression of cancer, which indicated the potential use of metals this kind of information in the early diagnosis, prognosis, or treatments of LC [103].

Pancreatic cancer (PaC) is one of the most lethal and aggressive malignancies with poor prognosis and survival, making early diagnosis very difficult. A recent urine metallomic study of PDAC (PDAC, the most common form of Pa) by Schilling et al. showed significant differences in several base metals (including significantly reduced Ca and Mg levels and higher levels of Cu and Zn), implying that metallomics is a promising approach to discover early diagnosis [104]. There is still a lack of metallomic analysis of PaC blood samples [93].

Metallomic studies have also reported several types of gastrointestinal cancers, including colorectal, gastric, and esophageal cancers. A metallomics study analyzed 18 elements in tumor and adjacent non-tumor paired samples from Spanish colorectal cancer (CRC) patients and found that accumulation of many essential elements such as Mn, Se, Cu, and Fe can be used as indicators of cancer progression [105]. A recent Asian cross-sectional study also assessed the levels of trace elements and heavy metals in cancerous and noncancerous tissues of CRC patients, and the results showed somewhat different changes in elemental profiles, and indicated that other metals (such as increased Zn and Cr and decreased Mn levels) may play a role in developing CRC [106]. Kozadi et al. compared the concentrations of trace elements in gastric cancer tissue and normal tissue, and found that compared with non-cancerous tissue, the content of Fe, Mg, and As in cancer tissue was higher, while the content of Cr, Cu, Ca, and Ni was lower [107]. Lin et al. used metallomics and machine learning methods to study the relationship between changes in serum elemental profiles and esophageal squamous cell carcinoma (ESCC), which indicated the potential application of metallomics in the diagnosis and prognosis of ESCC [108].

In addition to the aforementioned cancer types, metallomics has also been applied to several other cancers [93]. Burton et al., examining the urinary ionomer of women newly diagnosed with breast cancer, found significant increases in Cu and Pb in patients, and constructed a multivariate model with information on Cu, Pb, and patient age for breast cancer early detection of cancer [109]. Lee et al. examined the association of multiple metals in serum with gallbladder cancer (GBC) and gallstones, an important risk factor for GBC, as a strong link between the timing of metal exposure and the natural history and mechanisms of gallbladder cancer. Correlation provided cross-sectional evidence for GBC [110]. Golasik et al. analyzed the status of various essential and toxic elements in the hair and nails of laryngeal cancer patients and healthy controls and showed that the patient group had significantly lower levels of Zn and Fe, while the opposite was observed for heavy metals trend [111].

Currently, there are many studies on the effects of single-factor exposure on cancer, but there are fewer scholarly studies on mixed exposures to multiple metals. Cancer is a current hot issue in the health field, its etiology is definitely not simply determinable, and it is important to pay attention to all aspects of exposure pathways to better prevent cancer and reduce its incidence. People should further prove its etiology.

Obesity More than 2 billion people worldwide are overweight or obese, which means that approximately one-third of the global population suffers from overweight or obesity-related health problems. From 1979 to 2016, the number of obese girls worldwide rose from 5 to 50 million in 2016, and the number of obese boys rose from 6 to 76 million; the number of obese adult women increased from 69 to

390 million, and obese men increased from 31 to 281 million. An additional 213 million children and adolescents and 1.3 billion adults are overweight, with obesity precursors [112].

Wang, X., et al. explored the association between exposure to heavy metal mixtures and obesity and its complications in US adults in a study of NHANES data from 2003–2014 using the environmental Risk Score (ERS) to assess the association of heavy metal mixtures with obesity and its complications, with cumulative exposure to heavy metal mixtures being associated with obesity and its associated chronic diseases such as hypertension and T2DM disease [113]. Niehoff, N.M., et al. in a prospective sister cohort study in the United States with a sample size of 1221, examined the effect of 16 metals on body mass index, and essential metals were found to be associated with lower body mass index (BMI), whereas nonessential metals were associated with unhealthy BMI correlation [114]. Some studies have shown that Hg, manganese, and Co also have an effect on lipid metabolism in fat, which not only causes disease but also accelerates the disease process [115]. A large body of evidence suggests that population exposure to a mixture of heavy metals is more likely to cause obesity and its complications in the population, increasing the risk of disease. Yao Shuangshuang et al. explored the correlation between trace elements in the human body and obesity and metabolism by studying the differences in serum trace element content between obese and normal weight people. Vanadium, Fe, Co, selenium, and strontium were negatively correlated with body mass index, waist circumference, hip circumference, percentage of body fat content, and other metabolic-related indicators (fasting blood glucose, triglycerides, etc.), and serum Cr, manganese, and Cu were positively correlated with the above indexes ($P < 0.05$). The conclusion of this study is that serum trace elements are closely related to obesity, and may be involved in the metabolic process in vivo and have further influence on obesity and its complications [116].

In summary, many investigators and study design protocols preferentially use the blood and urine of subjects as biomarkers. Blood and urine, as an indicator of exposure over a short period of time, can reflect intraindividual exposure levels over a recent period of time relatively quickly but are not necessarily good at obtaining a true correlation between prolonged metal mixture exposure and disease. While obesity is a long-term process, the causes of obesity can be many, so the association between heavy metal mixtures for obesity can focus on the selection of hair, teeth, and nails and other biomarkers that can detect long-term exposure to better obtain the relationship between long-term exposure and disease. Subsequent studies should also consider whether heavy metals affect obesity directly by affecting metabolism and causing obesity or through other related pathways that cause imbalance in physiological levels and thus indirectly cause obesity.

Liver Diseases The liver participates in metabolic processes involving heavy metals [117]. Heavy metals can accumulate in the liver and damage it [118]. Total protein (TP), albumin (ALB), aspartate aminotransferase (AST), and alanine aminotransferase (ALT) are common biomarkers of liver damage [119]. Hepatocellular damage results in the release of intracellular AST and ALT into the blood, thus increasing the serum levels of these biomarkers [120]. TP and ALB are synthesized by the liver; a decrease in their levels indicates impaired liver function [121].

Evidence of an association between heavy metal exposure and liver function biomarkers continues to increase. In animal experiments, AST and ALT levels were noted to increase with exposure to Cr [122], Cd [123], Pb [124], and Mn [118a, 125]. Furthermore, in animals exposed to Cr and Pb, plasma TP levels were reported to be decreased [124c, 126], whereas in Mn-treated animals, they were increased, with ALB levels unchanged [125]. A Greek epidemiological study indicated that Cr is positively associated with TP and ALB in people with long-term exposure to Cr in water. A positive dose–response relationship has been discovered between Cr and ALT [127]. The Korea National Health and Nutrition Examination Survey (KNHANES) indicated that Cd raises the risk of increased AST and ALT levels [128]. Another group of researchers found a positive association between Cd and serum ALT and AST levels in 2953 Korean adults [129]. Similarly, Nakata et al. reported a positive relationship between Cd and ALT in a population living in mining areas of South Africa [130]. Pb is associated with elevated ALT and the prevalence of nonalcoholic fatty liver disease (NAFLD) [131]. In addition, childhood Pb exposure was associated with higher ALT level in young Mexican adults [132]; by contrast, Huang et al. reported a negative association between Pb and ALT [133]; Mn was noted to be related to increased liver enzyme levels and mortality associated with chronic liver disease [134]. Furthermore, in patients with liver cirrhosis, the average serum Mn level was significantly higher than that of the control group [135]. By contrast, Zhang et al. reported that Mn may be a protective factor for NAFLD in men [136].

In epidemiological studies exploring liver damage and heavy metals, researchers have generally focused on the effect of a single heavy metal. However, heavy metals in the environment are often not present alone. A key principle in the Framework for Metals Risk Assessment of the US environmental Protection Agency is that metals usually enter the environment as a mixture [137]. Multiple heavy metals have been reported to coexist in air, surface water, and soil [138], leading to individuals being exposed to multiple heavy metals simultaneously. In addition, heavy metals have interactive effects [139]. Thus, studying the effects of a single metal cannot reflect the real exposure status of a population. It's more valuable to study overall effects of Cr, Cd, Mn, and Pb on liver function.

Kidney Disease The number of people with chronic kidney disease(CKD) reached 697.5 million worldwide in 2017, including 132.3 million in China and 115.1 million in India [140]. CKD is responsible for 1.2 million deaths worldwide, a 41.5% increase from 1990, and an additional 1.4 million cardiovascular deaths attributable to renal impairment, accounting for 7.6% of cardiovascular deaths in 2017. The study predicts that the number of deaths due to CKD could increase to 2.2 million by 2040 and to 4 million in the worst-case scenario [140].

Heavy metals primarily contribute to kidney disease because a variety of low-molecular-weight proteins can specifically bind certain heavy metals, such as Cd, Hg, and Pb, which have a strong effect on the kidney. Tsai, C.-C., et al in

2018 in a study on the exposure of chronic kidney disease patients with mixed exposure to heavy metals in northern Taiwan found that the incidence of chronic kidney disease (indexed by HR) was associated with several heavy metals such as Pb, Cd, As, and Hg, and that patients living in polluted areas had an increased risk of chronic kidney disease, CVD, and hypertension, and that excess Zn and Ni were associated with advanced. There is an additive effect of the negative effects of nephropathy [141]. The kidneys do not easily metabolize these metals quickly, and the mixture of multiple metals has a synergistic effect, reducing glomerular filtration capacity and decreasing kidney function. Additionally, exposure to Pb, Cd, Hg, and As plays a negative role in the progression of kidney disease and greatly increases the risk of chronic kidney disease, more so than with single heavy metals. Sanders et al. conducted a study on the relationship between mixed exposure to heavy metals and biomarkers of kidney injury in 2709 adolescents in the National Health and Nutrition Examination Survey (NHANES) from 2009 to 2014. The results showed that the urinary mixture of As, Cd, Pb, and Hg had a positive association with glomerular filtration rate (EGFR), urine albumin, and blood urea nitrogen (BUN) [142]. Similarly, Luo et al. also found a positive correlation between the mixture of Co, Pb, Cd, and methyl Hg (meHg) and the prevalence of CKD and ACR (albumin-to-creatinine ratio) in 1435 elderly people of NHANES [143]. Meanwhile, evidence showed that there were positive interactions between Pb and Cd/Co. Furthermore, a retrospective cohort study of CKD patients in Taiwan also observed that residents living in areas with high soil heavy metal concentrations had an increased risk of end-stage renal disease (ESRD) [141]. Prodanchuk et al. analyzed the concentrations of about 20 trace elements in whole blood of patients with ESRD and found that the levels of most of the examined elements were significantly increased in ESRD patients (especially non-essential and toxic trace elements), while the essential elemental selenium was significantly reduced, implying that monitoring and elimination of accumulated toxic elements and increased selenium intake should be considered in ESRD patients undergoing hemodiafiltration [144].

The above studies show that the damage of heavy metal mixed exposure to kidney is not limited to early functional index damage, but reflected in aggravating and promoting the deterioration of renal disease. However, kidney diseases are preventable and treatable to a great extent, which should be given more attention in global health policy-making.

Cardiovascular Disease Cardiovascular disease (CVD) is the leading cause of mortality worldwide and the leading cause of disability [145]. CVD is the leading cause of death in the Chinese population [146], and approximately 40% of total deaths in China are due to cardiovascular disease.

Chowdhury, R., et al.'s 2018 meta-analysis found a positive association between multiple heavy metal mixtures and cardiovascular disease, with exposure to As, Pb, Cd, and Cu increasing the risk of cardiovascular disease and coronary heart disease [147]. Yang, A.-M., et al. and Leone, N., et al. found an increased risk of cardiovascular disease with exposure to Zn, Cu, Cr, Co, Mn, Se, Ni, and W in both population and animal studies [148]. In 2018, the British academic Chowdhury, R., et al. found that population exposure to As, Pb, Cd, and Cu increased the risk of cardiovascular disease and coronary heart disease, but there is no evidence that Hg modifies the risk of cardiovascular morbidity [147]. In 2019, Chinese and American scholars Zhao, D., et al. found a weak positive association between W and CVD morbidity and mortality when urinary molybdenum levels were low and a strong positive association between W and CVD morbidity and mortality when urinary molybdenum levels were high in a study on a mixture of molybdenum and W [149]. In addition, urinary Cd was found to be associated with an increased risk of ischemic stroke, but when serum concentrations of Zn were low, the risk of stroke was increased [149]. The researchers thus speculated that Zn in serum may have a role in inhibiting the harmful effects of Cd in humans and reducing the chance of ischemic strokes. In 2003, Choe, S.Y., et al. found that toxic heavy metals (e.g. As, Cd, Pb, and Hg) and some essential metals (e.g. Co, Cu, Ni, and Se) may also increase the risk of CVD in the population by affecting the endocrine system and disrupting hormone levels in the body [150]. Domingo-Relloso, A., et al. found a positive association between urinary Cu, Zn, Cd, Cr, and V and cardiovascular morbidity in a study of cardiovascular morbidity in a sample size of 1171 Spanish adults without cardiovascular disease, with antimony and Cd being the main metal ions involved [151]. DeVille, N.V., et al., in a prospective cohort study in the United States, examined the relationship between 11 metal mixtures and cardiometabolic risk in children, and found prenatal levels of selected metals increased cardiometabolic risk in children aged four to six years [152].

Hypertension Globally, the prevalence of hypertension in the population is still on the rise. Although treatment and control rates for hypertension have improved significantly in recent years, treatment and control rates are still at low levels. A survey by British academic Sarki, A.M., et al. in 2015 showed that on average, one in three people in developing countries will suffer from hypertension [153]. In recent years, it has been found that elevated ambient concentrations of pollutants such as $PM_{2.5}$, PM_{10}, SO_2, and O_3 may lead to an increased risk of developing hypertension.

Wu, W., et al. found in a 2015 cross-sectional study of 823 middle-aged and older adults in Wuhan, China, that environmental exposure to vanadium, Fe, Zn, selenium, and Hg may increase the risk of hypertension or raise blood pressure levels and that exposure to Hg may be associated with elevated diastolic blood pressure levels [154]. The same team found in a case–control study in Wuhan that exposure to environmental V, Fe, Co, Ni, Cu, Zn, Rb, Sr, Cd, and Hg may be associated with the prevalence of hypertension: urinary Fe, Co, Ni, Cu, Zn, Sr, and Hg levels were positively associated with the risk of hypertension, but urinary V, Rb, and Cd levels were negatively associated with the risk of hypertension [155]. Xu, J., et al., in a sister cohort study in the United States, examined the risk of mixed exposure to 10 metals in air and hypertension, As, Co, Cd, Cr, Pb, and Mn were found to be positively associated with the prevalence of hypertension, whereas Ni, Se, Hg, and Sb were negatively associated with the prevalence of hypertension [156]. Castiello, F., et al. found in a cohort study of 133 Spanish adolescents with combined exposure to toxic metals that exposure to metal mixtures, especially As and Cd, Hg, As, Cd, and Cr combined, may affect their hormone levels and may lead to elevated blood pressure in male adolescents [157]. It is suggested that exposure to metal mixtures may have an impact on the risk of hypertension in adolescents.

Most of the cardiovascular system studies have been conducted in the middle-aged and elderly population, whose incidence and prevalence of cardiovascular diseases and hypertension are inherently higher than those of the general population due to aging of body organs and declining physical functions. The risk of morbidity in middle-aged and elderly people is related to several factors, so we need to investigate the magnitude of the weight of cardiovascular system diseases caused by heavy metal mixtures. Several existing experiments have demonstrated that the incidence of cardiovascular disease and hypertension in adolescents in heavy metal-contaminated areas is also increasing year by year, which provides strong evidence for the existence of an association between metal mixtures and cardiovascular system risk and suggests that further research on the adolescent population is needed. Therefore, further attention needs to be paid in the future to the exposure of metal mixtures in the adolescent population and the relationship between exposure to metal mixtures and target outcomes, and more experiments need to be designed to investigate the extent to which metal mixtures actually have an effect on cardiovascular disease in humans.

Dyslipidemia Dyslipidemia refers to abnormalities in the metabolism of lipoproteins, including triglycerides (TG), total cholesterol (TC), low-density lipoprotein cholesterol (LDL-C), and high-density lipoprotein cholesterol (HDL-C), and commonly occurs in the older population [158]. Dyslipidemia is a major risk factor for cardiovascular disease (CVD). Over the past two decades, the incidence and prevalence of dyslipidemia have gradually increased in the older population in China [159]. The Chinese National Nutrition Examination Survey indicated that the overall prevalence of dyslipidemia in Chinese adults increased from 18.60% to 40.40% from 2002 to 2012. The incidence of dyslipidemia is increasing not only in China but also in other countries [160]. The Global Burden of Disease Study that was conducted in 204 countries and territories from 1990 to 2019 reported that elevated LDL-C levels are a major risk factor for disability-adjusted life years globally in older populations [161].

The development of modern industry and the advancement of technology have led to the release of many toxic heavy metals; these metals adversely affect health and cause diseases through different exposure pathways, including direct oral ingestion, dermal absorption, and air inhalation [162]. The World Health Organization (WHO) has listed Hg, As, Pb, and Cd as the top 10 pollutants and metals that adversely affect health [163]. Numerous epidemiological studies have reported an association between exposure to heavy metals and various adverse health outcomes, including dyslipidemia [158, 164]. To our knowledge, few studies have investigated the relationship between mixed heavy metal exposure and dyslipidemia, and the findings of these studies are inconclusive. Although some studies have reported an association between heavy metal exposure and CVD, no study has demonstrated the relationship between mixed heavy metal exposure and dyslipidemia [165]. Several epidemiological studies have reported a significant association between single heavy metal exposure and dyslipidemia. A cross-sectional study conducted in China indicated that high Mn exposure reduced the risk of high TG among workers [166]. Another study conducted in Taiwan observed that an increase in blood Pb levels caused a significant decrease in HDL-C levels and an increase in LDL-C levels [167]. Two NHANES studies have suggested that an increase in urinary Cd levels led to a decrease in HDL-C levels and an increase in LDL-C levels [168]. A meta-analysis demonstrated a negative relationship between urinary Cd and HDL-C levels [169]. A 2011–2014 US NHANES study reported that increases in blood Pb and urinary Cd levels caused a decrease in HDL-C levels [170]. A Korean study demonstrated that blood Pb level was significantly positively correlated with TC level ($p < 0.05$) and significantly negatively correlated with HDL-C level ($p < 0.05$) [171]. Epidemiological studies conducted in the United States, Ghana, Ethiopia, and China have all reported that urinary, serum, and nail Cr cause alterations in animal lipoproteins and a significant relationship between elevated urinary Cr levels and elevated HDL-C levels [172]. However, a study conducted in the United States indicated that the TG level gradually increased with an increase in the Cr level. By contrast, other studies conducted in China and Ethiopia have reported that an increase in the urinary Cr level caused a decrease in the TG level [172a–c].

Given the potential hazards of metal mixtures to human health, studies have suggested paying more attention to exposure to heavy metal mixtures rather than single metals in real-life scenarios. A prospective cohort study conducted in Hubei Province, China, reported that the incidence of dyslipidemia increased with increasing plasma levels of aluminum, As, strontium, and vanadium [164b]. Zhu et al. indicated that combined exposure to Cd, strontium (Sr), and Pb was a risk factor for dyslipidemia [158]; this finding is consistent with that of the NHANES study [164c]. However, several studies have not observed a relationship between exposure to multiple urinary metals and dyslipidemia (mainly elevated TG and decreased HDL-C levels [170, 173]. Although many studies have investigated the effect of exposure to metal mixtures on dyslipidemia, few studies have evaluated the relationship between exposure to metals and health outcomes. Zhao et al. reported that Cr exposure was related to lipid metabolism in the older population. The authors observed a negative additive interaction between urinary Cr and urinary Cd for body mass index (BMI) and abdominal obesity but no interaction between urinary Cr and urinary Pb or Mn [174]. A cross-sectional study that included 637 individuals from China identified the interactive effect of Zn and Cd on LDL-C but did not explore the interaction direction [175]. The Korea National Health and Nutrition Examination Survey (KNHANES) indicated that the cumulative effects of exposure to heavy metal mixtures may be additive or synergistic; however, they did not observe any specific interaction [176]. A study reported that an increase in the co-exposure dose of Pb and Cd would reduce the prevalence of dyslipidemia [170]. Therefore, determining the potential interaction between metals is crucial for preventing dyslipidemia.

Neuromas study observed a potential synergistic effect of metal–metal on LDL-C. Zhu et al. reported that the combined exposure of Pb and Cd increased the prevalence of dyslipidemia [158]. Although the interaction between the two metals was noted, the interaction direction remains unclear. Guo et al. reported a significant interaction between Pb and Cd [177]. The possible reason is that both Cd and Pb cause oxidative stress, resulting in a decrease in the MDA level. Simultaneously, Pb and Cd interfere with and change the levels of Zn, Cu, and Fe in blood and tissues [178], thus interfering with normal lipid metabolism.

Essential metals in the body may weaken the toxicity of some heavy metals. Thus, when the levels of some essential metals in the population are high, the potential interaction between heavy metals would not be observed. For example, Cd would interfere with Ca^{2+} homeostasis [117a] when the Ca level in the population is high, and the interaction between Cd and other metals may not be observed. Similarly, Pb interferes with the utilization of Cu and inhibits the expression of Pb when the Cu level in the population is high [179]. Previous studies investigated the combined effect of heavy metal exposure as well as the interaction of different metal pairs on dyslipidemia, and observed potential synergistic and antagonistic effects of heavy metals in their relationship with dyslipidemia. In the future, these results should be confirmed in longitudinal studies by using repeated measurement data. Future epidemiological studies investigating the effect of metals on human health must consider the single, combined, and interaction effects of multiple pollutants on health.

Other Diseases Recently, metallomic studies of several other diseases have also been reported. For example, Ilyas et al. analyzed the concentrations of trace metals in the blood and scalp hair of patients with ischemic heart disease (IHD) and showed significant differences in the elemental profiles in the two tissues, suggesting that metal imbalances may contribute to better understanding the occurrence of ischemic heart disease and preventing IHD [180]. In addition, sepsis is one of the major problems threatening human health worldwide, with extremely high morbidity and mortality. Li Rui et al. summarized the research progress on the correlation between the occurrence and development of sepsis and micronutrients, and it was found that micronutrients (trace elements and vitamins), as essential components or cofactors of enzymes in the metabolic process, play an important role in the occurrence and development of sepsis by regulating immunity and anti-oxidation [181].

6.3.2.2 Toxic-Element-Related Diseases

With the process of modern urbanization and the rapid rise of industrialization, it has promoted the development of the national economy, but it has also led to environmental pollution, releasing heavy metals such as As, Pb, Hg, and Cr of toxic waste, smoke, and fumes. And most heavy metal pollution is serious, long term, and irreversible. The most common heavy metals as pollutants include As, Pb, Hg, Cr, Zn, and Cd. [182]. Sharma et al. summarized those heavy metals such as Hg, as, and Pb have toxic effects on the kidney and nervous system, further leading to mental disorders, weakness, headache, abdominal colic, diarrhea, and anemia. A detailed description of all eight heavy metals (as, Pb, Hg, Cr, Zn, CD, Cu, and Ni), including their pollution sources (natural and man-made), uses, and adverse effects on human health can be found in [183].

As and Related Diseases In nature, As exists in two forms, organic and inorganic. As is a carcinogen of human skin and lung cancer confirmed by the International Agency for Research on Cancer (IARC). Long-term exposure can lead to systemic chronic poisoning mainly by skin pigment loss and/or hyperpigmentation, palm and toe keratosis, and cancelation. As poisoning seriously affects the patient's life. The common forms of As are As(III), As(V), monomethyl As acid (MMA), dimethyl As acid (DMA), arecoline (AsC), reobtained (AsB), etc. Its toxicity decreases in turn. In general, the toxicity of inorganic As is greater than that of organic As [184]. As poisoning can cause abdominal pain, destruction of red blood cells (hemolysis), shock, melanosis, keratosis, hyperkeratosis, back, nonpitting edema, gangrene, and skin cancer. Studies have linked chronic, low-dose As exposure to skin, bladder, lung, kidney, and liver cancers [185].

Pb and Related Diseases Pb can act on various systems and organs in the body, mainly involving the nervous system, digestive system, blood and hematopoietic system, cardiovascular system, and kidneys. There are two types of Pb poisoning, acute (due to short-term intense exposure) and chronic (due to repeated low-level exposure over long periods of time). Acute Pb poisoning is mostly manifested as gastrointestinal symptoms, such as nausea, vomiting, and abdominal cramps, and a few symptoms of toxic encephalopathy may appear. Diagnosis and treatment of Pb poisoning depend on the amount of Pb in the blood, measured in micrograms of Pb per deciliter of blood [186].

Hg and Related Diseases Hg mainly has three forms: metallic Hg or Hg vapor, inorganic Hg or Hg salts, and organic Hg compounds. The toxicity of methyl Hg and dimethyl Hg is much higher than that of inorganic Hg. Hg is a poison that damages multiple organs. Hg poisoning refers to systemic diseases caused by exposure to metallic Hg, mainly in the central nervous system and oral cavity, and involving the respiratory tract, gastrointestinal tract, and kidneys [187]. Excessive exposure to Hg can lead to various diseases such as allergies, language disorders, neurological and renal-system-related diseases [188].

Cd and Related Diseases Cd is another toxic non-essential transition metal classified as a human carcinogen by the National Toxicology Program [185a]. Sources of human exposure to Cd include employment in the metals industry, the production of certain batteries, some electroplating processes, and the consumption of tobacco products. Cd poses a huge risk to human health because the body has a limited ability to respond to Cd exposure because the metal cannot be metabolized to less toxic species and is poorly excreted. Several studies have linked occupational exposure to Cd with lung cancer, as well as cancers of the prostate, kidney, liver, hematopoietic system, bladder, pancreas, and stomach. Fagerberg et al. reviewed evidence on local Cd and its association with altered innate immunity in smokers.

Exposure to Cd (Cd) through food and smoking is associated with an increased risk of atherosclerotic cardiovascular disease (ASCVD). It is concluded that Cd may cause smoking-related lung diseases, possibly altering redox balance and malfunction of macrophages. However, new studies on local Cd levels in long-term smokers and their association with pathology, as well as more in-depth studies of cellular and molecular mechanisms, are needed to elucidate the importance of Cd in smoking-related diseases [189].

Cr and Related Diseases Cr existing in nature mainly has two valence states: trivalent Cr [Cr(III)] and hexavalent Cr [Cr(VI)]. Among them, Cr(III) is considered to be a trace element needed by the human body, and it is involved in maintaining the normal life activities of the body. Cr(VI) was recognized as a human carcinogen as early as 1990 by the International Agency for Research on Cancer with carcinogenicity and mutagenicity. Cr(VI)-induced epigenetic changes play an important role in the cytotoxic and carcinogenic mechanisms. Epigenetics is the study of heritable changes in gene function that do not require changes in DNA sequence; common epigenetic phenomena include DNA methylation, histone modifications, noncoding RNA regulation, signaling pathways, genetic imprinting alterations, phosphorylation, RNA methylation, etc. The research on Cr(VI)-induced epigenetic changes mainly focuses on DNA methylation, histone modification, and non-coding RNA regulation [190].

6.3.2.3 Combined Toxicity of Multiple Heavy Metal Mixtures

Nervous system damage caused by heavy metal poisoning is extremely common in clinical practice, often endangering patients' lives and bringing a heavy burden to society and families. Fan Shuangyi taking several common heavy metal poisons as examples, combined with domestic and foreign research, reviewed the nervous system damage caused by heavy metal poisoning, and provided a key basis for early clinical prevention, diagnosis, treatment, and research [191]. Lin et al. selected 8 common heavy metals (Pb, Cd, Hg, Cu, Zn, Mn, Cr, Ni) that cause environmental pollution and studied the combined toxicity of different heavy metal mixtures to HL7702 cells [192].

Samuel conducted in vivo and in vitro experiments on the combined toxicity of multiple heavy metal mixtures. They have studied multiple heavy metal mixtures: Ni, Zn, Pb, Cr, Cd, Cu, Hg, and Mn systemic toxicity to HL7702 cells and Sprague Dawley (SD). It provided a scientific basis for exploring the combined toxic effects of various heavy metal mixtures and their molecular mechanisms and preventive measures [193].

Recent data suggest that As, Cd, Hg, and Pb exposures are associated with respiratory dysfunction and respiratory diseases (COPD, bronchitis). Observations by Skalny AV et al. confirm the role of heavy metal exposure in impaired mucociliary clearance, reduced barrier function, airway inflammation, oxidative stress, and apoptosis. There is an association between heavy metal exposure and severity of viral diseases, including influenza and respiratory syncytial virus. The latter can be considered as a consequence of the detrimental effects of metal exposure on adaptive immunity. Therefore, reducing toxic metal exposure may be considered as a potential tool to reduce susceptibility and severity of viral diseases affecting the respiratory system, including COVID-19 [194].

6.3.2.4 Genetic Diseases Associated with Metallomics

Inherited diseases refer to a class of diseases in which the disease-causing genes carried by parents are passed on to the next generation through genes and germ cells.

Cu transport promotion includes inherited Wilson disease and Menkes disease. The Wilson ATPase gene encodes a Cu-transporting P-type ATPase (ATP7B). ATP7B binds six Cu atoms at the N-terminus of the molecule. Defects in this protein lead to toxic accumulation of Cu in various tissues. Menkes disease is another inherited disorder of Cu metabolism characterized by impaired dietary Cu absorption and severe disturbance of intracellular Cu transport [195]. The Menkes ATPase gene encodes a Cu-transporting P-type ATPase (ATP7A) that is highly homologous to ATP7B. The protein also binds Cu in a 1:6 protein to Cu ratio. Defects in this protein lead to the classic clinical symptoms of Menkes disease. As a result, the chemical species and metallomic supply of Cu is prevented from reaching developmentally important Cu enzymes such as dopamine beta-hydroxylase, superoxide dismutase, and amine oxidase [195a, 196].

Fe(II) and Fe(III) transporters, e.g. hereditary hemochromatosis, are human diseases associated with abnormal metal metabolism in organisms. Hereditary hemochromatosis is a disorder of Fe metabolism in which Fe uptake by the digestive system is not downregulated, resulting in excess Fe uptake [195a]. The body continues to absorb high levels of dietary Fe regardless of the amount of Fe stored in the body. The major ferroportin in plasma is transferrin, which has two ferric Fe-binding sites. Several proteins are involved in cellular uptake of Fe, including transferrin receptors 1 and 2 and divalent metal transporter (DMT-1). In addition to intracellular Fe transport following endocytosis of transferrin-bound Fe, DMT-1 is also important in Fe absorption in the gut. DMT-1 transports Fe(II), but not Fe(III) [197]. Fe(II) is soluble under physiological conditions and can also diffuse across membranes. In contrast, Fe(III) is easily hydrolyzed in an aqueous environment, resulting in poorly soluble products. Regulation of gene products involved in Fe metabolism is influenced by two cytoplasmic Fe regulatory proteins (IRP-1 and IRP-2) with mRNA-binding capacity [195a, 198].

6.3.2.5 Application of Metallomics in Disease Treatment

Exploiting Intermetallic Interactions Using metallomic methods to study the interaction of Hg and selenium in living organisms is an effective strategy. Selenium is one of the trace elements necessary for human and animals to maintain life activities, and has a significant antagonistic effect on the toxicity of heavy metals such as Hg. A comprehensive and accurate understanding of the distribution and existence forms of Hg and selenium in living organisms, the absorption, migration, transformation, and accumulation process of Hg in living tissues, and the influence of selenium on these biological behaviors and effects of Hg is of practical significance for the control of environmental Hg pollution and Hg toxicity. It is also helpful to understand the interaction process and mechanism of Hg and selenium in organisms.

Li Min et al. studied the antagonistic effect of selenium on lead toxicity. On the one hand, selenium is an important component of the antioxidant system in the body, which can significantly improve the lipid peroxidation reaction induced by Pb poisoning and reduce the harm of Pb; on the other hand, it has a strong affinity with metals and can combine with lead to form metal–selenoprotein complexes in the body, thereby reducing the toxic effects of Pb. The results show that supplementing different forms of selenium can promote and hinder the excretion and absorption of Pb in the body, respectively, that is, it has different degrees of antagonism on Pb toxicity, indicating that selenium does antagonize Pb toxicity. It can not only prevent Pb poisoning, but also improve the body's antioxidant function and reduce the body's sensitivity to Pb [199].

There is mutual antagonism between Zn and Pb, and other elements such as Cu also antagonize the metabolism of Zn; some trace elements such as vitamin C and calcium can also play a certain role in combating Pb poisoning.

Jiao Shi et al. have studied Zn gluconate and sodium selenite alone and in combination to antagonize the toxic effects of silica on alveolar macrophages. Experiments have confirmed that Zn gluconate and sodium selenite can effectively antagonize the cytotoxic effect of silica dust in vitro, reduce the content of malondialdehyde in alveolar macrophages, reduce the activity of lactate dehydrogenase in the culture medium, and increase the activity of superoxide dismutase in macrophages, the combined effect of the two is more obvious.

Application of Metal-Containing Clinical Drugs Some clinical medicines contain metals, Fe preparations for the treatment of Fe-deficiency anemia, silver for antibacterial effects, lithium for the treatment of mania, As-containing drugs for the treatment of leukemia, anticancer drugs cisplatin or ruthenium metals, and gold for anti-rheumatism, bismuth drugs for the treatment of Helicobacter pylori, etc.

Zn has many physiological functions, such as promoting human growth and development; accelerating the healing of wounds and ulcers; improving the immunity of the elderly; delaying aging; enhancing appetite; improving digestive function; and maintaining normal sexual function. Zn has therapeutic effect on coronary heart disease, diabetes, cancer, blood disease, and viral cold, so Zn preparations are widely used in clinical.

Chen Baoquan conducted research on the anticancer drugs of selenium. Selenium plays a key role in maintaining normal physiological functions of the human body, has a wide range of biological activities, and exerts anti-tumor effects through various mechanisms. Selenium deficiency is associated with coronary heart disease, anemia, arthritis, diabetes, etc. A study by Zhang Zhihong used vitamin D and calcium combined with Zn gluconate oral liquid in the treatment of rickets in infants and young children has a significant effect, which can significantly improve the bone mineral density, blood Zn level, and blood calcium level of the children.

Cisplatin and successive generations of platinum-based anticancer drugs have demonstrated that metal complexes are very promising in anticancer therapy. Platinum drugs attack nuclear DNA, causing the DNA to twist. DNA damage caused by cisplatin is recognized by proteins that activate downstream events, leading to a major apoptotic response [200]. Since Pt anticancer agents have different chemical structures, such as cisplatin, carboplatin, oxaliplatin, and satrap Latin, metallomics studies clearly provide a way to improve our understanding of how these drugs behave in living organisms. This aspect was explored by using analytical methods for metallomic studies of antitumor Pt drugs [201].

6.3.2.6 Perspectives

Metallomics is an emerging "omics" discipline developed in recent years. Clinical metallomics not only helps to reveal the dynamic variation characteristics and interactions of different elements in the pathogenesis of complex diseases, but also provides new ideas for early clinical diagnosis, risk assessment, and efficacy evaluation of related diseases. However, it should be acknowledged that the current work on disease metallomics has only just begun, where the use of bioinformatics and web-based approaches is still limited. Since ion data provide a wealth of information to explore, there is an urgent need to develop new efficient algorithms for data processing and analysis. Furthermore, the use of metallomics has so far not been embraced by many other complex diseases, suggesting that this field is very promising.

In the future, with the availability of multi-omics data and the advent of new computational methods for high-throughput data analysis, the integration of metallomic and other omics data will provide valuable information to study the role of minerals and trace elements in life sciences. And it will promote the clinical translation of these achievements (e.g. etiology, diagnosis, and treatment). This could lead to major advances in our understanding of the dynamic network of elements and their relationship to the onset and progression of these diseases. Elemental analysis of single cells will also receive more attention.

List of Abbreviations

AAS	atomic absorption spectrometry
ACR	albumin-to-creatinine ratio
AD	Alzheimer's disease
AES	atomic emission spectrometry
AFS	atomic fluorescence spectrometry
Al	aluminum
ALB	albumin
ALS	amyotrophic lateral sclerosis
ALT	alanine aminotransferase
As	arsenic

AsB	reobtained
AsC	arecoline
ASCVD	atherosclerotic cardiovascular disease
AST	aspartate aminotransferase
BALF	bronchoalveolar lavage fluid
BMI	body mass index
BUN	blood urea nitrogen
CA	*A Cancer Journal for Clinicians*
Ca	calcium
CE	capillary electrophoresis
CKD	chronic kidney disease
Co	cobalt
COPD	chronic obstructive pulmonary disease
Cr	chromium
CRC	colorectal cancer
CSF	cerebrospinal fluid
Cu	copper
CV	cold vapor
CVD	cerebrovascular disease
DALYs	disability-adjusted life years
DMA	dimethyl As acid
DMT-1	divalent metal transporter
DNA	deoxyribonucleic acid
EGFR	glomerular filtration rate
ERS	environmental risk score
ESCC	esophageal squamous cell carcinoma
ESI-MS	electrospray ionization mass spectrometry
ESI-MS	electrospray mass spectrometry
ESI-MS/MS	electrospray tandem mass spectrometry
ESI-TOF-MS	electrospray time-of-flight mass spectrometry
ESRD	end-stage renal disease
ETAAS	electrothermal atomic absorption spectrometry
FAAS	flame atomic absorption spectrometry
Fe	ferrum
GBC	gallbladder cancer
GC	gas chromatography
GD	gestational diabetes
GE	gel electrophoresis
GF-AAS	graphite furnace atomic absorption spectrophotometer
HD	Huntington's disease
HDL-C	high-density lipoprotein cholesterol
HG	hydride generation
HPLC	high-performance liquid chromatography
HR	hazard ratio
IARC	International Agency for Research on Cancer

ICP-AES	inductively coupled plasma atomic emission spectroscopy
ICP-MS	inductively coupled plasma mass spectrometry
ICP-OES	inductively coupled plasma atomic emission spectrometry
ICWM	International Commission for Weights and Measures
IHD	ischemic heart disease
IRP-1	iron regulatory protein-1
IRP-2	iron regulatory protein-2
IUPAC	International Union of Pure and Applied Chemistry
KNHANES	Korea National Health and Nutrition Examination Survey
LC	lung cancer
LDL-C	low-density lipoprotein cholesterol
MALDI-MS	matrix-assisted laser desorption mass spectrometry
MCI	mild cognitive impairment
MDA	malonaldehyde
Mg	manganese
MMA	monomethyl As acid
Mo	molybdenum
NAA	neutron activation analysis
NAFLD	nonalcoholic fatty liver disease
NHANES	National Health and Nutrition Examination Survey
Ni	nickel
P	phosphorus
PaC	pancreatic cancer
Pb	lead
PC	prostate cancer
PD	Parkinson disease
PDAC	pancreatic ductal adenocarcinoma
Rb	rubidium
RNA	ribonucleic acid
SD	Sprague Dawley
Se	selenium
Sr	strontium
T1D	type I diabetes
T2D	type II diabetes
TC	total cholesterol
TG	triglycerides
TP	total protein
UV-Vis	ultraviolet–visible spectroscopy
V	vanadium
vasc	vascular components
VSG	volatile species generation
W	wolfram
Zn	zinc

References

1. Nordberg, M. and Nordberg, G.F. (2016). *J. Trace Elem. Med. Biol.* 38: 46.
2. Wu, W., Wu, P., Yang, F. et al. (2018). *Sci. Total Environ.* 630: 53.
3. Cai, L.M., Wang, Q.S., Luo, J. et al. (2019). *Sci. Total Environ.* 650: 725.
4. Tchounwou, P.B., Yedjou, C.G., Patlolla, A.K., and Sutton, D.J. (2012). *Exp. Suppl.* 101: 133.
5. WHO, 1996, Switzerland: Geneva: World Health Organization, www.who.org/.
6. Mahan, B., Chung, R.S., Pountney, D.L. et al. (2020). *Cell. Mol. Life Sci.* 77: 3293.
7. (a) H. Haraguchi, 2002; (b) Haraguchi, H. (2017). *Metallomics* 9: 1001. (c) H. H. a. H. Matsuura, 2003; (d) Haraguchi, H. (2004). *J. Anal. At. Spectrom.* 19: 5.
8. Mounicou, S., Szpunar, J., and Lobinski, R. (2009). *Chem. Soc. Rev.* 38: 1119.
9. Chen, B., Hu, L., He, B. et al. (2020). *TrAC, Trends Anal. Chem.* 126: 115875.
10. Gu, W. and Tong, Z. (2020). *Lab. Med.* 51: 116.
11. Engler, M. (1995). *Chem. Ing. Tech.* 67.
12. Yu, X., Yan, C., and Shen, X. (2006). *Guo Ji Jian Yan Yi Xue Za Zhi* 27: 2.
13. Wang, S., Chen, B., Liang, J. et al. (2016). *Guang Dong Yi Xue* 37: 3.
14. Li, Y., Jiang, Y., and Yan, X. (2007). Presented at *Symposium on organic analysis and bioanalysis. Chinese Chemical Society*.
15. B. Chen ICP-MS-based Coupling Techniques and Their Application in Elemental and Morphological Analysis of Biological Systems, Wuhan University, 2010.
16. Jiang, G. and He, B. (2005). *China Sci. Found.* 19: 5.
17. Peng, H. (2014). Application of Mass Spectrometry Coupling Techniques in Elemental Speciation Analysis. Wuhan University.
18. Wang, S. and Zheng, Q. (1999). *Di Fang Bing Tong Bao* 14: 3.
19. H. Shi. (2009). Development of a Portable Neutron Irradiation Dose Detector for Blood. Chengdu University of Technology.
20. (a) Gao, B., Li, Q., Zhou, H. et al. (2014). *Spectrosc. Spect. Anal.* 34: 5. (b) Liu, L.N. (2019). *Chem. Eng. Desig. Commun.*
21. Li, X. (2010). Study of Birth Defect Metabolomics and Metalomics Based on Chromatography-Mass Spectrometry Coupling Techniques. Shanghai Jiaotong University.
22. Maret, W. (2018). *Adv. Exp. Med. Biol.* 1055: 1.
23. (a) Santos, A.B.D., Kohlmeier, K.A., Rocha, M.E. et al. (2018). *J. Trace Elem. Med. Biol.* 47: 134. (b) Xu, Q. (2007). Research and Application of Plasma Spectroscopy Coupling Techniques in Elemental Speciation Analysis. Guangxi Normal University.
24. Yan, X. and Ni, Z. (2001). *Guang Pu Xue Yu Guang Pu Fen Xi* 1: 129.
25. Yan, D., Yang, L., and Wang, Q. (2008). *Anal. Chem.* 80: 6104.
26. Du, P. (2004). Application Research of Ionic Liquid Dispersive Liquid-Liquid Microextraction in Heavy Metals and Speciation Analysis. Central China Agricultural University.
27. Zhang, Y. (2017). *Adv. Exp. Med. Biol.* 1005: 63.

28 Chen, T.C., Parker, J.D., Clark, J. et al. (2018). *Vital Health Stat.* 2: 1.
29 Bocato, M.Z., Bianchi Ximenez, J.P., Hoffmann, C., and Barbosa, F. (2019). *J. Toxicol. Environ. Health B Crit. Rev.* 22: 131.
30 Haines, D.A., Saravanabhavan, G., Werry, K., and Khoury, C. (2017). *Int. J. Hyg. Environ. Health* 220: 13.
31 Eykelbosh, A., Werry, K., and Kosatsky, T. (2018). *Environ. Int.* 119: 536.
32 Karagas, M.R., Choi, A.L., Oken, E. et al. (2012). *Environ. Health Perspect.* 120: 799.
33 (a) Wang, R., Zhang, L., Chen, Y. et al. (2020). *Environ. Int.* 144: 106061.
(b) Chiudzu, G., Choko, A.T., Maluwa, A. et al. (2020). *J. Pregnancy* 2020: 9435972.
34 Hays, S.M., Becker, R.A., Leung, H.W. et al. (2007). *Regul. Toxicol. Pharm.* 47: 96.
35 Wei, Y., Zhu, J., and Nguyen, A. (2014). *Int. J. Environ. Health Res.* 24: 459.
36 IARC (1990). *IARC Monogr. Eval. Carcinog. Risks Hum.* 49: 1.
37 Rehman, K., Fatima, F., Waheed, I., and Akash, M.S.H. (2018). *J. Cell. Biochem.* 119: 157.
38 Navarro Silvera, S.A. and Rohan, T.E. (2007). *Cancer Causes Control* 18: 7.
39 Järup, L. and Elinder, C.G. (1993). *Br. J. Ind. Med.* 50: 598.
40 Al Osman, M., Yang, F., and Massey, I.Y. (2019). *Biometals* 32: 563.
41 (a) Fel, B.M., Zaky, E.A., El-Sayed, A.B. et al. (2015). *Behav. Neurol.* 2015: 545674; (b) Ye, B.S., Leung, A.O.W., and Wong, M.H. (2017). *Environ. Pollut.* 227: 234.
42 Reuben, A., Caspi, A., Belsky, D.W. et al. (2017). *JAMA* 317: 1244.
43 Rodrigues, E.G., Bellinger, D.C., Valeri, L. et al. (2016). *Environ. Health* 15: 44.
44 Sherief, L.M., Abdelkhalek, E.R., Gharieb, A.F. et al. (2015). *Medicine* 94: e740.
45 Absalon, D. and Slesak, B. (2010). *Sci. Total Environ.* 408: 4420.
46 Reyes, J.W. (2015). *Econ. Inquiry* 53: 1580.
47 Schumacher, L. and Abbott, L.C. (2017). *J. Appl. Toxicol.* 37: 4.
48 Otebhi, O.E. and Osadolor, H.B. (2016). *J. Appl. Sci. Environ. Manag.* 20: 5.
49 Milton, A.H., Smith, W., Rahman, B. et al. (2005). *Epidemiology* 16.
50 Agrawal, A. (2012). *Adv. Life Sci.* 2: 29.
51 Wang, Z., Xu, X., He, B. et al. (2019). *Ecotoxicol. Environ. Saf.* 169: 232.
52 Liu, Y., Yuan, Y., Xiao, Y. et al. (2020). *Ecotoxicol. Environ. Saf.* 189: 110006.
53 Yokel, R.A., Lasley, S.M., and Dorman, D.C. (2006). *J. Toxicol. Environ. Health B Crit. Rev.* 9: 63.
54 Liu, Y., Buchanan, S., Anderson, H.A. et al. (2018). *Environ. Res.* 160: 212.
55 Yang, L., Zhang, Y., Wang, F. et al. (2020). *Chemosphere* 245: 125586.
56 Bjørklund, G., Dadar, M., Mutter, J., and Aaseth, J. (2017). *Environ. Res.* 159: 545.
57 Liang, C.M., Wu, X.Y., Huang, K. et al. (2019). *Chemosphere* 218: 869.
58 Killin, L.O., Starr, J.M., Shiue, I.J., and Russ, T.C. (2016). *BMC Geriatr.* 16: 175.
59 (a) Mirnamniha, M., Faroughi, F., Tahmasbpour, E. et al. (2019). *Rev. Environ. Health* 34: 339; (b) Zemrani, B. and Bines, J.E. (2020). *Curr. Opin. Gastroenterol.* 36: 110.

60 Feigin, V., Nichols, E., Alam, T. et al. (2019). *Lancet Neurol.* 18 (5): 459–480.
61 Emmett, S.E. (1989). *Prog. Clin. Biol. Res.* 317: 1077.
62 Slanzi, A., Iannoto, G., Rossi, B. et al. (2020). *Front. Cell Dev. Biol.* 8: 328.
63 (a) Yegambaram, M., Manivannan, B., Beach, T.G., and Halden, R.U. (2015). *Curr. Alzheimer Res.* 12: 116; (b) Kawahara, M., Kato-Negishi, M., and Tanaka, K. (2017). *Metallomics* 9: 619.
64 Sussulini, A. and Hauser-Davis, R.A. (2018). *Adv. Exp. Med. Biol.* 1055: 21.
65 Gerhardsson, L., Lundh, T., Minthon, L., and Londos, E. (2008). *Dement. Geriatr. Cogn. Disord.* 25: 508.
66 González-Domínguez, R., García-Barrera, T., and Gómez-Ariza, J.L. (2014). *Metallomics* 6: 292.
67 Koseoglu, E., Koseoglu, R., Kendirci, M. et al. (2017). *J. Trace Elem. Med. Biol.* 39: 124.
68 (a) Lane, D.J.R., Ayton, S., and Bush, A.I. (2018). *J. Alzheimer's Dis.* 64: S379; (b) Bonda, D.J., Lee, H.G., Blair, J.A. et al. (2011). *Metallomics* 3: 267.
69 Ciavardelli, D., Consalvo, A., Caldaralo, V. et al. (2012). *Metallomics* 4: 1321.
70 Zheng, L., Zhu, H.Z., Wang, B.T. et al. (2016). *Sci. Rep.* 6: 39290.
71 van Eersel, J., Ke, Y.D., Liu, X. et al. (2010). *Proc. Natl. Acad. Sci. U.S.A.* 107: 13888.
72 Forte, G., Bocca, B., Senofonte, O. et al. (1996). *J. Neural Trans.* 2004 (111): 1031.
73 Zhao, H.W., Lin, J., Wang, X.B. et al. (2013). *PLoS One* 8: e83060.
74 Sanyal, J., Ahmed, S.S., Ng, H.K. et al. (2016). *Sci. Rep.* 6: 35097.
75 McColgan, P. and Tabrizi, S.J. (2018). *Eur. J. Neurol.* 25: 24.
76 Muller, M. and Leavitt, B.R. (2014). *J. Neurochem.* 130: 328.
77 Squadrone, S., Brizio, P., Abete, M.C., and Brusco, A. (2020). *J. Trace Elem. Med. Biol.* 57: 18.
78 Sirabella, R., Valsecchi, V., Anzilotti, S. et al. (2018). *Front. Neurosci.* 12: 510.
79 De Benedetti, S., Lucchini, G., Del Bò, C. et al. (2017). *Biometals* 30: 355.
80 Oggiano, R., Solinas, G., Forte, G. et al. (2018). *Chemosphere* 197: 457.
81 Dubey, P., Thakur, V., and Chattopadhyay, M. (2020). *Nutrients* 12.
82 I. D. Federation, Vol. 2019, http://www.diabetesatlas.org/ 2019.
83 Sun, L., Yu, Y., Huang, T. et al. (2012). *PLoS One* 7: e38845.
84 Liu, G., Sun, L., Pan, A. et al. (2015). *Int. J. Epidemiol.* 44: 240.
85 Li, X.T., Yu, P.F., Gao, Y. et al. (2017). *Biomed. Environ. Sci.* 30: 482.
86 Badran, M., Morsy, R., Soliman, H., and Elnimr, T. (2016). *J. Trace Elem. Med. Biol.* 33: 114.
87 Marín-Martínez, L., Molino-Pagán, D., and López-Jornet, P. (2019). *Diabetes Res. Clin. Pract.* 157: 107871.
88 Forte, G., Bocca, B., Peruzzu, A. et al. (2013). *Biol. Trace Elem. Res.* 156: 79.
89 Peruzzu, A., Solinas, G., Asara, Y. et al. (2015). *Chemosphere* 132: 101.
90 Squitti, R., Negrouk, V., Perera, M. et al. (2019). *J. Trace Elem. Med. Biol.* 56: 156.
91 Roverso, M., Berté, C., Di Marco, V. et al. (2015). *Metallomics* 7: 1146.
92 Roverso, M., Di Marco, V., Badocco, D. et al. (2019). *Metallomics* 11: 676.

93 Zhang, Y., Xu, Y., and Zheng, L. (2020). *Int. J. Mol. Sci.* 21.
94 Ross, D.S., Burch, H.B., Cooper, D.S. et al. (2016). *Thyroid* 26: 1343.
95 Ferlay, J.S.I., Ervik, M., Dikshit, R. et al. (2015). *Int. J. Cancer J. Int. Cancer* 136: E359.
96 Fitzmaurice, C. and Balakrishnan, S. (2019). *JAMA Oncol.* 5 (12): 1749–1768.
97 Siegel, R.L., Miller, K.D., and Jemal, A. (2020). *CA Cancer J. Clin.* 70: 7.
98 (a) Chen, Y., Fan, Z., Yang, Y., and Gu, C. (2019). *Int. J. Oncol.* 54: 1143; (b) Koedrith, P., Kim, H., Weon, J.I., and Seo, Y.R. (2013). *Int. J. Hyg. Environ. Health* 216: 587.
99 Qayyum, M.A. and Shah, M.H. (2014). *Biol. Trace Elem. Res.* 162: 46.
100 Saleh, S.A.K., Adly, H.M., Abdelkhaliq, A.A., and Nassir, A.M. (2020). *Curr. Urol.* 14: 44.
101 Lee, K.Y., Feng, P.H., Chuang, H.C. et al. (2018). *Biol. Trace Elem. Res.* 182: 14.
102 Callejón-Leblic, B., Gómez-Ariza, J.L., Pereira-Vega, A., and García-Barrera, T. (2018). *Metallomics* 10: 1444.
103 Callejón-Leblic, B., Arias-Borrego, A., Pereira-Vega, A. et al. (2019). *Int. J. Mol. Sci.* 20.
104 Schilling, K., Larner, F., Saad, A. et al. (2020). *Metallomics* 12: 752.
105 Lavilla, I., Costas, M., Miguel, P.S. et al. (2009). *Biometals* 22: 863.
106 Sohrabi, M., Gholami, A., Azar, M.H. et al. (2018). *Biol. Trace Elem. Res.* 183: 1.
107 Kohzadi, S., Sheikhesmaili, F., Rahehagh, R. et al. (2017). *Chemosphere* 184: 747.
108 Lin, T., Liu, T., Lin, Y. et al. (2017). *BMJ Open* 7: e015443.
109 Burton, C., Dan, Y., Donovan, A. et al. (2016). *Clin. Chim. Acta* 452: 142.
110 Lee, M.H., Gao, Y.T., Huang, Y.H. et al. (2020). *Hepatology* 71: 917.
111 Golasik, M., Jawień, W., Przybyłowicz, A. et al. (2015). *Metallomics* 7: 455.
112 NCD Risk Factor Collaboration (NCD-RisC) (2017). *Lancet* 390 (10113): 2627–2642. https://doi.org/10.1016/S0140-6736(17)32129-3.
113 Wang, X., Mukherjee, B., and Park, S.K. (2018). *Environ. Int.* 121: 683.
114 Niehoff, N.M., Keil, A.P., O'Brien, K.M. et al. (2020). *Environ. Res.* 184: 109396.
115 Kawakami, T., Hanao, N., Nishiyama, K. et al. (2012). *Toxicol. Appl. Pharmacol.* 258: 32.
116 Fan, Y., Zhang, C., and Bu, J. (2017). *Nutrients* 9.
117 (a) Xu, S., Pi, H., Chen, Y. et al. (2013). *Cell Death Dis.* 4: e540; (b) Herman, D.S., Geraldine, M., and Venkatesh, T. (2009). *J. Hazard. Mater.* 166: 1410.
118 (a) Jiang, J., Wang, F., Wang, L. et al. (2020). *Biol. Trace Elem. Res.* 197: 254; (b) Elshazly, M.O., Morgan, A.M., Ali, M.E. et al. (2016). *J. Adv. Res.* 7: 413; (c) Vinodhini, R. and Narayanan, M. (2008). *Int. J. Environ. Sci. Technol.* 5: 179.
119 Green, R.M. and Flamm, S. (2002). *Gastroenterology* 123: 1367.
120 (a) Zhao, C., Li, L., Harrison, T.J. et al. (2009). *J. Gastroenterol.* 44: 139; (b) Robles-Diaz, M., Lucena, M.I., Kaplowitz, N. et al. (2014). *Gastroenterology* 147.
121 McGill, M.R. (2016). *EXCLI J.* 15: 817.
122 Mohamed, A.A.-R., El-Houseiny, W., El-Murr, A.E. et al. (2020). *Ecotoxicol. Environ. Saf.* 188: 109890.

123 (a) Athmouni, K., Belhaj, D., Hammi, K.M. et al. (2018). *Arch. Physiol. Biochem.* 124: 261; (b) Cao, Z., Fang, Y., Lu, Y. et al. (2017). *J. Pineal Res.* 62.

124 (a) Chen, C., Lin, B., Qi, S. et al. (2019). *Biol. Trace Elem. Res.* 191: 426; (b) Luo, T., Shen, M., Zhou, J. et al. (2019). *Environ. Toxicol.* 34: 521; (c) Alhusaini, A., Fadda, L., Hasan, I.H. et al. (2019). *Antioxidants* 8.

125 Hoseini, S.M., Hedayati, A., and Ghelichpour, M. (2014). *Ecotoxicol. Environ. Saf.* 107: 84.

126 Sadeghi, P., Savari, A., Movahedinia, A. et al. (2014). *Environ. Sci. Pollut. Res. Int.* 21: 6076.

127 (a) Sazakli, E., Villanueva, C.M., Kogevinas, M. et al. (2014). *Int. J. Environ. Res. Public Health* 11: 10125; (b) Wang, X., Bin, W., Zhou, M. et al. (2021). *J. Hazard. Mater.* 419: 126497.

128 (a) Kang, M.Y., Cho, S.H., Lim, Y.H. et al. (2013). *Occup. Environ. Med.* 70: 268; (b) Park, E., Kim, J., Kim, B., and Park, E.Y. (2021). *Chemosphere* 266: 128947.

129 Kim, D.W., Ock, J., Moon, K.W., and Park, C.H. (2021). *Int. J. Environ. Res. Public Health* 18.

130 Nakata, H., Nakayama, S.M.M., Yabe, J. et al. (2021). *Chemosphere* 262: 127788.

131 (a) Zhai, H., Chen, C., Wang, N. et al. (2017). *Environ. Health* 16: 93; (b) Wahlang, B., Appana, S., Falkner, K.C. et al. (2020). *Environ. Sci. Pollut. Res. Int.* 27: 6476.

132 Betanzos-Robledo, L., Cantoral, A., Peterson, K.E. et al. (2021). *Environ. Res.* 196: 110980.

133 Huang, R., Pan, H., Zhou, M. et al. (2021). *Environ. Res.* 201: 111598.

134 (a) Deng, Q., Liu, J., Li, Q. et al. (2013). *Environ. Health* 12: 30; (b) Spangler, J.G. (2012). *Int. J. Environ. Res. Public Health* 9: 3258.

135 Kobtan, A.A., El-Kalla, F.S., Soliman, H.H. et al. (2016). *Biol. Trace Elem. Res.* 169: 153.

136 Zhang, D., Wu, S., Lan, Y. et al. (2022). *Chemosphere* 287: 132316.

137 Fairbrother, A., Wenstel, R., Sappington, K., and Wood, W. (2007). *Ecotoxicol. Environ. Saf.* 68: 145.

138 (a) Kumar, V., Parihar, R.D., Sharma, A. et al. (2019). *Chemosphere* 236: 124364; (b) Yang, S., Liu, J., Bi, X. et al. (2020). *Ecotoxicol. Environ. Saf.* 197: 110628; (c) Yuan, X., Xue, N., and Han, Z. (2021). *J. Environ. Sci.* 101: 217.

139 (a) Bauer, J.A., Devick, K.L., Bobb, J.F. et al. (2020). *Environ. Health Perspect.* 128: 97002; (b) Oladipo, O.O., Ayo, J.O., Ambali, S.F., and Mohammed, B. (2016). *Toxicol. Mech. Methods* 26: 674.

140 Bikbov, B., Purcell, C.A., Levey, A.S. et al. (2020). *Lancet* 395: 709.

141 Tsai, C.-C., Wu, C.-L., Kor, C.-T. et al. (2018). *Nephrology* 23: 830.

142 Sanders, A.P., Mazzella, M.J., Malin, A.J. et al. (2019). *Environ. Int.* 131: 104993.

143 Luo, J. and Hendryx, M. (2020). *Environ. Res.* 191: 110126.

144 Prodanchuk, M., Makarov, O., Pisarev, E. et al. (2014). *Cent. Eur. J. Urol.* 66: 472.

145 Roth, G.A., Mensah, G.A., Johnson, C.O. et al. (2020). *J. Am. Coll. Cardiol.* 76 (25): 2982–3021.

146 Zhou, M., Wang, H., Zhu, J. et al. (2016). *Lancet* 387: 251.

147 Chowdhury, R., Ramond, A., O'Keeffe, L.M. et al. (2018). *BMJ* 362: k3310.
148 (a) Yang, A.-M., Lo, K., Zheng, T.-Z. et al. (2020). *Chron. Dis. Transl. Med.*; (b) Leone, N., Courbon, D., Ducimetiere, P., and Zureik, M. (2006). *Epidemiology* 17: 308.
149 Zhao, D., Domingo-Relloso, A., Kioumourtzoglou, M. et al. (2019). *Environ. Epidemiol.* 3: 465.
150 Choe, S.Y., Kim, S.J., Kim, H.G. et al. (2003). *Sci. Total Environ.* 312: 15.
151 Domingo-Relloso, A., Grau-Perez, M., Briongos-Figuero, L. et al. (2019). *Int. J. Epidemiol.* 48: 1839.
152 DeVille, N.V., Khalili, R., Levy, J.I. et al. (2020). *J. Exposure Sci. Environ. Epidemiol.* 31: 197.
153 Sarki, A.M., Nduka, C.U., Stranges, S. et al. (2015). *Medicine* 94: e1959.
154 Wu, W., Jiang, S., Zhao, Q. et al. (2018). *Environ. Pollut.* 233: 670.
155 Wu, W., Jiang, S., Zhao, Q. et al. (2018). *Sci. Total Environ.* 622-623: 184.
156 Xu, J., White, A.J., Niehoff, N.M. et al. (2020). *Environ. Res.* 191: 110144.
157 Castiello, F., Olmedo, P., Gil, F. et al. (2020). *Environ. Res.* 182: 108958.
158 Zhu, X., Fan, Y., Sheng, J. et al. (2021). *Biol. Trace Elem. Res.* 199: 1280.
159 (a) Zhang, M., Deng, Q., Wang, L. et al. (2018). *Int. J. Cardiol.* 260: 196; (b) Opoku, S., Gan, Y., Fu, W. et al. (2019). *BMC Public Health* 19: 1500; (c) Xi, Y., Niu, L., Cao, N. et al. (2020). *BMC Public Health* 20: 1068.
160 (a) Hernández-Alcaraz, C., Aguilar-Salinas, C.A., Mendoza-Herrera, K. et al. (2020). *Salud Pub. Mex* 62: 137; (b) Boo, S., Yoon, Y.J., and Oh, H. (2018). *Medicine* 97: e13713; (c) Gupta, R., Rao, R.S., Misra, A., and Sharma, S.K. (2017). *Ind. Heart J.* 69: 382.
161 G. R. F. Collaborators (2020). *Lancet (London, England)* 396 (10258): 1223.
162 Martín, J.A.R., De Arana, C., Ramos-Miras, J.J. et al. (2015). *Environ. Pollut.* 196: 156.
163 Donoso, G. and Melo, O. (2006). Water Quality Management in Chile: Use of Economic Instruments. In: *Water Quality Management in the Americas. Water Resources Development and Management.* (eds. A.K. Biswas, B. Braga, C. Tortajada, D.J. Rodriguez). Springer: Berlin, Heidelberg.
164 (a) Xu, H., Mao, Y., Xu, B., and Hu, Y. (2021). *J. Trace Elements Med. Biol.* 63: 126651. (b) Jiang, Q., Xiao, Y., Long, P. et al. (2021). *Chemosphere* 285: 131497. (c) Buhari, O., Dayyab, F.M., Igbinoba, O. et al. (2020). *Hum. Exp. Toxicol.* 39: 355.
165 (a) Cosselman, K.E., Navas-Acien, A., and Kaufman, J.D. (2015). *Nat. Rev. Cardiol.* 12: 627. (b) Alissa, E.M. and Ferns, G.A. (2011). *J. Toxicol.* 2011: 870125.
166 Luo, X., Liu, Z., Ge, X. et al. (2020). *BMC Public Health* 20: 874.
167 Yang, C.C., Chuang, C.S., Lin, C.I. et al. (2017). *J. Clin. Lipidol.* 11: 234.
168 (a) Obeng-Gyasi, E. (2020). *J. Environ. Sci. Health. Part A Toxic/Hazard. Subst. Environ. Eng.* 55: 726. (b) Xu, C., Weng, Z., Zhang, L. et al. (2021). *Ecotoxicol. Environ. Saf.* 208: 111433.
169 Gallagher, C.M. and Meliker, J.R. (2010). *Environ. Health Perspect.* 118: 1676.
170 Bulka, C.M., Persky, V.W., Daviglus, M.L. et al. (2019). *Environ. Res.* 168: 397.
171 Park, Y. and Han, J. (2021). *Int. J. Environ. Res. Public Health* 18.
172 (a) Xiao, L., Zhou, Y., Ma, J. et al. (2019). *Chemosphere* 215: 362. (b) Ngala, R.A., Awe, M.A., and Nsiah, P. (2018). *PLoS One* 13: e0197977. (c) Shun, C.H.,

Yuan, T.H., Hung, S.H. et al. (2021). *Environ. Sci. Pollut. Res. Int.* 28: 27966. (d) Wolide, A.D., Zawdie, B., Alemayehu, T., and Tadesse, S. (2017). *BMC Endocr. Disord.* 17: 64.
173 (a) Xu, Y., Wei, Y., Long, T. et al. (2020). *Chemosphere* 254: 126763. (b) Asgary, S., Movahedian, A., Keshvari, M. et al. (2017). *Chemosphere* 180: 540.
174 Zhao, M., Ge, X., Xu, J. et al. (2022). *Ecotoxicol. Environ. Saf.* 231: 113196.
175 Li, Z., Xu, Y., Huang, Z. et al. (2021). *Chemosphere* 264: 128505.
176 Moon, S.S. (2014). *Endocrine* 46: 263.
177 Guo, C.H., Chen, P.C., Lin, K.P. et al. (2012). *Environ. Toxicol. Pharmacol.* 33: 288.
178 (a) Turgut, S., Polat, A., Inan, M. et al. (2007). *Ind. J. Pediatr.* 74: 827; (b) Ahamed, M., Singh, S., Behari, J.R. et al. (2007). *Clin. Chim. Acta* 377: 92.
179 Klevay, L.M. (2006). *Cell Mol. Biol.* 52: 11.
180 Ilyas, A. and Shah, M.H. (2017). *Biol. Trace Elem. Res.* 180: 191.
181 Anschau, V., Dafré, A.L., Perin, A.P. et al. (2013). *Parasitol. Res.* 112: 2361.
182 Tak, H.I., Ahmad, F., and Babalola, O.O. (2013). *Rev. Environ. Contam. Toxicol.* 223: 33.
183 Sharma, B., Singh, S., and Siddiqi, N.J. (2014). *Biomed Res. Int.* 2014: 640754.
184 Yu, X., Liu, C., Guo, Y., and Deng, T. (2019). *Molecules* 24.
185 (a) Arita, A. and Costa, M. (2009). *Metallomics* 1: 222; (b) Salnikow, K. and Zhitkovich, A. (2008). *Chem. Res. Toxicol.* 21: 28.
186 Gao, Q.Q., Zhang, H.D., Zhu, B.L. et al. (2022). *Chinese J. Ind. Hygiene Occupat. Dis.* 40: 57.
187 Wallace, D.R., Taalab, Y.M., Heinze, S. et al. (2020). *Cells* 9.
188 Kuang, X.Y., Feng, Y.M., Zhang, X.T. et al. (2011). *Zhonghua lao dong wei sheng zhi ye bing za zhi = Zhonghua laodong weisheng zhiyebing zazhi = Chinese J. Ind. Hygiene Occupat. Dis.* 29: 376.
189 Fagerberg, B. and Barregard, L. (2021). *J. Int. Med.* 290: 1153.
190 Nickens, K.P., Patierno, S.R., and Ceryak, S. (2010). *Chem. Biol. Interact.* 188: 276.
191 Yang, J. and Ma, Z. (2021). *Ecotoxicol. Environ. Saf.* 213: 112034.
192 Lin, X., Gu, Y., Zhou, Q. et al. (2016). *J. Appl. Toxicol.* 36: 1163.
193 Luo, S., Terciolo, C., Bracarense, A. et al. (2019). *Environ. Int.* 132: 105082.
194 Skalny, A.V., Lima, T.R.R., Ke, T. et al. (2020). *Food Chem. Toxicol.* 146: 111809.
195 (a) Kulkarni, P.P., She, Y.M., Smith, S.D. et al. (2006). *Chemistry* 12: 2410; (b) de Bie, P., Muller, P., Wijmenga, C., and Klomp, L.W. (2007). *J. Med. Genet.* 44: 673.
196 La Fontaine, S., Burke, R., and Giedroc, D.P. (2016). *Metallomics* 8: 810.
197 Shindo, M., Torimoto, Y., Saito, H. et al. (2006). *Hepatol. Res.* 35: 152.
198 Cairo, G. and Recalcati, S. (2007). *Expert Rev. Mol. Med.* 9: 1.
199 Li, M., Gao, J.Q., and Li, X.W. (2005). *J. Hygiene Res.* 34: 375.
200 Tan, C.P., Lu, Y.Y., Ji, L.N., and Mao, Z.W. (2014). *Metallomics* 6: 978.
201 Esteban-Fernández, D., Moreno-Gordaliza, E., Cañas, B. et al. (2010). *Metallomics* 2: 19.

7

Matermetallomics

Qing Li[1], Zhao-Qing Cai[1], Wen-Xin Cui[1], and Zheng Wang[1,2]

[1]Chinese Academy of Sciences, Shanghai Institute of Ceramics Center of Inorganic Mater. Analysis & Testing, 585 Heshuo Road, Shanghai 201899, P. R. China
[2]University of Chinese Academy of Sciences, Center of Materials Science and Optoelectronics Engineering, 19 Yuquan Road, Beijing 100049, P. R. China

7.1 The Concept of Matermetallomics

7.1.1 Introduction

With the in-depth development of biology, biochemistry, and life science, the terms genomics, transcriptomics, proteomics, and metabolomics were coined during the past decades. The suffix "-ome" refers to the entirety of, for example, genes, proteins, or metabolites, in a regarded system, while "-omics" corresponds to relevant scientific research to obtain the overall qualitative and quantitative information of each object. The ambitious goal of these "-omics" fields is to integrate genome, transcriptome, proteome, and metabolome data in order to expand our knowledge of organisms or ecosystems [1–4]. The integration and interpretation of large data sets can improve the understanding of channel functions and regulatory networks.

Metals play an important role in many life processes; on the one hand, they can be essential, on the other hand, metals can be toxic. In analogy to genome, transcriptome, proteome, and metabolome, the term metallome was coined to describe the entirety of metal and metalloid compounds in an organism or its parts (cells, body fluids, or tissues). The study of the metallome and thus the related scientific investigations necessary to acquire and integrate metallome data were denominated by the term metallomics. At present, metallomics has become a frontier interdisciplinary subject and has attracted extensive attention in the academic community [5]. The research on metallomics has made rapid progress, and both basic theories and technical methods are constantly improving.

Material is the basis for manufacturing. Materials innovation has always been the core of disruptive technical revolution. To accelerate the process from discovery to the application of new materials, the then US President Barack Obama announced the Materials Genome Initiative (MGI) in 2011 and clearly stated that the goal of

Applied Metallomics: From Life Sciences to Environmental Sciences, First Edition.
Edited by Yu-Feng Li and Hongzhe Sun.
© 2024 WILEY-VCH GmbH. Published 2024 by WILEY-VCH GmbH.

the MGI is to discover, develop, manufacture, and deploy advanced materials at twice the speed and for a fraction of the cost of what is currently possible. MGI has quickly aroused the positive response of scientists in the field of materials, and a new round of material revolution began. The essence of MGI is that materials design is conducted by up-front simulations and predictions, followed by key validation experiments. It combines big data and high-throughput screening to achieve efficient materials discovery and parameter optimizations, and greatly reduces the developmental costs for new materials. Thus, the quantitative relationships between composition, structure, process, properties, and performance are of great significance for research and development of new materials, which is also an important content of materials genomics [6].

Recently, this technology has been successfully applied to some novel materials, such as energy materials, composite materials, alloy materials, catalytic materials, polymer materials, and biomaterials [7–12]. Studies have shown that the composition as genes of materials is an important link in the chain of processing, structure, properties, and performance, which is worthy of attention and discussion.

Materials can be divided into metallic materials, inorganic nonmetallic materials, polymer materials, and composite materials depending on their chemical composition. Metallic elements play an important role and influence in these materials. Metallic materials are mainly composed of metal elements, including metallic simple substances and alloys. In the periodic table, about 50 kinds of metallic elements are labeled as metallic simple substances, and almost all metallic elements and some non-metallic elements have been prepared into alloys. These are important materials used in the fields of aerospace, energy, mechatronics, and automobile because of their good thermal conductivity, strength, hardness, and ductility. Studies also show that these properties are directly related to the type and proportion of metals. Besides, inorganic nonmetallic materials, polymer materials, and composite materials involve metallic elements, such as rare-earth-doped laser crystals, metal organic frameworks (MOF), and Mxenes. Metallic elements can be used as matrix, dopant, crosslinker, and impurity, which cover almost all of the metallic elements in the periodic table as shown in Figure 7.1. As the matrix of metallic materials,

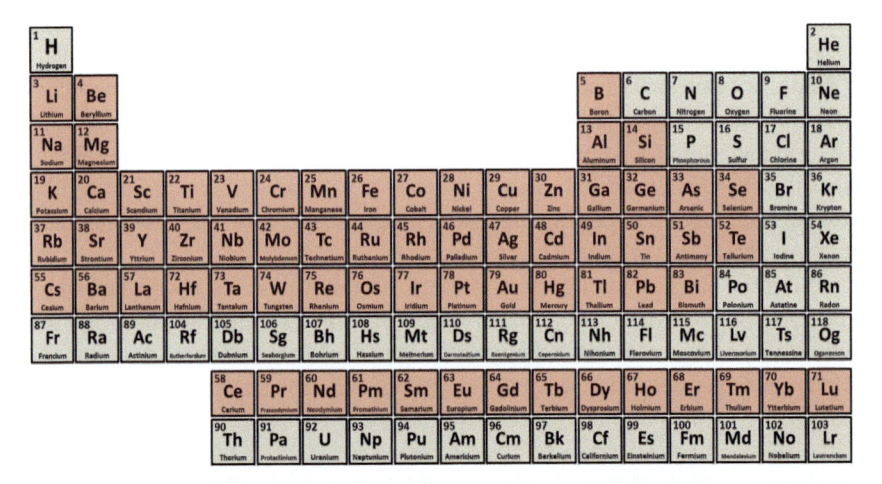

Figure 7.1 Elements already used for doping. (Almost all metallic and semi-metallic elements have been used for doping in the periodic table). Source: Qing Li.

the importance of metal elements is self-evident. Different metallic compositions produce different properties and performance, which is also one of key focuses of researchers.

7.1.2 Metallic Elements as Dopant

Doping is the purposeful incorporation of a small number of other elements or compounds into a material to change its composition, and making materials possess specific electrical, magnetic, and optical properties. At present, almost all of the discovered metallic elements have been used for doping, such as in crystals, ceramics, organic compounds, and other substances. The doped elements cover almost all of the metallic elements and semi-metallic elements (B, Si, As, Se, Te, etc.) in the periodic table of elements. Table 7.1 lists the doping metallic elements in typical materials and related properties.

Doping improves the properties of materials by changing their composition or structure and is widely used in materials science. It was found that advanced materials with different properties of optical, magnetic, and electrical properties have been obtained by doping metallic elements. The doped elements include almost all metallic elements and semi-metallic elements, and the forms of doping involve ions, atoms, isotopes, valence states, nanoparticles, bimetallic/polymetallic doping. At the same time, doping of non-metallic elements has been rarely reported.

Metallic element doping can effectively change the photoluminescence (PL) characteristics of the metal halide perovskite (the general structure is ABX_3) [13]. It has been reported that $CoFe_2O_4$ nanoparticles have been doped with more than 20 kinds of metal elements to achieve applications in different fields [14]. GaN-based magnetic semiconductor materials incorporated with different metallic elements changed different degrees of magnetic performance. The non-magnetic dopant Cu has a 0.70 µB/atom, and the magnetic moment of Ag doped GaN was larger than that of Pd doped because the N sites contribute 0.42 µB, which is much greater than any other 4d-metal-doped GaN. In addition, there was a unique observation reported as a colossal magnetic moment in REEs-doped GaN [15]. Atomically precise noble metal (mainly silver and gold) nanoclusters are an emerging category of promising functional materials. Without doping, the magnetic response of noble metal clusters is generally weak. Doping with ferromagnetic metals, such as Fe, Co, and Ni, is a promising pathway to introduce magnetic properties into such clusters [16].

In addition to the type of doped metal elements, closely related to the properties of the materials, the content of the doped elements also greatly affects the properties of the materials. Shobana et al. [17] found that the increase in Y^{3+} content in yttrium doped cobalt ferrite prepared by sol–gel combustion can increase the crystallite size and decrease the conductivity of the nanoparticles. The valence state or dopant form is also related to the properties of materials. The dopants are mainly in the form of species (ion, atom, isotope) nanoparticles and compounds. For example, when adding metal halide salts (e.g. $SbCl_3$, $BiCl_3$, VCl_3, $NiCl_2$, $ZnCl_2$, $SnCl_2$, $SnCl_4$, $PbCl_2$, and $CuCl_2$) as dopants to $CsPbCl_3$ nanocrystals, the difference in the photoluminescence quantum yield (PLQY) enhancements was not due to the doping of these different doping metallic ions, but because of the varied ability of these metal compounds to release active chloride ions for surface passivation [18, 19].

Table 7.1 Doping and properties of some representative materials.

Material	Doping metal elements	Related performance
Metal halide perovskites	Sb, Bi, Sn, Pb, transition metals (e.g. V, Ni, Cu, Zn), lanthanide metals (e.g. Sm, Dy, Er, Yb), and alkali metals (e.g. Li, Na, K, Rb)	Optoelectronic properties (e.g. absorption band gap, PL emission, and quantum yield [QY]) and stabilities Power conversion efficiency (PCE), the reproducibility, and stability
Ceramic–magnetic nanoparticles	Mg, Al, In, 3d metals (e.g. Ti, V, Cr, Mn), 4d metals (e.g. Y, Ag, Cd), rare-earth metals (e.g. La, Ce, Pr, Nd)	Electrical Structural Optical characteristics
Nitride ceramic phosphor	Rare-earth metals (e.g. Eu, Ce, Tb, Er, Gd)	Optical characteristics
Zinc oxide semiconductor	Transition metals (e.g. Mn, Cu, Fe) Noble metals (e.g. Au, Ag, Pd, Pt)	Photocatalytic properties (energy levels mobility, conductivity, and optical and magnetic properties) Gas-sensing properties
Graphite carbon nitride	Alkali metals (e.g. Na, K), transition metals (e.g. Fe, Cd, Co, Mo), rare-earth metals (e.g. Ce, Eu, Se, Y)	Photocatalytic properties (band gap, interlayer resistance)
Silicon carbide ceramics	Transition metals (e.g. Fe, Ti)	Band diagram Conduction Spin-related-features
Electrolyte ceramics	Cu, Sb, Nb, Ta, Mn	Oxygen ion conductivity
Noble metal nanoclusters	Fe, Co, Ni	Ferromagnetic properties
GaN-based magnetic semiconductor materials	3d metals (e.g. Fe, Co, Ni, Cu), 4d metals (e.g. Ag, Pd), rare-earth metals (e.g. Eu, Gd, Ce)	Ferromagnetic properties
Structural ceramics	Rare-earth metals (e.g. La, Y, Ce)	Compactness properties Thermal conductivity Mechanical properties Electrical properties
Conductive polymer	Transition metals (e.g. Fe, Co, Ni)	Electrochemical activity

For ferric-chloride-doped polyvinyl alcohol (Fe: PVA) polymer films, the decrease of the refractive index in the visible range induced by UV exposure was related to the reduced oxidation state of the doping metal ($Fe^{3+} \rightarrow Fe^{2+}$) [20]. Structural ceramics are widely used in daily life, and also have been extended to aerospace, integrated circuit, energy, and the environmental protection fields, such as alumina, silicon nitride, and silicon carbide. In order to improve the compactness, thermal conduction, and reduced sintering temperature, rare-earth oxides are often doped, such as La/Y-Al_2O_3 [21].

Bimetallic/polymetallic doping has been tried because single-metal doping sometimes cannot achieve satisfactory results. Bimetallic doping caused shifts of the absorption and emission band maxima in the luminescence spectra and the appearance of optical properties unattainable for mono-doped materials [22]. Moreover, such systems often demonstrate the sensitization effect which allows a considerable enhancement of the luminescence intensity and thus avoids the limitations imposed by the concentration quenching effect. Experimental and theoretical studies have proven that incorporation of co-dopants advances the photocatalytic properties of graphite gallium nitride (g-C_3N_4) photocatalyst more efficiently [23].

7.1.3 Metallic Elements as Impurities

Actually, in addition to doping by introducing metal elements into the material, the impurities of the raw materials and the contamination accompanying the process (such as the diffusion from crucible material and matrix material) will also affect the final quality and performance of the materials. Appropriate impurity species and contents are very helpful for the improvement of material properties. Cobalt and copper oxides have shown more important contribution to the observed electroactivity, and their presence as impurities in carbon nanotubes (CNTs) should be considered in the evaluation of their electrochemical response [24]. It is considered that the ferromagnetism of some fullerene polymers and graphite, such as hard-carbon-phase samples, is not related to fullerene at all [25]. Most of the works published previously as evidence of ferromagnetism in fullerene polymers synthesized at high-pressure high-temperature (HPHT) conditions can be explained by contamination with magnetic impurities. Formation of iron carbide (Fe_3C) due to the reaction of metallic iron with fullerene molecules explains the observed Curie temperature of close to 500 K and the levels of magnetization reported for "magnetic carbon."

However, the existence of some impurities with the content exceeding a certain limit often causes adverse effects, and the impurity removal process usually should be carried out. Metallic impurities such as iron, chromium, and titanium can reduce the carrier diffusion length in polysilicon. In polysilicon cells, copper affected not only the composition of the substrate, but also the composition of the emitter and, more importantly, it affected the performance of the cell [26]. Pumera et al. [27] investigated the electrochemical response of CNTs containing different amounts of impurities toward the reduction of an important biomarker, hydrogen peroxide, and the oxidation of an important impurity marker. They found that the borderline between being redox active/inactive for iron-based impurities was in the middle-ppm range.

7.1.4 Metallic Elements as Crosslinkers

Metal coordination is a special non-covalent interaction with both high bond energy and dynamic characteristics, for which metal ions are widely used as crosslinkers to assemble and regulate materials [28] and to broaden the properties and applications prospects of materials. For example, crosslinking can fix the structure of polymer micelles and improve the stability of polymer micelles, and the use of some chemicals can be avoided in the formation of non-covalent crosslinking with metal ions (e.g. Eu^{3+}, Ru^{2+}, and Zn^{2+}), which has been applied in the fabrication of metal coordination crosslinked polymer micelles. Based on the unique advantages of metal ion crosslinking, such as forming a reversible network and simply tuning material properties, it has also been applied to polyethylene glycol (PEG)-based links, crosslinked with metal ions (e.g. V^{3+}, Fe^{3+}, and Al^{3+}) as materials for 3D extrusion printing, thus facilitating the adaptation of the system to the requirements of the printing process broad range of the printing parameters and application with remarkable flexibility [29]. All in all, metal crosslinking is a good way to adjust and assemble materials, which has been successfully applied in many fields.

The specific mechanisms of the relationship between doping, crosslinking, or impurity elements and properties are still unclear and worthy of further study, especially with regard to the MGI. The materials genome can be compared with biological genomes, and one may conclude that: at any moment, the performance of a specific material depends on its chemical composition (inherent property stored in its genome), which determines the properties of materials and its environment (external interactions–processing–conditions during usage). Thus, it is important to materials science that the distribution, species, and content type of doping or impurity elements, and the effects on the resulting properties of the materials be evaluated.

Metals play an important role in life process. Metallomics was coined to describe the interactions and functional connections of entirety of metal and metalloid compounds in an organism or its parts. In analogy to metallomics, the term matermetallomics was coined to describe the roles and functions of all metallic elements and parts semi-metallic elements to structure, properties, and performances of materials. It is therefore important to specify that a metallomics study implies:

(a) A focus on metallic or semi-metallic elements (e.g. B, Si, Ge, As, Se, Sb, Po, and At) in the context of materials science. The metallic or semi-metallic elements as matrix, dopants, linkers, impurities, and other components containing metals and semi-metals in the material system are collectively referred to as matermetallome. It is not recommended to extend the term to non-metals, such as sulfur or phosphorus.

(b) A correlation of the element concentration mapping or element speciation with materials. This correlation may be statistical (the distribution, content, and species of elements coincide with the presence of a particular property or character), structural (the interaction between the elements and microstructure), or functional (the presence of elements is the result of process adjustment).

(c) A systematic, comprehensive, or global approach. If a metal species does not explain its significance and contribution to materials science, it is not matermetallomics!

In summary, matermetallomics aims to answer the following question: what are the interactions and functional connections between metal ions and material properties, properties, and structures? This includes a global identification and quantification of matermetallome as well as their link to the material structure–activity relationship.

Matermetallomics is not an isolated research field; it has to be regarded as subcategory of metallomics. The study of the matermetallome requires specific analytical strategies and matermetallomics is strongly interrelated to metallomics and material science.

7.2 The Analytical Techniques in Matermetallomics

In matermetallomics, many existing analytical techniques have been widely used, while more advanced analytical techniques are also needed to explore the new research fields. Some representative analytical methods in materials analysis are listed in Table 7.2. The analysis techniques are splitted into the following sections based on the research subjects.

7.2.1 Element Imaging Analysis

In material studies, elemental maps are of great importance to improvements of manufacturing and processing techniques such as deposition, diffusion, or segregation processes, and coating or combustion procedures, which enables localization of chemical elements including trace metals at micron and even at nanometer level. In this chapter, we present four complementary element-specific imaging techniques: laser-induced breakdown spectroscopy (LIBS) and laser ablation inductively coupled plasma mass spectrometry (LA-ICP-MS) based on laser plasma, secondary ion mass spectrometry (SIMS), transmission electron microscopy coupled with energy-dispersive X-ray spectroscopy (TEM/X-EDS), and synchrotron-based X-ray fluorescence (SXRF). The common principle is that a high-energy beam scans the surface of a solid material, and characteristic spectrum of excited elements or ionized ion intensity signals are recorded by mass spectrometer or spectrometer.

Table 7.2 Representative analysis methods in material analysis.

Analyte	Analysis method	Comments	References
Cr, Co, Ni, and Cu	LA-ICP-MS	Quantitative analysis of Cr, Co, Ni, and Cu in metallic materials without complex processing	[30]
Ratio of Fe(II)/Fe(III)	XPS	Analyze the ratio of Fe (II)/Fe (III) in 1045 steel and J55 steel	[31]
Cr, Fe, Ni	SIMS	Depth-profiling analysis of Cr, Fe, Ni in oxidized steel samples	[32]
Th and U	ICP-MS	Detect ultra-trace Th and U in copper	[33]
Sc, Y, La, Ce, Pr, Nd, Sm, Eu, Gd, Tb, Dy, Ho, Er, Tm, Yb, Lu, Th	Glow discharge mass spectrometry (GD-MS)	Qualitative analysis of 72 impurity elements in high-purity copper powder	[34]
Cd, Cr, Mn, Mo, Pb, V, and Zn	Electrothermal vaporization (ETV)-ICP-AES	Determination of Cd, Cr, Mn, Mo, Pb, V, and Zn in 2.0–2.5 mg aliquots of nickel foam samples without pretreatment	[35]
Cr, Mn, Fe, Co, Ni, Cu, Zn, As, and Pb at trace level	Total reflection-XRF	Quantify trace elements in light and middle distillates (gasoline, racing, and jet fuel)	[36]
Sulfur and iron valence state	Wavelength dispersion (WD)-XRF	Determine the content of ferrous iron and to estimate sulfur valence state in coal concentrates and ashes from the pressed pellet without additional sample preparation	[37]
Lithium isotopic	MC-ICP-MS	Determination of lithium isotopes in coal to help to track atmospheric haze and polluted water in the environment	[38]
C, H, Ni, S, and V	LIBS&LA-ICP-AES	LIBS and LA-ICP-AES were used simultaneously for the elemental analysis of asphaltene samples using minimum sample pretreatment	[39]
Cr, Hg, and Pb	LIBS	Quantitative analysis of plastics by LIBS was used to measure Cr, Hg, and Pb	[40]
The ^{137}Cs and ^{40}K radio isotopes	High-purity germanium gamma spectrometry (HPGe)	The HPGe method was used to detect the concentration of ^{137}Cs and ^{40}K isotope in edible salt	[41]
Pd, Pt, and Rh	High-resolution continuum source graphite furnace (HR-CS)-GFAAS	HR-CS-GFAAS detected the Pd, Pt, and Rh (PGMs) in spent automobile catalysts without chemical separation	[42]

Gas concentrations below a tungsten surface	LIBS-LAMS	LIBS-LAMS was used to measure the depth-dependent concentration of gaseous substances under the surface of tungsten	[43]
$^{235}U/^{238}U$ ratio	NanoSIMS	The micro-scale isotopic heterogeneity of nuclear fuel pellets was characterization by nano-scale secondary ion mass spectrometry (NanoSIMS)	[44]
Two Cm isotopic ratios	AMS	Two Cm isotopic ratios ($^{244}Cm/^{246}Cm$ and $^{245}Cm/^{246}Cm$) were determined to date irradiated in nuclear fuels	[45]
Cu, Zr, Ag, W	FIB-TOF-SIMS	High spatial resolution of elemental distribution was determined	[46]
Cu, Se, In, Ga	GD-OES	GD-OES was used for quantitative depth-profiling of copper indium gallium sulfur selenide (CIGS) thin films	[47]
Mn, Cu, Fe	Electron probe microanalyzer (EPMA)	EPMA was used to measure the partial substitution of iron in the TiFe-system	[48]
Al, Si, K, Ca, Ti, Mn, Fe, Cu, Zn, and Sr	PIXE	Multi-elements and non-destructive analysis on the young leaves of Neem	[49]
Na, Mg, Al	PIGE	In situ quantitative analysis of four main elements (silicon, sodium, magnesium, and aluminum) in the soda lime glass sample	[50]
Sc	NMR	Solid-state NMR (^{45}Sc MAS NMR) was used to measure the intramolecular charge transfers, which confirmed clearly the increase of electron density around the Sc^{3+} species after Sc-EBTC interacted with the analytes	[51]
Complexation of terpolymer and Ga ions	HPLC	HPLC was used to characterize the complexation of the terpolymer and Ga ions	[52]
Na	MALDI-TOF-MS	MALDI-TOF MS spectra showed the ionization with H^+ and Na^+ were almost the same to the theoretical calculated values, confirmed chemical structure of octa carboxyl polyhedral oligomeric silsesquioxane	[53]
Microstructure of $La_2CaB_{10}O_{19}$ crystal	Raman spectrometer	In situ Raman spectroscopy was applied to obtain the microstructure information on $La_2CaB_{10}O_{19}$ crystal and its growth solution	[54]
Titanate	FTIR	FTIR measured the typical stretching vibrations band of Ti–O, explained the molecular structure of titanate nanofibers	[55]

In order to generate an image, the signal intensities for each element are converted into a color or grayscale intensity plot where each image is a visual representation of the relative distribution of one chemical element.

7.2.1.1 Laser Ablation Inductively Coupled Plasma Mass Spectrometry (LA-ICP-MS)

LA-ICP-MS is a spatially resolved analytical technique mainly measuring metallic elements in solid materials. The process of LA-ICP-MS analysis involves four processes: (i) the sample is placed inside a gas-tight chamber with a special geometry and size, which is connected with the ICP-MS torch. The chamber is continuously flushed with argon or helium, which is used as carrier gas. (ii) A pulsed laser (nanoseconds to femtoseconds pulse duration) is used to ablate the surface of the sample, which results in the formation of sample aerosols. (iii) The aerosols transported into the ICP torch via the continuous carrier gas flow are atomized and ionized. (iv) Atomic cations are separated and measured by mass-to-charge ratio in MS.

Most commercial LA-ICP-MS systems apply solid-state Nd:YAG laser operated at different wavelengths (mostly 193, 213, or 266 nm) allowing a spatial resolution down to 2 µm [56]. The application of near-field laser ablation allows even a spatial resolution in the nanometer range. Also, examples of the application of excimer laser have been described, which causes less fractionation effects during the ablation process. In addition, many studies have been carried out on the sampling cell to reduce fractionation effect or improve imaging resolution, such as a low dispersion sample chamber (tube cell), which allows improvement of the imaging capabilities by reduction of the single LA shot duration to 30 ms (full width at 1% maximum) [57]. The two most important parameters in element-specific imaging are spatial resolution and element sensitivity, which are also contradictory. In order to achieve high spatial resolution, it is necessary to decrease the laser spot diameter, resulting in delivering less material to the ICP-MS and weaken MS signal intensity. In addition, another reason is that the existence of non-thermal equilibrium photochemical or photothermal process makes the ablation pit often larger than the laser spot in the ablation process. Similarly, high sensitivity requires enlarging laser spot size to deliver more material to the ICP-MS; however, this will be at the detriment of spatial resolution.

LA-ICP-MS is capable of being a quantitative technique through the application of matrix-matched calibration standards. Certified reference materials (CRMs) for tissues are readily available [58], although these tend to have a limited range of elements certified and it is rare to have more than one CRM available for each tissue type leading to single- or two-point calibrations, which have limited accuracy across the entire concentration range.

While two-dimensional imaging of the distribution of elements offers clear benefit over 1D and single-point analysis, it would be beneficial to observe the whole material in a 3D distribution to the study of material defects and other mechanisms. This has been achieved by Hare et al. resulting in a quantitative three-dimensional, high-resolution (HR) virtual reconstruction of metal-ion distributions in mouse brain [58].

7.2.1.2 Laser-Induced Breakdown Spectroscopy (LIBS)

LIBS, another laser-assisted technique used for elemental analysis, allows to complement drawbacks of LA-ICP-MS. When a high-power laser pulse strikes a material, the high electric field generated at the focal spot causes an electric breakdown and leads to a certain amount of mass ablation. The evaporated mass expands in the form of a plasma plume and compresses the surrounding atmosphere by releasing a shockwave. The plasma plume is a high-temperature dense collection of electronically excited atoms and ions of the ablated mass, whose radiation emission should be efficiently collected by a detection system. LIBS yields a unique list of advantages such as real-time, standoff analysis capabilities; quasi nondestructive ablation; simultaneous all-element detection; safe operation; and necessitating only optical access with little or no sample preparation [59–63]. Its disadvantages are poor stability, high sensitivity (up to µg/g level). LIBS has been successfully applied to the analysis of materials of interest in many different fields, ranging from industrial diagnostics [61, 64, 65], environmental protection [66, 67], bio-medicine [68], forensic analysis [69], cultural heritage and archaeology [70], and so on. The characteristics of speed, portability, capability of in situ remote analysis without sample treatment, and ease of use of the equipment have made LIBS one of the fastest growing analytical techniques in the last century.

LIBS enables extremely fast imaging experiments with pixel acquisition rates in the kHz range and a spatial resolution down to some µm with valuable und numerous developments and applications published by the group of Vincent Motto-Ros [71]. The lack of need for the transport of ablated matter also eliminates carry-over and wash-out effects, and transport efficiency is not an issue. The ability to measure almost every element of the periodic table also including the elements H, C, N, O, and F which are not easily accessible by ICP-MS is a further remarkable benefit of LIBS that are recognized in elemental imaging studies. Compared to alkali and earth alkali elements, which provide best detection limits (DLs), the sensitivity of non-metals is reduced. Moreover, LIBS spectra may also provide molecular information, which is useful especially in terms of polymer characterization and capabilities for stand-off analysis. In the work of Veber et al. LIBS was used as an analytical tool to investigate elemental distributions of Ca and Zr in lead-free piezoelectric crystals grown by the micro-pulling down technique [72]. Longitudinal LIBS analysis was in good agreement with electron-probe micro-analysis (EPMA) analysis and was able to reveal inhomogeneities of Zr especially at high pulling speeds. Additionally, cross-sections of pulled fibers revealed elemental segregation at the core.

7.2.1.3 Secondary Ion Mass Spectrometry (SIMS)

SIMS is an analytical technique that relies on the sputtering of ions from a solid surface by focused positive or negative primary ion beams and the subsequent analysis of the produced secondary ions by a mass spectrometer under high vacuum. Different combinations of primary ion sources (e.g. Ga^+, Bi^+, Cs^+, O^-, C_{60}^+) and mass spectrometers (quadrupole, time of flight, double sector) enable a wide range of applications for this technique. The primary ion beam (either Cs^+ or O^-) is scanned over the surface of the sample, and the secondary ions are directed toward a double-sector analyzer. This double sector is composed of an electrostatic field that homogenizes the kinetic energy and focuses the ions beam and a magnetic field that will deflect the ions according to their mass over charge ratios (m/z). Numerous other components (lenses, deflectors, diaphragms, hexapole) are used in the secondary ion path to shape and focus the ion beam. Finally, the ions are counted using seven parallel detectors allowing simultaneous detection, which is a requisite for high-precision isotopic ratios.

Thus, SIMS can be described as a scanning ion microprobe that can analyze at the sub-micrometer scale and produce elemental and isotopic images in 2D and 3D. Since its introduction in the 1990s [73, 74], the high spatial resolution of SIMS opened numerous new research possibilities [75]. SIMS can generally be divided into static SIMS and dynamic SIMS, and the latter can be used for depth profiling. During depth profiling, dynamic SIMS continuously detects all substances being stripped, which can provide information on the elemental and isotopic composition of the sample at each time. The depth limits can reach about one atomic layer. According to the analysis needs, depths of more than 10 µm samples can be analyzed continuously. Static SIMS can analyze thermally unstable and non-evaporating organic materials.

SIMS has so far been mainly used in five different fields of science, namely biology, geology, cosmochemistry, geochemistry, and material sciences. In geology, the lateral resolution of SIMS enables the determination of chemical gradient along single olivine grains for volcanology studies, the U–Pb dating of small zircons, isotopic ratios in very small and rare samples such as meteorites [76] and presolar grains [77]. In material science, the high lateral resolution is very useful to study, for example, element migration at grain boundaries [78] of alloys or electrochemical properties of electrodes for batteries [79]. Corrosion processes can also be dynamically monitored using enriched stable isotopes (^{18}O, ^2H) as shown by Yardley et al. [80].

7.2.1.4 TEM/X-EDS

A TEM/X-EDS is based on high-voltage electron emission allowing imaging of the texture and structure combined with elemental analysis of solid samples. The principle for imaging samples using an electron source and electromagnetic lenses is that an accelerated electron probe (between 80 and 300 keV) is focused on a sample – transparent to electron beam (thickness should be less than 100 nm depending on its nature and accelerating voltage) – by a set of electromagnetic lens. The objective lens allows imaging of the texture (direct imaging) but also of the structure (diffraction pattern); the interaction between the electron beam and the sample generates among others the production of X-photons issued from a small volume of the sample. The total thickness of the prepared sample sections should not be more than 100 nm for inorganic and 150 nm for organic/biological samples, respectively, which are placed on TEM grids (Cu, Ni, Au, etc.) with a diameter of 3.2 mm. The TEM/X-EDS technique generates graphs of the morphology, size, phase distribution, atomic structure, and elemental composition of the sample. TEM/X-EDS can analyze any single point of interest on the sample with high spatial resolution nm-μm level. TEM in HR or lattice fringes (LF) modes can achieve very fine resolution down to 0.2 nm, which can even provide details at atomic scale. However, these technologies can obtain 2D element imaging maps, while 3D imaging maps are difficult to obtain directly. It is usually analyzed after cutting materials into sections, used by pre-processing technology, such as focused ion beam (FIB) technology.

7.2.1.5 Synchrotron Radiation X-Ray Fluorescence Spectrometry (SR-XRF)

Synchrotron radiation X-ray fluorescence spectrometry (SR-XRF) is an X-fluorescence spectral analysis technology with synchrotron radiation X-ray as the excitation light source. Synchrotron radiation X-ray fluorescence analysis includes synchrotron radiation XRF for microarea and trace element analysis, synchrotron radiation total reflection X-ray fluorescence (SR-TXRF) for surface and film analysis, and synchrotron radiation X-ray fluorescence scanning and imaging methods for three-dimensional nondestructive analysis (such as X-ray fluorescence CT, X-ray fluorescence full field imaging, confocal X-ray fluorescence, and grazing out X-ray fluorescence). X-ray fluorescence spectrometry identifies elements by measuring the characteristic X-ray emission wavelength or energy of elements. Due to the use of high brightness light source, SR-XRF has the advantages of nondestructive, in situ microarea, micro- and trace-element concentration, and spatial distribution analysis compared with ICP-MS and other technologies. Combined with X-ray diffraction, absorption. and imaging technology, SR-XRF can obtain the chemical composition, morphology, structure, and imaging information of trace substances at the same time, and has been widely used in fields of biomedicine, environment, archaeology, geology, and material science.

Each technique has its specificity, strengths, and limitations. Most important parameters are lateral resolution and sensitivity, but also sample preparation and stability, range of detectable elements and potential for isotopic measurements, elemental speciation, and 3D imaging should be taken into account. For example, LA-ICP-MS provides excellent sensitivity in the low microgram per kilogram range, but the spatial resolution is usually between 4 and 20 μm, in the best case about 1 μm, which is not enough for highly resolved lattice imaging. In contrast, TEM/X-EDS provide resolutions down to 0.2 nm for structural and textural imaging and down to 5 nm for elemental mapping, but with a low sensitivity of about 1 g/kg. SIMS and SXRF show lateral resolution below 50 nm and sensitivities from the milligram per kilogram down to the microgram per kilogram range, depending on the element. In view of the range of detectable types, LA-ICP-MS and SIMS allow isotope ratio analysis, which is not possible with both X-ray-based techniques. In contrast, SXRF can additionally provide information on chemical bonding by the use of XAS. In the view of all these developments, it can be expected that elemental imaging will become an indispensable tool in future matermetallomics studies.

7.2.2 Quantitative and Qualitative Analysis

Analytical techniques providing quantitative and qualitative analysis information about the composition of samples can be classified into three main groups according to the type of particle detected: optical spectrometry, where the intensity of either non-absorbed photons (absorption) or emitted photons (emission and fluorescence) is detected as a function of photon energy (in most cases, plotted against wavelength); mass spectrometry (MS), where the number of atomic ions is determined as a function of their mass-to-charge ratio; and electron spectroscopy, where the number of electrons ejected from a given sample is measured according to their kinetic energy, which is directly related to the bonding energy of the corresponding electron in a given atom. Among them, the first two techniques are widely used in quantitative and qualitative analysis of elements. In this chapter, we present two complementary element-specific imaging techniques: atomic emission spectrometry and ICP MS.

For the main application of material composition analysis, the quantitative or qualitative analysis of dopants and impurities in materials, the analysis techniques mainly include inductively coupled plasma atomic emission spectrometry (ICP-AES), flame or graphite furnace atomic absorption spectrometry (FAAS/GFAAS), atomic fluorescence spectrometry (AFS) and ICP-MS. For these techniques, a dissolved sample is usually employed in the analysis to form a liquid spray which is delivered to an atomizer (e.g. a flame or plasma). Concerning optical spectrometry, techniques based on photon absorption, photon emission, and fluorescence will be described, while for MS particular attention will be paid to the use of an ICP as the ionization source.

7.2.2.1 Inductively Coupled Plasma Atomic Emission Spectrometry (ICP-AES)

Electrically generated plasmas produce flame-like atomizers with significantly higher temperatures and less reactive chemical environments compared with flames. The plasmas are energized with high-frequency electromagnetic fields (radio-frequency or microwave energy) or with direct current (dc). By far the most common plasma used in combination with AES for analytical purposes is the ICP.

The main body of an ICP consists of a quartz torch made of three concentric tubes and surrounded externally by an induction coil that is connected to a radiofrequency generator commonly operating at 27 MHz. An inert gas, usually argon, flows through the tubes. The spark from a Tesla coil is used first to produce "seed" electrons and ions in the region of the induction coil. Subsequently the plasma forms, provided that the flow patterns are adequate inside the torch, giving rise to high-frequency currents and magnetic fields inside the quartz tube. The induced current heats the support gas to a temperature of the order of 7000–8000 K and sustains the ionization necessary for a stable plasma. Usually, an aerosol from the liquid sample is introduced through the central channel transported by an argon flow. The high temperatures and the relative long residence time of the atoms in the plasma (2–3 ms) lead to nearly complete solute vaporization and high atomization efficiency. Accordingly, although matrix and inter-element effects should be relatively low, it has been observed that sometimes they are significant. Further, the high excitation capacity of this source gives rise to very rich spectra, so a careful assessment of potential spectral interferences is essential. On the other hand, the ICP emission frequently has an important background due to bremsstrahlung (i.e. continuous radiation produced by the deceleration of a charged particle, such as an electron, when deflected by another charged particle, such as an atomic nucleus) and to electron–ion recombination processes.

In addition to radiofrequency-generated plasma, other plasmas which have been used in combination with AES are the following:

Microwave-induced plasma (MIP) is initiated by providing "seed" electrons. The electrons oscillate in the microwave field and gain sufficient energy to ionize the support gas by collisions. MIPs operate at 2450 MHz and at substantially lower powers than ICP devices. The low-power levels do not provide plasmas of sufficient energy to get an efficient desolvation of solutions and, hence, MIPs have been used mostly with vapor-phase sample introduction. Sample introduction difficulties have been primarily responsible for the lower popularity of MIPs compared with ICPs.

Direct current plasma (DCP): this is produced by a dc discharge between electrodes. DCPs allow the analysis of solutions. Experiments have shown that although excitation temperatures can reach 6000 K, sample volatilization is not complete because residence times in the plasma are relatively short (this can be troublesome with samples containing materials that are difficult to volatilize). A major drawback is the contamination introduced by the electrodes.

The performance comparison of these analysis techniques is summarized in the table shown as below, including DLs, linear ranges, precision, versatility, and sample throughput.

7.2.2.2 Inductively Coupled Plasma Mass Spectrometry (ICP-MS)

ICP-MS is ordinarily performed using ICP as ion source and either a quadrupole or a scanning sector-field mass spectrometer as an analyzer. The remarkable attributes of such a combination, being the gold standard in terms of element-specific detection, include: very low detection limits for many elements; availability of isotopic information in a relatively simple spectrum; acceptable levels of accuracy and precision.

Quadrupole ICP-MS Quadrupole ICP-MS represents the simplest instrumental setup that has been used for multi-element analysis or as an element species detector when using hyphenated approaches. However, due to the presence of interfering argides, hydrides, carbides, nitrides, and oxides, which are naturally formed in argon plasma operated under normal laboratory conditions or due to the matrix constituents of samples, accurate multi-element determination was always challenging. Chemical matrix removal or mathematical corrections have been applied to account for specific interferences.

Pure quadrupole ICP-MS are getting rare in most laboratories, due to the introduction of collision/reaction cell inductively coupled plasma mass spectrometry (CC-ICP-MS) or the dynamic reaction cell inductively coupled plasma mass spectrometry (DRC-ICP-MS).

Such instruments consist of a multipole ion guide located in front of the quadrupole mass filter (e.g. a quadru, hexa, or octopole), which can be pressurized with a specific gas at a flow rate of several milliliters per minute. The pressurized ion guide allows the utilization of gas-phase chemistry or physical processes (e.g. kinetic energy discrimination [KED]), which result in the often dramatic reduction of specific interferences. By utilizing cell gases such as H_2, He, O_2, or NH_3, most-polyatomic interferences can be separated from the targeted ions or minimized to an insignificant level utilizing different gas-phase mechanisms.

Triple-Quadrupole ICP-MS In comparison to the already available DRC-ICP-MS, which utilizes an rf/dc quadrupole reaction cell, which can be pressurized with a reactive gas, in order to promote specific ion–molecule reactions and which also features an adjustable DRC band pass, allowing the partial suppression of new interferences produced through sequential reactions within the cell [81], this new instrumental setup includes two real independent quadrupole mass filters connected via an octopole collision/reaction cell, which allows an improved interference reduction as well as a better control of the gas-phase reactions or even completely new detection schemes, to handle specific interferences. Overall, this new instrumental setup provides various mechanisms to allow the interference-free detection of most elements of the periodic table with high sensitivity and lowest backgrounds due to the unique MS/MS approach.

Further ICP-based Mass Spectrometry Besides the already mentioned ICP-MS techniques, a few examples on the application of further instrumental setup such as HR double-focusing sector-field ICP-MS (HR-ICP-SF-MS) and multi-collector sector-field ICP-MS (MC-ICP-MS) as possible analyzer can be found in the literature. The interference problems have been overcome with the availability of HR-ICP-SF-MS, which uses a magnetic sector as well as an electrostatic sector field and different beam apertures to physically separate the mass of the targeted analyte ions from the interfering polyatomic species that have been formed inside the plasma. Different resolution settings up to 10 000 can be achieved with such instrumentation allowing the interference-free analysis of many elements. However, every improvement in resolution will cause a loss in sensitivity (e.g. about 99% reduction at a high resolution of 10 000). In consequence, trace analytes could not be detected any more. During the past decade, MC-ICP-MS has gained much attention, as it allows, for example, the accurate determination of isotopic ratios or the fully simultaneous acquisition of the entire m/z range covered by the elements of the periodic table. The technological developments strongly promote applications, such as marine geochemistry [82], geochronology [83, 84], cosmochemistry [85], or provenance studies [86, 87]. Such setups provide a mass resolving power (edge resolution) of up to 10 000, which is required in particular when analyzing interfered isotopes.

7.2.2.3 X-Ray Fluorescence (XRF)
Elemental analysis by means of X-ray fluorescence (XRF) spectrometry is based on the element-specific electromagnetic radiation induced as a consequence of

inner-shell ionization. The characteristic X-ray fluorescence is functional for several analytical methods. Besides the XRF fluorescence spectrometry, also the EPMA and the proton-induced X-ray emission (PIXE), are based on the X-ray fluorescence process. XRF spectrometry is ideal for the direct analysis of solid samples, but can also investigate fluid samples. Firstly, this method allows the rapid qualitative screening of unknown samples, without any particular sample preparation. Secondly, it is possible to perform the fully automated quantitative analysis of large sample sets. Thirdly, the availability of portable XRF systems is a further advantage for on-site measurements. And the rapidity and robustness of XRF methods for the quantitative analysis made the method the workhorse in many productive sectors. Further figures of merit are the "standard-less" analysis of samples in a non-destructive mode, and detection down to 0.01%. However, the main restriction on XRF techniques is the impossibility of measuring elements with atomic numbers lower than 10. In addition, matrix effects can influence the sensitivity to one specific element in different matrices, which makes the quantification complex.

Two XRF instrumentation architectures exist for the extraction of the element-specific information, namely a wavelength-dispersive (WD-XRF) or an energy-dispersive (ED-XRF) one. In the WD-XRF, the probing of the sample is accomplished with X-ray radiation of photon energy up to 100 keV. The fluorescence wavelength separation takes places by means of a crystal analyzer, according to the law of Bragg. ED-XRF relies on a solid-state detector to resolve spectral peaks of characteristic X-rays. In ED-XRF spectrometers, all of the elements in the sample are excited simultaneously, and an energy-dispersive detector in combination with a multi-channel analyzer is used to simultaneously collect the fluorescence radiation emitted from the sample and then separate the different energies of the characteristic radiation from each of the different sample elements. In comparison to ED-XRF, the WD-XRF instruments are at the time of writing more efficient in terms of fluorescence-line resolution but less detection sensitive in terms of concentration. XRF radiation in solid materials has a slightly larger penetration depth of approx. 1–100 µm, but also the elemental analysis using WD-XRF is limited to a shallow region, depending on matrix and photon energy.

7.2.2.4 GD Optical Emission Spectroscopy (GD-OES) and GD Mass Spectrometry (GD-MS)

The generation of a GD takes place typically in a low-pressure chamber through which argon flows continuously. The device consists of a grounded anode and a cathode (the sample). An electric current ionizes the discharge gas, forming a plasma and yielding argon ions which are attracted toward the sample surface producing the sputtering (removal) of atoms, electrons, and ions. The atoms of the analyte are excited and ionized through collisions in the plasma and, therefore, measurement by OES and MS is feasible.

GD optical emission spectroscopy (GD-OES) and GD mass spectrometry (GD-MS) are mainly applied in the materials sciences where they are used routinely for bulk and surface analysis. For direct analysis of conducting solids, GD-MS is virtually unrivaled for the trace analysis of impurities in high-purity materials. On the other hand, for bulk analysis of less pure metals, GD-OES competes with spark emission spectroscopy and X-ray techniques, but in many cases GD-OES exhibits fewer matrix effects or gives lower DLs. The success of GD-OES and GD-MS comes up from their interesting analytical performance as well as from some working features, such as low running costs (a low flow rate of argon is typically used as plasma gas) and high sample throughput (seconds to minutes per specimen) thanks to the fast sputtering rate (in the order of μm/min) and the lack of ultrahigh vacuum requirements in the discharge chamber.

For conducting materials, dc-GD-MS gives the best limits of detection in the range far below 1 ng/g, followed only by LA-ICP-MS. But to give a fairer assessment, the figure changes immediately once non-conducting materials such as glasses, geological samples, or ceramics become of interest, in particular if they are relatively thick. Non-conducting material can be only analyzed directly with RF-powered discharges and the generator power is usually coupled capacitively to the plasma, so that the plasma power decreases with increasing thickness. In last decades, some strategies have proved their interest in the analysis of nonconductors with dc GDs, such as the use of a secondary cathode [88–90], or by combining the surface-coating method and tantalum carrier method, eight in general terms it can be affirmed that dc-GDs applications mostly address depth profiling of conductive materials.

7.2.3 Metal Speciation Analysis

Metal elements present different valence states or forms in materials, which will contribute to the properties and performances of materials to varying degrees. Metal speciation analysis technologies include molecular spectroscopy technology (such as infrared, Raman spectroscopies), nuclear magnetic resonance (NMR) technology, and chromatography technology. The FTIR method is a non-destructive analysis and free from the restriction of the sample's physical state, but it is suitable for qualitative analysis, rarely for quantitative analysis. NMR is the most frequently used method for molecular structure characterization, which can obtain sample structure, composition, and kinetic information under non-destructive conditions with highly specific and good repeatability, but its low sensitivity limits its application in low doped materials. Gas/liquid chromatography-MS is mainly applied to identify unknown organic substances containing metallic elements, but the analysis time is generally longer than with other MS-based techniques. In this chapter, we will mainly introduce Raman spectroscopy and X-ray photo electron spectroscopy (XPS).

7.2.3.1 Raman Spectroscopy

Raman spectroscopy measures inelastically scattered light upon irradiating a sample with a visible or ultraviolet laser. Raman active vibrations are those that feature a change in the polarizability in the material/molecule. Raman spectroscopy can obtain the microstructure information of materials, such as bond length, coordination number information, cluster information with different coordination numbers, and quantitatively obtain the corresponding content of the microstructure types. A limitation to this technique is the small fraction of photons that are inelastically scattered (1 out of every 10^6) and thus, techniques to enhance this such as resonance Raman (RR) [91] or surface-enhanced Raman spectroscopy (SERS) [92], tip-enhanced Raman spectroscopy (TERS) [93] are often employed. What's more, it can be used for in situ detection at high temperature. The application of Raman spectroscopy in matermetallomics is widely used for analyzing metal complexes, such as MOF, and also used for the analysis of different metal compounds, such as uranium oxides.

7.2.3.2 X-Ray Photo Electron Spectroscopy (XPS)

XPS is considered a powerful technique for analyzing chemical state of elements on samples surface with good depth and lateral resolution. In XPS, a source of photons in the X-ray energy range is used to irradiate the sample. Superficial atoms emit electrons (called photoelectrons) after the direct transfer of energy from the photon to a core-level electron. Photo electrons are subsequently separated according to their kinetic energy and counted. The kinetic energy will depend on the energy of the original X-ray photons (the irradiating photon source should be monochromatic) and also on the atomic and, in some cases, the molecular environment from which they come. This, in turn, provides important information about oxidation states and chemical bonds because the stronger the binding to the atom, the lower is the photo electron kinetic energy. XPS is widely used in the analysis of chemical components on the surface of materials because of its following advantages: (i) XPS can analyze all elements with 0.1% of DL at the micro-area surface, except H and He; (ii) XPS can obtain the corresponding chemical valence information; (iii) XPS can measure the chemical composition distribution of the layer of 1–10 nm on the sample surface; (iv) XPS is a less destructive material analysis method.

7.2.4 Techniques Providing Depth Information

As mentioned above, many imaging technologies can realize the two-dimensional element distribution map of materials, such as LA-ICP-MS, LIBS, SEM, and SXRF. In materials science, depth profiling is also an important tool to reveal elements distribution information in layers. Techniques with spatially resolved information capabilities include: SIMS, glow discharge optical emission spectrometry (GD-OES), or mass spectrometry (GD-MS). SIMS has been introduced above and will not give unnecessary details here. GD-OES has become a widely used technique for compositional depth profiling of thin films.

Application of GD-OES depth profiling can be divided in two areas: (i) quantitative determination of layer thickness, and (ii) qualitative and quantitative measurement of major and minor compositions of layers, whereas GD-MS is mostly applied in this field for trace and ultra-trace compositions of layers. GD-OES offers fairly good DLs (mg/kg) and high sample throughput. Further, the sputtering and excitation processes are rather separated, giving rise to minimal matrix effects as compared with other direct solids analysis techniques, and this simplifies the quantitation of depth profiles. However, problems in depth quantitation still remain; for example for the analysis of multilayered samples containing layers of very different composition (from one layer to another, the composition of a given element can change from a very low concentration to almost 100%; therefore, if a highly sensitive emission line is chosen, self-absorption can occur). Moreover, algorithms need to be developed to correct for the effect of some light elements (such as hydrogen, nitrogen, and oxygen) which can produce serious effects in the calibration curves.

For depth profile and surface analysis [94], especially of thick coatings, GD-OES does not have much competition due to the low total cost of the instrumentation, the ease of sample handling, and the speed of analysis. Nowadays, most commercial GD-OES instruments are used to characterize ever thinner layers, even extremely thin atomic layers. For very thin layers, AES, SIMS, and secondary neutral mass spectrometry (SNMS) are the competing methods. And these methods are always preferred due to the possible depth resolution, the information on the lateral distribution of elements, and the structural information, although they are more costly and require longer analysis times.

7.3 The Application of Matermetallomics in Material Science and the Perspectives

In the last decades, the measurement of sample composition, additive levels, and elemental impurities has become important in the field of materials science. Primary goal of these efforts is to maintain or even improve the intended chemical, physical, or mechanical product properties.

7.3.1 Matermetallomics in Semiconductor Materials

As a third-generation semiconductor material, GaN has attracted widespread attention of researchers in the field of photocatalysis because of its good photoelectric and photocatalytic properties [95]. However, GaN has a large band gap (about 3.390 ev), which greatly limits its application. Doping is one of the effective means to regulate the band gap [96]. It is found that doping GaN with Mo, Cu, V, Mn, Ni, and other elements [97–101] will form lattice defects, hinder the recombination of electron hole pairs, and reduce the band gap of the doped system. Metal-doped semiconductors can improve the response ability of the system to visible light, but the stability of metal-doped systems is poor; therefore, Si, B, C, O, and other elements [102–105] are also used to dope GaN, which reduces the band gap of the system, some impurity energy levels appear, increase the electron transition probability improving photocatalytic activity of the system. However, the non-metallic is inactive and has strong repulsion with the same properties of elements, resulting in great difficulty in doping. Through theoretical simulation and experimental research, it is found that double-doping of metal and non-metallic elements is conducive to improve the photocatalytic activity. Castiglia et al. [106] found that the incorporation of Mg–H can inhibit electron hole pair recombination and has better stability and catalytic activity than a single-doping semiconductor.

Thermal instability of $In_xGa_{1-x}N$ quantum wells (QWs) is an obstacle to construct efficient blue and green LEDs and laser diodes. Structural degradation of QWs with indium content above 15% becomes severe at temperatures above 930 °C leading to formation of extended non-radiative areas within the active region. Lachowski et al. [107] showed a method to overcome this problem by using heavy Si doping of the GaN barrier layers. In the method, the presence of silicon atoms increases the energy barrier for gallium vacancies migration. This effectively reduces possibility of diffusion of gallium vacancies from the n-type layer to the active region. As a result, improved thermal stability of QWs was achieved and significant degradation was not observed up to temperatures of 980 °C in comparison to 930 °C for the undoped structure.

Gladkov et al. studies [108] showed that the changes of luminescence properties with doping concentration, temperature and excitation power, the intrinsic correlation between infrared and blue emissions with the doped Fe ions were explored in Fe-doped GaN. The infrared luminescence will be suppressed at higher doping concentrations, while the blue emission intensity will increase significantly. And the these findings were characterized by SIMS and TEM.

7.3.2 Matermetallomics in Artificial Crystal Materials

Inorganic nonmetallic materials are one of the three major contemporary materials. Among them, the class of materials developed from traditional silicate materials has been widely applied in optics [109, 110], aerospace [111], and others because they have high-temperature resistance, high hardness, and corrosion resistance. However, it is important to note that the properties of inorganic nonmetallic materials are significantly influenced by impurities, even at trace concentrations [112, 113]. For example, their light transmission performance can be significantly affected by impurities such as Fe and Pb introduced improperly during the preparation process [114]. Further, the properties of such materials can be improved by elemental doping [115–117]. For example, Nd^{3+}-doped yttrium aluminum garnet (YAG) is currently the most commonly used type of solid-state lasers. It is therefore of practical significance to conduct accurate quantitative analysis of the elemental concentrations in inorganic nonmetallic materials.

To obtain the elemental impurities and dopants distribution maps, LA-ICP-MS with direct solid sampling and low risk of contamination was applied to YAG analysis. However, the lack of appropriate standards or CRMs is considered to be significant a drawback of quantitative analysis by the LA-ICP-MS technique. A dried-droplet calibration approach based on external calibration was developed in the quantitative analysis of samples by LA-ICP-MS [118]. The linear correlation coefficients of the calibration curves ranged from 0.9936 to 0.9999 without any internal standard correction. Moreover, the results agree well with the results obtained by ICP-AES or inductively coupled plasma atomic MS and certified values.

PbF_2 is a promising Cherenkov radiator for electromagnetic calorimeters. An important criterion for evaluating the performance of a Cherenkov radiator is its transmission, especially that observed in the UV region of the electromagnetic spectrum [119]. However, transmission loss, which can occur during the growth, machining, storage, and transport of the crystals, is a major problem encountered during the manufacture of this type of material. Furthermore, trace elements, such as Mg, Lu, Cr, and Fe, can be accumulated during the growth of crystal and influence its properties, including transmission [120]. It is important to make sure that these trace elements are distributed evenly in the materials produced; this is because uneven distribution of trace elements during crystal growth can render the crystal fragile and diaphanous, which in turn can affect the structural properties of the material. Therefore, a method for determining the concentrations and distributions of trace elements in PbF_2 crystals would be of great value to the production of these types of material [121]. Zhang et al. developed a procedure for preparing matrix-matched calibration standards for the quantitative imaging of multiple trace elements in PbF_2 crystals by LA-ICP-MS. The analysis showed good agreement with the results observed by established ICP-MS methods, following acid dissolution of the samples. Finally, the element distributions and transmission curves of a PbF_2 sample with non-transparent and transparent sections were visualized. The distribution images, in conjunction with the transmission curves, suggested that the enrichment of Mg, Al, Rh, Cs, and Bi atoms in the non-transparent section of the sample could explain the loss in transmission observed for that section.

Acknowledgments

This work was financially supported by the National Natural Science Foundation of China (No. 52203302), and the Scientific Research Project of Science and Technology of Shanghai (No. 21142202000), the Science and Technology Innovation Project of Shanghai Institute of Ceramics, and the Shanghai Technical Platform for Testing and Characterization on Inorganic Materials (No. 19DZ2290700).

List of Abbreviations

AFS	atomic fluorescence spectrometry
CC-ICP-MS	collision/reaction cell inductively coupled plasma mass spectrometry
CRMs	certified reference materials
CT	computed X-ray tomography
dc-GDs	direct current glow discharge
DCP	direct current plasma
DLs	detection limits
DRC-ICP-MS	dynamic reaction cell inductively coupled plasma mass spectrometry

ED-XRF	energy-dispersive X-ray fluorescence
EPMA	electron-probe micro-analysis
ETV	electrothermal vaporization
FAAS	flame or graphite furnace atomic absorption spectrometry
FIB-TOF-SIMS	focused ion beam time of flight secondary ion mass spectrometry
FTIR	Fourier transform infrared spectrometry
GD-MS	glow discharge mass spectrometry
GD-OES	glow discharge optical emission spectroscopy
GFAAS	graphite furnace atomic absorption spectrometry
HPGe	high-purity germanium gamma spectrometry
HPLC	high-performance liquid chromatography
HR-CS-GFAAS	high-resolution continuum source graphite furnace
HR-ICP-SF-MS	high-resolution double-focusing sector-field ICP-MS
ICP-AES	include inductively coupled plasma atomic emission spectrometry
ICP-MS	include inductively coupled plasma mass spectrometry
KED	kinetic energy discrimination
LA-ICP-MS	laser ablation inductively coupled plasma mass spectrometry
LIBS	laser-induced breakdown spectroscopy
MALDI-TOF-MS	matrix-assisted laser desorption ionization time-of-flight mass spectrometry
MC-ICP-MS	multi-collector sector-field ICP-MS
MGI	materials Genome Initiative
MIP	microwave-induced plasma
MOF	metal organic frameworks
Nd:YAG	neodymium yttrium aluminum garnet
NMR	nuclear magnetic resonance
PIGE	proton excitation γ X-ray emission spectrometry
PIXE	proton-induced X-ray emission
QWs	quantum wells
SEM	scanning electron microscopy
SERS	surface-enhanced Raman spectroscopy
SIMS	secondary ion mass spectrometry
SNMS	secondary neutral mass spectrometry
SR-TXRF	synchrotron radiation total reflection X-ray fluorescence
SXRF	synchrotron-based X-ray fluorescence
TEM/X-EDS	transmission electron microscopy coupled with energy-dispersive X-ray spectroscopy
TERS	tip-enhanced Raman spectroscopy
WD-XRF	wavelength-dispersive X-ray fluorescence
XAS	X-ray absorption
XPS	X-ray photo electron spectroscopy
XRF	X-ray fluorescence

References

1 Waldron, K.J., Rutherford, J.C., Ford, D., and Robinson, N.J. (2009). *Nature* 460: 823.
2 Wesenberg, D., Krauss, G.J., and Schaumloffel, D. (2011). *Int. J. Mass Spectrom.* 307: 46.
3 Jamieson, E.R. and Lippard, S.J. (1999). *Chem. Rev.* 99: 2467.
4 Karasawa, T. and Steyger, P.S. (2015). *Toxicol. Lett.* 237: 219.
5 Mounicou, S., Szpunar, J., and Lobinski, R. (2009). *Chem. Soc. Rev.* 38: 1119.
6 Li, Y., Liu, L., Chen, W., and An, L. (2018). *Sci. Sin. Chim.* 48: 243.
7 Breneman, C.M., Brinson, L.C., Schadler, L.S. et al. (2013). *Adv. Funct. Mater.* 23: 5746.
8 Gaultois, M.W., Sparks, T.D., Borg, C.K.H. et al. (2013). *Chem. Mater.* 25: 2911.
9 Olivares-Amaya, R., Amador-Bedolla, C., Hachmann, J. et al. (2011). *Energy Environ. Sci.* 4: 4849.
10 Wang, B.B., Zhou, L.L., Xu, K.L., and Wang, Q.S. (2017). *Ind. Eng. Chem. Res.* 56: 47.
11 Kuenemann, M.A. and Fourches, D. (2017). *Mol. Inform.* 36: https://doi.org/10.1002/minf.201600143.
12 Wilmer, C.E., Leaf, M., Lee, C.Y. et al. (2012). *Nat. Chem.* 4: 83.
13 Lu, C.-H., Biesold-McGee, G.V., Liu, Y. et al. (2020). *Chem. Soc. Rev.* 49: 4953.
14 Sharifianjazi, F., Moradi, M., Parvin, N. et al. (2020). *Ceram. Int.* 46: 18391.
15 Shakil, M., Hussain, A., Zafar, M. et al. (2018). *Chin. J. Phys.* 56: 1570.
16 Ghosh, A., Mohammed, O.F., and Bakr, O.M. (2018). *Acc. Chem. Res.* 51: 3094.
17 Shobana, M.K., Nam, W., and Choe, H. (2013). *J. Nanosci. Nanotechnol.* 13: 3535.
18 Behera, R.K., Das Adhikari, S., Dutta, S.K. et al. (2018). *J. Phys. Chem. Lett.* 9: 6884.
19 Chen, J.-K., Ma, J.-P., Guo, S.-Q. et al. (2019). *Chem. Mater.* 31: 3974.
20 Bulinski, M., Kuncser, V., Plapcianu, C. et al. (2004). *J. Phys. D. Appl. Phys.* 37: 2437.
21 Thompson, A.M., Soni, K.K., Chan, H.M. et al. (1997). *J. Am. Ceram. Soc.* 80: 373.
22 Akhmadullin, N.S., Shishilov, O.N., and Kargin, Y.F. (2020). *Russ. Chem. Bull.* 69: 825.
23 Hasija, V., Raizada, P., Sudhaik, A. et al. (2019). *Appl. Mater. Today* 15: 494.
24 Zhang, M., Yuan, Z., Song, J., and Zheng, C. (2010). *Sensors Actuators B Chem.* 148: 87.
25 Talyzin, A.V. and Dzwilewski, A. (2007). *J. Nanosci. Nanotechnol.* 7: 1151.
26 Kveder, V., Kittler, M., and Schroter, W. (2001). *Phys. Rev. B* 63: 115208.
27 Pumera, M. and Miyahara, Y. (2009). *Nanoscale* 1: 260.
28 Liu, Z.P., He, W.J., and Guo, Z.J. (2013). *Chem. Soc. Rev.* 42: 1568.
29 Wlodarczyk-Biegun, M.K., Paez, J.I., Villiou, M. et al. (2020). *Biofabrication* 12: 035009.
30 Makino, Y., Kuroki, Y., and Hirata, T. (2019). *J. Anal. At. Spectrom.* 34: 1794.

31 Wongpanya, P., Saramas, Y., Chumkratoke, C., and Wannakomol, A. (2020). *J. Pet. Sci. Eng.* 189: 106965.
32 Misnik, M., Konarski, P., Zawada, A., and Azgin, J. (2019). *Nucl. Instrum. Meth. B* 450: 153.
33 Arnquist, I.J., di Vacri, M.L., and Hoppe, E.W. (2020). *Nucl. Instrum. Meth. A* 965: 167323.
34 Zhang, J.Y., Zhou, T., Cui, Y.J. et al. (2020). *J. Anal. At. Spectrom.* 35: 2712.
35 Harrington, K., Al Hejami, A., and Beauchemin, D. (2020). *J. Anal. At. Spectrom.* 35: 461.
36 Cinosi, A., Siviero, G., Monticelli, D., and Furian, R. (2020). *Spectrochim. Acta B* 164: 105749.
37 Chubarov, V.M., Amosova, A.A., and Finkelshtein, A.L. (2020). *Spectrochim. Acta B* 163: 126.
38 He, M.Y., Luo, C.G., Lu, H. et al. (2019). *J. Anal. At. Spectrom.* 34: 1773.
39 Oropeza, D., Gonzalez, J., Chirinos, J. et al. (2019). *Appl. Spectrosc.* 73: 540.
40 Liu, K., Tian, D., Xu, H.Y. et al. (2019). *Anal. Methods* 11: 4769.
41 Caridi, F., Messina, M., Belvedere, A. et al. (2019). *Appl. Sci. (Basel)* 9: 2882.
42 Eskina, V.V., Dalnova, O.A., Filatova, D.G. et al. (2020). *Spectrochim. Acta B* 165: 105784.
43 Shaw, G., Garcia, W., Hu, X.X., and Wirth, B.D. (2020). *Phys. Scr.* T171: 014041.
44 Kips, R., Weber, P.K., Kristo, M.J. et al. (2019). *Anal. Chem.* 91: 11598.
45 Christl, M., Guerin, N., Totland, M. et al. (2019). *J. Radioanal. Nucl. Chem.* 322: 1611.
46 Priebe, A., Utke, I., Petho, L., and Michler, J. (2019). *Anal. Chem.* 91: 11712.
47 Kodalle, T., Greiner, D., Brackmann, V. et al. (2019). *J. Anal. At. Spectrom.* 34: 1233.
48 Dematteis, E.M., Cuevas, F., and Latroche, M. (2021). *J. Alloys Compd.* 851: 156075.
49 Sharma, V., Anand, A., and Singh, B. (2020). *J. Radioanal. Nucl. Chem.* 323: 291.
50 Elayaperumal, M., Vedachalam, Y., Loganathan, D. et al. (2021). *Biol. Trace Elem. Res.* 199: 3540.
51 Zhan, D.Y., Saeed, A., Li, Z.X. et al. (2020). *Dalton Trans.* 49: 17737.
52 Gorshkov, N.I., Murko, A.Y., Gavrilova, I.I. et al. (2019). *Dokl. Chem.* 485: 91. https://doi.org/10.3390/polym12122889.
53 Zhang, W.Y., Zhang, W.C., Pan, Y.T., and Yang, R.J. (2021). *J. Hazard. Mater.* 401: e54378. 123439.
54 Liu, S.S., Zhang, G.C., Feng, K. et al. (2020). *Cryst. Growth Des.* 20: 6604.
55 Luo, Q.Y., Huang, X.H., Luo, Y. et al. (2021). *Chem. Eng. J.* 407: 20220144. 127050.
56 Gonzalez, J., Mao, X.L., Roy, J. et al. (2002). *J. Anal. At. Spectrom.* 17: 1108.
57 Wang, H.A.O., Grolimund, D., Giesen, C. et al. (2013). *Anal. Chem.* 85: 10107.
58 Hare, D., Austin, C., and Doble, P. (2012). *Analyst* 137: 1527.
59 Afgan, M.S., Sheta, S., Hou, Z.Y. et al. (2019). *J. Anal. At. Spectrom.* 34: 2385.
60 Bauer, A.J.R. and Buckley, S.G. (2017). *Appl. Spectrosc.* 71: 553.

61 Noll, R., Fricke-Begemann, C., Connemann, S. et al. (2018). *J. Anal. At. Spectrom.* 33: 945.
62 Wang, Z., Yuan, T.B., Hou, Z.Y. et al. (2014). *Front. Phys.* 9: 419.
63 Fortes, F.J., Moros, J., Lucena, P. et al. (2013). *Anal. Chem.* 85: 640.
64 Bulajic, D., Cristoforetti, G., Corsi, M. et al. (2002). *Spectrochim. Acta B* 57: 1181.
65 Lorenzetti, G., Legnaioli, S., Grifoni, E. et al. (2015). *Spectrochim. Acta B* 112: 1.
66 Yamamoto, K.Y., Cremers, D.A., Ferris, M.J., and Foster, L.E. (1996). *Appl. Spectrosc.* 50: 222.
67 Kasem, M.A., Russo, R.E., and Harith, M.A. (2011). *J. Anal. At. Spectrom.* 26: 1733.
68 Corsi, M., Cristoforetti, G., Hidalgo, M. et al. (2003). *Appl. Opt.* 42: 6133.
69 Tofanelli, M., Pardini, L., Borrini, M. et al. (2014). *Spectrochim. Acta B* 99: 70.
70 Botto, A., Campanella, B., Legnaioli, S. et al. (2019). *J. Anal. At. Spectrom.* 34: 81.
71 Jolivet, L., Leprince, M., Moncayo, S. et al. (2019). *Spectrochim. Acta B* 151: 41.
72 Veber, P., Bartosiewicz, K., Debray, J. et al. (2019). *CrystEngComm* 21: 3844.
73 Slodzian, G., Daigne, B., Girard, F. et al. (1992). *Biol. Cell.* 74: 43.
74 Slodzian, G., Daigne, B., and Girard, F. (1992). *Microsc. Microanal. Microstruct.* 3: 99.
75 Hoppe, P., Cohen, S., and Meibom, A. (2013). *Geostand. Geoanal. Res.* 37: 111.
76 Duprat, J., Bardin, N., Engrand, C. et al. (2014). *Meteorit. Planet. Sci.* 49: A103.
77 Kodolanyi, J., Hoppe, P., Groner, E. et al. (2014). *Geochim. Cosmochim. Acta* 140: 577.
78 Christien, F., Downing, C., Moore, K.L., and Grovenor, C.R.M. (2013). *Surf. Interface Anal.* 45: 305.
79 Chung, E.H., Han, H.J., Khan, F.N. et al. (2013). *J. Ceram. Process. Res.* 14: 304.
80 Yardley, S.S., Moore, K.L., Ni, N. et al. (2013). *J. Nucl. Mater.* 443: 436.
81 Baranov, V.I. and Tanner, S.D. (1999). *J. Anal. At. Spectrom.* 14: 1133.
82 Xue, Z.C., Rehkamper, M., Schonbachler, M. et al. (2012). *Anal. Bioanal. Chem.* 402: 883.
83 Pickering, R., Kramers, J.D., Partridge, T. et al. (2010). *Quat. Geochronol.* 5: 544.
84 Ireland, T.R. (2013). *Rev. Sci. Instrum.* 84: 011101.
85 Blichert-Toft, J., Zanda, B., Ebel, D.S., and Albarede, F. (2010). *Earth Planet. Sci. Lett.* 300: 152.
86 Degryse, P., Shortland, A., De Muynck, D. et al. (2010). *J. Archaeol. Sci.* 37: 3129.
87 Albarede, F., Desaulty, A.M., and Blichert-Toft, J. (2012). *Archaeometry* 54: 853.
88 Muniz, R., Lobo, L., Kerry, T. et al. (2017). *J. Anal. At. Spectrom.* 32: 1306.
89 Heras, L.A.D., Hrnecek, E., Bildstein, O., and Betti, M. (2002). *J. Anal. At. Spectrom.* 17: 1011.
90 Schelles, W. and VanGrieken, R.E. (1996). *Anal. Chem.* 68: 3570.
91 Nurrohman, D.T. and Chiu, N.F. (2021). *Nanomaterials (Basel)* 11: 216.
92 Bousiakou, L.G., Gebavi, H., Mikac, L. et al. (2019). *Croat. Chem. Acta* 92: 479.
93 Zenobi, R. and Deckert, V. (2000). *Angew. Chem. Int. Edit.* 39: 1746.

94 Lobo, L., Fernandez, B., and Pereiro, R. (2017). *J. Anal. At. Spectrom.* 32: 920.
95 Shur, M. (2019). *Solid State Electron.* 155: 65.
96 He, T.H., Zeng, X.S., and Rong, S.P. (2020). *J. Mater. Chem. A* 8: 8383.
97 Rezaei-Sameti, M. and Moradi, F. (2017). *J. Incl. Phenom. Macrocycl. Chem.* 88: 209.
98 Lin, L., Zhu, L.H., Zhao, R.Q. et al. (2018). *New J. Chem.* 42: 9393.
99 Yao, G.R., Fan, G.H., Zheng, S.W. et al. (2012). *Opt. Mater.* 34: 1593.
100 Khan, M.J.I., Liu, J., Hussain, S. et al. (2020). *Optik* 208: 164529.
101 Xiao, G., Wang, L.L., Rong, Q.Y. et al. (2017). *Physica B* 524: 47.
102 Rummukainen, M., Oila, J., Laakso, A. et al. (2004). *Appl. Phys. Lett.* 84: 4887.
103 Koller, C., Pobegen, G., Ostermaier, C. et al. (2017). *Appl. Phys. Lett.* 111: 031101.
104 Li, H., Zhang, L.W., Cai, X.L. et al. (2018). *Mater. Res. Express* 5: 1–10.
105 Arakawa, Y., Ueno, K., Imabeppu, H. et al. (2017). *Appl. Phys. Lett.* 110: 101002.
106 Castiglia, A., Carlin, J.F., and Grandjean, N. (2011). *Appl. Phys. Lett.* 98: 213505.
107 Lachowski, A., Grzanka, E., Grzanka, S. et al. (2022). *J. Alloys Compd.* 900: 163519.
108 Gladkov, P., Humlicek, J., Hulicius, E. et al. (2010). *J. Cryst. Growth* 312: 1205.
109 Cao, J.K., Xu, D.K., Hu, F.F. et al. (2018). *J. Eur. Ceram. Soc.* 38: 2753.
110 Mao, X.J., Shimai, S.Z., Dong, M.J., and Wang, S.W. (2007). *J. Am. Ceram. Soc.* 90: 986.
111 Yang, H., Zhang, L.J., Guo, X.Z. et al. (2011). *Ceram. Int.* 37: 2031.
112 Gao, J.Q., Chen, J., Liu, G.L. et al. (2012). *Int. J. Appl. Ceram. Tec* 9: 847.
113 Wang, X.J., Xie, J.J., Wang, Z.J. et al. (2018). *Ceram. Int.* 44: 9514.
114 Esposito, L., Piancastelli, A., Costa, A.L. et al. (2011). *Opt. Mater.* 33: 713.
115 Gong, H., Tang, D.Y., Huang, H., and Ma, J. (2009). *J. Am. Ceram. Soc.* 92: 812.
116 Gan, Q.J., Jiang, B.X., Zhang, P.D. et al. (2018). *J. Inorg. Mater.* 33: 107.
117 Ikesue, A., Kinoshita, T., Kamata, K., and Yoshida, K. (1995). *J. Am. Ceram. Soc.* 78: 1033.
118 Guo, L.Q., Li, Q., Chen, Y.R. et al. (2020). *J. Anal. At. Spectrom.* 35: 1441.
119 Nikolskaya, O.K. and Demyanets, L.N. (1994). *Inorg. Mater.* 30: 1097.
120 Yonezawa, T., Matsuo, K., Nakayama, J., and Kawamoto, Y. (2003). *J. Cryst. Growth* 258: 385.
121 Zhang, G., Wang, Z., Li, Q. et al. (2016). *Talanta* 154: 486.

8

Archaeometallomics*

Li Li[1], Yue Zhou[1,2], Sijia Li[1,2], Lingtong Yan[1], Heyang Sun[1], and Xiangqian Feng[1]

[1] Chinese Academy of Sciences, Institute of High Energy Physics, CAS-HKU Joint Laboratory of Metallomics on Health and Environment, & Beijing Metallomics Facility, 19B Yuquan Road, Shijingshan District, Beijing 100049, P.R. China
[2] University of Chinese Academy of Sciences, Beijing 100049, P.R. China

8.1 The Concept of Archaeometallomics

Metallomics is a research field that can provide a systematic understanding of the metal uptake, source of origin, role, and excretion in biological systems [1, 2]. Although metallomics was originally focused on the essential roles of metal and metalloids in biological systems, the concept of metallomics has been extended to the study of metals and metalloids in material sciences, such as matermetallomics [3].

Relics from human activities such as ancient paintings, stone tools, wooden and jade wares, bronzes, ceramics, etc. contain metals and metalloids substances. The relics can help to condense and reflect on the developmental process of ancient history. Using scientific techniques to extract the information contained in cultural relics has enhanced the research in archaeology and achieved valuable results. In addition, because of the rarity of the samples, various means of nondestructive analysis or micro-damage analysis are widely used to obtain the respective physical and chemical information.

We propose the concept of archaeometallomics as a tool to systematically study the role of trace elements in cultural relics and clarify the role and function of the metal elements. Archaeometallomics can facilitate the understanding of the origin, the processing technology used, and to verify the authenticity or falsification of cultural relics such as for ancient ceramics, ancient bronzes, and ancient paintings.

---

* This chapter has been modified to feature as Reviews: (1) Li, L., Yan L.T., Sun H.Y., et al. (2021). Archaeometallomics as a tool for studying ancient ceramics. *At. Spectrosc.* 42(5): 247–253; (2) Zhou, Y., Yan, L.T., Li, L., Sun, H.Y., and Feng, X.Q. (2021). The progress of the application of X-ray spectroscopy in the nondestruction analysis of relics. *Spectrosc. Spect. Anal.* 41(5): 1329–1335.

Applied Metallomics: From Life Sciences to Environmental Sciences, First Edition.
Edited by Yu-Feng Li and Hongzhe Sun.
© 2024 WILEY-VCH GmbH. Published 2024 by WILEY-VCH GmbH.

8.2 The Analytical Techniques in Archaeometallomics

Due to the particularity of cultural relics, nondestructive analysis techniques or micro-damage analysis are always desired. Neutron activation analysis (NAA), X-ray fluorescence analysis (XRF), X-ray absorption fine structure spectroscopy (XAFS), Laser Ablation Inductively Coupled Plasma Mass Spectrometry (LA-ICP-MS), Laser-induced Breakdown Spectroscopy (LIBS), Atomic Absorption Spectroscopy (AAS), X-ray diffraction (XRD), and Neutron diffraction are often used to analyze the content, valence, and structure of metallic elements in cultural relics such as for ancient ceramics, ancient bronzes, and ancient paintings [4–7].

8.2.1 Neutron Activation Analysis (NAA)

NAA uses neutrons, charged particles, and high-energy photons with certain energy to bombard the sample. Radioactive nuclides are generated, which can be used for the quantitative analysis of elements in the sample [8]. Compared with other elemental analysis methods, it has many advantages: first, high sensitivity, high accuracy, and precision; second, multi-element analysis, which can give the contents of dozens of elements, especially trace elements; third, the sample size is small, not easy to stain, and not affected by reagent blank. The disadvantage is that this method cannot detect elements that cannot be activated by neutrons, nor can the elements with short half-life be measured. The earliest application of the neutron activation method to ceramic archaeology was in 1954, when Sayre and Dodson at Brookhaven Laboratory of Princeton University, United States, was first studied for the origin of ancient ceramics [9]. For more than half a century, experts at home and abroad have used NAA to carry out a large number of studies on the origin, age, and production process of ancient ceramics [10, 11]. Neutron activation has gradually become a widely accepted method for trace element analysis in Chinese ceramic archaeology.

8.2.2 X-Ray Fluorescence Analysis (XRF)

According to the different excitation sources, the XRF technology is available as X-ray fluorescence (using X-ray tube as the excitation source), synchrotron radiation X-ray fluorescence (SRXRF), and proton-induced X-ray fluorescence (PIXE), etc. [12]. As a non-destructive analysis technique, XRF is one of the most commonly used methods in cultural analysis and archaeological research. The sample is bombarded with photons, electrons, protons, particles, or other ions at a certain energy in which the inner shell (K, L, or M shell) electrons are excited, causing electron transitions in the shell and emitting characteristic X-rays (or auger electrons) of the element. By determining the wavelength (or energy) of characteristic X-rays, it can determine the elements in the sample. The percentage content of an element in the sample can be obtained by measuring the intensity of the characteristic X-ray and adopting the appropriate method for calibration and correction [13]. XRF is often used in conjunction with other methods, such as XAFS and XRD, for elemental analysis of regions of interest [14, 15]. Confocal μ-XRF developed on the basis of the original XRF in recent years has also been used in the field of cultural analysis and archaeology [16].

8.2.3 Laser Ablation Inductively Coupled Plasma Mass Spectrometry (LA-ICP-MS)

LA-ICP-MS is a powerful analytical technology that enables highly sensitive elemental and direct isotopic analysis of solid samples [17]. The principle of LA-ICP-MS is to focus the laser beam on the surface of the sample to melt and vaporize it. The carrier gas (He or Ar) will send the sample particles (aerosols) to the plasma for ionization and then mass filtration through the mass spectrometry system. Finally, the receiver will be used to detect ions with different mass charge ratios. In recent years, the micro-elemental and isotope ratio analysis technology of LA-ICP-MS has been further developed and is widely used in geology, metallurgy, environment, biology, chemistry, materials, archaeology, and other fields.

8.2.4 Laser-induced Breakdown Spectroscopy (LIBS)

In LIBS, a high-energy laser is applied to the sample to form a laser spot on the surface of the sample, which causes the sample to excite and glow. The light is then analyzed by the spectral and the monitoring systems, and the elemental composition and the content of the sample are obtained. In recent years, great progress has been made in the theory and application of LIBS, especially in the fields of materials, soil, biology, environment, metallurgy, medicine, ancient art, and painted cultural relics [18].

8.2.5 Atomic Absorption Spectroscopy (AAS)

AAS is based on a sample of measured elemental vapor-phase ground-state atoms by narrowband characteristic of the atoms of the radiation of the light source to produce resonance absorption. Its absorbance within a certain range and the vapor phase is proportional to the measured element of the ground state of atomic concentration. So, the element content in the sample is measured [19]. AAS has been widely used in various fields, such as geology, metallurgy, machinery, chemical industry, agriculture, food, light industry, biological medicine, environmental protection, and material science.

8.2.6 X-Ray Absorption Fine Structure Spectroscopy (XAFS)

When X-rays pass through an object, they are absorbed by the object and their intensity is changed. This strength decay obeys exponential law. On the high-energy side near the absorption limit, the absorption curve presents a fine structure of up and down fluctuations. By measuring, analyzing, and calculating this fine structure, much information can be obtained about the arrangement of atoms around the absorption atoms [20]. XAFS is usually divided into two parts: Extended X-ray

Absorption Fine Structure (EXAFS) and X-ray Absorption Near Side Structure (XANES) [21]. The characteristics of the EXAFS spectrum reflect the near-range coordination conditions around the absorption atoms, while the XANES spectrum is basically compared with the standard reference, and the valence state and coordination environment are qualitatively determined according to the deviation of the absorption edge position, and the height and position of the edge front peak. For example, in the study of ancient ceramics, μ-XANES and μ-XRF are used together to reveal information about metal elements in glazes. Dejoie et al. used synchrotron radiation μ-XANES and μ-XRF to analyze Fe crystals in the glaze when studying the crystals in the black glaze porcelain of the Jian kiln, and determined that the Fe crystals in the surface layer were Fe_3O_4 [22]. In the analysis of metal relics, μ-XANES can be used to determine the oxidation state of metal and its structure [23].

8.2.7 X-Ray Diffraction (XRD)

When X-rays interact with matter, energy conversions are generally divided into three parts, one part is scattered, one part is absorbed, and one part continues to propagate in the original direction through matter. When the wavelength of the scattered X-ray is the same as that of the incident X ray, the diffraction phenomenon will occur in the crystal. The characteristic diffraction pattern of each crystal material is compared with the standard diffraction pattern, and the phase in the sample can be identified by using the principle of three strong peaks [24]. For amorphous materials, because they do not have long-range ordered structure, XRD has limited ability to detect amorphous materials, which can provide other important information such as average atomic spacing or molecular spacing. Compared with conventional XRD, SR-XRD has higher signal-to-noise ratio and angular resolution. As an important method to study crystal structure, XRD has been widely used in the analysis of cultural relics and archaeological samples [25]. In recent years, grazing incidence X-ray diffraction (GIXRD), developed from XRD technology, has been used in cultural and archaeological samples. Liu et al. studied Fe crystals in purple gold glaze samples in the Qing Dynasty unearthed from the Forbidden City, and GIXRD was used to analyze and confirm the ε-Fe_2O_3 crystals [26].

8.2.8 Neutron Diffraction

The neutron source emits a beam and irradiates on the sample. The interaction between the neutron beam and the sample leads to anisotropic scattering of neutrons, and the elastic scattering beam produces diffraction, which is received by the detector through the wire harness device, and the detector is connected to the data acquisition system to obtain the results [27]. Neutron diffraction can be used to analyze the composition of mineral or metallic phases, the crystal structure, and microstructure of each component phase. Neutron diffraction, as a nondestructive testing technology, was not applied to cultural relics until about 2000. Compared with XRD, neutron diffraction can easily penetrate the thick coating and corrosion layer on the surface to obtain the internal structure information of cultural relics. It can detect the magnetic sequence of magnetic minerals such as hematite and magnetite in cultural relics [28]. Huang Wei et al. used neutron diffraction technology to conduct non-destructive analysis on phase structure, alloy composition, and corrosion composition of three iron coins in Song Dynasty. The results showed that the main phases of iron coins were ferrite and cementite (including Fe_3P), and the corrosion products were mainly goethite [29].

8.3 The Application of Archaeometallomics in Archaeological Science

8.3.1 The Application of Archaeometallomics in Ancient Ceramics

Ancient ceramics are among the most outstanding and important historical and cultural heritage of the Chinese nation. They are not only the special carrier of human society and culture but also the crystallization of science and technology of ancient craftsmen. The developmental process used in ancient ceramics contains very rich scientific technology and social and cultural connotations. Tracing the origin, identifying the authenticity or falsification, understanding the development, and evolution of porcelain technology are important topics [4, 30, 31].

8.3.1.1 Archaeometallomics in Studying the Origin and Dating of Ancient Ceramics

Ancient ceramics were generally made with local materials, so their production and raw material of origin are the same. In the same region, the clay used to make the body and glaze for the ceramic has certain trace metallic elements in common, and these trace elements are different from other regions, referred to as "fingerprint elements." Fingerprint elements in clay do not change when they are made and fired, and the fingerprint information is retained in the ceramic. The composition and content of chemical elements in ancient ceramics are used to find the common points and differences, and then the origin of ancient ceramics can be identified [32].

Energy-Dispersive X-ray Fluorescence (EDXRF) was used by Ma et al. [33] to analyze the characteristics of the different kilns used for the production of Tang Sancai wares. Some metal elements became the fingerprint elements of different Tang Sancai kiln sites (Figure 8.1). The fingerprint elements provide valuable scientific criteria for provenance identification for Tang Sancai pottery of unknown origin.

Fantuzzi et al. [34] analyzed Late Roman amphorae from four kiln sites located in the Guadalquivir River basin (Spain) by using a combination of instrumental analytical techniques, including thin-section optical microscopy, XRD, and XRF. The contents of some metallic elements (Ca, Al, V, Cr, etc.) established the origin characteristics (Figure 8.2). The results not only contribute to new evidence on the study of oil amphora production in this region, but also serves as a basis for the identification and sourcing of these amphorae and, consequently, for a better understanding of the trade networks during the Late Roman period. Prinsloo et al. [35] re-dated the

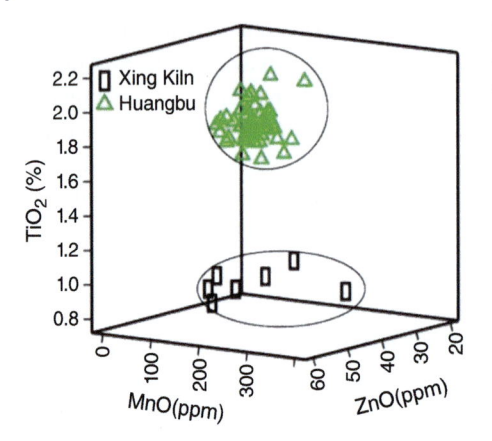

Figure 8.1 Distribution of TiO$_2$, MnO, and ZnO for Tang Sancai bodied specimens from Huangbu Kiln and Xing Kiln [33].

Chinese celadon shards excavated on Mapungubwe Hill, a thirteenth century Iron Age site in South Africa, using Raman spectroscopy, XRF, and XRD. According to the ratio of alkaline earth metals to alkali metals, the date of the Chinese celadon shards was possibly the period of the Yuan (1279–1368 AD) or even the early Ming (1368–1644 AD) dynasty. These results have an impact on the chronology of the history of the region and therefore calls for further research of a comparative nature for other Chinese celadon shards excavated at archaeological sites in Africa, in addition to additional carbon dating of the Mapungubwe hill area. Blagoev et al. [36] analyzed excavated ceramic fragments to obtain their chemical composition by using ns-LIBS and XRF. Combining different methods and comparing the obtained results, it provided complementary information regarding white-clay ceramic production and the complete chemical characterization of the examined artifacts.

The lead and strontium isotopic composition in pottery is an important feature for identifying the origin of ancient pottery, in addition to the analysis of the main and trace elemental content of the body and the glaze. Zhang et al. [37, 38] used MS to

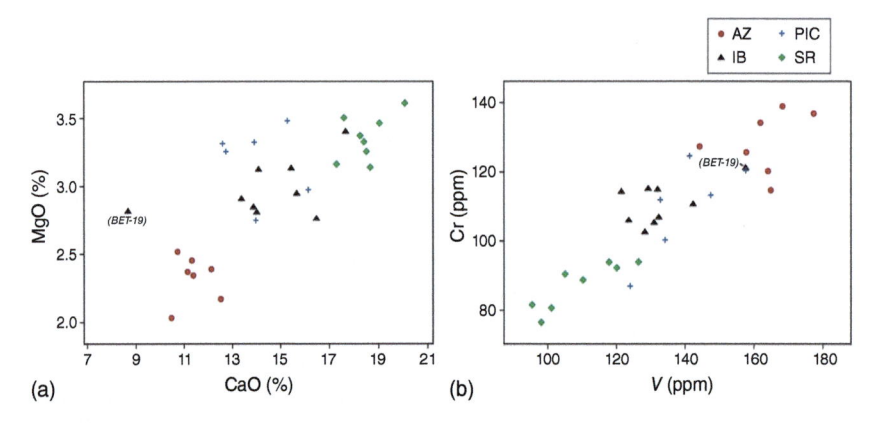

Figure 8.2 Binary diagrams, using normalized data, of (a) CaO vs MgO and (b) V vs Cr for the amphora samples analyzed. An indication of the kiln site for each individual is given [34].

analyze the lead isotopic composition of ancient pottery samples from the neolithic sites of Jiahu and Xishan in the Henan province. Their study showed that the lead isotopic composition can be used to identify the origin of some ancient pottery samples. The Pb isotopic composition of the ancient pottery from Xishan and Hualing was found similar to the Liangzhu culture, but the Sr isotopic composition was significantly different. Strontium isotope analysis significantly enhances the ability of using isotope composition to identify the origin of ancient pottery.

8.3.1.2 Archaeometallomics in Studying the Color Mechanism and Firing Technology of Ancient Ceramics

In the process of high-temperature firing and because of the inherent law of metal elements at different temperatures and different firing atmospheres, rich and colorful glazes are produced. By researching the structure of a metal's color elements, the ceramic firing technology can be deduced, which provides a scientific basis for the restoration of the original in the traditional manufacturing technology. For example, Mn, Fe, Cu, and Co in the glaze at different temperatures and different firing atmospheres contribute to the colorful glaze of ancient Chinese ceramics. The distribution of metal elements and their adjacent atomic structures is closely related to the provenance, dating, and processing technology [39]. It was found that green glaze and red glaze contained Cu as the coloring agent; celadon, white, and black porcelain used Fe as the coloring agent, while blue and white porcelain used Co as the coloring agent, etc. [40–42]. However, the same coloring element in the ceramic glaze color can show different colors and depends on the relationship with the content of the ceramic formula and the firing process [43].

Coutinho et al. [44] used EDXRF to analyze the elemental composition of blue and white porcelain fragments unearthed in a monastery in Portugal but from the late Ming Dynasty, and found that the element (Mn, Fe, Co) content ratio of the blue and dark blue regions in blue and white porcelain was different. In the dark blue region, the ratio of Mn/Co is higher and the ratio of Fe/Co is lower. It suggested that ancient potters probably used both Co pigments as shown in Figure 8.3a,b.

XAFS is suitable for the study of the fine structure of metal elements in ancient ceramic-colored materials and the state of colored metal elements, to discuss the color mechanism and firing technology. XAFS research was carried out by Matsunaga et al. [41] on a pottery billet in Turkey where the clay of the vessel was from the vicinity of the dig site. It was found that the absorption edge energy of the Fe element spectrum increased gradually with the color of the unearthed pottery changing from gray to brown to orange. Gray pottery is mainly fired in a reductive atmosphere or a weak oxidizing atmosphere, while yellow pottery is fired in an oxidizing atmosphere. Barilaro et al. [45] collected a series of painted pottery fragments excavated in Sicily, Italy, which were identified as products of different historical periods based on the morphological samples, ranging from the eighteenth century BC to the sixteenth century AD. Through comprehensive analysis of the K-edge near the edge and extended-edge structure of Cu in the sample, and comparison with the fine structure spectrum of CuO and Cu_2O in the standard sample, it was confirmed that the composition of the green pigment in the painted pottery is CuO (Figure 8.4).

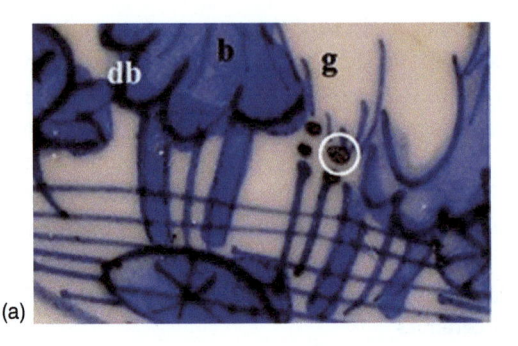

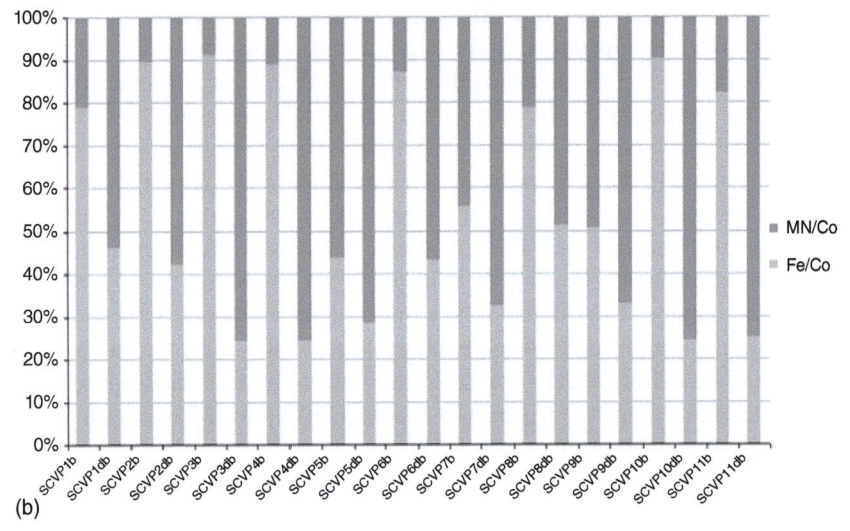

Figure 8.3 (a): The detail of samples SCP4 showing an example of the analyzed areas (g) glaze, (b) blue and (db) dark blue [44]. (b): The bar chart with the ratio of Fe/Co and Mn/Co oxides for blue (b) and dark blue (db) area [44].

Archaeometallomics is proposed as a tool to systematically study the role of trace elements in ancient ceramics, which can facilitate the understanding of the origin, the technology used, and the authenticity or falsification. With the continuous development and improvement of various analytical techniques, there are a variety of analytical techniques available today to comprehensively study the metal elements in the body, glaze, and color of ancient ceramics as well as their origin.

8.3.2 The Application of Archaeometallomics in Metal Cultural Relics

As a symbol of civilization, metal cultural relics such as bronze ware, iron ware, gold ware, and lead ware have been found all over the world, and relevant research has been widely valued by scholars. Take bronze ware as an example. Most bronze ware is an alloy of Cu, Sn, and Pb. The ratio of these three metal ingredients is not constant. The different proportions of the various components reflect the function, origin, and age of the bronze ware [46]. Therefore, the use of natural scientific means to study the composition and microstructure of metal in ancient metal relics and corrosion

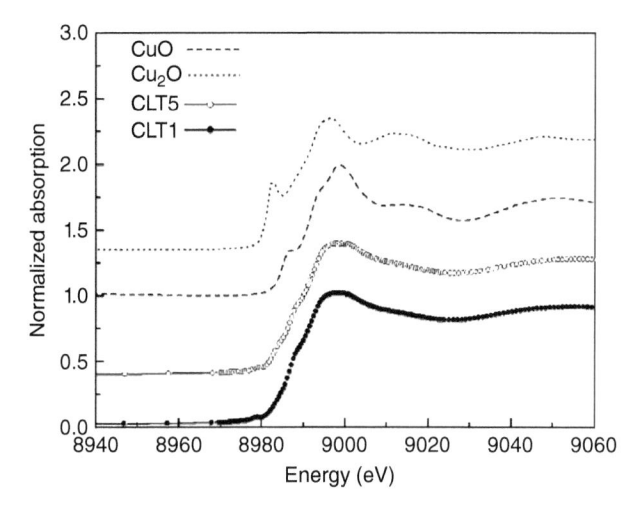

Figure 8.4 Cu K-edge XANES spectra of the samples CLT1 (−●−) and CLT5 (−o−). CuO (cupric oxide) (- - -) and Cu_2O (cuprous oxide) (···) compounds, used as standards, are also reported [45].

layer is of great importance to their craft, origin, and age, which is conducive to achieve the purpose of protecting metal relics. Archaeometallomics is suitable for studying the origin, casting process, corrosion, and protection of metal relics.

8.3.2.1 Archaeometallomics in Studying the Origin of Metal Cultural Relics

The study of mineral materials and origin of metal cultural relics is one of the key points and difficulties in metallurgical archaeology, which can not only reflect the origin and inheritance of metallurgical technology, but also reflect the political, cultural, economic and trade, transportation, and other deep-rooted problems of the society at that time. The research method of archeology has been used for a long time, which is greatly influenced by subjective factors of researchers. In recent years, scholars at home and abroad have determined the origin of metal cultural relics from the scientific analysis of lead isotope, trace element tracing method, and the residue on the surface of cultural relics [47, 48]. For example, in the research of origin tracing of bronze ware, X-ray fluorescence spectrum can be used to analyze the content information of main and trace elements of residual clay cores in bronze ware, and trace back the casting place of bronze ware according to the regional characteristics of element composition [49, 50].

Luo et al. [49] analyzed the composition of mud core in unearthed bronze vessels flocking together in Yunxian county, Hubei province, with X-ray fluorescence spectrum, and prepared samples by glass slice method. The results showed that the content of CaO, Na_2O, MgO, K_2O, and MnO in the residual mud core was low, while the content of SiO_2 was high. The element composition was similar to the characteristics of red soil in south China, while different to the characteristics of the Yellow River basin loess. According to the judgment, the unearthed bronze is not likely to be cast in the north, and should be Chu State after casting to ancient Jun country. Ma et al. [50] used XRF and ICP-AES to characterize the major elements, trace elements,

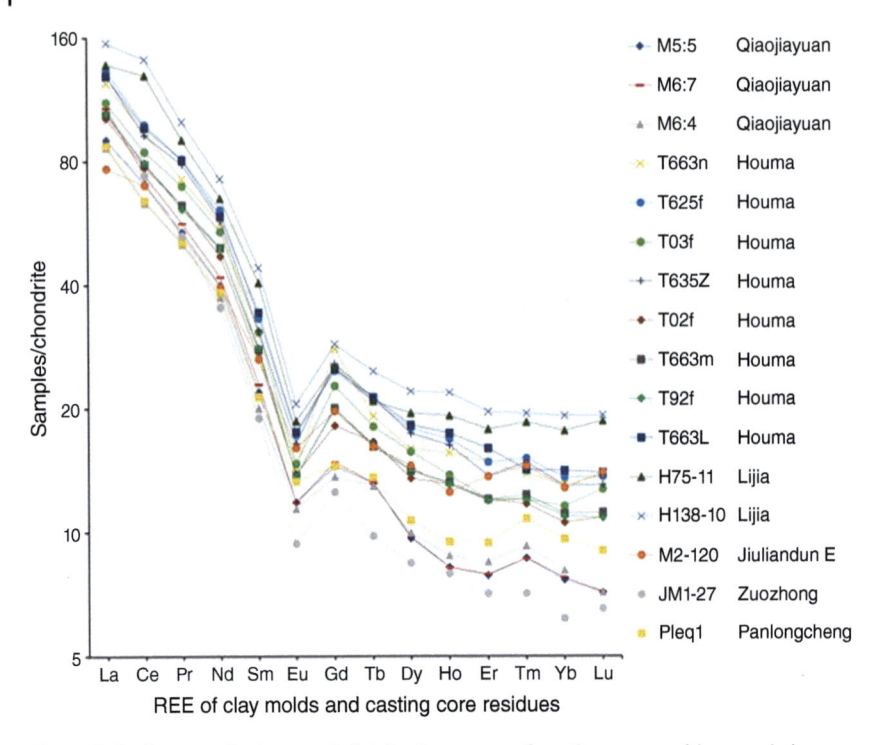

Figure 8.5 Rare earth elements' distribution curve of casting core residues and clay molds [50].

and rare-earth elements in the casting core residues of the Qiaojiayuan bronze ritual vessels. The results show that the samples from Qiaojiayuan well resembled with those from Panlongcheng and Zuozhong, but differ from the casting core residues or clay molds manufactured in Northern China, and they are not consistent with the local soil geochemical characteristics as shown in Figure 8.5. It can be further speculated that the Qiaojiayuan bronzes were first cast and finished in the Chu State before being transported to the area of Jun.

8.3.2.2 Archaeometallomics in Studying the Manufacturing Technology of Metal Cultural Relics

In metal cultural relics, modern analytical methods are mostly used to study metal materials and production techniques, such as metallographic microscopy and scanning electron microscopy to observe the structure of metal elements in cultural relics [51]. By analyzing the content information of metal elements in cultural relics by NAA, XRF, and PGAA, we can have a further understanding of the making technology of ancient alloy [52]. The matrix-phase composition of different parts of metal relics was detected by neutron diffraction [53].

Fan et al. [54] used μ-XRF on synchrotron radiation to analyze the Jiangzhai brass (4700–4000 BC) unearthed in Shaanxi Province. By comparing with the simulated samples, it was found that the distribution of Zn and Pb in the unearthed Jiangzhai brass alloy was consistent with that in solid reduced copper, but significantly different from that in smelting copper. Therefore, it is speculated that the unearthed Jiangzhai brass alloy products were made by solid-state reduction. Hu Fei et al. [55] study the element composition of metal wire in the early of Western Zhou period unearthed from Dazhixiezi site in Hubei province of China by dispersive X-ray fluorescence spectrometer. The results show that metal wire is tin-lead alloy. Its chemical composition is similar to that of low-temperature solder from the late of Western Zhou Dynasty to the early of Warring States period. It speculated that they might be the "wire" of bronze ware during the welding process.

Neutrons can penetrate the sample and nondestructively analyze crystal structure, composition, and other information to infer the manufacturing process (casting, stamping, and bunching). Stephen E. Nagler et al. [56] analyzed the Composition and Minting of Ancient Judaean "Biblical" Coins by time-of-Flight Neutron Diffraction. The results of XRF and ND-TOF analysis were compared. Pb is preferentially forced to the surface during striking, and it is easily smeared on surfaces during polishing if care is not taken. Bronze phases with a higher Sn-to-Cu ratio are harder and more resistant to abrasion, so natural wear or polishing of the surface can cause Sn levels to be higher than in the bulk composition.

8.3.2.3 Archaeometallomics in Studying the Corrosion of Metal Cultural Relics

Metal cultural relics are buried underground all year long before being unearthed. It is easy to have different types of corrosion on the surface or the broken place. Especially for unearthed the metal cultural relics, harmful corrosion products are easy to appear on the basis of the original corrosion products due to environmental changes, resulting in corrosion expansion and even perforation [57]. Modern analysis techniques can be used to figure out the chemical composition and type of corrosion product. It is also possible to analyze the chemical composition of the metal ware inside, in corrosion layers, and in the burial environment [58, 59]. The migration and diffusion of metal elements during corrosion were investigated by the analysis techniques. The above research can reveal the mechanism of corrosion and provide basis and guidance for the protection and repair of metal cultural relics.

The material and workmanship of a metal mask of the Palace Museum from the Liao Dynasty was researched for understanding the types of disease products and their causes [60]. A portable X-ray fluorescence spectrometer (P-XRF) was used for nondestructive composition analysis of the mask, and it was found that the mask was made of silver – copper – tin alloy. By means of XRF, XRD, and micro-laser Raman, the corrosion products on the surface of the mask were determined, including "harmful rust" such as chlorite. Combined with the traditional technology of metal relic restoration, the mask was protected and restored. It improves the scientific nature of the protection and restoration of metal cultural relics, solves some problems that are difficult to overcome with traditional technology, and provides a reference case for the protection and restoration of similar cultural relics.

8.3.3 The Application of Archaeometallomics in Ancient Painting

Painting cultural relics, such as colored and painted cultural relics, ancient oil paintings, and paper cultural relics, are of great artistic and historical value. The origin, aging mechanism, and authenticity identification of cultural relics have always been concerned by people. Metal elements play an important role in the coloring of the pigments for painting cultural relics. The information of metal elements can help to study the aging mechanism, production technology, and authenticity identification of painting cultural relics. The microscopic chemical structure information around the metal atoms can reveal the micro-mechanism of pigment aging effect.

8.3.3.1 Archaeometallomics in Studying the Aging Mechanism of Painting Cultural Relics

Due to a variety of reasons, most ancient painting collections have been damaged by aging to varying degrees, such as the change of pigment color, local cracking and peeling of the paint layer, and obvious protrusion in the part of the picture [61]. The composition of pigments in painting cultural relics is complex. People often mix several pigments and use them together. In addition, the pigments contain organic substances, which make it easy for complex chemical reactions to occur in the preservation process of the pigment layer, causing damage to the cultural relics [62]. For example, metal soap is a common aging product in oil paintings. It comes from the reaction of fatty acids contained in the painting material with metal ions such as Pb, Zn, and Cu in the pigment, and is an important cause of fission aging of oil paintings [63]. It is not only important for the protection and restoration of paintings of historical value, but also of great significance for the protection and restoration of traditional cultural relics [64, 65].

Chen-wiegart et al. [62] scanned and analyzed the area where lead soap appeared on a fifteenth century oil painting with SRXRF equipment with a spatial resolution of micron magnitude, and the results of element distribution showed that the content of lead element was higher in the central area of lead soap, while the content of tin element was higher in the surrounding area of lead soap, as shown in Figure 8.6. They also found the difference of L3 near-edge structure between Pb pigments and lead soaps by XAFS. Eleanor et al. [66] studied the degradation mechanism of synthetic pigment group cyan blue by X-ray absorption fine structure spectrum. Comparing the group of cyanine before and after degradation almost edge structure of Al element K absorption edge, it was found that the differences between the two confirmed the degradation mechanism is due to the aluminum atoms from the tetrahedron coordination from, has formed six ligand-containing aluminum compounds and out from the pigment, resulting in the degradation of the paint. Ferreira et al. [67] found in an analysis of two nineteenth-century paintings that the surface had an unusual granular texture, given by the application of metallic particles and the presence of large translucent green agglomerates. They investigate the composition and origin of the metal particles and green agglomerates. A combination of bulk analysis (GC–MS and FTIR) and analytical microscopy techniques (X-ray tomographic microscopy, light microscopy, and SEM–EDX) were used. The results showed that the metal particles are composed of a lower-quality brass (copper/zinc alloy) and ground from foil. The current appearance resulted from the reaction of the brass and its corrosion products with the fatty acids in the surrounding paint/varnish matrix.

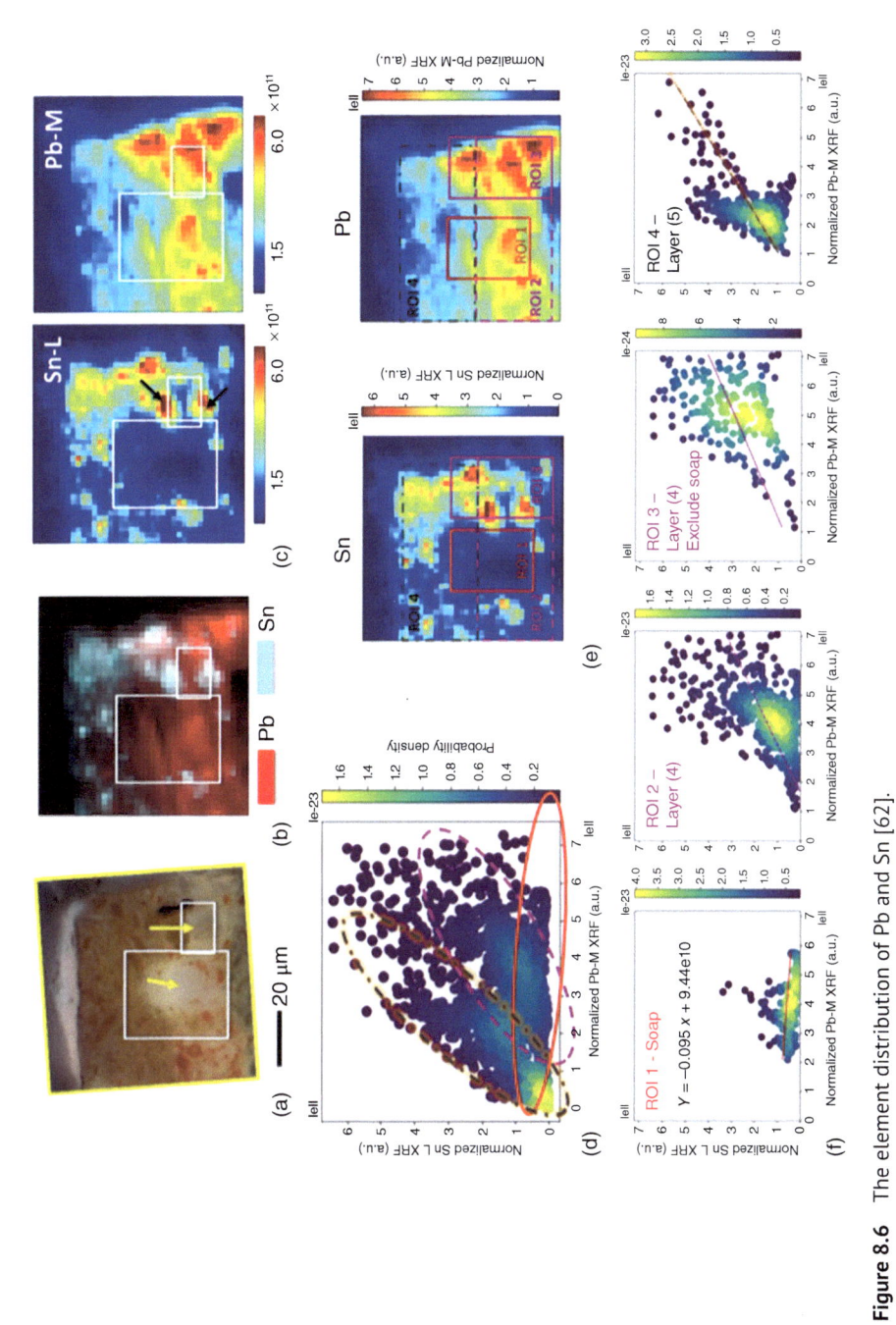

Figure 8.6 The element distribution of Pb and Sn [62].

This has led to the formation of agglomerates of zinc and copper carboxylates, the latter responsible for their green color.

8.3.3.2 Archaeometallomics in Studying the Authenticity Identification of Painting Cultural Relics

Under normal circumstances, scientists use X-ray fluorescence spectroscopy to conduct *in situ* analysis of painting cultural relics to obtain the composition information of metal elements in a specific area, and then combine with other experimental means, such as Raman spectroscopy and Fourier transform infrared spectroscopy XRD, to determine the type of pigments [68, 69]. Certain types of pigments were only used at certain times, so identifying the pigments used in artifacts can identify the age and authenticity of their production.

In order to distinguish the age information of an oil painting, Lehmann et al. [70] used microbeam X-ray fluorescence spectrometer to analyze the element composition of pigments in some tiny areas of the painting, and found metallic element zinc in the pigments at the bottom of the cracks in the picture, and trace metallic element titanium in some positions on the surface of the picture, as shown in Figure 8.7. The pigment zinc white was mainly used in the nineteenth century, while titanium white was only used in the twentieth century, so it was confirmed that the painting was a copy produced in the nineteenth century and restored in the twentieth century. Wei et al. [16] used synchrotron radiation confocal μ-XRF to analyze the element depth profile of the spotted bamboo color paintings in qianlong Garden of the

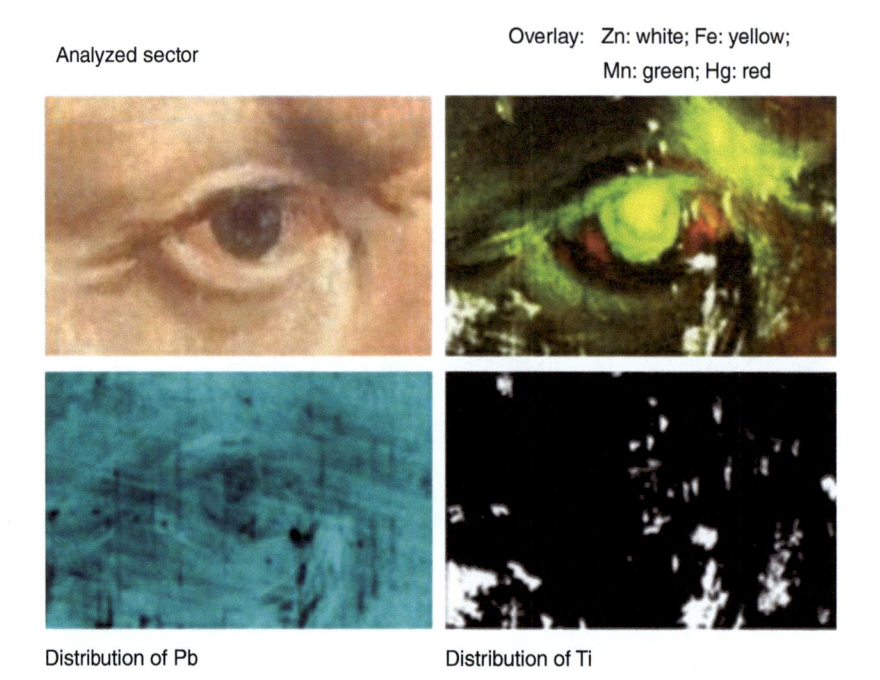

Figure 8.7 Distribution of elements in the eye region, measured by μ-XRF, measurement matrix 256 × 200 pixel (3,4 × 2,5 cm), 1000 msec measurement time for each pixel [70].

Imperial Palace, and mainly detected four elements: Ca, Fe, As, and Pb. The painting is divided into Ca and Pb paint layers, As paint layers, and Pb and Fe paint layers. The delamination and restoration of bamboo flower pigment were determined.

8.4 Summary and Perspectives

The analysis of metal elements is of great significance in many types of cultural relic research, and the analysis combined with archaeology greatly promotes the depth and breadth of cultural relic related research. The composition analysis of metal elements in cultural relics plays an important role in the study of ancient ceramics, bronzes, and paintings. The micro-neighbor structure of specific metal elements in cultural relics can be used to study the mechanism of corrosion and aging of cultural relics and technological restoration, providing important reference value for the protection and restoration of cultural relics.

Acknowledgments

The project was supported by National Natural Science Foundation of China (NSFC, grant numbers 12075259, 12075260 and 12175260).

List of Abbreviations

AAS	atomic absorption spectroscopy
EDXRF	energy-dispersive X-ray fluorescence
EXAFS	extended X-ray absorption fine structure
FTIR	Fourier transform infrared spectroscopy
GC–MS	gas chromatography mass spectrometry
GIXRD	grazing incidence X-ray diffraction
ICP-AES	inductively coupled plasma atomic emission spectrometer
LA-ICP-MS	laser ablation inductively coupled plasma mass spectrometry
LIBS	laser-induced breakdown spectroscopy
MS	mass spectrometry
NAA	neutron activation analysis
ND-TOF	time-of flight neutron diffraction
PGAA	prompt gamma activation analysis
PIXE	proton-induced X-ray fluorescence
P-XRF	portable X-ray fluorescence spectrometer
SEM–EDX	scanning electron microscopy coupled with energy dispersive X-ray spectroscopy
SR-XRD	synchrotron radiation X-ray diffraction
SRXRF	synchrotron radiation X-ray fluorescence
XAFS	X-ray absorption fine structure spectroscopy

XANES   X-ray absorption near side structure
XRD   X-ray diffraction
XRF   X-ray fluorescence analysis

References

1 Haraguchi, H. (2004). Metallomics as integrated biometal science. *J. Anal. At. Spectrom.* 19: 5–14.
2 Li, Y.F., Sun, H.Z., Chen, C.Y., and Chai, Z.F. (2016). *Metallomics*. Beijing: Science Press.
3 Li, Q., Cai, Z., Fang, Y., and Wang, Z. (2021). Matermetallomics: concept and analytical methodology. *Atom. Spectrosc.* 42 (5): 238–246.
4 Feng, S.L., Xu, Q., Feng, X.Q. et al. (2005). Application of nuclear analysis techniques in ancient Chinese porcelain. *Nucl. Phys. Rev.* 22 (1): 131–134.
5 Yan, L.T., Zhou, Y., Ma, B. et al. (2022). The coordination structure and optical property of iron ions in the traditional lead glazes. *J. Mol. Struct.* 1257 (5): 132593.
6 Fabrizi, L., Turo, F.D., Medeghini, L. et al. (2019). The application of non-destructive techniques for the study of corrosion patinas of ten Roman silver coins: the case of the medieval Grosso Romanino. *Microchem. J.* 145: 419–427.
7 Monico, L., Janssens, K., Hendriks, E. et al. (2015). Evidence for degradation of the chrome yellows in Van Gogh's Sunflowers: a study using noninvasive in situ methods and synchrotron-radiation-based X-ray techniques. *Angew. Chem. Int. Ed.* 54: 1–6.
8 Chai, Z.F. (1982). *Basis of Activation Analysis*. Beijing: Atomic Energy Press.
9 Sayre, E.V. and Dodson, R.W. (1957). Neutron activation study of mediterranean potsherds. *Am. J. Archaeol.* 61: 35–41.
10 Hein, A., Mommsen, H., and Maran, J. (1999). Element concentration distributions and most discriminating elements for provenancing by neutron activation analyses of ceramics from Bronze Age sites in Greece. *J. Archaeol. Sci.* 26: 1053–1058.
11 Lei, Y. (2007). Instrument neutron activation analysis of Tang Tricolour ware and a study of its trajectory of development in the region of capitals. *Palace Museum J.* 3: 86–127. 158.
12 Ma, G.Z. and Yao, H.Y. (1985). Progress of X-ray fluorescence spectroscopy in China. *Spectrosc. Spect. Anal.* 5: 140.
13 Ji, A., Tao, G.Y., Zhuo, S.J., and Luo, L.Q. (2003). *X-Ray Fluorescence Spectrum Analysis*. Beijing: Science Press.
14 Wei, L., Wang, L.Q., Zhou, T. et al. (2012). Application and progress of the nondestructive spectral technology used in polychrome ceramic relics analysis. *Spectrosc. Spectr. Anal.* 32 (2): 481–485.
15 Yan, L.T., Liu, M., Sun, H.Y. et al. (2020). A comparative study of typical early celadon shards from Eastern Zhou and Eastern Han dynasty (China). *J. Archaeol. Sci. Rep.* 33: 102530.

16 Wei, X.J., Lei, Y., Sun, T.X. et al. (2008). Elemental depth profile of faux bamboo paint in Forbidden City studied by synchrotron radiation confocal μ-XRF. *X-Ray Spectrom.* 37 (6): 595–598.

17 Gray, A.L. (1985). Solid sample introduction by laser ablation for inductively coupled plasma source mass spectrometry. *Analyst* 110: 551–556.

18 Chen, J.Z., Wang, J., Ju, G.J. et al. (2016). Recent development and application of laser induced breakdown spectroscopy. *Chin. Sci. Bull.* 61: 1086–1098.

19 Xu, W.Y., Li, X.R., and Guo, B. (2015). A review of detection techniques by atomic absorption spectrometry. *Gansu Agric. Sci. Technol.* 11: 76–78.

20 Wang, Q.W. and Liu, W.H. (1994). *X-Ray Absorption Fine Structure and its Application*. Beijing: Science Press.

21 Koningsberger, D.C., Mojet, B.L., van Dorssen, G.E., and Ramaker, D.E. (2000). XAFS spectroscopy; fundamental principles and data analysis. *Top. Catal.* 10: 143–155.

22 Dejoie, C., Sciau, P., Li, W.D. et al. (2014). Learning from the past: rare epsilon-Fe_2O_3 in the ancient black–glazed Jian (Tenmoku) wares. *Sci. Rep.* 4: 4941.

23 Gaowei, M.J., Liu, Y.Z., Chu, W.S. et al. (2009). Sn-L3 EDGE and Fe K edge XANES spectra of the surface layer of ancient Chinese black mirror Heiqigu. *Nucl. Tech.* 32 (9): 662–666.

24 Yang, T.Y., Wen, W., Yin, G.Z., and Li, X.L. (2015). Introduction of the X-ray diffraction beamline of SSRF. *Nucl. Sci. Tech.* 26 (2): 5–9.

25 Jonynaitė, D., Senvaitienė, J., Kiuberis, J. et al. (2009). XRD characterization of cobalt-based historical pigments and glazes. *Chemija* 20 (1): 10–18.

26 Liu, Z., Jia, C., Li, L. et al. (2018). The morphology and structure of crystals in Qing Dynasty purple-gold glaze excavated from the Forbidden City. *J. Am. Ceram. Soc.* 101 (11): 5229–5240.

27 Loongc, K., Scherillo, A., and Festa, G. (2017). Scattering techniques: small- and wide- angle neutron diffraction. In: *Neutron Methods for Archaeology and Cultural Heritage* (ed. K. Nikolay and F. Giulia), 183–207. Cham: Springer International Publishing Switzerland.

28 Kockelmann, W. and Kirfel, A. (2004). Neutron diffraction studies of archaeological objects on ROTAX. *Phys. B Condens. Matter* 350 (1–3): 581–585.

29 Huang, W., Kockelmann, W., Gordfrey, E. et al. (2010). Non-destructive phase analysis of Song Dynasty iron coins by TOF neutron diffraction. *Acta Sci. Nat. Univ. Pekin.* 46 (2): 245–250.

30 Feliu, M.J., Edreira, M.C., and Martı́n J. (2004). Application of physical–chemical analytical techniques in the study of ancient ceramics. *Anal. Chim. Acta* 502 (2): 241–250.

31 Li, B.P., Zhao, J.X., Collerson, K.D., and Greig, A. (2003). Application of ICP-MS trace element analysis in study of ancient Chinese ceramics. *Chin. Sci. Bull.* 2003 (48): 1219–1224.

32 Li, L., Xie, G.X., Feng, S.L. et al. (2013). Provenance research by INAA on ancient Chinese white porcelain excavated from the Maojiawan site of Beijing. *J. Archaeol. Sci.* 40: 1449–1453.

33 Ma, B., Liu, L., Feng, S.L. et al. (2014). Analysis of the elemental composition of Tang Sancai from the four major kilns in China using EDXRF. *Nucl. Instrum. Meth. A* 319: 95–99.

34 Fantuzzi, L. and Ontiveros, M.C. (2019). Amphora production in the Guadalquivir valley (Spain) during the Late Roman period: petrographic, mineralogical, and chemical characterization of reference groups. *Archaeol. Anthropol. Sci.* 11: 6785–6802.

35 Prinsloo, L.C., Wood, N., Loubser, M. et al. (2005). Re-dating of Chinese celadon shards excavated on Mapungubwe Hill, a 13th century Iron Age site in South Africa, using Raman spectroscopy, XRF and XRD. *J. Raman Spectrosc.* 36: 806–816.

36 Blagoev, K., Grozeva, M., Malcheva, G., and Neykova, S. (2013). Investigation by laser induced breakdown spectroscopy, X-ray fluorescence and X-ray powder diffraction of the chemical composition of white clay ceramic tiles from Veliki Preslav. *Spectrochim. Acta B* 79–80: 39–43.

37 Zhang, X., Ma, L., Chen, J.F. et al. (2003). Discrimination of the production sites of ancient pottery by using lead isotopic composition. *Nucl. Tech.* 26: 693–696.

38 Zhang, X., Chen, J.F., Ma, L. et al. (2004). Pb and Sr isotopic compositions of ancient pottery: a method to discriminate production sites. *Nucl. Tech.* 27: 201–206.

39 Yu, Y.B., Zhang, M.L., Wu, J. et al. (2011). Application of X-ray absorption fine structure in the study of ancient ceramics. *Jiangsu Ceram.* 44 (1): 12–14.

40 Tian, S.B., Liu, Y.Z., Zhang, M.L. et al. (2009). A preliminary study on coloring mechanism of Jun copper red glaze. *Nucl. Tech.* 32 (6): 413–418.

41 Matsunaga, M. and Nakai, I. (2004). A study of the firing technique of pottery from Kaman-Kalehöyük, Turkey, by synchrotron radiation-induced fluorescence X-ray absorption near-edge structure (XANES) analysis. *Archaeometry* 46: 103–114.

42 Wang, L.H. and Wang, C.S. (2011). Co speciation in blue decorations of blue-and-white porcelains from Jingdezhen kiln by using XAFS spectroscopy. *J. Anal. At. Spectrom.* 26: 1796–1801.

43 Tanaka, I., Mizoguchi, T., and Yamamoto, T. (2005). XANES and ELNES in ceramic science. *J. Am. Ceram. Soc.* 88 (8): 2013–2029.

44 Coutinho, M.L., Muralha, V.S.F., Mirao, J., and Veiga, J.P. (2014). Non-destructive characterization of oriental porcelain glazes and blue underglaze pigments using μ-EDXRF, μ-Raman and VP-SEM. *Appl. Phys. A Mater. Sci. Process.* 114: 695–703.

45 Barilaro, D., Crupi, V., Majolino, D., and Venuti, V. (2007). Decorated pottery study: analysis of pigments by x-ray absorbance spectroscopy measurements. *J. Appl. Phys.* 101: 064909.

46 Miao, J.M. and Song, C.Z. (1991). X-ray nondestructive testing technology for bronze ware. *Palace Museum J.* 03: 93–96.

47 Brill, R.H. and Wampter, J.M. (1967). Isotope studies of ancient lead. *Am. J. Archaeol.* 71 (1): 63–77.

48 Tylecote, R.F., Ghaznavi, H.A., and Boydell, P.J. (1977). Partitioning of trace elements between the ores, fluxes, slags and metal during the smelting of copper. *J. Archaeol. Sci.* 4 (4): 305–333.

49 Luo, W.G., Qin, Y., Huang, F.C., and Wang, C.S. (2010). Study on the microstructure and alloy technology of bronzes excavated from the ancient Jun district. *Stud. Hist. Nat. Sci.* 29 (3): 329–338.

50 Ma, D., Luo, W.G., Qin, Y. et al. (2020). Study on the casting cores to identify the manufacturing place of Chinese bronze vessels excavated in the Qiaojiayuan tombs from Spring and Autumn period. *Archaeol. Anthropol. Sci.* 12: 203.

51 Liu, Y., Yuan, S.X., and Zhang, X.M. (2000). Research on the corrosion of bronze wares excavated from Tianma-Qucun site of Jin state in Zhou Dynasty. *Sci. Conserv. Archaeol.* 12 (2): 9–18.

52 Maróti, B., Révay, Z., Szentmiklósi, L. et al. (2018). Benchmarking PGAA, in-beam NAA, reactor-NAA and handheld XRF spectrometry for the element analysis of archeological bronzes. *J. Radioanal. Nucl. Chem.* 317: 1151–1163.

53 Siano, S., Bartoli, L., Santisteban, J.R. et al. (2006). Non-destructive investigation of bronze artefacts from the Marches National Museum of Archaeology using neutron diffraction. *Archaeometry* 48 (1): 77–96.

54 Fan, X.P., Harbottle, G., Gao, Q. et al. (2012). Brass before bronze? Early copper-alloy metallurgy in China. *J. Anal. At. Spectrom.* 27: 821–826.

55 Hu, F., Qin, Y., Qu, Y., and Luo, Y.B. (2018). Preliminary study on metal wire of early Western Zhou Dynasty unearthed from Xiezidi Site in Daye City, Hubei Province. *Nonferrous Metals: Extract. Metall.* 12: 76–79.

56 Nagler, S.E., Stoica, A.D., Stoica, G.M. et al. (2019). Time-of-flight neutron diffraction (TOF-ND) analyses of the composition and minting of ancient Judaean "Biblical" coins. *J. Anal. Methods Chem.* 2019: 6164058. https://doi.org/10.1155/2019/6164058.

57 Ma, J.Y., Liang, H.G., and Wang, J.L. (2012). The formation of corrosion products on the bronze of Western Zhou Dynasty, excavated from Ouhai, Zhejiang province. *Sci. Conserv. Archaeol.* 24 (2): 84–89.

58 Robbiola, L., Blengino, J.M., and Fiaud, C. (1998). Morphology and mechanisms of formation of natural patinas on archaeological Cu-Sn alloys. *Corros. Sci.* 40 (12): 2083–2111.

59 Mezzi, A., de Caro, T., Riccucci, C. et al. (2013). Unusual surface degradation products grown on archaeological bronze artefacts. *Appl. Phys. A Mater. Sci. Process.* 113: 1121–1128.

60 Qu, L., Gao, F., Liu, J.Y. et al. (2018). Study of metal artifacts reparation using modern technology and traditional technology: taking Liao dynasty metal mask in the palace museum as an example. *Museum* (02): 119–127.

61 Cianchetta, H., Colantoni, I., Talarico, F. et al. (2012). Discoloration of the smalt pigment: experimental studies and ab initio calculations. *J. Anal. At. Spectrom.* 27 (11): 1941–1948.

62 Chen-Wiegart, Y.C.K., Catalano, J., Williams, G.J. et al. (2017). Elemental and molecular segregation in oil paintings due to lead soap degradation. *Sci. Rep.* 7 (1): 11656.

63 Eumelen, G.J.A.M., Bosco, E., Suiker, A.S.J. et al. (2019). A computational model for chemo-mechanical degradation of historical oil paintings due to metal soap formation. *J. Mech. Phys. Solids* 132: 103683.

64 Van Driel, B.A., Van den Berg, K.J., Gerretzen, J., and Dik, J. (2018). The white of the 20th century: an explorative survey into Dutch modern art collections. *Herit. Sci.* 6: 16.

65 Hermans J.J., Keune K., Van Loon A., Stols-Witlox M.J.N., Corkery R.W. and Ledema P.D.2014). The synthesis of new types of lead and zinc soaps: a source of information for the study of oil paint degradation. ICOM-CC 17th Triennial Conference: 17–19 September 2014, Melbourne, Australia (pp. 1603). International Council of Museums.

66 Cato, E., Borca, C., Huthwelker, T., and Ferreira, E.S.B. (2016). Aluminium X-ray absorption near-edge spectroscopy analysis of discoloured ultramarine blue in 20th century oil paintings. *Microchem. J.* 126: 18–24.

67 Ferreira, E., Gros, D., Wyss, K. et al. (2015). Faded shine .... The degradation of brass powder in two nineteenth century painting. *Herit. Sci.* 3: 24.

68 Wang, B., Yu, H., Bovyn, G., and Caen, J. (2017). Identification of pigments from "Zhenhai Temple", a Qing Dynasty export oil painting. *Sci. Conserv. Archaeol.* 29 (2): 82–88.

69 Stratulat, L., Geba, M., and Salajan, D. (2019). Village from Muscel by Ion Marinescu Valsan State of conservation and the chromatic palette. *Rev. Chim.* 69 (12): 3464–3468.

70 Lehmann, R., Schmidt, H.-J., Costa, B.F.O. et al. (2016). 57Fe Mössbauer, SEM/EDX, p-XRF and µ-XRF studies on a Dutch painting. *Hyperfine Interact.* 237: 69.

9

Metallomics in Toxicology

Ruixia Wang[1,3], *Ming Gao*[1,3], *Jiahao Chen*[2], *Mengying Qi*[2], *and Ming Xu*[1,2,3]

[1]*Chinese Academy of Sciences, Research Center for Eco-Environmental Sciences, State Key Laboratory of Environmental Chemistry and Ecotoxicology, No. 18, Shuangqing Road, Haidian District, Beijing 100085, China*
[2]*University of Chinese Academy of Sciences, School of Environment, Hangzhou Institute for Advanced Study, No. 1 Xiangshan Rd, Xihu District, Hangzhou 310024, China*
[3]*University of Chinese Academy of Sciences, College of Resources and Environment, No. 1 Yanqihu East Rd, Huairou District, Beijing 100049, China*

9.1 Metallomic Research on the Toxicology of Metals

All living things require essential transition metals that participate in a number of biochemical reactions necessary to maintain the normal physiological activities, such as iron in red blood cells and manganese in mitochondria. On the other side of the coin, exposure to non-essential heavy metals is inevitable as a result of various geologic processes and anthropogenic activities, for example, oral exposure to mercury in seafood or platinum drugs used in cancer treatment, which may have a negative impact on health. Thus, whether metabolic dyshomeostasis of essential metals or excessive intake of non-essential metals can be toxic has to be studied. To this end, there is always a strong need for researchers to figure out how metals and living matter interact, and to disclose why heavy metals interfere with normal life process.

Chemically, at the molecular level, the bioactivity of metals is determined by their atomic characteristics, resulting in their benefits or hazards to life. Therefore, in theory, there should be some underlying rules that govern the metal–biomolecular interactions. However, the live cell has a complex running system and most of its parts are still not well understood yet, making it impossible to get a full picture of the toxicological process of essential or non-essential metals. Hence researchers constantly look for new approaches to elaborate how biometals play a part in the process of life. Nearly two decades ago, the terms "metallome" and "metallomics" were proposed to integrate the research field related to bio-trace elements as part of omics-science, in a similar manner to "genome" and "genomics," as well as "proteome" and "proteomics" (Figure 9.1) [1]. Since then, metallomics has

Applied Metallomics: From Life Sciences to Environmental Sciences, First Edition.
Edited by Yu-Feng Li and Hongzhe Sun.
© 2024 WILEY-VCH GmbH. Published 2024 by WILEY-VCH GmbH.

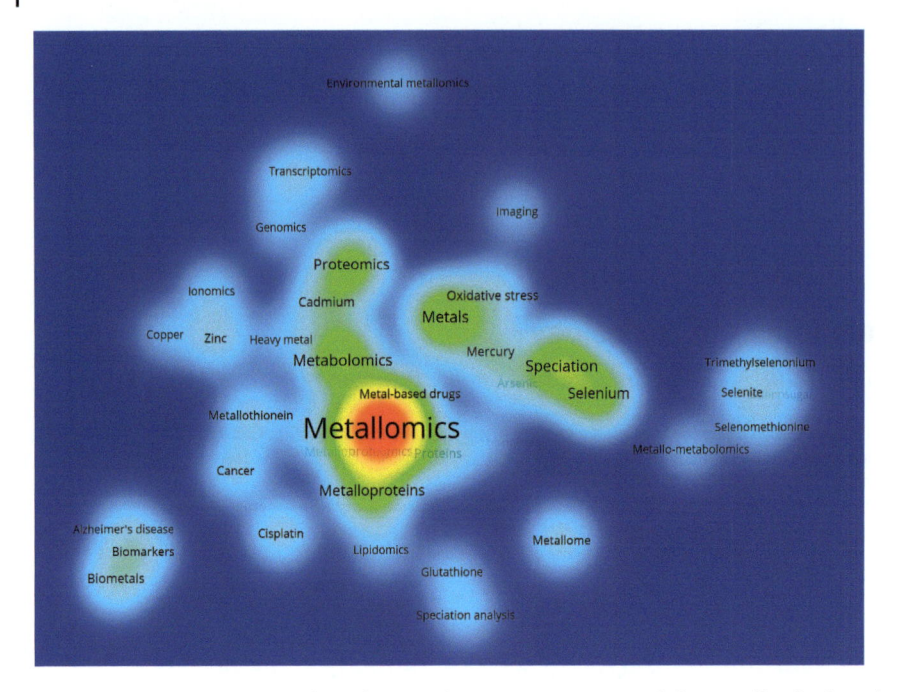

Figure 9.1 Density visualization of keyword co-occurrence network for metallomics based on the total link strength from January 2004 to December 2021. Only the keywords with at least five records are shown in the network using VOSviewer. Source: Adapted from Web of Science.

evolved into a multidisciplinary field, for example, on the aspect of toxicology, referred as "toxicometallomics" [2]. Metallomics is currently being applied to gain a deep insight into the metal toxicity in many different disciplines, and the latest progresses in this field have been reviewed thoroughly in several important literatures [3–5].

In biological scenarios, metals in the form of free ions or other species will come into contact with diverse kinds of biomolecules like nucleic acids, proteins, lipids, and polysaccharides. These biomolecules may directly react with metals via covalent or non-covalent interactions, resulting in the formation of multifarious metal-containing species. When a metal is incorporated into a biomolecule, conformational or functional transitions may happen, thus giving rise to changes of the molecular activity and enabling the biomolecule to play its physiological roles or impairing its normal functions. As a result, the downstream signaling pathways that are responsible for the detrimental effects of metals would be activated. However, the interactions between metals and biomolecules within cells are highly dynamic and unpredictable, most of which are still a mystery to researchers. There are many knowledge gaps that need to be filled, especially with regard to how metals induce detrimental effects and undermine health. In the scope of toxicometallomics, we here present some recent progresses and efforts to understand the interplay between heavy metals and life (Figure 9.2), illustrating how heavy metals work at the molecular, cellular, and organ levels.

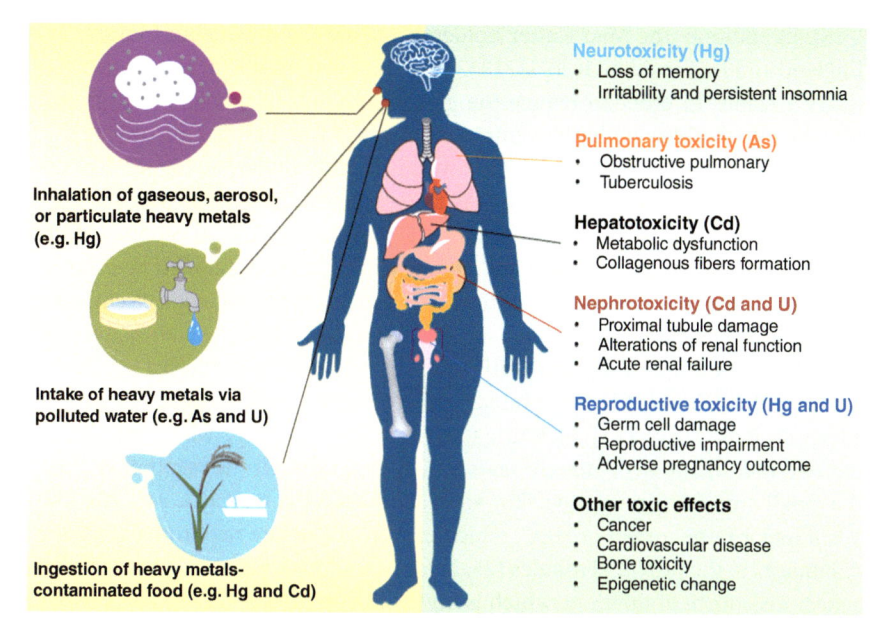

Figure 9.2 Primary exposure routes and health risks of heavy metals. Heavy metals can enter the human body through inhalation of gaseous and aerosol heavy metals (e.g. mercury), drinking water (e.g. arsenic and uranium), or food ingestion (e.g. mercury and cadmium) or dermal contact (Left), and induce health problems through impairing the kidneys, bones, liver, brain, lungs, and reproductive system (Right). Source: Ming Xu.

9.2 Recent Progresses in Understanding the Health Effects of Heavy Metals

9.2.1 Mercury, Oxidative Stress, and Cell Death

Mercury (Hg) emitted by natural processes and anthropogenic activities has a residence time up to a year in the atmosphere, and can be transported over long distances before finally depositing to terrestrial and marine ecosystems [6]. The anthropogenic emission of mercury is estimated to be 2,220 metric tons per year in 2018 [7], mainly from artisanal small-scale gold mining (ASGM, 37.7%) [8–10], coal combustion (21%) [11, 12], metal smelting (15%) [13, 14], and cement production (11%) [15]. A wide variety of mercury species exist in the environment, such as elemental mercury (Hg^0), inorganic mercury (e.g. Hg^{2+}), and organic mercury (e.g. methyl mercury (MeHg)), undergoing complicated biogeochemical transformation processes. In particular, mercury is well known to have an ability to biomagnify through food chains, and bring health risks to human being, considered by World Health Organization (WHO) as one of the top ten chemicals of major public health concern [16]. For example, in 1956, Minamata disease occurred in the residents living around Minamata City, south-west region of Japan's Kyushu Island, who ingested fish and shellfish contaminated by MeHg in industrial wastewater discharged from the Chisso Corporation, leading to the death of 1,784 people since then

[17, 18]. In addition, the Mad Hatter Epidemic during the nineteenth Century era in England and Iraq incident in 1971–1972 is well known for the mercury poisoning [18]. As a result, in order to reduce the global pollution of mercury and protect human health, the Minamata Convention on Mercury was adopted in 2013 [19].

Mercury is one of the most toxic heavy metals to human health, highly depending on its chemical and physical form. For example, in the case of gaseous Hg^0, its main exposure pathway is inhalation [20, 21], especially for the occupationally exposed population such as ASGM and mercury miners. After inhalation, gaseous Hg^0 can be quickly distributed within the body through lung and blood circulation, and even permeate through the blood–brain barrier (BBB) into the central nervous system (CNS), because of its highly diffusibility and lipid lipophilicity [22]. In the circulating blood, inhaled Hg^0 can rapidly diffuse into erythrocytes and be oxidized to mercuric ions (Hg^{2+}) [23]. Mercuric ions show low lipophilic with a half-life of about two months in vivo, and are mainly absorbed through the respiratory tract, and a small amount through the skin and gastrointestinal tract [23, 24]. The kidney is a major target organ of Hg^{2+}, which can induce severe nephrotoxic injuries and damages to the proximal tubules [25, 26]. In addition, MeHg is considered to be the most toxic form of mercury, which is highly lipophilic and easy to be adsorbed through the gastrointestinal tract. The major source of MeHg exposure in humans is regarded to be fish consumption over the past years, but recent studies suggest that rice intake may be another important dietary source of human MeHg exposure in South and Southeast Asia [27–29]. MeHg can easily pass through the BBB and placental barrier, exhibiting more pronounced toxic effects in infants and fetus than adults. Furthermore, MeHg exposure can induce neurotoxic effects related to the adult neurodegenerative diseases, because MeHg preferentially accumulates in the CNS [23, 30].

Oxidative stress is a key pathway to initiate the toxic effects and health risks of mercury (Figure 9.3), though the underlying mechanisms are still not fully elucidated yet. According to the hard–soft–acid–base principle, soft acidic metal ions such as Hg^{2+} have a strong affinity to covalently bind to the soft bases such as thiol groups (–SH) in biomolecules (e.g. cysteine and serum albumin), forming the Hg–S bond in vivo that is implicated in the transport and transformation of mercury through Rabenstein's reactions [31, 32]. On the other hand, there are thousands of cysteine residue-containing proteins that are mercury reactive in a proteome, offering a large number of –SH binding sites for mercury [33]. Binding of mercury to the thiols in antioxidants would inhibit the intracellular antioxidant defense system and disrupt redox homeostasis through the over-production of reactive oxygen species (ROS), resulting in the increase of oxidative stress (Table 9.1). For example, as a nonenzymatic antioxidant, intracellular glutathione (GSH) mainly exists as its free thiol form, participating in cellular defense against oxidative stress and heavy metals. Many studies have reported the formation of GS–Hg complex in vitro, tightly related to the abnormal GSH level in the Hg-exposed mammals [44–48]. Metallothionein (MT) is another low-molecular-weight, thiol-rich protein involved in the detoxification of mercury through playing the antioxidative role and forming the Hg–MT complexes in cells. In particular, MT in vertebrates contains at least 20 cysteinyl

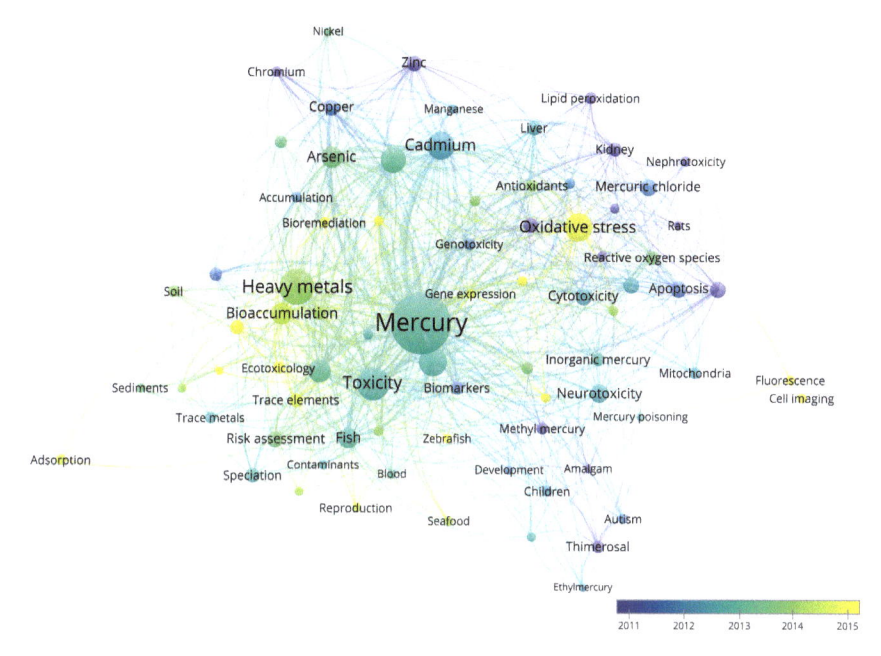

Figure 9.3 Network visualization map of keyword co-occurrence for mercury toxicity from 2000–2021. The colors represent the average time of keyword occurrence. Only the keywords with at least 25 records are shown in the network using VOSviewer. Source: Adapted from Web of Science.

thiols; hence, more than 1 mercury atom may be incorporated into 1 MT molecule, forming the Hg–MT complexes by binding 7 Hg^{2+} per MT molecule (Hg_7–MT) or 20 MeHg ions per MT molecule (($MeHg)_{20}$–MT) as reported [40, 49]. In this way, MT is able to protect against mercury poisoning by arresting and inactivating free forms of mercury in vivo.

Likely, selenol group (–SeH) is another high-affinity mercury-binding site in biomolecules, typically existing in the selenopeptides/selenoproteins. The selenium-mercury interactions have been extensively studied regarding the mercury toxicity in the past years [50–52]. For example, George et al. found nanoparticulate mercuric selenide in the brain tissues collected from human populations exposed to MeHg [53]. Chen et al. reported that, in mercury-polluted area, supplement of organic selenium could increase the urine excretion of mercury and decrease the oxidative damages in local residents after the long-term mercury exposure [54]. In rats after $HgCl_2$ administration, selenium-binding protein 1 was identified to be the most markedly upregulated protein in the kidneys [55]. In comparison with thiol group, selenol group is more chemically active and has a stronger capability of mercury binding with the binding energy of 9.96 ev [56, 57]. In vitro tests have demonstrated that selenol group in selenocysteine or selenoprotein could directly bind to mercury [58]. The formation of Hg–Se bond will inevitably affect the expression and activity of selenoenzymes such as glutathione peroxidase (GPx) and thioredoxin reductase (TrxR), and generate ROS because selenol participates in the redox reactions as

Table 9.1 Mercury-binding sites in representative proteins involved in oxidative stress.

Protein	Residue	Species	References
Thioredoxin reductase	Cys497	$HgCl_2$	[34]
	Sec498		
Glutathione reductase	Cys45	$HgCl_2$	[35]
	Cys50		
Glutathione peroxidase 1	Sec49	MeHg	[36]
	Cys52		
Sorbitol dehydrogenase	Cys44	MeHg	[37]
	Cys129		
	Cys119		
	Cys164		
Na,K-ATPase	Cys64	$HgCl_2$	[38]
	Cys136	MeHg	
Manganese superoxide dismutase	Cys196	MeHg	[39]
Metallothionein	Cys6, Cys8	MeHg	[40, 41]
	Cys14, Cys16		
	Cys20, Cys22		
	Cys25, Cys27		
	Cys30, Cys34		
	Cys35, Cys37		
	Cys38, Cys42		
	Cys45, Cys49		
	Cys51, Cys62		
	Cys64, Cys65		
Copper chaperone ATX1	Cys15	$HgCl_2$	[42, 43]
	Cys18		

reported [59–61]. For example, Chen et al. reported that, in mercury miners, both expression and enzymatic activity of GPx greatly increased accompanied by an elevation of selenium concentration in serum, suggesting the antioxidative role of GPx on eliminating the oxidative stress induced by Hg in vivo [62]. In the zebra-seabreams exposed to waterborne MeHg, the level of TrxR was found to decrease in the brain (~75%) and liver (~40%) [61].

Mitochondria are the primary organelles for the production of intracellular ROS, and calcium homeostasis is essential for mitochondrial function. Excessive calcium influx will boost oxidative phosphorylation, lead to an increase of ROS production, and eventually promote oxidative damages [63, 64]. Both inorganic and organic mercury may affect the cytosolic free calcium level [65]. For example, it was found that $HgCl_2$ enhanced the calcium entry through calcium channel, while

MeHg could increase calcium level through both calcium channel and calcium mobilization [65]. The mechanism behind opens a question that will need to be investigated and answered in the future.

Excessive ROS and severe oxidative damage will activate the signaling pathway for the regulation of cell death, which is a major cause of mercury-induced toxicity. For example, mercury was found to induce apoptosis dose-dependently in human liver cells through activating the caspase-3-mediated pathway [66–68]. During this process, caspases acted as crucial mediators of apoptosis which were implicated in the induction, transduction, and amplification of intracellular apoptotic signals in the cells exposed to mercury [69]. In addition, the long-term exposure of mercury can also lead to necrosis [70, 71]. Bhattacharya et al. reported that, after exposed to $HgCl_2$ at low-dose, the cell death mode shifted from apoptosis to necrosis via TNF-α-RIP3-caspase-8 pathway in hepatocytes [72]. Lin-Shiau et al. demonstrated that the induction of necrosis or apoptosis strongly depended on the dose of MeHg, since a high concentration of MeHg could lead to necrosis via a rapid rising of calcium ions, while a low concentration of MeHg could induce apoptosis through the activation of endonuclease [73]. Moreover, some recent literature reported that mercury could induce other types of programmed cell death [74, 75]. For instance, Lee et al. found that mercury inhibited the activation of NLRP3 inflammasomes leading to pyroptosis in macrophages [74], while Li et al. showed that mercury up-regulated the genes implicated in ferroptosis and necroptosis in the brain [75]. These findings suggest that there is a complicated regulatory network of cell death in response to mercury, which is highly desired to be disclosed in the future research.

9.2.2 Arsenic and Lung Cancer

Naturally occurring arsenic (As) exists as the form of sulfide minerals in the earth crust. Various geological processes can release arsenic into the air, water, and soil. Meanwhile, growing anthropogenic activities such as mining, coal consumption, oil extraction, semiconductor industry, and pesticide metallurgy cause the pollution of arsenic in the ecosystem, bringing potential health risks to human beings. Ingestion of arsenic in drinking water or crops and inhalation of arsenic-contaminated smoke during mining or smelting are regarded as the primary exposure routes. After ingestion or inhalation, arsenic will accumulate in organs and tissues, including liver, kidneys, lungs, bladder, nerves, and muscles. Initial clinical features of acute arsenic poisoning include stomach pain, diarrhea, dizziness, headache, and nausea. Severe poisoning may lead to trembling, breathing difficulty, convulsions, collapse, and, in some cases, death [76]. Chronic arsenic exposure is associated with direct myocardial injury, cardiac arrhythmias, cardiomyopathy, even diabetes and other cardiovascular dysfunctions. Besides, melanosis, hyperpigmentation, and keratinized skin lesions are also considered as the pathologic hallmarks of chronic arsenic exposure [77, 78]. Arsenic is classified as a class I carcinogen by the International Agency for Research on Cancer (IARC), and there are increasing epidemiological studies demonstrating the relationships between chronic arsenic exposure and skin, lung, bladder, and kidney cancer [79].

Lung cancer is the leading cause of cancer death worldwide, accounting for approximately 11.4% of the total cancer burden in 2020 [80, 81]. Lung cancer is historically classified into small-cell lung cancer (SCLC) and non-small-cell lung cancer (NSCLC). Thereinto, the latter accounts for 80~85% cases and can be further divided into different subtypes such as adenocarcinoma, adenosquamous carcinoma, squamous cell carcinoma, and large-cell carcinoma. The overall five-year survival rate for lung cancer is very low, varying from 6.7% to 26.3% according to the cancer types, because most of patients are diagnosed at a metastatic stage [80, 81]. A variety of environmental stresses is known to be correlated to the occurrence and metastasis of lung cancer [82, 83]. There are a lot of epidemiological studies reporting that chronic exposure to high level of arsenic (>100 μg/L) in drinking water is significantly associated with the incidence rates of lung cancer [84–87]. Thus, arsenic in drinking water was determined as a cause of lung cancer by IARC in 2004. However, a few controversial epidemiological studies suggest that there is still lack of sufficient evidences that chronic exposure to arsenic in drinking water at the low concentration ranging from 10 to 100 μg/L accounts for the development of lung cancer [88, 89].

The prevailing arsenic species are inorganic arsenate (As(V)) and arsenite (As(III)), the latter of which is much more toxic to mammals [90]. The pathological mechanisms for lung cancer induced by arsenic have been extensively investigated in vivo and in vitro. It is reported that the biotransformation of inorganic arsenic led to an increase of methylated metabolites and ROS production, giving rise to cell cycle progression, oxidative stress, and protein methylation [91, 92]. The stability and integrity of genome can be disrupted by arsenic through inducing DNA damage by reactive radicals, or interfering with DNA repair processes by affecting the expression of DNA damage response proteins [93]. Furthermore, arsenic exposure can alter the activation of P53-, Nrf2-, EGFR-, hedgehog (HH)-, TGF-β-, and NF-κB-mediated signaling pathways, involving in cell proliferation, differentiation, and epithelial-to-mesenchymal transition [94]. In particular, the expression of both oncogenic and tumor-suppressive genes at the epigenetic level may be affected by arsenic, including DNA methylation, non-coding RNA, histone modification, and RNA methylation [95, 96]. For example, Breda et al. found that arsenic exposure altered the DNA methylation profiling of whole genome to modulate the expression of target genes including p53, which plays a critical role in the promotion and progression of lung cancer [96]. Wang et al. demonstrated that arsenic activated the SOCS3-AKT/ERK pathway by increasing the histone modification of H3K9me2 involving in the cell transformation of lung cancer [97]. In addition, we recently revealed that arsenic exposure increased the somatic mutation rate of lung cancer cells by activating FTO-APOBEC3B pathway in a m6A-dependent manner [98].

9.2.3 Epigenetic Effects of Cadmium

Cadmium (Cd) is considered as one of the most ubiquitous and hazardous heavy metals, which is widely applied for industrial and consumer products such as batteries, pigments, electroplating, and plastic stabilizers [99, 100]. The etiopathogenesis of several diseases, such as kidney diseases, hypertension, diabetes, and carcinogenesis, has been reported to be associated with the long-term exposure of cadmium [101, 102]. The most severe form of chronic cadmium poisoning is Itai-itai disease, first reported in 1955 in Jinzu river basin in Toyama Prefecture, Japan, characterized by cardiovascular disease, osteoporosis and nephrotoxicity [103]. In 1993, IARC classified cadmium as a carcinogen. Therefore, the health risks and underlying toxicological mechanisms of cadmium exposure are persistently concerned by the public and researchers.

The half-life of cadmium within the human body can reach up to 30 years, resulting in the deposition of cadmium in the skeletal, gastrointestinal, cardiovascular, and renal systems [104, 105]. Most of cadmium is complexed by MT in vivo, responsible for accumulating cadmium in organs and protecting against cadmium-induced toxicity through the formation of Cd–MT complex [106, 107]. The primary mechanisms of cadmium-induced toxicity and resistance include the expression of stress response genes, generation of oxidative and endoplasmic reticulum stress, inhibition of DNA repair, and induction of cell death [108]. However, the factors initiating the cadmium-induced adverse effects are still not fully elucidated at the molecular level yet. It is reported that transcription factors such as nuclear factor-kappa B (NF-κB) and activator protein 1 (AP-1) could regulate the activation of numerous signaling pathways upon cadmium exposure to trigger diversity of deleterious effects [109]. In addition, cadmium exposure is able to cause DNA damages such as DNA mutations, DNA strand breaks, and chromosomal aberrations, or inhibit the expression or activity of DNA repair proteins, which are directly involved in the carcinogenesis and cell death induced by cadmium [110].

Recently, toxic effects of cadmium at the epigenetic level have attracted increasing attention [111]. Some epidemiological studies suggest that there is a potential association between blood cadmium level and epigenetic alterations [112, 113]. For example, in adult non-pregnancy women in Argentina, the concentration of blood cadmium was found to be positively associated with the DNA methylation in long interspersed nuclear element 1 (LINE-1), a transposable repetitive sequence whose hypomethylation represents the alterations of gene expression and genomic instability [114]. However, no significant relationship between blood cadmium and DNA methylation was found based on a 127 mother–child paired cohort study in Bangladesh [115]. On the contrary, there have more CpG site-specific hypermethylation in cord blood DNA. Another epidemiologic study also reported that the level of cadmium was significantly correlated with LINE-1 hypomethylation in the blood of pregnancy woman in United States [116]. Moreover, the expression of some specific microRNA (miRNA) or long non-coding RNA (lncRNA) in blood has been demonstrated to be tightly correlated with the occupational exposure of cadmium [117, 118].

Epigenetic alterations after exposure to cadmium have been observed at the mechanistic levels of DNA methylation, histone modification, and non-coding RNA [119]. For example, Arita and Costa found that cadmium exposure silenced the gene expression of tumor suppressors such as p53 and p16 through affecting the levels of DNA methylation [120]. Further mechanistic investigation suggested that cadmium exposure changed the expression level or enzymatic activity of DNA methyltransferases (DNMT1 and DNMT3b) and dioxygenase (Tet1 and Tet2), partly depending on the production of ROS [121, 122]. In addition, the alterations of miRNA expression have been demonstrated to be associated with cadmium-induced cytotoxicity and carcinogenesis, because miRNA can bind to a complementary sequence of mRNA and thereby regulate mRNA degradation and protein translation, allowing the cadmium-responsive miRNAs to play diverse roles in response to cadmium stress through regulating the expression of their target genes [123]. For example, in the human bronchial epithelial cells exposed to cadmium, the expression of miR-181a-2-3p significantly declined, leading to the enhanced inflammatory response and inflammasome activation by regulating the expression of Toll-like receptor 4 or sequestosome 1 genes [124]. In the rat liver cells exposed to cadmium, the expression of miR-155 was found to upregulate the level of autophagy, through suppressing the activation of Rheb/mTOR signaling pathway [125]. Recent advances also suggest that LncRNAs may play a crucial role as pivotal regulators under cadmium stress. For example, an increase of lncRNA MALAT1 was observed in the lungs of cadmium-exposed rats, as well as in the blood of workers exposed to cadmium [126]. LncRNA MT1DP was found to be involved in cadmium-induced cell death and oxidative stress through interacting with RhoC, miR-214, and miR-365, respectively [127, 128]. In addition, LncRNA MT1DP secreted by the liver after cadmium exposure could be packed into exosome and transferred to the kidneys to aggravate nephrotoxicity [129]. Post-translational modifications of histones and m6A RNA methylation can also be affected by cadmium. For example, alterations of global H3K4me3 and H3K9me2 in immortalized normal human bronchial epithelial cells have been detected after cadmium exposure through inhibiting the activity of histone demethylases KDM5A and KDM3A [130]. Under cadmium stress, an increase of m6A demethylase ALKBH5 could erase the m6A methylation of PTEN mRNA in human normal lung epithelial cells responsible for the cadmium-induced malignant transformation [131].

9.2.4 Nephrotoxicity of Uranium in Drinking Water

Uranium (U) is a naturally occurring actinide and ubiquitous in the earth's crust and bodies of water. Because of the geologic processes and anthropogenic activities, increasing attention is paid to the uranium contamination in groundwater on the world scale, which is a major source of drinking water [132–137]. For example, in the Datong Basin area of Shanxi, China, the level of uranium in the local shallow groundwater was reported to reach up to 288 µg/L [138, 139]. In the US California High Plains and Central Valley aquifers, where 1.9 million people live, the mean concentrations of uranium in groundwater were reported to be 128 and 71 µg/L, respectively [133]. On the other hand, anthropogenic activities including mining, fertilizer use, nuclear facility, and military activities all aggravate the uranium contamination in groundwater [140]. To ensure the safety of drinking water, a guideline concentration of 30 µg/L for uranium is recommended by the WHO, primarily based on its chemotoxicity rather than radiotoxicity [141]. In addition, many countries have set their own guidelines for uranium in drinking water (Table 9.2).

Table 9.2 Guideline values for uranium in drinking water.

Nation	Guideline value (µg/l)
Australia	17
Canada	20
Germany	10
United States	30
China	30

Once ingested, uranium in the form of uranyl ions (UO_2^{2+}) will enter the bloodstream, form complexes with bicarbonate (~50%), transferrin (~30%), and red blood cells (~20%), and then rapidly accumulate in organs and tissues like the kidneys [142, 143]. Numerous epidemiological and animal studies have documented that the kidney is the main target organ of soluble uranium in drinking water, which may cause nephrotoxicity (Figure 9.4) [144]. Contrary to common sense, there is increasing evidence suggesting that the toxicity of natural uranium in drinking water mainly originates from its chemotoxicity, rather than radiotoxicity, because of the relatively long-term exposure to low-dose natural uranium in some populations. For instance, Zamora et al. reported that a chronic ingestion of uranium in drinking water ranging from 2 to 781 µg/L affected the renal function and the proximal tubule of residents using the water from private wells in a Canadian community [145]. Kurttio et al. reported that the exposure of uranium in drinking water at a median concentration of 28 µg/L (6~135 µg/L) was significantly correlated with the secretion of calcium and phosphate in 325 persons who had used the drilled wells for drinking water in southern Finland, suggesting that even low uranium

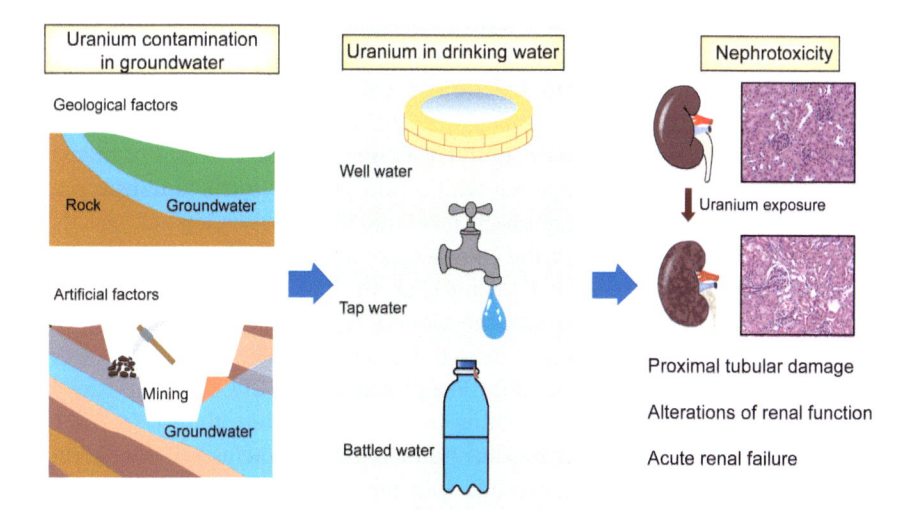

Figure 9.4 Nephrotoxic risk of uranium in drinking water. Source: Ming Xu.

concentration in drinking water could cause nephrotoxic risks [146]. Weaver et al. found that a significant association between urine uranium and creatinine clearance and NAG level existed in 684 lead workers exposed to uranium [147]. In particular, animal studies have demonstrated that the long-term exposure of uranium in drinking water led to renal damages, including apoptosis and cell loss in the proximal tubules, alterations of renal function, and acute renal failure [148–151].

Chemotoxicity is the primary reason of detrimental effects in the kidneys induced by natural uranium, in comparison with enriched uranium and depleted uranium. The main toxicological mechanisms of natural uranium include: (i) uranium is able to disrupt cellular redox homeostasis and lead to oxidative stress through perturbing the antioxidant defense system including superoxide dismutase, GPx, and catalase, and the elevation of intracellular ROS will boost lipid peroxidation and damage proximal tubular plasma membranes [152]. Meanwhile, abnormal alterations of metabolic enzymes in response to uranium stress are also closely related to renal injury [153]. At the subcellular level, the dose-dependent generation of ROS in renal cells exposed to uranium appears to be responsible for the collapse of mitochondrial membrane potential, release of cytochrome C, and dysfunction of mitochondrial respiratory chain and tricarboxylic acids cycle [154–158]. (ii) apart from radioactivity, uranium can induce DNA damage and genotoxicity in renal cells via chemical reactions, from either free radical production, or direct binding of uranyl ions to the negatively charged phosphate backbone of DNA, leading to DNA break or the formation of DNA adduct [159–161]. Thiébault et al. demonstrated that uranium could dose-dependently induce the DNA double-strand breaks in normal rat kidney proximal cells [157]. In addition, Stearns et al. showed that, though low concentrations of uranium (<10 µM) were regarded to be non-toxic to human embryonic kidney cells or normal human keratinocytes, concentrations of uranyl ions in the low micromolar range inhibited DNA repair through disruption of zinc finger domains of specific target DNA repair proteins in kidney cells, which might be associated with DNA repair deficiency in uranium-exposed human populations [162]. (iii) uranyl ions are high affinity to the carboxyl, phosphoryl, and amide groups in the amino acid side chain of protein molecule, probably leading to the structural and functional impairments of enzymes [163]. For example, more than 20% of uranyl ions have been found to be complexed with serum proteins, such as transferrin, albumin, and fetuin-A, which might participate into the translocation and toxicity of uranium in vivo [164]. However, the knowledge of interactions between uranium and nephric proteins is quite limited. Vidaud et al. previously reported the presence of uranium-protein complexes within the kidneys of rat exposed to uranium in drinking water [165]. Another study reported that, uranyl ions might interact with heat shock protein 90 (Hsp90), or ERO1-like protein alpha (ERO1A) in proximal tubular cells, affecting the protein folding in endoplasmic reticulum [166, 167]. Besides, uranium is found to bind to some iron regulatory proteins like transferrin and ferritin, which is probably correlated with the iron disorder observed in the kidneys, though the underlying mechanisms remain unclear yet [168–170]. (iv) uranium is reported to induce renal inflammatory responses in the kidneys through activating the NF-κB pathway and upregulating three inflammatory genes (*TNF-α*, *iNOS*, and *COX-2*) [171]. In another transcriptomic study, the overexpression of some inflammatory genes in the kidneys of uranium-administrated mice, such as *Opn*, *Pecam*, and *Gal-3*, was determined [172]. Moreover, the gene expression of intercellular and vascular cell adhesion molecules (ICAM and VCAM) in the kidneys was found to be upregulated after exposure to uranium, which are involved in the recruitment of inflammatory cells [173]. (v) uranium may elicit irreversible cell damage and induce kidney cell death. It is reported that uranium could induce kidney cell death via mitochondrial- and non-mitochondrial apoptotic pathways through the regulation of MAPK phosphorylation and activation of caspase-3, caspase-8, and caspase-9 pathways [149, 174]. Additionally, FasR/caspase-8 pathway was also found to be implicated in the uranium-induced apoptosis in renal cells [156]. Moreover, the high expression level of LC3-II indicated the formation of autophagosome in the mouse kidneys after uranium exposure [175, 176]. Very recently, the features of pyroptosis were identified in the uranium-treated rat kidney cells, including the overexpression of NLRP3 and cleaved caspase-1, upregulation of *GSDMD* gene, release of mature IL-18 and IL-1β, and leakage of LDH [177].

9.3 Knowledge Gaps, Challenges, and Perspectives

There is still a long way to go for metallomic research on the toxicology of heavy metals, elaborating their essential or non-essential roles in life. A major knowledge gap concerned in this field is what are the biomolecular targets of heavy metals. If the aim is to elaborate the toxicological mechanisms behind the detrimental effects of a heavy metal at the molecular level, the aforementioned query must be addressed first. Without this crucial knowledge, it is challenging to understand why some heavy metals can exert negative impacts and how to develop efficient detoxification means following heavy metal exposure. However, identification of the biomolecular targets of a heavy metal is still not an easy task at present. On the one hand, in organisms, interactions between heavy metals and biomolecules are unpredictable and changeable, requiring the development of advanced techniques to precisely detect the spatiotemporal changes of metal-binding biomolecules at any moment. But, for instance, there are thousands of proteins in a proteome, in charge of life activities, serving as the regulator or ligand of metals. A better description of the network of metal–protein interactions needs more specific approaches for high-throughput screening the protein targets and biomarkers of heavy metals. In order to uncover the regulatory pathways of heavy metal toxicity, multi-omics approaches will give more detailed information on the gene expression, protein functionality, and metabolic profile. Furthermore, another knowledge gap is, when one wants to uncover the fate and stress pathway of a heavy metal in vivo, the premise is that all basic issues on the biological level should be well addressed. Unfortunately, there are many black boxes in regard to biological processes in life, which need to be uncovered in the future. Hence future developments of more efficient means based on the metallomic strategies, as well as advancements in chemical and biological fields, are highly expected to bring more opportunities to expand the toxicological knowledge in medical, health, and environmental sciences.

Acknowledgments

This work was supported by the National Natural Science Foundation of China (21922611, 22176201, 22376211) and Youth Innovation Promotion Association of Chinese Academy of Sciences (2019042).

List of Abbreviations

Akt	protein kinase B
ALKBH5	alkB homolog 5
AP-1	activator protein 1
APOBEC3B	apolipoprotein B mRNA editing catalytic polypeptide-like 3B
As	arsenic
ASGM	artisanal small-scale gold mining
BBB	blood–brain barrier
Bcl-2	B-cell lymphoma-2
Cd	cadmium
CNS	central nervous system
COX-2	cyclooxygenase-2
DNMT1	DNA methyltransferase 1
DNMT3b	DNA methyltransferase 3b
EGFR	epidermal growth factor receptor
ERK	extracellular-signal-regulated kinase
ERO1A	ERO1-like protein alpha
FTO	fat mass and obesity-associated
Gal-3	lectin, galactose binding, soluble 3
GPx	glutathione peroxidase
GSDMD	gasdermin D
GSH	glutathione
H3K4me3	trimethylated histone H3 on lysine 4
H3K9me2	dimethylated histone H3 on lysine 9
H3K9me2	H3 lysine 9-dimethylation
Hg	mercury
Hg^0	elemental mercury
Hg^{2+}	mercurous ion
HH	hedgehog
Hsp90	heat shock protein 90
IARC	International Agency for Research on Cancer
ICAM	intercellular cell adhesion molecules
IL-18	interleukin (IL)-18

IL-1β	interleukin (IL)-1 beta
iNOS	inducible nitric oxide synthase
KDM3A	lysine-specific demethylase 3A
KDM5A	lysine-specific demethylase 5A
LC3-II	LC3-phosphatidylethanolamine conjugate
LDH	lactate dehydrogenase
LINE-1	long interspersed nuclear element 1
lncRNA	long non-coding RNA
m6A	N6-methyladenosine
MALAT1	metastasis-associated lung adenocarcinoma transcript 1
MAPK	mitogen-activated protein kinase
MeHg	methyl mercury
miR-155	miRNA 155
miR-181a-2-3p	miRNA 181a-2-3p
miRNA	microRNA
MT	metallothionein
MT1DP	metallothionein 1D pseudogene
mTOR	mammalian target of rapamycin
NAG	N-acetyl-β-D-glucosaminidase
NF-κB	nuclear factor-kappa B
NLRP3	nucleotide-binding oligomerization domain (NOD)-like receptor family pyrin domain containing 3
Nrf2	nuclear factor erythroid 2-related factor 2
NSCLC	non-small-cell lung cancer
Opn	secreted phosphoprotein 1
P53	tumor suppressor encoded by TP53
Pecam	platelet/endothelial cell adhesion molecule
PTEN	phosphatase and tensin homolog deleted on chromosome 10
Rheb	Ras homolog enriched in brain
RhoC	Ras homolog gene family, member C
RIP3	receptor-interacting protein 3
ROS	reactive oxygen species
SCLC	small-cell lung cancer
–SeH	selenol group
–SH	thiol groups
SOCS3	suppressor of cytokine signaling 3
Tet1	ten-eleven translocation methylcytosine dioxygenase 1
Tet2	ten-eleven translocation methylcytosine dioxygenase 2
TGF-β	transforming growth factor β
TNF-α	tumor necrosis factor α
TrxR	thioredoxin reductase
U	uranium
UO_2^{2+}	uranyl ions
VCAM	vascular cell adhesion molecules
WHO	World Health Organization

References

1 Haraguchi, H. (2017). Metallomics: the history over the last decade and a future outlook. *Metallomics* 9 (8): 1001–1013.
2 Harrington, C.F., Clough, R., Drennan-Harris, L.R. et al. (2011). Atomic spectrometry update. Elemental speciation. *J. Anal. At. Spectrom.* 26 (8): 1561–1595.
3 Li, Y., Sun, H., Chen, C., and Chai, Z. (2016). *Metallomics*. Beijing: Science Press.
4 Arruda, M.A.Z. (2018). *Metallomics: The Science of Biometals*, vol. 1055. Springer.
5 Makino, Y.; Ohara, S.; Yamada, M.; Mukoyama, S.; Hattori, K.; Sakata, S.; Tanaka, Y.; Suzuki, T.; Shinohara, A.; Matsukawa, T., Metallomics: recent analytical techniques and applications. 2017.
6 Dastoor, A., Angot, H., Bieser, J. et al. (2022). Arctic mercury cycling. *Nat. Rev. Earth Environ.* 3 (4): 270–286.
7 Amano, K. and Ntiri-Asiedu, A. (2020). Mercury emission from the aluminium industry: a review. *MOJ Ecol. Environ. Sci.* 5 (3): 129–135.
8 Gerson Jacqueline, R., Topp Simon, N., Vega Claudia, M. et al. (2020). Artificial lake expansion amplifies mercury pollution from gold mining. *Sci. Adv.* 6 (48): eabd4953.
9 Esdaile, L.J. and Chalker, J.M. (2018). The mercury problem in artisanal and small-scale gold mining. *Chemistry* 24 (27): 6905–6916.
10 Saalidong, B.M. and Aram, S.A. (2022). Mercury exposure in artisanal mining: assessing the effect of occupational activities on blood mercury levels among artisanal and small-scale goldminers in ghana. *Biol. Trace Elem. Res.* 200 (10): 4256–4266.
11 Tang, S., Feng, X., Qiu, J. et al. (2007). Mercury speciation and emissions from coal combustion in Guiyang, Southwest China. *Environ. Res.* 105 (2): 175–182.
12 Lee, S.-S. and Wilcox, J. (2017). Behavior of mercury emitted from the combustion of coal and dried sewage sludge: the effect of unburned carbon, Cl, Cu and Fe. *Fuel* 203: 749–756.
13 Zhang, L., Wang, S., Wu, Q. et al. (2012). Were mercury emission factors for Chinese non-ferrous metal smelters overestimated? Evidence from onsite measurements in six smelters. *Environ. Pollut.* 171: 109–117.
14 Ye, X., Hu, D., Wang, H. et al. (2015). Atmospheric mercury emissions from China's primary nonferrous metal (Zn, Pb and Cu) smelting during 1949–2010. *Atmos. Environ.* 103: 331–338.
15 Liu, Z., Wang, D., Peng, B. et al. (2017). Transport and transformation of mercury during wet flue gas cleaning process of nonferrous metal smelting. *Environ. Sci. Pollut. Res.* 24 (28): 22494–22502.
16 WHO Mercury and Health, (2017). https://www.who.int/news-room/fact-sheets/detail/mercury-and-health.
17 Harada, M. (1995). Minamata disease: methylmercury poisoning in Japan caused by environmental pollution. *Crit. Rev. Toxicol.* 25 (1): 1–24.
18 Crichton, R.R. (2012). Chapter 23 – Metals in the environment. In: *Biological Inorganic Chemistry*, 2e (ed. R.R. Crichton), 433–445. Oxford: Elsevier.

19 Minamata Convention on Mercury, (2013). https://www.mercuryconvention.org/en.
20 Snow, M.A., Darko, G., Gyamfi, O. et al. (2021). Characterization of inhalation exposure to gaseous elemental mercury during artisanal gold mining and e-waste recycling through combined stationary and personal passive sampling. *Environ. Sci. Processes Impacts* 23 (4): 569–579.
21 He, L., Liu, F., Zhao, J. et al. (2021). Temporal trends of urinary mercury in Chinese people from 1970s to 2010s: a review. *Ecotoxicol. Environ. Saf.* 208: 111460.
22 Cariccio, V.L., Samà, A., Bramanti, P., and Mazzon, E. (2019). Mercury involvement in neuronal damage and in neurodegenerative diseases. *Biol. Trace Elem. Res.* 187 (2): 341–356.
23 Aschner, M. and Aschner, J.L. (1990). Mercury neurotoxicity: mechanisms of blood-brain barrier transport. *Neurosci. Biobehav. Rev.* 14 (2): 169–176.
24 Genchi, G., Sinicropi, M.S., Carocci, A. et al. (2017). Mercury exposure and heart diseases. *Int. J. Environ. Res. Public Health* 14 (1): 74.
25 Clarkson, T. and Magos, L. (2006). The toxicology of mercury and its chemical compounds. *Crit. Rev. Toxicol.* 36: 609–662.
26 Zalups, R.K. (2000). Molecular interactions with mercury in the kidney. *Pharmacol. Rev.* 52 (1): 113.
27 Liu, M., Zhang, Q., Cheng, M. et al. (2019). Rice life cycle-based global mercury biotransport and human methylmercury exposure. *Nat. Commun.* 10 (1): 1–14.
28 Feng, X., Li, P., Qiu, G. et al. (2008). Human exposure to methylmercury through rice intake in mercury mining areas, Guizhou Province, China. *Environ. Sci. Technol.* 42 (1): 326–332.
29 Horvat, M., Nolde, N., Fajon, V. et al. (2003). Total mercury, methylmercury and selenium in mercury polluted areas in the province Guizhou, China. *Sci. Total Environ.* 304 (1-3): 231–256.
30 Carocci, A., Rovito, N., Sinicropi, M.S., and Genchi, G. (2014). Mercury toxicity and neurodegenerative effects. *Rev. Environ. Contam. Toxicol.* 229: 1–18.
31 Ajsuvakova, O.P., Tinkov, A.A., Aschner, M. et al. (2020). Sulfhydryl groups as targets of mercury toxicity. *Coord. Chem. Rev.* 417: 213343.
32 Bridges, C.C. and Zalups, R.K. (2017). Mechanisms involved in the transport of mercuric ions in target tissues. *Arch. Toxicol.* 91 (1): 63–81.
33 Gould, N.S., Evans, P., Martínez-Acedo, P. et al. (2015). Site-specific proteomic mapping identifies selectively modified regulatory cysteine residues in functionally distinct protein networks. *Chem. Biol.* 22 (7): 965–975.
34 Carvalho, C.M., Lu, J., Zhang, X. et al. (2011). Effects of selenite and chelating agents on mammalian thioredoxin reductase inhibited by mercury: implications for treatment of mercury poisoning. *FASEB J.* 25 (1): 370–381.
35 Picaud, T. and Desbois, A. (2006). Interaction of glutathione reductase with heavy metal: the binding of Hg (II) or Cd (II) to the reduced enzyme affects both the redox dithiol pair and the flavin. *Biochemistry* 45 (51): 15829–15837.
36 Burk, R.F. and Hill, K.E. (2010). Chapter 4.13 – Glutathione peroxidases. In: *Comprehensive Toxicology*, 2e (ed. C.A. McQueen), 229–242. Oxford: Elsevier.

37 Kanda, H., Toyama, T., Shinohara-Kanda, A. et al. (2012). S-Mercuration of rat sorbitol dehydrogenase by methylmercury causes its aggregation and the release of the zinc ion from the active site. *Arch. Toxicol.* 86 (11): 1693–1702.

38 Zichittella, A., Shi, H., and Argüello, J. (2000). Reactivity of cysteines in the transmembrane region of the Na, K-ATPase α subunit probed with Hg^{2+}. *J. Mem. Biol.* 177 (3): 187–197.

39 Kumagai, Y., Homma-Takeda, S., Shinyashiki, M., and Shimojo, N. (1997). Alterations in superoxide dismutase isozymes by methylmercury. *Appl. Organomet. Chem.* 11 (8): 635–643.

40 Xu, M., Yang, L., and Wang, Q. (2013). Chemical interactions of mercury species and some transition and noble metals towards metallothionein (Zn7MT-2) evaluated using SEC/ICP-MS, RP-HPLC/ESI-MS and MALDI-TOF-MS. *Metallomics* 5 (7): 855–860.

41 Boulanger, Y., Goodman, C.M., Forte, C.P. et al. (1983). Model for mammalian metallothionein structure. *Proc. Natl. Acad. Sci.* 80 (6): 1501–1505.

42 Serre, L., Rossy, E., Pebay-Peyroula, E. et al. (2004). Crystal structure of the oxidized form of the periplasmic mercury-binding protein MerP from *Ralstonia metallidurans* CH34. *J. Mol. Biol.* 339 (1): 161–171.

43 Alvarez, H.M., Xue, Y., Robinson, C.D. et al. (2010). Tetrathiomolybdate inhibits copper trafficking proteins through metal cluster formation. *Science* 327 (5963): 331–334.

44 Mah, V. and Jalilehvand, F. (2010). Glutathione complex formation with mercury(II) in aqueous solution at physiological pH. *Chem. Res. Toxicol.* 23 (11): 1815–1823.

45 Oram, P.D., Fang, X., Fernando, Q. et al. (1996). The formation constants of mercury(II)–glutathione complexes. *Chem. Res. Toxicol.* 9 (4): 709–712.

46 Xu, M., Yan, X., Xie, Q. et al. (2010). Dynamic labeling strategy with 204Hg-isotopic methylmercurithiosalicylate for absolute peptide and protein quantification. *Anal. Chem.* 82 (5): 1616–1620.

47 Woods, J.S. and Ellis, M.E. (1995). Up-regulation of glutathione synthesis in rat kidney by methyl mercury: relationship to mercury-induced oxidative stress. *Biochem. Pharmacol.* 50 (10): 1719–1724.

48 Kobal, A.B., Prezelj, M., Horvat, M. et al. (2008). Glutathione level after long-term occupational elemental mercury exposure. *Environ. Res.* 107 (1): 115–123.

49 Manceau, A., Bustamante, P., Haouz, A. et al. (2019). Mercury (II) binding to metallothionein in *Mytilus edulis* revealed by high energy-resolution XANES spectroscopy. *Chem. A Eur. J.* 25 (4): 997–1009.

50 Falnoga, I. and Tušek-Žnidarič, M. (2007). Selenium–mercury interactions in man and animals. *Biol. Trace Elem. Res.* 119 (3): 212–220.

51 Gerson, J.R., Walters, D.M., Eagles-Smith, C.A. et al. (2020). Do two wrongs make a right? Persistent uncertainties regarding environmental selenium–mercury interactions. *Environ. Sci. Technol.* 54 (15): 9228–9234.

52 Luque-Garcia, J.L., Cabezas-Sanchez, P., Anunciação, D.S., and Camara, C. (2013). Analytical and bioanalytical approaches to unravel the selenium–mercury antagonism: a review. *Anal. Chim. Acta* 801: 1–13.

53 Korbas, M., O'Donoghue, J.L., Watson, G.E. et al. (2010). The chemical nature of mercury in human brain following poisoning or environmental exposure. *ACS Chem. Neurosci.* 1 (12): 810–818.

54 Li, Y.-F., Dong, Z., Chen, C. et al. (2012). Organic selenium supplementation increases mercury excretion and decreases oxidative damage in long-term mercury-exposed residents from Wanshan, China. *Environ. Sci. Technol.* 46 (20): 11313–11318.

55 Lee, E.K., Shin, Y.-J., Park, E.Y. et al. (2017). Selenium-binding protein 1: a sensitive urinary biomarker to detect heavy metal-induced nephrotoxicity. *Arch. Toxicol.* 91 (4): 1635–1648.

56 Wall, A., Caprile, C., Franciosi, A. et al. (1986). Bonding and stability in narrow-gap ternary semiconductors for infrared applications. *J. Vac. Sci. Technol. A* 4 (4): 2010–2013.

57 Melnick, J.G., Yurkerwich, K., and Parkin, G. (2010). On the chalcogenophilicity of mercury: evidence for a strong Hg–Se bond in [TmBut] HgSePh and its relevance to the toxicity of mercury. *JACS* 132 (2): 647–655.

58 Xu, M., Yang, L.-M., and Wang, Q.-Q. (2015). Se-Hg dual-element labeling strategy for selectively recognizing selenoprotein and selenopeptide. *Chin. J. Anal. Chem.* 43 (9): 1265–1271.

59 Franco, J.L., Posser, T., Dunkley, P.R. et al. (2009). Methylmercury neurotoxicity is associated with inhibition of the antioxidant enzyme glutathione peroxidase. *Free Radical Biol. Med.* 47 (4): 449–457.

60 Glaser, V., Nazari, E.M., Müller, Y.M.R. et al. (2010). Effects of inorganic selenium administration in methylmercury-induced neurotoxicity in mouse cerebral cortex. *Int. J. Dev. Neurosci.* 28 (7): 631–637.

61 Branco, V., Canário, J., Holmgren, A., and Carvalho, C. (2011). Inhibition of the thioredoxin system in the brain and liver of zebra-seabreams exposed to waterborne methylmercury. *Toxicol. Appl. Pharmacol.* 251 (2): 95–103.

62 Chen, C., Yu, H., Zhao, J. et al. (2006). The roles of serum selenium and selenoproteins on mercury toxicity in environmental and occupational exposure. *Environ. Health Perspect.* 114 (2): 297–301.

63 Brookes, P.S., Yoon, Y., Robotham, J.L. et al. (2004). Calcium, ATP, and ROS: a mitochondrial love-hate triangle. *Am. J. Phys. Cell Physiol.* 287 (4): C817–C833.

64 Ryan, K.C., Ashkavand, Z., and Norman, K.R. (2020). The role of mitochondrial calcium homeostasis in Alzheimer's and related diseases. *Int. J. Mol. Sci.* 21 (23): 9153.

65 Tan, X.X., Tang, C., Castoldi, A.F. et al. (1993). Effects of inorganic and organic mercury on intracellular calcium levels in rat T lymphocytes. *J. Toxicol. Environ. Health* 38 (2): 159–170.

66 Sutton, D.J., Tchounwou, P.B., Ninashvili, N., and Shen, E. (2002). Mercury induces cytotoxicity and transcriptionally activates stress genes in human liver carcinoma (HepG2) cells. *Int. J. Mol. Sci.* 3 (9): 965–984.

67 Renu, K., Chakraborty, R., Myakala, H. et al. (2021). Molecular mechanism of heavy metals (lead, chromium, arsenic, mercury, nickel and cadmium) – induced hepatotoxicity – a review. *Chemosphere* 271: 129735.

68 Sutton, D.J. and Tchounwou, P.B. (2006). Mercury-induced externalization of phosphatidylserine and caspase 3 activation in human liver carcinoma (HepG2) cells. *Int. J. Environ. Res. Public Health* 3 (1): 38–42.

69 Fan, T.-J., Han, L.-H., Cong, R.-S., and Liang, J. (2005). Caspase family proteases and apoptosis. *Acta Biochim. Biophys. Sin.* 37 (11): 719–727.

70 Adams, D.H., Sonne, C., Basu, N. et al. (2010). Mercury contamination in spotted seatrout, *Cynoscion nebulosus*: an assessment of liver, kidney, blood, and nervous system health. *Sci. Total Environ.* 408 (23): 5808–5816.

71 Branco, V., Ramos, P., Canário, J. et al. (2012). Biomarkers of adverse response to mercury: histopathology versus thioredoxin reductase activity. *J. Biomed. Biotechnol.* 2012: 359879.

72 Chatterjee, S., Banerjee, P.P., Chattopadhyay, A., and Bhattacharya, S. (2013). Low concentration of $HgCl_2$ drives rat hepatocytes to autophagy/apoptosis/necroptosis in a time-dependent manner. *Toxicol. Environ. Chem.* 95 (7): 1192–1207.

73 Kuo, T.C. and Lin-Shiau, S.Y. (2004). Early acute necrosis and delayed apoptosis induced by methyl mercury in murine peritoneal neutrophils. *Basic Clin. Physiol. Pharmacol.* 94 (6): 274–281.

74 Ahn, H., Kim, J., Kang, S.G. et al. (2018). Mercury and arsenic attenuate canonical and non-canonical NLRP3 inflammasome activation. *Sci. Rep.* 8 (1): 13659–13659.

75 Zhang, Y., Lu, Y., Zhang, P. et al. (2022). Brain injury induced by mercury in common carp: novel insight from transcriptome analysis. *Biol. Trace Elem. Res.* 201: 403–411.

76 Ratnaike, R.N. (2003). Acute and chronic arsenic toxicity. *Postgrad. Med. J.* 79 (933): 391–396.

77 Rahman, M.M., Ng, J.C., and Naidu, R. (2009). Chronic exposure of arsenic via drinking water and its adverse health impacts on humans. *Environ. Geochem. Health* 31 (Suppl 1): 189–200.

78 Yu, S., Liao, W.T., Lee, C.H. et al. (2018). Immunological dysfunction in chronic arsenic exposure: from subclinical condition to skin cancer. *J. Dermatol. Sci.* 45 (11): 1271–1277.

79 Argos, M., Ahsan, H., and Graziano, J.H. (2012). Arsenic and human health: epidemiologic progress and public health implications. *Rev. Environ. Health* 27 (4): 191–195.

80 Blandin Knight, S., Crosbie, P.A., Balata, H. et al. (2017). Progress and prospects of early detection in lung cancer. *Open Biol. J.* 7 (9): 170070.

81 Bade, B.C. and Dela Cruz, C.S. (2020). Lung cancer 2020: epidemiology, etiology, and prevention. *Clin. Chest Med.* 41 (1): 1–24.

82 Shankar, A., Dubey, A., Saini, D. et al. (2019). Environmental and occupational determinants of lung cancer. *Transl. Lung Cancer Res.* 8 (Suppl 1): S31–S49.

83 Boffetta, P. and Nyberg, F. (2003). Contribution of environmental factors to cancer risk. *Br. Med. Bull.* 68: 71–94.

84 Celik, I., Gallicchio, L., Boyd, K. et al. (2008). Arsenic in drinking water and lung cancer: a systematic review. *Environ. Res.* 108 (1): 48–55.
85 Smith, A.H. and Smith, M.M.H. (2004). Arsenic drinking water regulations in developing countries with extensive exposure. *Toxicology* 198 (1-3): 39–44.
86 Ferreccio, C., González, C., Milosavjlevic, V. et al. (2000). Lung cancer and arsenic concentrations in drinking water in Chile. *Epidemiology* 673–679.
87 Smith, A.H., Goycolea, M., Haque, R., and Biggs, M.L. (1998). Marked increase in bladder and lung cancer mortality in a region of Northern Chile due to arsenic in drinking water. *Am. J. Epidemiol.* 147 (7): 660–669.
88 Ren, C., Zhou, Y., Liu, W., and Wang, Q. (2021). Paradoxical effects of arsenic in the lungs. *Environ. Health Preventative Med.* 26 (1): 80.
89 Wild, P., Bourgkard, E., and Paris, C. (2009). Lung cancer and exposure to metals: the epidemiological evidence. *Methods Mol. Biol.* 472: 139–167.
90 Minatel, B.C., Sage, A.P., Anderson, C. et al. (2018). Environmental arsenic exposure: from genetic susceptibility to pathogenesis. *Environ. Int.* 112: 183–197.
91 Hubaux, R., Becker-Santos, D.D., Enfield, K.S. et al. (2013). Molecular features in arsenic-induced lung tumors. *Mol. Cancer* 12: 20.
92 Hirano, S. (2020). Biotransformation of arsenic and toxicological implication of arsenic metabolites. *Arch. Toxicol.* 94 (8): 2587–2601.
93 Muenyi, C.S., Ljungman, M., and States, J.C. (2015). Arsenic disruption of DNA damage responses-potential role in carcinogenesis and chemotherapy. *Biomolecules* 5 (4): 2184–2193.
94 Ozturk, M., Metin, M., Altay, V. et al. (2022). Arsenic and human health: genotoxicity, epigenomic effects, and cancer signaling. *Biol. Trace Elem. Res.* 200 (3): 988–1001.
95 Ansari, J., Shackelford, R.E., and El-Osta, H. (2016). Epigenetics in non-small cell lung cancer: from basics to therapeutics. *Transl. Lung Cancer Res.* 5 (2): 155–171.
96 van Breda, S.G., Claessen, S.M., Lo, K. et al. (2015). Epigenetic mechanisms underlying arsenic-associated lung carcinogenesis. *Arch. Toxicol.* 89 (11): 1959–1969.
97 Wang, Z., Yang, P., Xie, J. et al. (2020). Arsenic and benzo[a]pyrene co-exposure acts synergistically in inducing cancer stem cell-like property and tumorigenesis by epigenetically down-regulating SOCS3 expression. *Environ. Int.* 137: 105560.
98 Gao, M., Qi, Z., Feng, W. et al. (2022). m6A demethylation of cytidine deaminase APOBEC3B mRNA orchestrates arsenic-induced mutagenesis. *J. Biol. Chem.* 298 (2): 101563.
99 Rafati Rahimzadeh, M., Rafati Rahimzadeh, M., Kazemi, S., and Moghadamnia, A.A. (2017). Cadmium toxicity and treatment: an update. *Casp. J. Int. Med.* 8 (3): 135–145.
100 Godt, J., Scheidig, F., Grosse-Siestrup, C. et al. (2006). The toxicity of cadmium and resulting hazards for human health. *J. Occup. Med. Toxicol.* 1: 22.
101 Byber, K., Lison, D., Verougstraete, V. et al. (2016). Cadmium or cadmium compounds and chronic kidney disease in workers and the general population: a systematic review. *Crit. Rev. Toxicol.* 46 (3): 191–240.

102 Rinaldi, M., Micali, A., Marini, H. et al. (2017). Cadmium, organ toxicity and therapeutic approaches: a review on brain, kidney and testis damage. *Curr. Med. Chem.* 24 (35): 3879–3893.

103 Uno, T., Kobayashi, E., Suwazono, Y. et al. (2005). Health effects of cadmium exposure in the general environment in Japan with special reference to the lower limit of the benchmark dose as the threshold level of urinary cadmium. *Scand. J. Work Environ. Health* 31 (4): 307–315.

104 Shi, Z., Taylor, A.W., Riley, M. et al. (2018). Association between dietary patterns, cadmium intake and chronic kidney disease among adults. *Clin. Nutr.* 37 (1): 276–284.

105 Huang, L., Liu, L., Zhang, T. et al. (2019). An interventional study of rice for reducing cadmium exposure in a Chinese industrial town. *Environ. Int.* 122: 301–309.

106 Klaassen, C.D., Liu, J., and Diwan, B.A. (2009). Metallothionein protection of cadmium toxicity. *Toxicol. Appl. Pharmacol.* 238 (3): 215–220.

107 Klaassen, C.D. and LIU, J. (1998). Metallothionein transgenic and knock-out mouse models in the study of cadmium toxicity. *J. Toxicol. Sci.* 23 (SupplementII): 97–102.

108 Rani, A., Kumar, A., Lal, A., and Pant, M. (2014). Cellular mechanisms of cadmium-induced toxicity: a review. *Int. J. Hyg. Environ. Health* 24 (4): 378–399.

109 Tokumoto, M., Lee, J.Y., and Satoh, M. (2019). Transcription factors and downstream genes in cadmium toxicity. *Biol. Pharm. Bull.* 42 (7): 1083–1088.

110 Giaginis, C., Gatzidou, E., and Theocharis, S. (2006). DNA repair systems as targets of cadmium toxicity. *Toxicol. Appl. Pharmacol.* 213 (3): 282–290.

111 Wang, B., Li, Y., Shao, C. et al. (2012). Cadmium and its epigenetic effects. *Curr. Med. Chem.* 19 (16): 2611–2620.

112 Vilahur, N., Vahter, M., and Broberg, K. (2015). The epigenetic effects of prenatal cadmium exposure. *Curr. Environ. Health Rep.* 2 (2): 195–203.

113 Martinez-Zamudio, R. and Ha, H.C. (2011). Environmental epigenetics in metal exposure. *Epigenetics* 6 (7): 820–827.

114 Hossain, M.B., Vahter, M., Concha, G., and Broberg, K. (2012). Low-level environmental cadmium exposure is associated with DNA hypomethylation in Argentinean women. *Environ. Health Perspect.* 120 (6): 879–884.

115 Kippler, M., Engstrom, K., Mlakar, S.J. et al. (2013). Sex-specific effects of early life cadmium exposure on DNA methylation and implications for birth weight. *Epigenetics* 8 (5): 494–503.

116 Boeke, C.E., Baccarelli, A., Kleinman, K.P. et al. (2012). Gestational intake of methyl donors and global LINE-1 DNA methylation in maternal and cord blood: prospective results from a folate-replete population. *Epigenetics* 7 (3): 253–260.

117 Bollati, V., Marinelli, B., Apostoli, P. et al. (2010). Exposure to metal-rich particulate matter modifies the expression of candidate microRNAs in peripheral blood leukocytes. *Environ. Health Perspect.* 118 (6): 763–768.

118 Moawad, A.M., Hassan, F.M., Sabry Abdelfattah, D., and Basyoni, H.A.M. (2021). Long non-coding RNA ENST00000414355 as a biomarker of cadmium exposure regulates DNA damage and apoptosis. *Toxicol. Ind. Health* 37 (12): 745–751.

119 Venza, M., Visalli, M., Biondo, C. et al. (2014). Epigenetic effects of cadmium in cancer: focus on melanoma. *Curr. Genomics* 15 (6): 420–435.

120 Arita, A. and Costa, M. (2009). Epigenetics in metal carcinogenesis: nickel, arsenic, chromium and cadmium. *Metallomics* 1 (3): 222–228.

121 Benbrahim-Tallaa, L., Waterland, R.A., Dill, A.L. et al. (2007). Tumor suppressor gene inactivation during cadmium-induced malignant transformation of human prostate cells correlates with overexpression of de novo DNA methyltransferase. *Environ. Health Perspect.* 115 (10): 1454–1459.

122 Hirao-Suzuki, M., Takeda, S., Sakai, G. et al. (2021). Cadmium-stimulated invasion of rat liver cells during malignant transformation: evidence of the involvement of oxidative stress/TET1-sensitive machinery. *Toxicology* 447: 152631.

123 Wallace, D.R., Taalab, Y.M., Heinze, S. et al. (2020). Toxic-metal-induced alteration in miRNA expression profile as a proposed mechanism for disease development. *Cells* 9 (4): 901.

124 Kim, J., Kim, D.Y., Heo, H.-R. et al. (2019). Role of miRNA-181a-2-3p in cadmium-induced inflammatory responses of human bronchial epithelial cells. *J. Thorac. Dis.* 11 (7): 3055–3069.

125 Zou, H., Wang, L., Zhao, J. et al. (2021). MiR-155 promotes cadmium-induced autophagy in rat hepatocytes by suppressing Rheb expression. *Ecotoxicol. Environ. Saf.* 227: 112895.

126 Huang, Q., Lu, Q., Chen, B. et al. (2017). LncRNA-MALAT1 as a novel biomarker of cadmium toxicity regulates cell proliferation and apoptosis. *Toxicol. Res.* 6 (3): 361–371.

127 Gao, M., Chen, M., Li, C. et al. (2018). Long non-coding RNA MT1DP shunts the cellular defense to cytotoxicity through crosstalk with MT1H and RhoC in cadmium stress. *Cell Discovery* 4: 5.

128 Gao, M., Li, C., Xu, M. et al. (2018). LncRNA MT1DP aggravates cadmium-induced oxidative stress by repressing the function of Nrf2 and is dependent on interaction with miR-365. *Adv. Sci.* 5 (7): 1800087.

129 Gao, M., Dong, Z., Sun, J. et al. (2020). Liver-derived exosome-laden lncRNA MT1DP aggravates cadmium-induced nephrotoxicity. *Environ. Pollut.* 258: 113717.

130 Xiao, C., Liu, Y., Xie, C. et al. (2015). Cadmium induces histone H3 lysine methylation by inhibiting histone demethylase activity. *Toxicol. Sci.* 145 (1): 80–89.

131 Li, L., Zhou, M., Chen, B. et al. (2021). ALKBH5 promotes cadmium-induced transformation of human bronchial epithelial cells by regulating PTEN expression in an m6A-dependent manner. *Ecotoxicol. Environ. Saf.* 224: 112686.

132 Steffanowski, J. and Banning, A. (2017). Uraniferous dolomite: a natural source of high groundwater uranium concentrations in northern Bavaria, Germany? *Environ. Earth Sci.* 76 (15): 1–11.

133 Nolan, J. and Weber, K.A. (2015). Natural uranium contamination in major US aquifers linked to nitrate. *Environ. Sci. Technol. Lett.* 2 (8): 215–220.

134 Coyte, R.M., Jain, R.C., Srivastava, S.K. et al. (2018). Large-scale uranium contamination of groundwater resources in India. *Environ. Sci. Technol. Lett.* 5 (6): 341–347.

135 Nriagu, J., Nam, D.-H., Ayanwola, T.A. et al. (2012). High levels of uranium in groundwater of Ulaanbaatar, Mongolia. *Sci. Total Environ.* 414: 722–726.

136 Godoy, J.M., Ferreira, P.R., Souza, E.M.D. et al. (2019). High uranium concentrations in the groundwater of the Rio de Janeiro State, Brazil, Mountainous Region. *J. Braz. Chem. Soc.* 30: 224–233.

137 Zhaobo, C., Fengmin, Z., Weidong, X., and Yuehui, C. (2000). Uranium provinces in China. *Acta Geol. Sin.* 74 (3): 587–594.

138 Wu, Y., Wang, Y., and Xie, X. (2014). Occurrence, behavior and distribution of high levels of uranium in shallow groundwater at Datong basin, northern China. *Sci. Total Environ.* 472: 809–817.

139 Wu, Y., Wang, Y., and Guo, W. (2019). Behavior and fate of geogenic uranium in a shallow groundwater system. *J. Contam. Hydrol.* 222: 41–55.

140 Ma, M., Wang, R., Xu, L. et al. (2020). Emerging health risks and underlying toxicological mechanisms of uranium contamination: lessons from the past two decades. *Environ. Int.* 145: 106107.

141 Ansoborlo, E., Lebaron-Jacobs, L., and Prat, O. (2015). Uranium in drinking-water: a unique case of guideline value increases and discrepancies between chemical and radiochemical guidelines. *Environ. Int.* 77: 1–4.

142 Ansoborlo, É., Amekraz, B., Moulin, C. et al. (2007). Review of actinide decorporation with chelating agents. *C.R. Chim.* 10 (10-11): 1010–1019.

143 Deng, B., Heyi, W., Jiang, S., and Ma, J. (2011). The biodistribution of uranium in mice. *Environ. Chem.* 30 (7): 1247–1252.

144 Arzuaga, X., Rieth, S.H., Bathija, A., and Cooper, G.S. (2010). Renal effects of exposure to natural and depleted uranium: a review of the epidemiologic and experimental data. *J. Toxicol. Environ. Health, Part B* 13 (7-8): 527–545.

145 Zamora, M.L., Tracy, B., Zielinski, J. et al. (1998). Chronic ingestion of uranium in drinking water: a study of kidney bioeffects in humans. *Toxicol. Sci.* 43 (1): 68–77.

146 Kurttio, P., Auvinen, A., Salonen, L. et al. (2002). Renal effects of uranium in drinking water. *Environ. Health Perspect.* 110 (4): 337–342.

147 Shelley, R., Kim, N.S., Parsons, P.J. et al. (2014). Uranium associations with kidney outcomes vary by urine concentration adjustment method. *J. Exposure Sci. Environ. Epidemiol.* 24 (1): 58–64.

148 Homma-Takeda, S., Kokubo, T., Terada, Y. et al. (2013). Uranium dynamics and developmental sensitivity in rat kidney. *J. Appl. Toxicol.* 33 (7): 685–694.

149 Yi, J., Yuan, Y., Zheng, J., and Hu, N. (2018). Hydrogen sulfide alleviates uranium-induced kidney cell apoptosis mediated by ER stress via 20S proteasome involving in Akt/GSK-3β/Fyn-Nrf2 signaling. *Free Radical Res.* 52 (9): 1020–1029.

150 Goldman, M., Yaari, A., Doshnitzki, Z. et al. (2006). Nephrotoxicity of uranyl acetate: effect on rat kidney brush border membrane vesicles. *Arch. Toxicol.* 80 (7): 387–393.

151 Kobayashi, S., Nagase, M., Honda, N., and Hishida, A. (1984). Glomerular alterations in uranyl acetate-induced acute renal failure in rabbits. *Kidney Int.* 26 (6): 808–815.

152 Banday, A.A., Priyamvada, S., Farooq, N. et al. (2008). Effect of uranyl nitrate on enzymes of carbohydrate metabolism and brush border membrane in different kidney tissues. *Food Chem. Toxicol.* 46 (6): 2080–2088.

153 Dublineau, I., Souidi, M., Gueguen, Y. et al. (2014). Unexpected lack of deleterious effects of uranium on physiological systems following a chronic oral intake in adult rat. *Biomed Res. Int.* 2014: 181989.

154 Shaki, F., Hosseini, M.-J., Ghazi-Khansari, M., and Pourahmad, J. (2012). Toxicity of depleted uranium on isolated rat kidney mitochondria. *Biochim. Biophys. Acta (Bba)* 1820 (12): 1940–1950.

155 Shaki, F. and Pourahmad, J. (2013). Mitochondrial toxicity of depleted uranium: protection by beta-glucan. *Iran. J. Pharm. Res.* 12 (1): 131.

156 Hao, Y., Ren, J., Liu, C. et al. (2014). Zinc protects human kidney cells from depleted uranium-induced apoptosis. *Basic Clin. Physiol. Pharmacol.* 114 (3): 271–280.

157 Thiébault, C., Carriere, M., Milgram, S. et al. (2007). Uranium induces apoptosis and is genotoxic to normal rat kidney (NRK-52E) proximal cells. *Toxicol. Sci.* 98 (2): 479–487.

158 Hu, Q., Zheng, J., Xu, X.N. et al. (2022). Uranium induces kidney cells apoptosis via reactive oxygen species generation, endoplasmic reticulum stress and inhibition of PI3K/AKT/mTOR signaling in culture. *Environ. Toxicol.* 37 (4): 899–909.

159 Asic, A., Kurtovic-Kozaric, A., Besic, L. et al. (2017). Chemical toxicity and radioactivity of depleted uranium: the evidence from in vivo and in vitro studies. *Environ. Res.* 156: 665–673.

160 Yellowhair, M., Romanotto, M.R., Stearns, D.M., and Lantz, R.C. (2018). Uranyl acetate induced DNA single strand breaks and AP sites in Chinese hamster ovary cells. *Toxicol. Appl. Pharmacol.* 349: 29–38.

161 Yazzie, M., Gamble, S.L., Civitello, E.R., and Stearns, D.M. (2003). Uranyl acetate causes DNA single strand breaks in vitro in the presence of ascorbate (vitamin C). *Chem. Res. Toxicol.* 16 (4): 524–530.

162 Cooper, K.L., Dashner, E.J., Tsosie, R. et al. (2016). Inhibition of poly (ADP-ribose) polymerase-1 and DNA repair by uranium. *Toxicol. Appl. Pharmacol.* 291: 13–20.

163 Lin, Y.-W. (2020). Uranyl binding to proteins and structural-functional impacts. *Biomolecules* 10 (3): 457.

164 Basset, C., Averseng, O., Ferron, P.-J. et al. (2013). Revision of the biodistribution of uranyl in serum: is fetuin-A the major protein target? *Chem. Res. Toxicol.* 26 (5): 645–653.

165 Frelon, S., Guipaud, O., Mounicou, S. et al. (2009). In vivo screening of proteins likely to bind uranium in exposed rat kidney. *Radiochim. Acta* 97 (7): 367–373.

166 Prat, O., Berenguer, F., Malard, V. et al. (2005). Transcriptomic and proteomic responses of human renal HEK293 cells to uranium toxicity. *Proteomics* 5 (1): 297–306.

167 Dedieu, A., Bérenguer, F., Basset, C. et al. (2009). Identification of uranyl binding proteins from human kidney-2 cell extracts by immobilized uranyl affinity chromatography and mass spectrometry. *J. Chromatogr. A* 1216 (28): 5365–5376.

168 Vidaud, C., Gourion-Arsiquaud, S., Rollin-Genetet, F. et al. (2007). Structural consequences of binding of $UO2^{2+}$ to apotransferrin: can this protein account for entry of uranium into human cells? *Biochemistry* 46 (8): 2215–2226.

169 Xu, M., Frelon, S., Simon, O. et al. (2014). Development of a non-denaturing 2D gel electrophoresis protocol for screening in vivo uranium-protein targets in *Procambarus clarkii* with laser ablation ICP MS followed by protein identification by HPLC–Orbitrap MS. *Talanta* 128: 187–195.

170 Donnadieu-Claraz, M., Bonnehorgne, M., Dhieux, B. et al. (2007). Chronic exposure to uranium leads to iron accumulation in rat kidney cells. *Radiat. Res.* 167 (4): 454–464.

171 Zheng, J., Zhao, T., Yuan, Y. et al. (2015). Hydrogen sulfide (H2S) attenuates uranium-induced acute nephrotoxicity through oxidative stress and inflammatory response via Nrf2-NF-κB pathways. *Chem. Biol. Interact.* 242: 353–362.

172 Taulan, M., Paquet, F., Argiles, A. et al. (2006). Comprehensive analysis of the renal transcriptional response to acute uranyl nitrate exposure. *BMC Genomics* 7 (1): 1–14.

173 Bontemps, A., Conquet, L., Elie, C. et al. (2019). In vivo comparison of the phenotypic aspects and molecular mechanisms of two nephrotoxic agents, sodium fluoride and uranyl nitrate. *Int. J. Environ. Res. Public Health* 16 (7): 1136.

174 PD, R. and Arun, A.B. (2019). Role of PI3K-Akt and MAPK signaling in uranyl nitrate-induced nephrotoxicity. *Biol. Trace Elem. Res.* 189 (2): 405–411.

175 Bontemps-Karcher, A., Magneron, V., Conquet, L. et al. (2021). Renal adaptive response to exposure to low doses of uranyl nitrate and sodium fluoride in mice. *J. Trace Elem. Med. Biol.* 64: 126708.

176 Hao, Y., Huang, J., Ran, Y. et al. (2021). Ethylmalonic encephalopathy 1 initiates overactive autophagy in depleted uranium-induced cytotoxicity in the human embryonic kidney 293 cells. *J. Biochem. Mol. Toxicol.* 35 (3): e22669.

177 Zheng, J., Hu, Q., Zou, X. et al. (2022). Uranium induces kidney cells pyroptosis in culture involved in ROS/NLRP3/caspase-1 signaling. *Free Radical Res.* 1–13.

10

Pathometallomics: Taking Neurodegenerative Disease as an Example

Xiubo Du[1,2], Xuexia Li[1,2], and Qiong Liu[1,2]

[1] Shenzhen Key Laboratory of Microbial Genetic Engineering, Shenzhen Key Laboratory of Marine Biotechnology and Ecology, College of Life Sciences and Oceanography, Shenzhen University, Xueyuan Avenue, Nanshan District, Shenzhen 518060, China
[2] Shenzhen-Hong Kong Institute of Brain Science-Shenzhen Fundamental Research Institutions, Xueyuan Avenue, Nanshan District, Shenzhen, 518060, China

10.1 Introduction to Pathometallomics

10.1.1 The Concept and Scope of Pathometallomics

Metallomics and metalloproteomics have emerged in recent years to study metals and metalloids within an organ or tissue or cell. As an integrated biometal science in symbiosis with genomics and proteomics [1], metallomics is designated to elucidate the interactions and functional connection of metals (or metalloids) with proteins and other biomolecules [2]. Metalloproteomics is a subdiscipline of metallomics, which aims to systematically identify large sets of metal-associated proteins (including metalloproteins and metal-binding proteins) and to analyze their regulation, modification, interaction, structural assembly, and function, as well as their involvement in disease states and physiological processes [3]. Pathometallomics is a concept derived from metallomics and metalloproteomics, designated for the study of metals (or metalloids) and their associated proteins under pathological conditions. The scope of pathometallomics encompasses all-round changes of metal and metalloid species present in diseased cells or tissues, including dysregulation of metal homeostasis, disturbed interaction of metals with biomolecules, and dysfunction of metal-associated proteins in disease states.

Dysregulation of metal homeostasis in biological processes has a critical impact on a wide range of diseases such as neurodegenerative diseases. Revealing metal dyshomeostasis and its molecular mechanism under pathological conditions is essential to regulate the disturbed biological systems and to develop drugs for potential treatment of the disease. Neurodegenerative diseases are characterized by the progressive degeneration of neurons. Increasing evidence shows that the disruption of metal homeostasis is associated with a number of neurodegenerative diseases including Alzheimer's disease (AD), Parkinson's disease (PD), Huntington's disease

Applied Metallomics: From Life Sciences to Environmental Sciences, First Edition.
Edited by Yu-Feng Li and Hongzhe Sun.
© 2024 WILEY-VCH GmbH. Published 2024 by WILEY-VCH GmbH.

(HD), amyotrophic lateral sclerosis (ALS), and autism spectrum disorder (ASD) [4]. A focus of pathometallomics in neurodegenerative diseases is to identify novel proteins (or other biomolecules) and pathways contributing to metal dyshomeostasis in brain.

In the past decades, there has been an increasing interest in the application of pathometallomics to find out dysregulation of metal homeostasis and its impacts on the fields of biology and medicine. Comprehensive reviews can be found in metallomics and metalloproteomics for disease processes [5] and clinical research [6] in neurodegeneration [7], which belong to the scope of pathometallomics. Meanwhile, the use of metallomics approaches has also been reported recently to unveil the action mechanism of potential drugs, including metallodrugs [8] and metal-associated drugs, for neurodegenerative diseases [9]. In this chapter, we introduce a new concept of pathometallomics and an outline of methodologies, focus on its major development and application in neurodegenerative diseases. Important research findings are highlighted in neuronal pathometallomics that are derived from dysregulation of metal homeostasis in central nervous system (CNS), dysfunction of metals-associated biomolecules in diseased processes, and action mechanism of metal-associated drugs. Finally, the challenges and perspectives are discussed for pathometallomics in brain health and diseases.

10.1.2 Brief Introduction to Methodologies for Pathometallomics

As an emerging interdisciplinary area of research, the methodologies in pathometallomics follow the steps of metallomics and metalloproteomics that have received growing attention for the study of neurodegenerative diseases. Here, we summarize the methodologies for pathometallomics, mainly introduce the technologies advance rapidly in recent years, including inductively coupled plasma mass spectrometry (ICP-MS), and X-ray fluorescence (XRF) and absorption spectroscopy. These advanced methodologies could be utilized rapidly to expand our knowledge of how metals and metalloids function in diseased brains and how metal-associated drugs play the roles in neurodegenerative diseases.

ICP-MS is one of the most frequently used techniques in pathometallomics due to its simultaneous and accurate determination of metals and non-metals in biological samples [10]. The detection limits of ICP-MS for multiple metals or metalloids can reach very low concentration (such as part per trillion), which satisfy the requirement for measuring trace elements in biological samples, especially those present in brain regions [8c, e, 11]. ICP-MS has also been successfully applied to large-scale pathometallomics studies of human samples (e.g. blood, urine, hair), which demonstrate the power of this technique to investigate metabolism and status of trace metals and metalloids under disease conditions. In addition, ICP-MS can be modified to implement multiple functions such as species analysis, in situ detection, and isotope analysis of biological trace elements.

Metalloproteins or metal/metalloid-binding proteins can be detected by ICP-MS-based hyphenated techniques, such as HPLC-ICP-MS, LC-ICP-MS, CE-ICP-MS, and GE-ICP-MS. High-performance liquid chromatography (HPLC), liquid chromatography (LC), or capillary electrophoresis (CE) and gel electrophoresis (GE) are used to separate metalloproteins or metal/metalloid-binding proteins from other biomolecules. HPLC-ICP-MS has a wide range of applications in quantitatively characterizing different species and valency of metal or metalloid compounds. LC-ICP-MS can be used for metalloproteomics study. However, the relatively low separating resolution of LC requires multidimensional chromatographic steps [12]. CE-ICP-MS is of limited use in analysis of metalloproteins in real biological samples owing to its insufficient detection limits [13]. Using 2D non-denaturing polyacrylamide gel electrophoresis (PAGE), GE-ICP-MS is successfully applied to separate and analyze the proteomes of Fe-rich *Ferroplasma acidiphilum* [14]. Recently, column-type continuous-flow GE-ICP-MS has been developed for the analysis of metalloproteomics [15].

Laser ablation coupled with ICP-MS (LA-ICP-MS) is a powerful technique for in situ analysis and imaging multiple elements in solid-phase biological samples (e.g. brain slides, or cells grown on glass slides), which offers high spatial resolution, excellent sensitivity, and feasibility for quantitative imaging. It has been used to determine the elements distribution (mapping or imaging) in thin-tissue sections of brain [16]. Neurological studies focusing on monitoring the changes in concentration of Mn, Fe, Cu, and Zn occurring in PD rats were recently performed [17]. Furthermore, a method for in vivo droplet-based sample collection combined with a specifically designed analytical platform was developed for sample analysis. The time-resolved brain-fluid samples are analyzed using LA-ICP-MS [18].

Different isotopes of a metal or metalloid in biological samples can be identified and quantitatively measured by multicollector-ICP-MS (MC-ICP-MS) [19]. The natural abundance of heavy and light stable isotopes varies between tissues and metabolites due to isotopic effects in biological processes, that is, isotope discriminations occur between heavy and light isotopic forms during the processes involved in enzyme or transporter activity [20]. The metabolic deregulation associated with many diseases leads to alterations in metabolic fluxes, resulting in changes in isotope abundance that can be identified easily with current isotope ratio technologies such as MC-ICP-MS. Currently, application of MC-ICP-MS for the early detection of AD is still at an initial stage.

Immobilized metal affinity chromatography (IMAC) is another widely used and powerful technique for separation and enrichment of metalloproteins on the basis of differential binding affinities for immobilized metals (e.g. Cu(II), Zn(II), As(III), Co(II), Ni(II), and Bi(III)) on a solid-phase support [21]. When IMAC is combined with protein identification techniques (e.g. MALDI-MS), extensive metalloproteins, even those in low abundance, can be profiled and identified [22].

With the advances of nuclear analytical techniques, the utilization of synchrotron radiation for high-resolution elemental mapping in tissues, cellular compartments, or even individual cells has received growing interest [23]. Synchrotron-radiation-based XRF uses characteristic X-rays emitted from an atom that has been excited by the absorption of high-energy X-rays or gamma rays for elemental analysis [24]. Recently, micro-XRF and nano-XRF have been successfully used as a rapid screening and imaging technique for elemental quantification in biological samples [25].

X-ray absorption spectroscopy (XAS) is a valuable tool for probing the changes in metal centers' chemical environment, such as metal oxidation states in cells, the coordination motif of the probed metal, and the identity and number of adjacent atoms. As a noninvasive technique, XAS has been successfully used to monitor the biotransformation of metallodrugs in biological fluids [26]. In addition, a new 3D cryo correlative and quantitative methodology was developed through combining cryo–soft X-ray tomography and cryo-XRF tomography [27]. This original correlative and quantitative method can be readily extended to track different elements involved in biochemical processes in different organelles of human cells.

Recently, a direct, label-free nanoscale visualization of neuromelanin and associated metal ions in human brain tissue has been achieved using synchrotron scanning transmission X-ray microscopy (STXM), which illustrates the wider potential of STXM as a label-free spectromicroscopy technique applicable to biological samples [28].

10.2 Application of Pathometallomics in Neurodegenerative Diseases

10.2.1 Pathometallomics in Alzheimer's Disease

AD is a neurodegenerative disorder that mainly affects the elderly. In 2021, among people over 65 years old in the United States, the incidence of AD is as high as 11.5%, which is 1.5% higher than that in 2020. According to the *Chinese Alzheimer's Disease Report 2021*, the number of AD and other dementia patients reached 1.32 million, with the prevalence rate of 924.1/100 000, and the mortality rate of 22.5/100 000, which are all slightly higher than the global average levels. The cost of AD treatment brings a huge economic burden to society and families. As estimated, by 2030, the global cost of AD treatment and care will reach 2 trillion US dollars. The development of therapeutic drugs for AD has always been a global research hotspot. In recent decades, large-scale preclinical screening of new anti-AD drugs has been carried out worldwide, but the failure rate is as high as 99%. Before the approved of GV-971 by CFDA in 2019 and Biogen's amyloid-β (Aβ) monoclonal antibody drug Aducanumab by FDA in 2021, only five drugs were available for clinic use. The efficacy of these drugs is limited to the short-term symptomatic treatment of AD, and they are completely ineffective for the five-year course of AD. The clinical symptom of AD is the progressive loss of memory and cognitive functions, and the pathological hallmarks are the amyloid plaques formed by the extracellular aggregation of Aβ and neurofibrillary tangles (NFTs) caused by the intracellular accumulation of hyperphosphorylated tau proteins [29]. In addition, oxidative stress, inflammation, metal dyshomeostasis, mitochondrial disorder, etc. all participate in the development of AD.

10.2.1.1 Dysregulation of Metal Homeostasis in AD

Extensive studies have demonstrated that altered metal homeostasis in the brain is associated with the onset of AD and its pathological changes. Elevated concentrations of metal ions were found in Aβ plaques. In addition, some other pathological features of AD, including NFTs formation, neuronal loss, and reduced synaptic plasticity, are also associated with metal ions. Metal ions such as iron, copper, and zinc are involved in the regulation of synaptic activity and the biological function of metalloproteins in the brain, while calcium is a ubiquitous second messenger. Therefore, the homeostasis of metal ions is essential to maintain the normal physiological function of the brain, while its dyshomeostasis has been considered as one of the critical reasons causing AD.

Both elevated and reduced and as well as mis-localization of metal ions and some metalloproteins have been reported in AD cases. Table 10.1 summarizes the published results of various metal changes in AD. In 2011, Schrag et al. conducted a meta-analysis of the published data and demonstrated that there was no significant difference for the total Cu or Fe levels between AD brains and age-matched cohorts [57]. However, tremendous studies through ICP-MS, proton-induced X-ray emission, immunohistochemistry, and SR-µXRF have shown the quite higher levels of Cu, Fe, and Zn in senile plaques from AD patients and mouse models. Systemic Cu levels in both cerebrospinal fluid (CSF) and serum are elevated in AD patients [58], while systemic Zn is depleted in AD patients [59]. Although the brain is the main affected tissue of AD and a large body of studies demonstrated the close relationship between metal dyshomeostasis and AD pathogenesis, there are relatively few ionomic studies based on human brain tissues related with AD due to the limited availability of human brains. Furthermore, results of studies comparing brain metal levels between AD patients and the controls are somewhat heterogeneous [57], probably due to the experimental variabilities in sample processing and the intrinsic difference in the distribution of metals in various brain regions. For instance, Fe, Cu, and Zn were found hyper-accumulated in senile plaques from AD brains [60], with an increase of 2.8, 5.7, and 3.1 times, respectively; however, measurement of their levels in the whole-brain tissue has yielded inconsistent data [61]. In 2017, Xu et al. measured post mortem levels of Na, Mg, K, Ca, Mn, Fe, Cu, Zn, and Se in seven brain regions from 9 AD cases and 13 controls by ICP-MS. In the control group, metal levels significantly varied among different brain regions. AD-associated perturbations in most metals occurred in only a few: brain areas more severely influenced by degeneration commonly showed changes in more metals. Among all the elements studied, only copper levels were substantively decreased in all AD-brain regions, to 1/2–2/3 of the control values. These observations suggest that widespread brain-Cu deficiency may play critical roles in the pathogenesis of AD [45]. In 2020, the same research group evaluated the effects of key and potentially confounding variables including age, sex-matching, post mortem delay (PMD), and neuropathological stage on the metal concentrations in post mortem brains and found the brain metal levels were not affected by the above factors [62]. It should be noted that the regions of interest for study of brain metal metabolism should be selected carefully, since metal levels varied substantially among regions in both human and rat brains [45, 62].

Table 10.1 Pathometallomics in Alzheimer's disease.

Methods	Model	Tissue	Number of subjects (disease/control)	Elevated metals	Unchanged metals	Decreased metals	References
ICP-MS	Patients	Hippocampus	12/12	Al	Sn	Zn and Se	[30]
ICP-MS	Patients	Brain	11/10	—	Cd	Zn	[31]
ICP-MS	Patients	Brain	4/4	Sn, Al, and Mn in the parietal cortex	Cu and Ca in the parietal cortex; Al in the cerebellum	—	[32]
ICP-MS	Patients	Plasma and CSF	173/54	Mn and Hg in plasma	—	V, Mn, Rb, Sb, Cs, and Pb in CSF	[33]
ICP-MS	Mild to moderate AD patients, MCI[a] patients, and healthy controls	Serum	18/19/16	—	Zn among any of the three group regardless of gender	Zn in men with MCI	[34]
ICP-MS	Patients	CSF	21 (14 late-onset AD and 7 early-onset AD)/15	Cu and Zn in late-onset AD	Fe, Mg, and Mn in late-onset AD	—	[35]
ICP-MS	Patients	Blood and serum	80/130	Cd in blood; Pb and Hg in serum	—	Pb and Hg in blood	[36]
ICP-MS	3xTg-AD mice	Brains and cerebella	5/5	Al, Co, Ca, Fe, Cu, Na, and Mg in brains; Na, Ca, and Zn in cerebella	Mn in brain and cerebella; Al, Se, Fe, and Cu in cerebella	Li in brain; Cr in cerebella	[37]

ICP-MS	AD patients, MCI patients and healthy controls	Serum	30/16/30	Al in the different fractions (TOTAL, HMM[b] and LMM[c]) in MCI and AD); Fe in LMM in MCI and AD	Li, V, Cr, Co, Cu, Mo, Cd, and Pb in the different fractions (TOTAL, HMM, and LMM)	Mn in the different fractions (TOTAL and HMM) in MCI and AD	[38]
ICP-MS	Patients	Blood	15/10	—	Hg	Cu, Zn, Se, and Pb	[39]
ICP-MS	Moderate AD, severe AD, and healthy controls	Brain	15/30/28	Na in severe AD	Na in moderate AD K in severe AD	K in moderate AD	[40]
ICP-MS	Patients	Hair and serum	45/33	Cu and Mn in hair	Zn, Se, Cu, Fe, and Mg in serum; Fe and Mg in hair	Se and Zn in hair; Mn in serum	[41]
LA-ICP-MS	Patients	White and gray matter	4/5	Fe in gray matter	—	—	[42]
ICP-MS	Patients	Hair and nails	62/60	Na and K in hair	Na, Mg, Al, K, Ca, and Co in nail; Mg and Zn in hair	Mn, Fe, Cu, Zn, Cd, and Hg in nail Al, Ca, Mn, Fe, Co, Cu, Cd, Hg, and Pb in hair	[43]
ICP-MS	Patients	Plasma	92/161	Al, Cu, and Fe	Mg, Ca, Ti, V, Cr, Co, Ba, Pb, Cd, Se, Mo, and Sr	Mn, Li, and Zn	[44]
ICP-MS	Patients	Hippocampus and entorhinal cortex	9/13	Na, Mn, and Fe in Hippocampus; Na, Mg, Ca, Fe, and Zn in entorhinal cortex	Mg, K, Ca, Se, and Zn in Hippocampus; K, Mn, and Se in entorhinal cortex	Cu in Hippocampus and in entorhinal cortex	[45]

(Continued)

Table 10.1 (Continued)

Methods	Model	Tissue	Number of subjects (disease/control)	Elevated metals	Unchanged metals	Decreased metals	References
ICP-MS	Patients	Middle temporal gyrus and sensory cortex	9/13	Na, Mg, Fe, and Zn in middle temporal gyrus	K, Ca, Mn, and Se in middle temporal gyrus; Na, Mg, K, Ca, Mn, Fe, Zn, and Se in sensory cortex	Cu in middle temporal gyrus and sensory cortex	[45]
ICP-MS	Patients	Motor cortex and cingulate gyrus	9/13	—	Na, Mg, K, Ca, Mn, Fe, and Zn in motor cortex; Na, Ca, Mn, Fe, and Zn in cingulate gyrus	Cu and Se in motor cortex; Mg, K, Cu, and Se in cingulate gyrus	[45]
ICP-MS	Patients	Cerebellum	9/13	—	Na, Ca, and Fe in cerebellum	Mg, K, Mn, Cu, Zn, and Se in cerebellum	[45]
ICP-MS	AD patients and elderly	Erythrocytes	40/32	Fe and Cu	Se	—	[46]
LA-ICP-MS	Patients	Brain	4/4	Fe	—	—	[47]
ICP-MS	5xFAD and L66 mice	Brain	20/20 and 26/14	Fe in L66 mice; Zn in 5xFAD mice	Cu and Zn in L66 mice	Cu in 5xFAD mice	[48]
ICP-MS	Patients	Blood and urine	53/217	As in urine; Cr in blood	—	Se in blood	[49]

Technique	Subject	Sample	n	Findings			Ref.
ICP-MS	Patients	Plasma	42/43	Zn in male patients	Na, Mg K, Fe, Cu, Se, and Ca in male, female or all patients; Zn in female patients	—	[50]
ICP-MS	5xFAD mice	Hippocampus	6/6	Fe	—	—	[51]
ETAAS[d]	Patients	Brain	186/53	Al	—	—	[52]
TXRF[e]	Patients	Plasma	44/44	Ca and Zn	Cu and Se	Fe	[53]
XANES[f] and WDXRFS[g]	Patients	Hair	15/15	Ca	Mg	—	[54]
XRF and XAS	PSAPP, 5xFAD, CVN and Control mice	Brain	6/4/2/2	Fe in the cortex of the three AD mouse	—	—	[55]
XRF	B6C3-Tg(APPswe, PSEN1dE9)85 Dbo/J (PSAPP) mice	Brain	22/26	Fe in the cortex in 24 week-mice; Zn in the cortex in 56 week-mice	Cu in cortex; Zn and Cu in hippocampus; Fe in the hippocampus in 40 week-mice	Fe in the hippocampus in 24 week-mice and in 56 week-mice	[56]

a) MCI: mild cognitive impairment.
b) HMM: high molecular mass.
c) LMM: low molecular mass.
d) ETAAS: electrothermal atomic absorption spectrophotometry.
e) TXRF: total reflection X-ray fluorescence.
f) XANES: X-ray absorption near-edge spectroscopy.
g) WDXRFS: wavelength-dispersive X-ray fluorescence spectrometry.

Though the human-brain-tissue-based ionomic studies are relatively rare, a large number of ionomic studies have been performed in animal brains. In 2012, Wang et al. investigated the distribution profile and oxidation states of metals in APP/V717I transgenic mouse model and age-matched control by SR-µXRF [63]. Transition metals including Cu, Fe, and Zn were significantly elevated in AD mice and specifically enriched in the cortex and hippocampus, the major regions that senile plaques deposited. Fe and Ca were found to be increased with age in both AD and control mice, although they appeared significantly elevated in AD mice than control ones. By XANES analysis, Cu not Fe seemed to have redox properties in the AD brain. The authors further studied the co-localization of Aβ_{42} with metals Cu, Zn in the brains of the same APP transgenic mice by using the immunogold labeling technique combined with SR-µXRF, demonstrating the strong association of Cu and Zn with Aβ_{42} during plaque formation [64].

10.2.1.2 Metal-Associated Dysfunction in AD

A large amount of evidence shows that metal ions, including copper, iron, zinc, calcium, magnesium, and aluminum, are closely related to the pathogenesis of AD. First, erroneous deposition of metal ions (Zn^{2+}, Cu^{2+}, Fe^{3+}, etc.) or increased Ca^{2+} in different brain regions promotes the overproduction and aggregation of Aβ. Van et al. found the cortical iron reflects the severity of AD [65]. Elevated iron is associated with increased Aβ plaques amounts and it aggravates the neurotoxicity of Aβ by impeding its ordered aggregation [66]. Bush et al. reported that Zn^{2+} not only affects the processing of APP to generate Aβ [67], but also directly combines with Aβ and promotes the aggregation of Aβ to form senile plaques [68]. In addition to Zn^{2+}, some other metal ions, such as Cu^{+}, Cu^{2+}, and Al^{3+} can also bind to Aβ and induce the generation of Aβ oligomers, the known most toxic species of Aβ aggregates [69]. Secondly, metal ions promote tau phosphorylation and the formation of NFT. Studies have shown that Zn^{2+}, Cu^{2+}, Fe^{3+}, Ca^{2+}, and Al^{3+} all can promote the hyperphosphorylation of tau protein. Among them, Cu^{2+} directly combines with tau and NFT and aggravates the hyperphosphorylation of tau protein through modulation the activity of CDK5 and GSK-3β [70]. Zinc can directly bind to tau monomers and stimulate the phosphorylation of tau by the activation of GSK3β, ERK1/2, and c-Jun N-terminal kinase (JNK) [71]. Additionally, zinc also inactivates protein phosphatase 2A (PP2A), the main enzyme that catalyzes the dephosphorylation of phosphorylated-tau (p-tau), through Src-dependent pathway [72]. Thirdly, metal ions promote the generation of ROS and oxidative stress. Metal ions (Cu, Fe, Mn) induce the ROS production either by direct reaction with oxygen or by binding with Aβ or tau [73]. With the overload of iron or copper, oxidative stress occurs via Haber–Weiss and Fenton reaction that directly produce ROS. Both iron and copper can bind with APP and Aβ, produce ROS, including hydroxyl free radical, leading to mitochondrial dysfunction and cell death [74]. Fourthly, metal ions alter the synaptic plasticity of neural cells. For instance, the zinc transporter ZNT1, which is responsible for exporting Zn^{2+} from neurons, was recently identified as a novel postsynaptic protein and directly affected the synaptic plasticity of neural cells [75]. And the level changes of Zn^{2+}, Cu^{2+}, Fe^{3+}, Al^{3+}, Mg^{2+}, Ca^{2+}, or Al^{3+} all have been shown to impair synaptic plasticity, including both long-term potentiation (LTP) and long-term depression (LTD). Studies have found Ca^{2+}, Zn^{2+}, Mg^{2+}, and Cu^{2+} affected LTP mainly through the CAMKII/BDNF pathway, while Al^{3+} destroyed LTP through the ERK1/2/ARC pathway [76]. Fifthly, metal ions increase neurotoxicity. In addition to inducing ROS/NOS and Aβ aggregation which damage neurons, metal ions such as Zn^{2+} also trigger the expression of endoplasmic reticulum stress genes, including GADD34, GADD45, and p8, while Cu^{2+} and Fe^{3+} induced neurotoxicity through TNF-α pathway and Akt/Nrf2 pathway, respectively. Finally, metal ions accelerate the development of AD by over-activating autophagy and apoptosis. Although it has been reported that autophagy is inhibited in AD, over-activated autophagy can cause cell damage. For example, Ca^{2+} and Zn^{2+} were found to activate autophagy and apoptosis by inhibiting the PI3K/Akt/mTOR pathway. Elevation of Cu^{2+} promotes autophagy and apoptosis of glioma cells via ROS and JNK activation [77]. Studies have shown that iron exposure in the neonatal period increased hippocampal levels of ubiquitinated proteins and induced memory impairment [78].

10.2.1.3 Application of Metallomics in the Prognosis of AD

Changes of metal ions in the plasma and CSF were considered to be applied in the predication of AD and the progression of dementia. In 2020, a pilot study performed in Portugal revealed that higher levels of selenium (Se) and nickel (Ni) are related with lower cognitive decline in elders [79]. Consistently, many in vivo studies have proved that supplementation of Se, either organic or inorganic, ameliorates neuropathology and cognitive deficits of AD model mice [80]. Measuring the levels of 19 elements in the plasma and CSF, Gerhardsson et al. found compared with the healthy control, Mn and Hg in plasma were elevated, while the levels of V, Mn, Pb, and several other trace metals in CSF were decreased in AD patients [33]. Furthermore, the authors investigated the relationships between metal levels and the change of several well-known AD markers, including Aβ, total tau (T-tau), and p-tau in CSF. They found the level of Mn is positively correlated with T-tau and p-tau, while Cs is negatively correlated with T-tau [81]. González-Domínguez et al. investigated the change of serum ionomes with the development of neurodegeneration and found Fe, Cu, Zn, and Al are progressively changed with the progression of dementia [38]. In 2018, Xu et al. conducted a case–control study of eight essential elements (Na, K, Ca, Mg, Fe, Zn, Cu, and Se) in the plasma of sporadic AD patients and matched control cases. Although metal levels did not differ between cases and controls in all participants and in female group, Zn levels trended toward increase in AD than controls in male participants. Interestingly, co-regulated pairs including copper–sodium ($R_{control} = -0.03$, $R_{AD} = 0.65$; $p = 0.009$), and copper–calcium ($R_{control} = -0.01$, $R_{AD} = 0.65$; $p = 0.01$) were found to be associated with AD, which could be considered as potential markers of AD status [50]. Very recently, Lin et al. assessed the associations of plasma metal levels with amnestic mild cognitive impairment (aMCI) and AD. Lower plasma levels of boron (B), bismuth (Bi), thorium (Th), and uranium (U) were found to correlate with an increase in disease severity. Higher baseline concentration of calcium was associated with slower cognitive decline, while higher baseline levels of B, zirconium, and Th were associated with rapid annual cognitive decline in the aMCI group. In AD group, higher baseline levels of manganese resulted in rapid annual cognitive decline. These studies suggest the great potential of plasma metals as in vivo biomarkers for AD [82]. In addition to the fluid-based studies, metal levels of other tissues such as hair and nail in AD patients at different clinical stages were also measured. Koseoglu et al. found alkali elements, especially Na levels, were accumulated in nails with the progression of the disease [43].

10.2.1.4 Metal Chelators as AD Therapeutics

Though tremendous efforts have been made worldwide, there is still no effective drug for the treatment of AD. Since zinc was first found to be enriched in the senile plaques of AD patients in 1994 [67], researchers have conducted extensive and deep studies to explore the function of transition metals in the nervous system, especially their roles in the pathogenesis of AD. Since perturbed metal homeostasis plays a critical role in the onset and progression of AD, the application of metal chelators to reduce metal overload in certain regions of AD brains, to relieve or even treat AD, has attracted an increased amount of attention.

Deferoxamine (DFO) and Deferiprone (DFP) The first clinical trial of iron chelator, i.e. deferoxamine (DFO), in the treatment of AD was carried out in 1991 [83]. In this single-blind study of 48 AD patients, DFO administration (125 mg intramuscularly twice daily, five days per week, for 24 months) slowed the clinical progression of AD-related dementia. Mechanism study revealed that DFO impeded APP holoprotein translation via targeting the iron response element (IRE) in the *APP* transcript [84]. Deferiprone (DFP), as an oral iron chelator previously applied in the treatment of thalassemia syndrome, shows anti-AD potential through reducing Aβ and phosphorylated tau levels but not ROS in the hippocampus of rabbits with cholesterol-enriched diet [85]. This iron chelator is undergoing a phase II clinical trial with prodromal for mild AD patients. Iron plays an important role in both the CNS and peripheral system. Consequently, attempts to design novel iron chelators should take into account the specific selectivity and penetrating ability of the candidates.

Clioquinol and PBT2 As the first generation derivative of 8-hydroxyquinolines (8HQ) and an effective metal (Cu, Zn, and Fe) chelating agent, clioquinol has undergone extensive translational and clinical studies. Clioquinol (CQ, PBT1) was a widely used parasiticide drug before 1970s, which was abandoned as it may cause myelo-optic neuropathy. Studies have found that CQ reduced the levels of APP and senile plaques in the brains of transgenic mouse models of AD and consequently improved the cognitive deficits of the mice [86]. Mechanism investigation suggests that CQ extracts metals from Aβ and prevents the aggregation of Aβ due to its high affinity toward iron, copper, and zinc [87]. In coincidence with the observations in mouse models of AD, a pilot phase II clinical trial with 36 moderate AD patients found that clioquinol administration slowed the cognitive decline of the patients [88]. Furthermore, the level of Aβ_{42} in the plasma of AD patients declined with the administration of CQ

[89]. PBT2 is the second-generation 8-OHQ derivative of CQ, which exhibited better BBB permeability than CQ. Studies with transgenic mouse model of AD documented that PBT2 reduced Aβ levels, restored dendritic spine density deficits, rescued LTP impairment, and improved cognitive performance [90]. In a three-month phase II clinical trial with 78 AD patients, PBT2 (250 mg/day) significantly reduced CSF Aβ burden and improved cognitive performance of the patients [91].

Cu-Specific Chelators Bernard Meunier et al. designed a series of N4-tetradentate quinolines based on the bis(8-amino) quinoline scaffold, which are specific for Cu^{2+} chelation but not Zn^{2+} with very high affinity and can cross the BBB. These chelators can efficiently extract Cu^{2+} from Cu^{2+}-Aβ complex [92], but with unsatisfactory solubility and druggability. Next, they designed a new series of N4-tetradentate copper chelators named TDMQ, based on the mono(8-amino) quinoline scaffold. By offering a square planar coordination suitable for Cu^{2+} but not Cu^{+}, these N4-tetradentate ligands efficiently extract Cu^{2+} from Cu^{2+}-Aβ complex and subsequently release copper which can be transferred to an apo-copper protein in the presence of glutathione. Furthermore, TDMQ chelators prevent the production of ROS induced by Cu-Aβ [93] and do not affect the activities of Cu, Zn-superoxide dismutase (SOD) and tyrosinase, two copper enzymes participating in the modulation of redox process in the brain [94].

In vivo studies with both transgenic 5XFAD mice and C57BL/6 mice received a single injection of human Cu-Aβ$_{1-42}$ in the lateral ventricles or in the hippocampus, revealed that oral administration of TDMQ20 efficiently inhibited the cognitive and behavioral impairment in these three different murine models of AD [9c].

Metalloproteins As a small molecular weight and Cys-rich protein, metallothioneins-3 (MT3) is predominately expressed within CNS and downregulated in AD [95]. Studies have demonstrated the multiple functions of MT3 in AD, including detoxification and storage of heavy metals, regulation the homeostasis of copper and zinc, modulation the endocytosis of Aβ by astrocytes [96]. By sustained release of Zn_7MT3 to the CNS of APP/PS1 mice, Tan et al. demonstrated that Zn_7MT3 substantially improved the cognitive deficits of the transgenic mouse model of AD by regulation the metal homeostasis, reducing amyloid plaque burden and oxidative stress [97]. Du et al. found a His-rich peptide derived from selenoprotein P named Selenop-H, bound metal ions (Zn^{2+}, Cu^{+}, and Cu^{2+}) with high affinities and suppressed metal-induced aggregation and neurotoxicity of Aβ and tau in vitro [69b, 98]. In vivo study revealed that Selenop-H improved the spatial learning and memory deficits, preserved neuron activities, and inhibited both tau pathology and Aβ aggregation in the mice model of AD. For the mechanism, the researchers demonstrated that Selenop-H activated TrkB signaling pathway and restored MT3 and ZnT3 deficiency and Zn^{2+} homeostasis in the mice brains [99].

Though some chelators were used clinically for the treatment of metal overload diseases such as Wilson's disease, still no metal chelator was approved for therapeutic use for AD in clinic. One of the primary reasons is the potential adverse effects caused by removal of essential metals from both CNS and periphery system.

Moreover, most chelators are hydrophilic, making them hard to pass BBB. Therefore, to develop an effective and suitable chelator for AD therapy, the following issues should be considered: (i) easy to cross the blood–brain barrier (BBB); (ii) limited side effects; (iii) specifically target to the metal overload areas (senile plaques or NFTs); and (iv) capable of transferring the chelated metals to other biological metalloproteins, such as transferrin and SOD.

10.2.2 Pathometallomics in Parkinson's Disease

PD is the second most prevalent neurodegenerative disease after AD in the middle-aged and elderly people. The main clinical manifestations of PD are resting tremor, bradykinesia, rigidity, and postural instability [100]. And the pathological features are the apoptosis of dopaminergic neurons in substantia nigra, and the accumulation of α-synuclein in the cytoplasm of residual neurons in substantia nigra to form Lewy bodies [101]. The prevalence rate of PD among people over 60 years old in China is about 1–3% [101]. At present, there are at least 4.94 million patients in China [102], and number of people affected by PD is expected to continue to increase due to population aging. However, the etiology of PD is poorly understood and many factors may contribute to the disease such as age, environment, autophagy, mitochondrial dysfunction, inflammation, and oxidative stress [103]. Numerous studies have shown an association between PD and metal imbalances such as copper, iron, zinc, aluminum, and manganese. Here, the pertinent literature is reviewed.

10.2.2.1 Dysregulation of Metal Homeostasis in PD

It was recognized very early that metals in the environment and human health are closely related. Arsenic trioxide, for example, is one of the most well-known toxic substances, but arsenic-containing compounds can also be used to treat human malignancies [104]. Therefore, it is not scientific to simply classify metals as completely harmful or beneficial in terms of their relevance to human health. The speciation and accumulation level of metal ions in body are closely related to their biological effects. A considerable research effort has gone into the relationship between metal ions and PD. However, the reports regarding metal status in patients of PD or PD models are mixed. These seemingly disparate results may reasonably be attributed to confounding factors such as differences in PD models, detection methods, and ethnicity, diet, and environment of PD patients. Table 10.2 summarizes the published results of various metal changes in PD.

Iron Iron is an essential trace element, but it also has toxic effects due to its redox activity and the generation of free radicals during the interconversion between Fe^{2+} and Fe^{3+}. Using XRF technique, numerous studies have shown that iron levels are elevated in the substantia nigra of PD patients [124, 126, 136, 137]. Similarly, Fe concentration is elevated in the olfactory bulb of PD patients [113]. The same phenomena have also been found in PD animal models. In the evaluation of animal models induced by 1-methyl-4-phenyl-1, 2, 3, 6-tetrahydropyridine (MPTP), increased Fe

Table 10.2 Pathometallomics in Parkinson's disease.

Methods	Model	Tissue	Number of subjects (disease/control)	Elevated metals	Unchanged metals	Decreased metals	References
ICP-MS	MPTP[a]-treated cynomolgus monkeys	Brain	6/2	Fe	Ca and Mn	—	[105]
LA-ICP-MS	MPTP-treated mice	Brain	18 or 19	Fe in the interpeduncular nucleus and hypothalamic	—	Cu in the periventricular zone and the fascia dentata	[17]
ICP-MS	Patients	Serum	325/304	—	Fe and Zn	Cu	[106]
ICP-MS	Patients	Brain	9/9	—	Na, Ca, and Fe	Mg, K, and Se in MCX[b]; K and Zn in HP[c]; K in MTG; Mn and Cu in most areas of brain	[107]
LA-ICP-MS and MALDI-IM-MS	6-OHDA-treated mice	Brain	—	Fe, Cu, and Mn	Zn	—	[108]
AAS[d] and FAAS[e]	Indian Patients	CSF and serum	50/60 for CSF, and 250/280 for serum	Ca, Cr, Pb, and Mg in CSF; Al, Ca, Mg, and Pb in serum	Cu in CSF; Cr, Co, Mn, and Zn in serum	Al, Co, Fe, Mn, and Zn in CSF; Cu and Fe in serum	[109]
ICP-MS and AAS	Patients	Blood and urine	52/70	Mn and Fe in blood; Fe in blood and urine	Cu and Zn in blood; Mn, Cu, and Zn in urine	—	[110]

(Continued)

Table 10.2 (Continued)

Methods	Model	Tissue	Number of subjects (disease/control)	Elevated metals	Unchanged metals	Decreased metals	References
ICP-AES[f] and SF-ICP-MS[g]	Patients	Urine, serum, blood, and CSF	26/13	Ca in urine, serum, and blood; Fe in urine and blood; Zn in blood	Mn; Zn in serum, urine, and CSF	Al in urine, serum, blood, and CSF; Cu and Mg in serum; Fe in CSF	[111]
ICP-AES and ICP-MS	Patients	Serum	45/42	Al, Cu, K, Mn, Mo, Na, P, V, Ca, Hg, Mg, Ni, and Pb	Cr, Ag, and Rb	Co, Fe, Se, Zn, and Cd	[112]
ICP-MS	Patients	Olfactory bulb	7/7	Fe and Na	Cu, K, Ca, Mg, Zn, Rb, Ni, Cr, Mn, Pb, and V	—	[113]
ICP-MS	Patients	CSF	75/68	—	Se	—	[114]
ICP-MS and HPLC-ICP-MS	Patients	Serum, urine, and hair	13/14	—	Cu, Zn, Fe, and Mn	—	[115]
ICP-MS	Patients	Hair	46/24	—	—	Ca, Mg, Sr, and Cd	[116]
ICP-MS	Patients	Serum	65/65	—	Fe, Cu, Hg, and Mn	—	[117]
ICP-AES and SF-ICP-MS	Patients	CSF	42/20	—	Al, Ba, Be, Bi, Ca, Cd, Cu, Hg, Li, Mg, Mn, Mo, Ni, Sb, Sr, Tl, V, W, Zn, and Zr	Co, Cr, Fe, Pb, and Sn	[118]
ICP-MS	Patients	CSF	20/15	Cu, Mn, and Zn	Fe and Mg	—	[35]
ICP-MS	Patients	Brain	1/3	Se in the globus pallidus, superior temporal gyrus, and frontal cortex	—	—	[119]

ICP-MS	6-OHDA-treated mice	Brain	24/24	Fe, Cu, Mn, and Zn	—	—	[120]
ICP-AES and SF-ICP-MS	Patients	CSF, blood, serum, urine, and hair	91/18	Ca, Cu, Fe, Mg, and Zn in blood; Ca and Fe in urine	Al, Ca, Cu, Mg, Mn, and Zn in CSF; Al and Mn in blood; Ca, Fe, Mg, Mn, and Zn in serum; Cu, Mg, and Zn in urine; Al, Ca, Cu, Mg, Mn, and Zn in hair	Fe in CSF; Al and Cu in serum; Al and Mn in urine; Fe in hair	[121]
ICP-MS	Patients	Plasma, whole blood, and urine	23/24	—	Sb, Ba, Cd, Cs, Co, Cr, Cu, Hg, Pb, Mo, Se, Tl, Sn, Zn, and U in plasma, whole blood, and urine	—	[122]
ICP-MS	Patients	Temporal cortex	10/10	—	Mn, Ni, and Cu in the temporal cortex	Fe in the temporal cortex	[123]
XRF	Patients	Substantia nigra	1/1	Fe	—	Cu	[124]
XRF and PIXE[h]	6-OHDA[i] lesioned rat	Substantia nigra	15/12	—	Fe in SNpe[j], SNpr[k], and VTA[l]	—	[125]
XRF	Patients	Substantia nigra pars compacta	4/4	Ca, Fe, and Zn in SN nerve cells; Zn and Rb in areas outside the nerve cells	—	—	[126]

(Continued)

Table 10.2 (Continued)

Methods	Model	Tissue	Number of subjects (disease/control)	Elevated metals	Unchanged metals	Decreased metals	References
PIXE and RBS[m]	FeCl$_2$-treated PMN[n] cells over expressing α-synuclein	Cell	10–18/per group	Ca and Fe	K, Cu, and Zn	—	[127]
XRF	MnCl$_2$-treated rats	Substantia nigra	12/16	Mn	Fe, Cu, and Zn	—	[128]
XRF	MnCl$_2$-treated rats	Brain	7/6	Mn	Cu and Fe	Zn in GP	[129]
AAS[o]	MnCl$_2$-treated rats	Plasma and CSF	8–10/per group	Mn in plasma and CSF; Fe in CSF	—	Fe in plasma	[130]
XRF and AAS	MnCl$_2$-treated rats	Hippocampus	4/4	Mn in almost all areas	—	Fe in DG/CA1	[131]
XRF and PIXE	Patients	Brian	—	Fe in NM-containing neurons in Substantia nigra	Cu in occipital cortex	Cu in Substantia nigra and locus coeruleus	[132]
SR-XRF	MnCl$_2$-treated rats	Hippocampus	20/3	—	P, K, Ca, Cu, Zn, and Mn in HF	Fe in HF; Fe and Zn in CA3/DG	[133]
XRF	MnCl$_2$-treated PC12 dopaminergic cells	—	—	Mn	Ca, Cu, and Zn	Fe	[134]

XRF	Rats with electrical stimulation of vagus nerve	Brain	—	K, Fe, and Zn in the left side of corpus striatum	[135]
XRF	Patients	Brain	—	Fe in SN	[136]
XRF	Patients	SN	—	Ca, Fe, Cu, Zn, and Se in SN	[137]
XRF			—	Ca, Zn, and Rb in substantia nigra of right hemisphere	[135]
XRF			—	Fe in most areas and Zn in all areas	[136]
XRF			—	—	[137]

a) MPTP: 1-methyl-4-phenyl-1, 2, 3, 6-tetrahydropyridine.
b) MCX: primary motor cortex
c) HP: hippocampus.
d) AAS: atomic absorption spectrophotometry.
e) FAAS: flame atomic absorption spectrophotometry.
f) ICP-AES: inductively coupled plasma atomic emission spectrometry.
g) SF-ICP-MS: sector field inductively coupled plasma mass. spectrometry (SF-ICP-MS).
h) PIXE: particle-induced X-ray emission.
i) 6-OHDA: 6-hydroxydopamine.
j) SNpc: SN pars compacta.
k) SNpr: pars reticulata.
l) VTA: ventral tegmental area.
m) RBS: micro-Rutherford backscattering spectrometry.
n) PMN: primary midbrain neurons.
o) AAS: atomic absorption spectrophotometry.

contents in the interpeduncular nucleus of mouse as well as in the brain of cynomolgus monkeys were found [17, 105]. In another study using rats with electrical stimulation of vagus nerve, Fe level in the left side of corpus striatum was increased [135]. Significant increases of iron concentrations were also observed in the brain tissues of 6-hydroxydopamine (6-OHDA)-treated mice [108, 120]. These above studies suggest that Fe levels are elevated in the brains of PD patients or PD animal models. Although there are some exceptions. One study found that Fe level is decreased in the temporal cortex of PD patients [123], and another found no change in the PD brains [107]. Moreover, completely different results were found in the brains of $MnCl_2$-treated rats, in which Fe levels were decreased or with no change [128, 129, 131, 133].

Studies on changes of iron levels in biological fluid and their association with PD have been conflicting. Increases in iron are consistently reported in the blood of PD patients [110, 111, 121]. Meanwhile, there are also several reports on reduced or unchanged iron levels within the serum [106, 109, 112]. Some studies examined iron levels in CSF, and trends toward decreased iron levels are usually observed [109, 111, 118, 121]. Inconsistent with that, one study found elevated iron levels in CSF of $MnCl_2$-treated rats [130], and another held the view that Fe level in CSF of PD patients was constant [35]. Additionally, most investigations demonstrated that the Fe concentrations in the urine of PD patients were increased [110, 111, 121]. To sum up, iron levels in the same biological matrices tend to be approximately the same, even in different PD models except the $MnCl_2$-treated rats. Thus, careful consideration may be required in the study of metal dyshomeostasis-related mechanisms or intervention drugs in PD when using $MnCl_2$-treated rats as the PD model.

Copper Copper ions, like iron ions, are also able to promote the production of reactive oxygen species through Fenton-like reactions under physiological conditions in the body. In addition, as cofactors and/or structural components of various enzymes, copper ions are involved in cellular respiration, free radical detoxification, iron metabolism, and neurotransmitter synthesis. In MPTP-treated mice, there was a specific decrease of Cu concentration in the periventricular zone and the fascia dentata within a short time after injection of MPTP [17]. Consistent with this model, Cu levels detected by ICP-MS were also reduced in most areas of brains in PD patients [107]. Likewise, through XRF and PIXE, the content of Cu in the substantia nigra of patients was decreased [124, 132], which is different from the case of iron. However, ICP-MS showed no change of Cu levels in the olfactory bulb and temporal cortex [113, 123]. Unlike other PD models, the content of Cu in the brain of $MnCl_2$-treated rats, especially in the substantia nigra, did not change significantly [128, 129]. Moreover, several studies found Cu levels were elevated in the brains of 6-OHDA-treated mice [108, 120].

Alterations of Cu levels in biofluids including CSF, serum, blood, and urine have also been reported. Significantly lower Cu level was been observed in serum from PD patients [106, 109, 111, 121], whereas one exact opposite conclusion was reached by another group [112]. Most studies have shown that the levels of copper ions in CSF, blood, or urine did not change significantly [109, 110, 115, 118, 121, 122] even though an increasing trend for Cu levels in the blood and CSF of PD patients has also been reported [35, 121].

Zinc Similar to iron in the brain, Zn level seems to be elevated in the brain of PD models. Studies show that the concentration of Zinc is higher both extracellularly and intracellularly in substantia nigra pars compacta [126]. Another study also observed elevated Zn level in the substantia nigra [137]. It was found that after electrical stimulation of the left vagus nerve, Zn in the left striatum of rats was increased, whereas it decreased in the substantia nigra of the right hemisphere [135]. The decrease of Zn content in PD brain was also reported by some literatures. By XFR, Popescu et al. observed a decrease in zinc levels in all regions of PD brains [136]. Through ICP-MS experiments, another study found that Zn level was decreased in the hippocampus [107]. XFR experiment indicated that the content of Zn in the globus pallidus and CA3/DG region of MnCl2-treated rats was decreased [129, 133].

Using ICP-MS for Zn quantification in patients, most studies have not observed changes of Zn level in the PD biofluids such as urine, serum, blood, or CSF [106, 115, 118, 122]. Bocca et al. found Zn was increased in blood, but didn't change in CSF, serum or urine [121]. Similar results were reported by Forte et al. [111]. Sanyal indicated there was no change of Zn in serum, while Zn level was reduced in CSF [109].

Dysfunction of Metal-associated Proteins in PD Amyloid formation by α-synuclein in brain cells is a hallmark of PD. It has been found that α-synuclein is a copper-binding protein, and Cu^{2+} mainly interacts with residues located in the N-terminal region of α-synuclein and effectively promotes α-synuclein aggregation at physiologically relevant concentrations [138]. The aggregation of α-synuclein is also influenced by intracellular iron ions. As α-synuclein has a strong affinity for both Fe^{2+} and Fe^{3+}, both valence states of iron ions promote α-synuclein aggregation in vitro [139]. Further studies showed that Fe^{2+} induces α-synuclein aggregation and neurotoxicity by inhibiting Nrf2/HO-1. In turn, the inhibition of Nrf2/HO-1 also leads to more α-syn aggregation and greater iron-induced toxicity [140]. Yang et al. found copper/iron ions accelerate prion-like transmission of α-synuclein fibrils via promoting cellular internalization of α-synuclein fibrils, intracellular α-synuclein aggregation, and subsequent release of mature fibrils into the extracellular space to induce further dissemination [141]. In addition, toxic metals arsenite and cadmium were reported to incorporate into α-synuclein amyloid fibers, alter the structure of amyloid, and aggravate α-synuclein toxicity [142].

Ceruloplasmin, a copper-containing protein with ferroxidase oxidase function, can oxidize toxic ferrous iron in cells to ferric form. It is asserted that the reduction of ceruloplasmin aggravates the brain iron deposition, since low serum ceruloplasmin levels in PD patients are correlated with abundant substantia nigra iron, which is a prominent pathophysiological feature of PD [143]. Furthermore, the ceruloplasmin ferritase activity in CSF and serum was significantly reduced in patients with idiopathic PD, which was hypothesized to result in the pathological features of the pro-oxidant iron accumulation [144]. In addition, ceruloplasmin knockout mice developed Parkinson's symptoms, which could be rescued by iron chelation, and ceruloplasmin alleviated substantia nigra iron deposition in PD model mice [144].

SOD is a kind of metal protein that catalyzes the disorientation of superoxide anion effectively and contains metal ions as cofactors. Superoxide dismutase 1 (SOD1) or CuZn-SOD is a homodimer containing copper and zinc, which exists in cytoplasm. SOD2 or Mn-SOD exists in the mitochondria in the form of tetramers. SOD3 or EC-SOD is localized in the extracellular space as tetramers containing copper and zinc [145]. Significantly lower SOD levels in serum have been reported in PD patients [146], while another study found no significant differences in serum SOD levels between PD controls [147]. Roy et al. pointed out that SOD1-specific activity was significantly higher in PD than in controls [148]. And Liu et al. found single-nucleotide polymorphisms of SOD1 and SOD2 are related to the genetic predisposition of PD [149].

The brain is considered to be the most important part of the body for selenium, as the brain retains this nutrient most preferentially, even in the presence of selenium deficiency. Therefore, the brain is rarely deficient in selenium. Fabian et al. analyzed selenium speciation in the CSF of PD patients and age-matched control. They found that selenoprotein P (SELENOP), human serum albumin-bound Se (Se-HSA), selenomethionine (Se-Met), and an unidentified se-compound didn't show significant differences between both groups. However, glutathione peroxidase (GPX), and the inorganic substances selenite and selenate were not detected in most PD and control samples [114]. Using a chronic PD mouse model, Zhang et al. investigated selenium transcriptome changes in five brain regions (cerebellum, substantia nigra, cortex, pons, and hippocampus). They found the GPX family levels were decreased in PD mouse brains. And in the substantia nigra, 17 selenoproteins levels were significantly reduced, whereas no selenoproteins were upregulated. Although some selenoproteins showed mixed patterns of up- and down-regulation in different brain regions, most selenoprotein transcriptomes were generally unchanged [150].

10.2.2.2 Application of Metallomics in the Prognosis of PD

Compared with other biomarker molecules such as RNA, protein, or metabolite, biological elements contents remain stable without the influence of the compounds degradation, temperature, or other factors. It is generally accepted that the accumulation of iron in the substantia nigra compacta is a major feature of PD. The changes of substantia nigra iron have been shown to be correlated with the severity of clinical PD, and the accumulation of substantia nigra iron distinguishes PD well from controls [151]. Meta-analysis studies revealed that zinc/iron levels in serum/plasma are inversely associated with PD risk, while elevated iron levels in CSF increase the risk of PD [152]. Wu et al. observed HAMD and HAMA scores were negatively correlated with serum iron level and positively correlated with serum transferrin level, indicating that serum iron and transferrin levels may be peripheral markers of PD [153]. Furthermore, serum copper levels are inversely associated with PD risk or clinical symptoms, and higher copper levels are associated with reduced PD risk [106]. In substantia nigra pars compacta, Fe, Ca, and Zn in nerve cells, and Fe and Cu outside the cells, distinguish well between PD and control [126]. These findings suggest that metal ions may play a role in the diagnosis of PD. As PD is a multifactorial neurodegeneration, a single-factor diagnosis may lead to large errors. Thus, combining changes in metal content with other biomarkers may improve the accuracy of PD diagnosis and prognosis, which needs to be further validated in the future investigations.

10.2.2.3 Application of Metallodrugs and Metalloproteins in the Treatment of PD

The correlation of metal ions with PD pathology suggests that metal chelators may be a promising therapeutic strategy. Multiple agents with iron-chelating features have been assessed preclinically in PD models, including DFO, 8HQ analogs, prochelators, and aroylhydrazones, while none of these compounds have progressed to clinical trial for PD [154]. In 2012, Kwiatkowski et al. explored the efficacy of DFP in patients with moderate PD. Before treatment, iron levels in the internal globi pallidi, dentate nuclei, substantia nigra, and red nuclei were accumulated. After treatment with DFP, the patients' clinical symptoms were relieved, activities of daily living were improved, and iron in bilateral dentate nucleus and substantia nigra were significantly decreased without obvious neurologic or hematological side effects [155]. In addition, some natural products also show obvious metal-chelating effect. Wang et al. demonstrated that ginkgetin inhibited the increase in the intracellular labile iron pool by chelating ferrous ions, down-regulating L-ferritin, and up-regulating transferrin receptor 1, thereby significantly improving the sensorimotor coordination in MPTP-induced PD mice [156]. Curcumin not only has the potential ability of chelating copper, but also has good antioxidant activity [157]. It has been reported that three-week oral curcumin treatment significantly improved behavioral disturbances, oxidative damage, and increased mitochondrial enzyme complex activity in rotenone-induced PD model mice [158]. Compared with drugs, these typical natural compounds have the advantages of anti-neurotoxicity and less side effects, which is worth further exploration.

10.2.3 Pathometallomics in Amyotrophic Lateral Sclerosis

ALS) has an incidence rate of 1.5 new cases in 100 000 individuals per year and a prevalence of five patients in 100 000 individuals [159]. There are two forms of ALS: sporadic (sALS) and familiar (fALS). sALS is the commonest form of the disease with no apparent pathological genetic background, accounting for 90% of the cases, while the remaining 10% cases belong to fALS. Approximately 20% of fALS is associated with inherited mutations in the Cu, Zn-SOD1 gene.

10.2.3.1 Dysregulation of Metal Homeostasis in ALS

ALS causes irreversible damage in humans, with the consequent loss of function of the central and peripheral motoneurons (MNs), leading to progressive paralysis of all muscles and eventually death. MNs in ALS show a number of changes including mitochondrial dysfunction, impairment of axonal transport, oxidative stress, excitotoxicity, protein aggregation, endoplasmic reticulum stress, and abnormal RNA processing [160]. In the last decade, metal dyshomeostasis has been proposed as the main trigger of the pathological cascade that brings to MNs loss [161]. Heavy metals are suggested to induce the onset of sALS [162]. In a small cohort investigation, the levels of several neurotoxic metals (such as Al, Cd, and Cu) were measured to be significantly increased in CSF but not in blood plasma of ALS patients, implying that accumulation of these toxic metals in the brain has an impact on the pathogenesis of ALS [163]. Another study analyzed the concentrations of a broad spectrum of metals in serum and whole blood of sALS patients and reported that higher concentrations of Se, Mn, and Al and lower serum concentration of As were associated with this disease [164].

The possible roles of metals and metalloids in the severity of ALS were investigated using the blood, urine, and hair of ALS patients at different stages. By analyzing 11 metals including Pb, Cd, Al, Hg, Mn, Fe, Cu, Zn, Se, Mg, and Ca in ALS patients, Pb, Cd, Al, and Hg were found detrimental to the function of the nervous system, whereas the other metals are essential in humans [159a, 165]. Further study showed that the levels of these elements were correlated with the progression of the illness, according to the ALS Functional Rating Scale-Revised (ALSFRSr). Elevated levels of lead may correlate with the severity of ALS, whereas Se may play a protective role [159a]. The possible involvement of Al and Mn in the severity of ALS required further study. No evidence was observed for an association between levels of Hg, Cd, Fe, and the severity of ALS. A major limitation of these studies is the relatively small number of patients analyzed, which could restrict the power of pathometallomic evaluation especially considering the high variability of trace amounts of metals or metalloids in biological systems [166]. Thus, definitive conclusions will require larger sample numbers and appropriate multiple test correction. Nevertheless, pathometallomics will be quite useful to explore the pathogenesis of ALS and to gain a comprehensive view of the complex interactions between metals/metalloids and the pathology of ALS.

10.2.3.2 Metal-Associated Dysfunction in ALS

Dysregulation of metal homeostasis in the pathology of ALS has been well investigated for several metals, i.e. Ca, Na, K, Cu, Zn, and Fe, with particular regard to these metal-associated proteins or pathways as putative targets for future therapeutic strategies [161].

Defects in cellular Ca^{2+} signaling are involved in the pathogenesis of ALS [167]. Deregulation of glutamate neurotransmission by increasing extracellular glutamate levels [168] may trigger Ca^{2+} entry, leading to altered Ca^{2+} homeostasis crucial for MN degeneration in both sALS and fALS [169] (P7; T1). An increase in cytosolic Ca^{2+}-buffering capacity protects vulnerable MNs from degeneration [170]. However, in ALS Ca^{2+}-buffering proteins (CaBPs) such as calbindin and parvalbumin [167b] are lost at an early stage of the disease in hypoglossal, spinal and low cranial MN populations, suggesting the association of intracellular Ca^{2+} homeostasis disturbance with disease progression [167b, 171]. Deregulation of intracellular Ca homeostasis has been reported in the terminal motor axons of subjects affected by ALS, as well as in the spinal MNs of ALS animals, in both sALS and fALS [172]. In fALS, Cu/Zn-SOD1 mutation causes mitochondrial Ca^{2+} overload and strong ROS generation, leading to MNs damage. Plasma membrane Na^+/Ca^{2+} exchanger isoform 3 (NCX3), a key protein in ALS pathogenesis, has a significant reduction in its expression and activity at muscular and neuronal levels. NCX3 thus may have crucial role in mediating the impairment of neuromuscular transmission [173], rendering it a putative druggable target in ALS.

In sALS, the phenotypes of increased persistent Na^+ conductance and reduced K^+ current are associated with alteration of axonal excitability. Membrane hyperexcitability due to the abnormality of Na^+ and K^+ conductance leads to muscle cramps and fasciculations, and promotes a neurodegenerative cascade mediated by Ca^{2+}-dependent processes. Axonal ion channel dysfunction evolves with disease progression and correlates with survival, thus representing a potential therapeutic biomarker in ALS [174]. Modulation of axonal Na^+ channel function in ALS results in amelioration of symptoms and stabilization of axonal excitability parameters. In fALS, the increase of Na^+ conductance has also been reported in transgenic SOD1 mice, although mechanisms of ectopic activity, such as cramp and fasciculations, and axonal degeneration still necessitate clarifications in patients. Currently, it is not clear whether any difference occurs within these processes between the subjects of fALS and sALS.

Cu ions may act as cofactor of pro-inflammatory cytokines in animal models [175], and its accumulation is extremely detrimental for the cells. Cu ions in the cuprous state (Cu^+) can induce oxidative stress directly catalyzing the formation of strong oxidants such as lipid hydroperoxides from hydrogen peroxide and hydroperoxides via a Fenton-like reaction [176]. Elevated cellular Cu levels may deplete glutathione levels, shifting the redox balance toward oxidizing environments, and hence, both enhancing the cytotoxic effect of ROS and allowing the metal to be more catalytically active, thus producing higher levels of ROS [177]. Alterations in intracellular Cu homeostasis could represent a possible mechanism responsible for the pathogenesis of ALS [178]. Cu dyshomeostasis is evident in G93A SOD1 mice before the onset of clinical symptoms, in a pre-symptomatic phase of the disease, suggesting its increase with pathological process and function as a hallmark of the pathology [179]. Notably, SOD1 mutants shift the Cu trafficking system toward Cu accumulation. In particular, the expression levels of the Cu importer 1 (CTR1) and the Cu efflux pump (ATP7A) increase and decrease, respectively, in the spinal cord of mutant animals. The expression level of metallothioneins (MT), a class of proteins with a very high affinity for Cu, is also augmented in the spinal cord of G93A SOD1 mice [180]. These altered expressions of proteins result in intracellular Cu accumulation and thus cytotoxicity in ALS. In human ALS cases, the involvement of Cu dyshomeostasis on ALS etiology remains to be elucidate.

Zn dyshomeostasis or Zn accumulation and oxidative injury may be another important contributor to ALS pathogenesis. ALS onset and progression depend on various interplaying processes that together lead to degeneration and atrophy of motor neurons. A significant increase of Zn levels was found in ALS patients [35]. In the spinal cords of G93A SOD1 transgenic mice, Zn accumulation and numerous MNs and astrocytes degeneration occurred with the appearance of ALS signs. Zn elevation in these cells induces lipid peroxidation and disrupts Zn^{2+} homeostasis by trigger Zn^{2+} release from MT and from G93A SOD1 [181]. Meanwhile, G93A SOD1 showed a weaker affinity for Zn and Zn-depleted SOD1 may also contribute to neurodegenerative process inducing nitrosactive stress with peroxynitrite and catalyzes nitration of protein tyrosine residues [182]. In addition, AMPA/kainate receptors, rather than NMDA glutamate receptors, may play a role allowing excessive Zn accumulation in MNs of ALS [183]. The expression of membrane Zn transporter ZnT6, which presents in the secretory pathway where Zn is required for correct folding and assembly of proteins, is significantly decreased in the spinal cords of sALS patients [184]. The reduced expression of ZnT6 may compromise Zn uptake in secretory pathway and, in turn, contribute to ER stress described in ALS [185].

Alteration of Fe homeostasis is also involved in neuronal cell death in ALS. Different pathogenetic mechanisms have been proposed to explain the abnormal accumulation of Fe in neurons and in glia observed in ALS mice. Alterations of proteins involved in both Fe influx and sensing of intracellular Fe concentrations lead to disruption of Fe balance and consequently oxidative stress and cell injury [186]. Blockage of anterograde axonal transport is also responsible for Fe accumulation in ventral motor neurons. Increased mitochondrial Fe load in neurons and glia also results in neurodegeneration in ALS [187].

The role of Se in ALS has been reported contradictory. Physiological damage has been attributed to Se. For instance, a case–control study of 41 ALS patients indicated a correlation between the disease and the consumption of well water rich in Se, although the possible effects of other pollutants were not excluded [188]. A cross-sectional study found higher levels of serum Se in ALS patients than in the controls [189]. In contrast, the protective role of Se in ALS has also been reported. Se was among the metals investigated in a population-based case–control study that analyzed the association between environmental exposure to trace elements and sALS, in which Se seemed to play a protective role and was lower in ALS patients than in the controls [190]. This result is supported by some other papers [159a, 191], demonstrating the defensive function of Se in antioxidant mechanism in CNS. Therefore, more studies are required to better understand the role of Se in ALS.

10.2.4 Pathometallomics in Autism Spectrum Disorder

Autism spectrum disorder (ASD) are a group of heterogeneous diseases that are likely generated under genetic background or caused by environmental factors or a combination of both [192]. The patients have been increased in prevalence up to 1 in 88 children [193], and within which around 90% ASD are highly heritable [194]. However, environmental factors also play a significant role in the pathogenesis through interaction with heritable factors. Individuals with ASD display specific behavioral features such as impairments in social interaction and communication, the presence of restricted and repetitive behaviors [193, 195]. In addition, these patients often experience sensory alterations and co-morbidities such as seizures, hyperactivity, sleep disorders, increased risk of allergies, gastrointestinal problems, and metabolic plus mitochondrial diseases, which commence early in life [196]. In terms of biochemical indices, ASD are characterized by elevated levels of mitochondrial dysfunction, lipid peroxidation, and decreased levels of antioxidants (including transferrin, ceruloplasmin, MT, and glutathione in blood serum) [197].

Metal dyshomeostasis has been consistently reported to play a role in the pathogenesis of ASD, demonstrating as essential metal deficiency, heavy metal toxicity, oxidative stress, inflammation, altered neurotransmitter synthesis, and neuronal dysfunction [198]. Many infantile ASD patients suffer from marginal to severe Zn and Mg deficiency and/or high toxic metal burdens. An imbalance in the Zn:Cu ratio has been reported in ASD, which plays a causative role in the mechanistic studies [199]. Meanwhile, some toxic metals or metalloids such as Pb, Cd, Hg, and As have been reported as environmental candidates causing ASD [200]. A recent study on 25 ASD children (clinically diagnosed according to Diagnostic and Statistical Manual of Mental Disorder [DMS-IV-TR] classification) and 25 neuro-typical (NT) children (controls) revealed that reduction in Zn and Mg levels with a concurrent increase in Pb was found in ASD children, which may be the basis of inadequate total antioxidant capacity (TAC) manifesting as increased malondialdehyde (MDA) and reduced total plasma peroxidase (TPP) levels in those patients [201].

Zn is a redox-inert metal that functions as an antioxidant in vivo through an indirect effect, acting as an inhibitor of NADPH oxidases [202]. A decrease in Zn and an increase in Cu result in higher levels of oxidative stress and thus the generation of ROS in brain. Apo MT binds to toxic metals with a high affinity and, as such, plays a pivotal role alongside Zn in the detoxification of toxic metals [203]. Metals such as Cd, Pb and Hg have been implicated in promoting intracellular ROS production. A deficiency in either Zn, Cu or Mn has an impact on oxidative stress by decreasing the expression levels and enzyme activity of SOD [204]. Zn, Cu and Mn also increase glutathione production, which maintains the redox state to protect cell function. It is widely known that ROS production is closely linked to pro-inflammatory processes. Increased inflammation has been reported in many human and animal studies in ASD [205]. Cd contributes to oxidative stress by promoting inflammation through the transcription factors NF-κB and AP-1 [206]. Zn can downregulate the production of inflammatory cytokines through NF-κB signaling [207]. Throughout fetal development, Zn deficiency results in increased inflammation and altered CNS development [208]. Zn also has pivotal function in postsynaptic scaffold formation by mediating proteins such as SHANK2 and SHANK3. These proteins are at the center of an ASD-linked postsynaptic signaling network [209].

The dyshomeostasis of metal ions such as Fe, Cu and Zn also typically hinders neurogenesis, due to their pathometallomic effects in various pathways [210]. Zn deficiency has shown to reduce proliferative development of neuronal stem cells [211]. Even marginal deficiency of Zn affects neurogenesis by altering the number of neurons and reducing neuronal specification [210b]. The altered metal composition may have significant knock-on effects on many biological processes that control proteostasis, gene expression, cell signaling, organ development, and metabolism.

However, an "infantile window" has been proposed in ASD, i.e. a critical time window in neurodevelopment and probably for treatment and prevention of ASD [192]. Therefore, early detection and intervention is likely to play a role in prevention of ASD. To achieve this purpose, it is also desired to introduce some innovative clinical tests for early estimation and treatment of ASD.

10.3 The Perspectives of Pathometallomics

The homeostasis of metal ions in the brains play critical roles in maintaining the normal functions; therefore, a disorder in their homeostasis can cause serious consequences. A common mechanism underlying the neurodegenerative disorders, including AD, PD, HD, and ALS, involves the misfolding of proteins that induces subsequent cytotoxicity. Metal ions dyshomeostasis not only promotes the misfolding of proteins and neurotoxicity, but can also lead to oxidative stress, apoptosis, autophagy disorder, mitochondrial dysfunction, lysosomal storage disorder, etc. Moreover, the metal deregulation-induced alterations can reversely aggravate metal misdistribution and deposition, forming a vicious cycle. The close connection between metal dyshomeostasis and neurodegenerative diseases pathogenesis opens doors for the predication and treatment of neurodegenerative diseases with metallomics and metal chelators, respectively. Though tremendous efforts have been made, still no metallomics-based approach or metal chelator was approved for the prediction or treatment of AD or other neurodegenerative disorders clinically. Elucidating the precise mechanisms in which metal homeostasis is affected in each disease of interest is crucial for the development of novel and efficient pharmacological agents. Furthermore, rebalancing metal homeostasis rather than chelation metals is considered as a more promising avenue in the drug development of AD and neurodegenerative diseases.

Acknowledgments

This work was financially supported by the Shenzhen Science and Technology Innovation Commission (JCYJ20200109110001818, JCYJ20220804182935001); Shenzhen-Hong Kong Institute of Brain Science-Shenzhen Fundamental Research Institutions (2023SHIBS0003).

List of Abbreviations

6-OHDA	6-hydroxydopamine
8HQ	8-hydroxyquinolines
AD	Alzheimer's disease
ALS	amyotrophic lateral sclerosis
ALSFRSr	ALS functional rating scale - revised
aMCI	amnestic mild cognitive impairment

ASD	autism spectrum disorder
ATP7A	Cu efflux pump
Aβ	aggregation of amyloid-β
B	boron
BBB	blood–brain barrier
Bi	bismuth
CaBPs	Ca^{2+}-buffering proteins
CE	capillary electrophoresis
CNS	central nervous system
CSF	cerebrospinal fluid
CTR1	Cu importer 1
DFO	deferoxamine
DFP	deferiprone
DMS-IV-TR	diagnostic and statistical manual of mental disorder
fALS	familiar ALS
GE	gel electrophoresis
GPX	glutathione peroxidase
HD	Huntington's disease
HPLC	high-performance liquid chromatography
ICP-MS	inductively coupled plasma mass spectrometry
IMAC	immobilized metal affinity chromatography
IRE	iron response element
JNK	c-Jun N-terminal kinase
LA-ICP-MS	laser ablation coupled with ICP-MS
LC	liquid chromatography
LTD	long-term depression
LTP	long-term potentiation
MC-ICP-MS	multicollector-ICP-MS
MDA	malondialdehyde
MNs	motoneurons
MPTP	1-methyl-4-phenyl-1, 2, 3, 6-tetrahydropyridine
MT	metallothioneins
MT3	metallothinoneins-3
NCX3	Na^+/Ca^{2+} exchanger isoform 3
NFTs	neurofibrillary tangles
Ni	nickel
NT	neuro-typical
PAGE	polyacrylamide gel electrophoresis
PD	Parkinson's disease
PMD	post mortem delay
PP2A	protein phosphatase 2A
p-tau	phosphorylated-tau
sALS	sporadic ALS
Se	selenium
Se-HSA	human serum albumin-bound Se

SELENOP	selenoprotein P
Se-Met	selenomethionine
SOD	superoxide dismutase
SOD1	superoxide dismutase 1
STXM	scanning transmission x-ray microscopy
TAC	total antioxidant capacity
Th	thorium
TPP	total plasma peroxidase
T-tau	total tau
U	uranium
XAS	X-ray absorption spectroscopy
XRF	X-ray fluorescence

References

1 Haraguchi, H. (2004). *J. Anal. Atom. Spectrom.* 19: 5.
2 Zhou, Y., Li, H.Y., and Sun, H.Z. (2022). *Annu. Rev. Biochem.* 91: 449.
3 Roberts, E.A. and Sarkar, B. (2014). *Curr. Opin. Clin. Nutr.* 17: 425.
4 Zhang, Y., Xu, Y.Z., and Zheng, L. (2020). *Int. J. Mol. Sci.* 21: 1–26. 8646.
5 Hare, D.J., Rembach, A., and Roberts, B.R. (2016). *Methods Mol. Biol.* 1303: 379.
6 Fu, D. and Finney, L. (2014). *Expert Rev. Proteomics* 11: 13.
7 Lothian, A., Hare, D.J., Grimm, R. et al. (2013). *Front. Aging Neurosci.* 5: 35.
8 (a) Wang, H., Zhou, Y., Xu, X. et al. (2020). *Curr. Opin. Chem. Biol.* 55: 171. (b) Wang, Y., Li, H., and Sun, H. (2019). *Inorg. Chem.* 58: 13673. (c) Zheng, L., Zhu, H.Z., Wang, B.T. et al. (2016). *Sci. Rep.* 6: 39290. (d) Iqbal, J., Zhang, K., Jin, N. et al. (2019). *ACS Chem. Neurosci.* 10: 2418. (e) He, Z., Zheng, L., Zhao, X. et al. (2022). *Biol. Trace Elem. Res.* 200: 3248.
9 (a) He, Z., Song, J., Li, X. et al. (2021). *J. Biol. Inorg. Chem.* 26: 551. (b) He, Z., Li, X., Han, S. et al. (2021). *Metallomics* 13. (c) Zhao, J., Shi, Q., Tian, H. et al. (2021). *ACS Chem. Neurosci.* 12: 140.
10 Xu, X., Wang, H., Li, H., and Sun, H. (2020). *Chem. Lett.* 49: 697.
11 Bulska, E. and Wagner, B. (2016). *Philos. Trans. A Math. Phys. Eng. Sci.* 374: 2079. 1–18.
12 Cvetkovic, A., Menon, A.L., Thorgersen, M.P. et al. (2010). *Nature* 466: 779.
13 Yin, X.-B., Li, Y., and Yan, X.-P. (2008). *Trends Anal. Chem.* 27: 554.
14 Ferrer, M., Golyshina, O.V., Beloqui, A. et al. (2007). *Nature* 445: 91.
15 Hu, L., Cheng, T., He, B. et al. (2013). *Angew. Chem. Int. Ed. Engl.* 52: 4916.
16 Becker, J.S., Zoriy, M., Becker, J.S. et al. (2007). *J. Anal. At. Spectrom.* 22: 736.
17 Matusch, A., Depboylu, C., Palm, C. et al. (2010). *J. Am. Soc. Mass Spectrom.* 21: 161.
18 Petit-Pierre, G., Colin, P., Laurer, E. et al. (2017). *Nat. Commun.* 8: 1239.
19 (a) Moynier, F., Borgne, M.L., Lahoud, E. et al. (2020). *Alzheimers Dement. (Amst.)* 12: e12112. (b) Moynier, F., Creech, J., Dallas, J., and Le Borgne, M. (2019). *Sci. Rep.* 9: 11894.

20 Tea, I., De Luca, A., Schiphorst, A.M. et al. (2021). *Metabolites* 11: 370. 1–17.
21 Abelin, J.G., Trantham, P.D., Penny, S.A. et al. (2015). *Nat. Protoc.* 10: 1308.
22 Block, H., Maertens, B., Spriestersbach, A. et al. (2009). *Methods Enzymol.* 463: 439.
23 McRae, R., Bagchi, P., Sumalekshmy, S., and Fahrni, C.J. (2009). *Chem. Rev.* 109: 4780.
24 Punshon, T., Guerinot, M.L., and Lanzirotti, A. (2009). *Ann. Bot.* 103: 665.
25 Streli, C., Rauwolf, M., Turyanskaya, A. et al. (2019). *Appl. Radiat. Isot.* 149: 200.
26 Aitken, J.B., Levina, A., and Lay, P.A. (2011). *Curr. Top. Med. Chem.* 11: 553.
27 Conesa, J.J., Carrasco, A.C., Rodríguez-Fanjul, V. et al. (2020). *Angew. Chem. Int. Ed. Engl.* 59: 1270.
28 Brooks, J., Everett, J., Lermyte, F. et al. (2020). *Angew. Chem. Int. Ed. Engl.* 59: 11984.
29 Lu, M., Pontecorvo, M.J., Devous, M.D. Sr., et al. (2021). *JAMA Neurol.* 78: 445.
30 Corrigan, F.M., Reynolds, G.P., and Ward, N.I. (1993). *Biometals* 6: 149.
31 Panayi, A.E., Spyrou, N.M., Iversen, B.S. et al. (2002). *J. Neurol. Sci.* 195: 1.
32 Srivastava, R.A. and Jain, J.C. (2002). *J. Neurol. Sci.* 196: 45.
33 Gerhardsson, L., Lundh, T., Minthon, L., and Londos, E. (2008). *Dement. Geriatr. Cogn. Disord.* 25: 508.
34 Dong, J., Robertson, J.D., Markesbery, W.R., and Lovell, M.A. (2008). *J. Alzheimers Dis.* 15: 443.
35 Hozumi, I., Hasegawa, T., Honda, A. et al. (2011). *J. Neurol. Sci.* 303: 95.
36 Lee, J.Y., Kim, J.H., Choi, D.W. et al. (2012). *Toxicol. Res.* 28: 93.
37 Ciavardelli, D., Consalvo, A., Caldaralo, V. et al. (2012). *Metallomics* 4: 1321.
38 González-Domínguez, R., García-Barrera, T., and Gómez-Ariza, J.L. (2014). *Metallomics* 6: 292.
39 Giacoppo, S., Galuppo, M., Calabrò, R.S. et al. (2014). *Biol. Trace Elem. Res.* 161: 151.
40 Graham, S.F., Nasarauddin, M.B., Carey, M. et al. (2015). *J. Alzheimers Dis.* 44: 851.
41 Koç, E.R., Ilhan, A., Zübeyde, A. et al. (2015). *Turk. J. Med. Sci.* 45: 1034.
42 Hare, D.J., Raven, E.P., Roberts, B.R. et al. (2016). *NeuroImage* 137: 124.
43 Koseoglu, E., Koseoglu, R., Kendirci, M. et al. (2017). *J. Trace Elem. Med. Biol.* 39: 124.
44 Guan, C., Dang, R., Cui, Y. et al. (2017). *PLoS One* 12: e0178271.
45 Xu, J., Church, S.J., Patassini, S. et al. (2017). *Metallomics* 9: 1106.
46 Vaz, F.N.C., Fermino, B.L., Haskel, M.V.L. et al. (2018). *Biol. Trace Elem. Res.* 181: 185.
47 Cruz-Alonso, M., Fernandez, B., Navarro, A. et al. (2019). *Talanta* 197: 413.
48 Solovyev, N., El-Khatib, A.H., Costas-Rodríguez, M. et al. (2021). *J. Biol. Chem.* 296: 100292.
49 Strumylaite, L., Kregzdyte, R., Kucikiene, O. et al. (2022). *Int. J. Environ. Res. Public Health* 19: 7309. 1–11.
50 Xu, J., Church, S.J., Patassini, S. et al. (2018). *Biometals* 31: 267.

51 Gurel, B., Cansev, M., Sevinc, C. et al. (2018). *J. Alzheimers Dis.* 61: 1399.
52 Lukiw, W.J., Kruck, T.P.A., Percy, M.E. et al. (2019). *J. Alzheimers Dis. Parkinsonism* 8: 457. 1–12.
53 Ashraf, A., Stosnach, H., Parkes, H.G. et al. (2019). *Sci. Rep.* 9: 3147.
54 Siritapetawee, J., Pattanasiriwisawa, W., and Sirithepthawee, U. (2010). *J. Synchrotron Radiat.* 17: 268.
55 Bourassa, M.W., Leskovjan, A.C., Tappero, R.V. et al. (2013). *Biomed. Spectrosc. Imaging* 2: 129.
56 Leskovjan, A.C., Kretlow, A., Lanzirotti, A. et al. (2011). *NeuroImage* 55: 32.
57 Schrag, M., Mueller, C., Oyoyo, U. et al. (2011). *Prog. Neurobiol.* 94: 296.
58 Squitti, R., Lupoi, D., Pasqualetti, P. et al. (2002). *Neurology* 59: 1153.
59 Baum, L., Chan, I.H.S., Cheung, S.K.K. et al. (2010). *Biometals* 23: 173.
60 (a) Pithadia, A.S. and Lim, M.H. (2012). *Curr. Opin. Chem. Biol.* 16: 67. (b) Lovell, M.A., Robertson, J.D., Teesdale, W.J. et al. (1998). *J. Neurol. Sci.* 158: 47.
61 (a) Akatsu, H., Hori, A., Yamamoto, T. et al. (2012). *Biometals* 25: 337. (b) Rembach, A., Hare, D.J., Lind, M. et al. (2013). *Int. J. Alzheimers Dis.* 2013: 623241.
62 Scholefield, M., Church, S.J., Xu, J. et al. (2020). *Metallomics* 12: 952.
63 Wang, H., Wang, M., Wang, B. et al. (2012). *Metallomics* 4: 289.
64 Wang, H.J., Wang, M., Wang, B. et al. (2012). *Metallomics* 4: 1113.
65 van Duijn, S., Bulk, M., van Duinen, S.G. et al. (2017). *J. Alzheimers Dis.* 60: 1533.
66 Liu, B., Moloney, A., Meehan, S. et al. (2011). *J. Biol. Chem.* 286: 4248.
67 Bush, A.I., Pettingell, W.H., Multhaup, G. et al. (1994). *Science* 265: 1464.
68 Damante, C.A., Osz, K., Nagy, Z. et al. (2009). *Inorg. Chem.* 48: 10405.
69 (a) Jiao, Y. and Yang, P. (2007). *J. Phys. Chem. B* 111: 7646. (b) Du, X., Wang, Z., Zheng, Y. et al. (2014). *Inorg. Chem.* 53: 1672.
70 Kitazawa, M., Cheng, D., and LaFerla, F.M. (2009). *J. Neurochem.* 108: 1550.
71 Kim, I., Park, E.J., Seo, J. et al. (2011). *Neuroreport* 22: 839.
72 Sun, X.Y., Wei, Y.P., Xiong, Y. et al. (2012). *J. Biol. Chem.* 287: 11174.
73 Tabner, B.J., Turnbull, S., El-Agnaf, O.M.A., and Allsop, D. (2002). *Free Radical Biol. Med.* 32: 1076.
74 (a) Rai, R.K., Chalana, A., Karri, R. et al. (2019). *Inorg. Chem.* 58: 6628. (b) Cheignon, C., Tomas, M., Bonnefont-Rousselot, D. et al. (2018). *Redox Biol.* 14: 450.
75 Sindreu, C., Bayes, A., Altafaj, X., and Perez-Clausell, J. (2014). *Mol. Brain* 7: 16. 1–7.
76 (a) Chen, T.J., Cheng, H.M., Wang, D.C., and Hung, H.S. (2011). *Toxicol. Lett.* 200: 67. (b) Nagasaki, N., Hirano, T., and Kawaguchi, S. (2014). *J. Physiol.-London* 592: 4891.
77 Sahni, S., Bae, D.H., Jansson, P.J., and Richardson, D.R. (2017). *Pharmacol. Res.* 119: 118.
78 Figueiredo, L.S., de Freitas, B.S., Garcia, V.A. et al. (2016). *Mol. Neurobiol.* 53: 6228.

79 Gerardo, B., Pinto, M.C., Nogueira, J. et al. (2020). *Int. J. Environ. Res. Public Health* 17: 6051. 1–18.
80 Du, X., Wang, C., and Liu, Q. (2016). *Curr. Top. Med. Chem.* 16: 835.
81 Gerhardsson, L., Blennow, K., Lundh, T. et al. (2009). *Dement. Geriatr. Cogn. Disord.* 28: 88.
82 Lin, Y.K., Liang, C.S., Tsai, C.K. et al. (2022). *J. Clin. Med.* 11: 3655. 1–14.
83 Crapper McLachlan, D.R., Dalton, A.J., Kruck, T.P. et al. (1991). *Lancet* 337: 1304.
84 Rogers, J.T., Randall, J.D., Cahill, C.M. et al. (2002). *J. Biol. Chem.* 277: 45518.
85 Prasanthi, J.R., Schrag, M., Dasari, B. et al. (2012). *J. Alzheimers Dis.* 30: 167.
86 (a) Cherny, R.A., Atwood, C.S., Xilinas, M.E. et al. (2001). *Neuron* 30: 665.
(b) Grossi, C., Francese, S., Casini, A. et al. (2009). *J. Alzheimers Dis.* 17: 423.
(c) Wang, T., Wang, C.Y., Shan, Z.Y. et al. (2012). *J. Alzheimers Dis.* 29: 549.
87 Opazo, C., Luza, S., Villemagne, V.L. et al. (2006). *Aging Cell* 5: 69.
88 Bush, A.I. (2008). *J. Alzheimers Dis.* 15: 223.
89 Ritchie, C.W., Bush, A.I., Mackinnon, A. et al. (2003). *Arch. Neurol.* 60: 1685.
90 (a) Adlard, P.A., Cherny, R.A., Finkelstein, D.I. et al. (2008). *Neuron* 59: 43.
(b) Adlard, P.A., Bica, L., White, A.R. et al. (2011). *PLoS One* 6: e17669.
91 Lannfelt, L., Blennow, K., Zetterberg, H. et al. (2008). *Lancet Neurol.* 7: 779.
92 Nguyen, M., Rechignat, L., Robert, A., and Meunier, B. (2015). *ChemistryOpen* 4: 27.
93 Zhang, W., Liu, Y., Hureau, C. et al. (2018). *Chemistry* 24: 7825.
94 Huang, J., Nguyen, M., Liu, Y. et al. (2019). *Eur. J. Inorg. Chem.* 1384–1388.
95 Tsuji, S., Kobayashi, H., Uchida, Y. et al. (1992). *EMBO J.* 11: 4843.
96 Lee, S.J., Seo, B.R., and Koh, J.Y. (2015). *Mol. Brain* 8: 84.
97 Xu, W., Xu, Q., Cheng, H., and Tan, X. (2017). *Sci. Rep.* 7: 13763.
98 (a) Du, X., Li, H., Wang, Z. et al. (2013). *Metallomics* 5: 861. (b) Du, X., Zheng, Y., Wang, Z. et al. (2014). *Inorg. Chem.* 53: 11221.
99 Yue, C., Shan, Z., Tan, Y. et al. (2020). *ACS Chem. Neurosci.* 11: 4098.
100 Bohnen, N.I., Yarnall, A.J., Weil, R.S. et al. (2022). *Lancet Neurol.* 21: 381.
101 Blesa, J., Foffani, G., Dehay, B. et al. (2022). *Nat. Rev. Neurosci.* 23: 115.
102 Zhang, X., Zhang, Y., Li, R. et al. (2020). *Aging (Albany NY)* 12: 9405.
103 (a) Liu, M., Jiao, Q., Du, X. et al. (2021). *Aging Dis.* 12: 2003. (b) Bjørklund, G., Hofer, T., Nurchi, V.M., and Aaseth, J. (2019). *J. Inorg. Biochem.* 199: 110717.
104 (a) Liu, J.X., Zhou, G.B., Chen, S.J., and Chen, Z. (2012). *Curr. Opin. Chem. Biol.* 16: 92. (b) Evens, A.M., Tallman, M.S., and Gartenhaus, R.B. (2004). *Leuk. Res.* 28: 891.
105 Li, S.J., Ren, Y.D., Li, J. et al. (2020). *Life Sci.* 240: 117091.
106 Kim, M.J., Oh, S.B., Kim, J. et al. (2018). *Parkinsonism Relat. Disord.* 55: 117.
107 Scholefield, M., Church, S.J., Xu, J. et al. (2021). *Front. Aging Neurosci.* 13: 641222.
108 Matusch, A., Fenn, L.S., Depboylu, C. et al. (2012). *Anal. Chem.* 84: 3170.
109 Sanyal, J., Ahmed, S.S., Ng, H.K. et al. (2016). *Sci. Rep.* 6: 35097.
110 Fukushima, T., Tan, X., Luo, Y. et al. (2013). *Fukushima J. Med. Sci.* 59: 76.

111 Forte, G., Bocca, B., Senofonte, O. et al. (2004). *J. Neural Transm. (Vienna)* 111: 1031.
112 Ahmed, S.S. and Santosh, W. (2010). *PLoS One* 5: e11252.
113 Gardner, B., Dieriks, B.V., Cameron, S. et al. (2017). *Sci. Rep.* 7: 10454.
114 Maass, F., Michalke, B., Willkommen, D. et al. (2020). *J. Trace Elem. Med. Biol.* 57: 126412.
115 Ajsuvakova, O.P., Tinkov, A.A., Willkommen, D. et al. (2020). *J. Trace Elem. Med. Biol.* 59: 126423.
116 Stefano, F., Cinzia, N., Marco, P. et al. (2016). *Toxics* 4: 27. 1–9.
117 Schirinzi, T., Martella, G., D'Elia, A. et al. (2016). *Parkinson's Dis.* 2016: 9646057.
118 Alimonti, A., Bocca, B., Pino, A. et al. (2007). *J. Trace Elem. Med. Biol.* 21: 234.
119 Ramos, P., Santos, A., Pinto, N.R. et al. (2015). *Biol. Trace Elem. Res.* 163: 89.
120 Tarohda, T., Ishida, Y., Kawai, K. et al. (2005). *Anal. Bioanal. Chem.* 383: 224.
121 Bocca, B., Alimonti, A., Senofonte, O. et al. (2006). *J. Neurol. Sci.* 248: 23.
122 McIntosh, K.G., Cusack, M.J., Vershinin, A. et al. (2012). *J. Toxicol. Environ. Health Part A* 75: 1253.
123 Yu, X., Du, T., Song, N. et al. (2013). *Neurology* 80: 492.
124 Carboni, E., Nicolas, J.D., Töpperwien, M. et al. (2017). *Biomed. Optics Express* 8: 4331.
125 Carmona, A., Roudeau, S., Perrin, L. et al. (2019). *Front. Neurosci.* 13: 1014.
126 Szczerbowska-Boruchowska, M., Krygowska-Wajs, A., and Adamek, D. (2012). *J. Phys. Condens. Matter* 24: 244104.
127 Ortega, R., Carmona, A., Roudeau, S. et al. (2016). *Mol. Neurobiol.* 53, 1925.
128 Robison, G., Sullivan, B., Cannon, J.R., and Pushkar, Y. (2015). *Metallomics* 7: 748.
129 Robison, G., Zakharova, T., Fu, S. et al. (2012). *PLoS One* 7: e48899.
130 Zheng, W., Zhao, Q., Slavkovich, V. et al. (1999). *Brain Res.* 833: 125.
131 Robison, G., Zakharova, T., Fu, S. et al. (2013). *Metallomics* 5: 1554.
132 Davies, K.M., Bohic, S., Carmona, A. et al. (2014). *Neurobiol. Aging* 35: 858.
133 Daoust, A., Barbier, E.L., and Bohic, S. (2013). *NeuroImage* 64: 10.
134 Carmona, A., Devès, G., Roudeau, S. et al. (2010). *ACS Chem. Neurosci.* 1: 194.
135 Szczerbowska-Boruchowska, M., Krygowska-Wajs, A., Ziomber, A. et al. (2012). *Neurochem. Int.* 61: 156.
136 Popescu, B.F., George, M.J., Bergmann, U. et al. (2009). *Phys. Med. Biol.* 54: 651.
137 Chwiej, J., Fik-Mazgaj, K., Szczerbowska-Boruchowska, M. et al. (2005). *Anal. Chem.* 77: 2895.
138 (a) Bortolus, M., Bisaglia, M., Zoleo, A. et al. (2010). *J. Am. Chem. Soc.* 132: 18057. (b) Rasia, R.M., Bertoncini, C.W., Marsh, D. et al. (2005). *Proc. Natl. Acad. Sci. U. S. A.* 102: 4294.
139 Peng, Y., Wang, C., Xu, H.H. et al. (2010). *J. Inorg. Biochem.* 104: 365.
140 He, Q., Song, N., Jia, F. et al. (2013). *Int. J. Biochem. Cell Biol.* 45: 1019.
141 Li, Y., Yang, C., Wang, S. et al. (2020). *Int. J. Biol. Macromol.* 163: 562.
142 Lorentzon, E., Horvath, I., Kumar, R. et al. (2021). *Int. J. Mol. Sci.* 22: 11455. 1–16.

143 (a) Jin, L., Wang, J., Zhao, L. et al. (2011). *Brain* 134: 50. (b) Zhao, X., Shao, Z., Zhang, Y. et al. (2018). *Brain Behav.* 8: e00995.

144 Ayton, S., Lei, P., Duce, J.A. et al. (2013). *Ann. Neurol.* 73: 554.

145 Zelko, I.N., Mariani, T.J., and Folz, R.J. (2002). *Free Radic. Biol. Med.* 33: 337.

146 Wang, Z., Liu, Y., Gao, Z., and Shen, L. (2021). *Clin. Lab.* 67: 1184–1189.

147 Medeiros, M.S., Schumacher-Schuh, A., Cardoso, A.M. et al. (2016). *PLoS One* 11: e0146129.

148 Roy, A., Mondal, B., Banerjee, R. et al. (2021). *J. Neuroimmunol.* 354: 577545.

149 Liu, C., Fang, J., and Liu, W. (2019). *J. Integr. Neurosci.* 18: 299.

150 Zhang, X., Ye, Y.L., Zhu, H. et al. (2016). *PLoS One* 11: e0163372.

151 (a) Guan, X., Xu, X., and Zhang, M. (2017). *Neurosci. Bull.* 33: 561. (b) Tuite, P. (2016). *Transl. Res.* 175: 4. (c) Uchida, Y., Kan, H., Sakurai, K. et al. (2020). *Mov. Disord.* 35: 1396.

152 (a) Du, K., Liu, M.Y., Zhong, X., and Wei, M.J. (2017). *Sci. Rep.* 7: 3902. (b) Jiménez-Jiménez, F.J., Alonso-Navarro, H., García-Martín, E., and Agúndez, J.A.G. (2021). *Eur. J. Neurol.* 28: 1041.

153 Xu, W., Zhi, Y., Yuan, Y. et al. (2018). *J. Neural Transm. (Vienna)* 125: 1027.

154 Devos, D., Cabantchik, Z.I., Moreau, C. et al. (2020). *J. Neural Transm. (Vienna)* 127: 189.

155 Kwiatkowski, A., Ryckewaert, G., Jissendi Tchofo, P. et al. (2012). *Parkinsonism Relat. Disord.* 18: 110.

156 Wang, Y.Q., Wang, M.Y., Fu, X.R. et al. (2015). *Free Radic. Res.* 49: 1069.

157 Gromadzka, G., Tarnacka, B., Flaga, A., and Adamczyk, A. (2020). *Int. J. Mol. Sci.* 21: 9259. 1–35.

158 Khatri, D.K. and Juvekar, A.R. (2016). *Pharmacol. Biochem. Behav.* 150–151: 39.

159 (a) Oggiano, R., Solinas, G., Forte, G. et al. (2018). *Chemosphere* 197: 457. (b) Ingre, C., Roos, P.M., Piehl, F. et al. (2015). *Clin. Epidemiol.* 7: 181.

160 Mancuso, R. and Navarro, X. (2015). *Prog. Neurobiol.* 133: 1.

161 Sirabella, R., Valsecchi, V., Anzilotti, S. et al. (2018). *Front. Neurosci.* 12: 510.

162 (a) Johnson, F.O. and Atchison, W.D. (2009). *Neurotoxicology* 30: 761. (b) Sutedja, N.A., Veldink, J.H., Fischer, K. et al. (2009). *Amyotroph. Lateral Scler.* 10: 302.

163 Roos, P.M., Vesterberg, O., Syversen, T. et al. (2013). *Biol. Trace Elem. Res.* 151: 159.

164 De Benedetti, S., Lucchini, G., Del Bò, C. et al. (2017). *Biometals* 30: 355.

165 (a) Bocca, B., Forte, G., Oggiano, R. et al. (2015). *J. Neurol. Sci.* 359: 11. (b) Forte, G., Bocca, B., Oggiano, R. et al. (2017). *Neurol. Sci.* 38: 1609.

166 Caroli, S., Alimonti, A., Coni, E. et al. (1994). *Crit. Rev. Anal. Chem.* 24: 363.

167 (a) Siklós, L., Engelhardt, J.I., Alexianu, M.E. et al. (1998). *J. Neuropathol. Exp. Neurol.* 57: 571. (b) Jaiswal, M.K. (2013). *Front. Cell Neurosci.* 7: 199.

168 Lin, C.L., Kong, Q., Cuny, G.D., and Glicksman, M.A. (2012). *Future Med. Chem.* 4: 1689.

169 (a) Plaitakis, A. and Caroscio, J.T. (1987). *Ann. Neurol.* 22: 575. (b) Trotti, D., Aoki, M., Pasinelli, P. et al. (2001). *J. Biol. Chem.* 276: 576. (c) Trotti, D., Rolfs, A., Danbolt, N.C. et al. (1999). *Nat. Neurosci.* 2: 848.

170 von Lewinski, F. and Keller, B.U. (2005). *Trends Neurosci.* 28: 494.
171 Mühling, T., Duda, J., Weishaupt, J.H. et al. (2014). *Front. Cell. Neurosci.* 8: 353.
172 Jaiswal, M.K. (2017). *Front. Cell. Neurosci.* 10, 295 %@ 1662: 295. 1–14.
173 (a) Casamassa, A., La Rocca, C., Sokolow, S. et al. (2016). *Glia* 64: 1124.
 (b) Anzilotti, S., Brancaccio, P., Simeone, G. et al. (2018). *Cell Death Dis.* 9: 206.
174 Park, S.B., Kiernan, M.C., and Vucic, S. (2017). *Neurotherapeutics* 14: 78.
175 Brewer, G.J. (2009). *Expert Opin. Investig. Drugs* 18: 89.
176 Halliwell, B. (2006). *J. Neurochem.* 97: 1634.
177 Jomova, K. and Valko, M. (2011). *Toxicology* 283: 65.
178 Tokuda, E., Okawa, E., Watanabe, S. et al. (2013). *Neurobiol. Dis.* 54: 308.
179 Tokuda, E., Watanabe, S., Okawa, E., and Ono, S. (2015). *Neurotherapeutics* 12: 461.
180 Gong, Y.H. and Elliott, J.L. (2000). *Exp. Neurol.* 162: 27.
181 Kim, J., Kim, T.Y., Hwang, J.J. et al. (2009). *Neurobiol. Dis.* 34: 221.
182 Puttaparthi, K., Gitomer, W.L., Krishnan, U. et al. (2002). *J. Neurosci.* 22: 8790.
183 Goto, J.J., Zhu, H., Sanchez, R.J. et al. (2000). *J. Biol. Chem.* 275: 1007.
184 Kaneko, M., Noguchi, T., Ikegami, S. et al. (2015). *J. Neurosci. Res.* 93: 370.
185 Ito, Y., Yamada, M., Tanaka, H. et al. (2009). *Neurobiol. Dis.* 36: 470.
186 Blasco, H., Vourc'h, P., Nadjar, Y. et al. (2011). *J. Neurol. Sci.* 303: 124.
187 Jeong, S.Y., Rathore, K.I., Schulz, K. et al. (2009). *J. Neurosci.* 29: 610.
188 Vinceti, M., Bonvicini, F., Rothman, K.J. et al. (2010). *Environ. Health* 9: 77.
189 Nagata, H., Miyata, S., Nakamura, S. et al. (1985). *J. Neurol. Sci.* 67: 173.
190 Bergomi, M., Vinceti, M., Nacci, G. et al. (2002). *Environ. Res.* 89: 116.
191 Pitts, M.W., Byrns, C.N., Ogawa-Wong, A.N. et al. (2014). *Biol. Trace Elem. Res.* 161: 231.
192 Stanton, J.E., Malijauskaite, S., McGourty, K., and Grabrucker, A.M. (2021). *Front. Mol. Neurosci.* 14: 695873.
193 (a) Weintraub, K. (2011). *Nature* 479: 22. (b) Pinto, D., Pagnamenta, A.T., Klei, L. et al. (2010). *Nature* 466: 368.
194 Huguet, G., Ey, E., and Bourgeron, T. (2013). *Annu. Rev. Genomics Hum. Genet.* 14: 191.
195 Yasuda, H. and Tsutsui, T. (2013). *Int. J. Environ. Res. Public Health* 10: 6027.
196 Lord, C., Brugha, T.S., Charman, T. et al. (2020). *Nat. Rev. Dis. Primers* 6: 5.
197 Chauhan, A., Chauhan, V., Brown, W.T., and Cohen, I. (2004). *Life Sci.* 75: 2539.
198 (a) Vela, G., Stark, P., Socha, M. et al. (2015). *Neural Plast.* 2015: 972791.
 (b) Ha, H.T.T., Leal-Ortiz, S., Lalwani, K. et al. (2018). *Front. Mol. Neurosci.* 11: 405. (c) Pangrazzi, L., Balasco, L., and Bozzi, Y. (2020). *Int. J. Mol. Sci.* 21: 972791. 1–15.
199 Bjorklund, G. (2013). *Acta Neurobiol. Exp. (Wars)* 73: 225.
200 (a) Palmer, R.F., Blanchard, S., and Wood, R. (2009). *Health Place* 15: 18.
 (b) Majewska, M.D., Urbanowicz, E., Rok-Bujko, P. et al. (2010). *Acta Neurobiol. Exp. (Wars)* 70: 196. (c) Jakovcevski, M. and Akbarian, S. (2012). *Nat. Med.* 18: 1194. (d) Jin, Y.H., Clark, A.B., Slebos, R.J. et al. (2003). *Nat. Genet.*

34: 326. (e) Arita, A. and Costa, M. (2009). *Metallomics* 1: 222. (f) Ciesielski, T., Weuve, J., Bellinger, D.C. et al. (2012). *Environ. Health Perspect.* 120: 758.
201 Omotosho, I.O., Akinade, A.O., Lagunju, I.A., and Yakubu, M.A. (2021). *J. Neurodev. Disord.* 13: 50.
202 Marreiro, D.D., Cruz, K.J., Morais, J.B. et al. (2017). *Antioxidants (Basel)* 6: 24. 1–9.
203 (a) Macedoni-Lukšič, M., Gosar, D., Bjørklund, G. et al. (2015). *Biol. Trace Elem. Res.* 163: 2. (b) Mostafa, G.A., Bjørklund, G., Urbina, M.A., and Al-Ayadhi, L.Y. (2016). *Metab. Brain Dis.* 31: 593.
204 Meguid, N.A., Dardir, A.A., Abdel-Raouf, E.R., and Hashish, A. (2011). *Biol. Trace Elem. Res.* 143: 58.
205 (a) Theoharides, T.C., Stewart, J.M., Panagiotidou, S., and Melamed, I. (2016). *Eur. J. Pharmacol.* 778: 96. (b) Freitas, B.C., Mei, A., Mendes, A.P.D. et al. (2018). *Front. Pediatr.* 6: 394.
206 Yang, Z., Yang, S., Qian, S.Y. et al. (2007). *Toxicol. Sci.* 98: 488.
207 Foster, M. and Samman, S. (2012). *Nutrients* 4: 676.
208 Sauer, A.K., Hagmeyer, S., and Grabrucker, A.M. (2016). *Nutritional Deficiency*. Croatia: InTech.
209 (a) Grabrucker, A.M. (2014). *Dev. Neurobiol.* 74: 136. (b) Grabrucker, S., Jannetti, L., Eckert, M. et al. (2014). *Brain* 137: 137. (c) Grabrucker, A.M. (2016). Zinc in the Developing Brain. In: *Nutrition and the Developing Brain*. CRC Press. (d) Grabrucker, A.M. (2020). *The History of Metals in Autism Spectrum Disorders, Biometals in Autism Spectrum Disorders*, 25–41. (e) Grabrucker, A.M. (2020). *Biometals in Autism Spectrum Disorders*, 149. Academic Press.
210 (a) Wang, H., Abel, G.M., Storm, D.R., and Xia, Z. (2019). *Toxicol. Sci.* (b) Kumar, V., Kumar, A., Singh, K. et al. (2021). *Eur. J. Nutr.* 60: 55.
211 Pfaender, S., Föhr, K., Lutz, A.K. et al. (2016). *Neural Plast.* 2016: 3760702.

11

Oncometallomics: Metallomics in Cancer Studies

Xin Wang[1], Chao Li[2], and Yu-Feng Li[3]

[1]*Anhui Medical University, School of Basic Medical Sciences, Department of Chemistry, No.81, Meishan Road, Hefei, Anhui 230032, China*
[2]*Anhui Medical University, The Second Affiliated Hospital, Department of Oncology, No. 678, Furong Road, Hefei, Anhui 230601, China*
[3]*Chinese Academy of Sciences, Institute of High Energy Physics, CAS-HKU Joint Laboratory of Metallomics on Health and Environment, No. 19B, Yuquan Road, Beijing 100049, China*

11.1 Introduction to Oncometallomics

Cancer is a serious threat to human health. It was reported that there were 19.3 million new cancer cases and 10 million cancer deaths worldwide in 2020 [1]. The causes of cancer may be related with chemical carcinogens, physical carcinogens, viral and bacterial infections, genetic factors, age, gender, etc. [2, 3]. Till now, the pathogenesis of most cancer is still unclear; therefore, early diagnosis and timely treatment are the most effective methods for controlling cancer [4].

At present, the common diagnostic methods for cancer include medical imaging examination, tumor marker examination, endoscopy examination, and histopathological examination. However, the above methods are insufficient for early diagnosis of cancer [5, 6]. For example, imageology/endoscopy usually detects cancer when it is in the middle or advanced stage [7], histopathological examination is the gold standard for cancer diagnosis, but it is time-consuming and highly subjective [8]. Therefore, the development of new diagnostic methods is very important for early diagnosis and treatment of cancer.

As an emerging research field, metallomics aims to systematically investigate the interactions and functional relationship between metals/metalloids and their species with genes, proteins, metabolites, and other biomolecules within organisms [9, 10]. Metallomics has been applied in different research field, leading to the branches of metallomics, such as nanometallomics, environmetallomics, and agrometallomics.

Elemental homeostasis is maintained in healthy people, while diseases like cancer may cause or be caused by elemental dyshomeostasis. The application of metallomics in cancer studies can be used for cancer screening and drug development, and here it is proposed to be called oncometallomics. Oncometallomics aims

Applied Metallomics: From Life Sciences to Environmental Sciences, First Edition.
Edited by Yu-Feng Li and Hongzhe Sun.
© 2024 WILEY-VCH GmbH. Published 2024 by WILEY-VCH GmbH.

to understand the occurrence, development, and treatment of cancers from the view of metallomics together with many other medical and clinical techniques. The samples for oncometallomics studies are mainly tissues, blood, hair, nail samples. The general procedure for oncometallomics study is shown in Figure 11.1.

Like other branches of metallomics, in oncometallomics, the quantification of the metallome can be carried out by using inductive-coupled plasma optical emission spectroscopy (ICP-OES), inductive-coupled plasma mass spectrometry (ICP-MS), and atomic absorption spectrophotometer (AAS), which require the digestion of biological samples. X-ray fluorescence analysis (XRF) provides another

Collection of biological samples
(tissue, blood, saliva, hair, nail, etc.)

⬇

Sample pretreatment
(denaturation, digestion, centrifugation, etc.)

⬇

High-throughput quantitative detection of elements
(XRF, SRXRF, ICP-OES, ICP-MS, AAS, etc.)

⬇

Statistical analysis
- Normalization of the data
- Analysis of differentially changed elements (Wilcoxon rank sum test, Mann–Whitney U test, t test, etc.)
- The other methods: PCA, logistic regression analysis, clustering, etc.

⬇

Oncometallomics network analysis
- Construction and analysis of network on elemental correlation
- Analysis of elemental correlation changes

⬇

Construction of metallome-based models for phenotypes classification/prediction
(decision tree, SVM, PLS-DA, etc.)

⬇

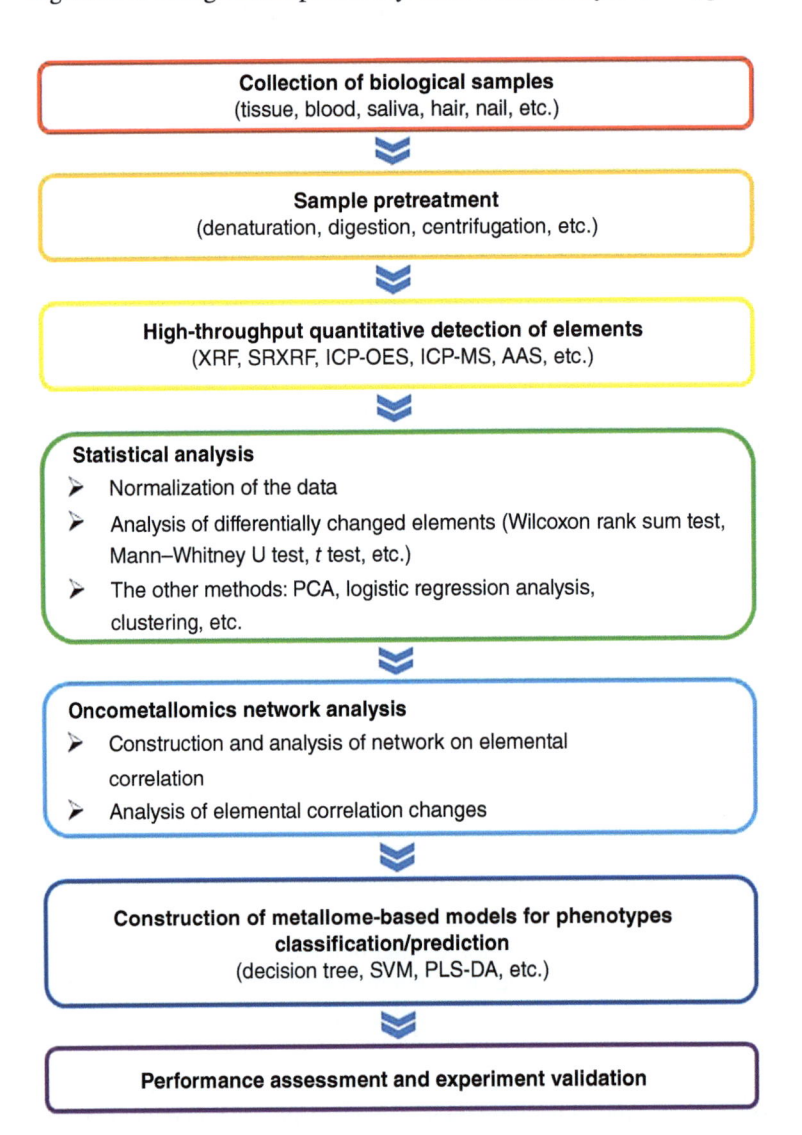

Performance assessment and experiment validation

Figure 11.1 A schematic diagram for oncometallomics [10–12]. Source: Adapted from [10, 13, 14]. Xin Wang.

multi-elemental method for oncometallomics, which does not require the digestion of samples (Figure 11.1) [10]. Moreover, synchrotron radiation X-ray fluorescence analysis (SRXRF) has the merit of low detection limit (ng/g), easy sample pretreatment, and the capacity of detecting most elements at a time [10].

As for data analysis, the machine learning algorithms, deep learning algorithms, and artificial intelligence are very important tools in oncometallomics. For example, principal component analysis (PCA) is an unsupervised classifier which can perform dimension reduction on high-dimension data [13, 14], while nonlinear support vector machine (SVM) and linear partial least squares discriminant analysis (PLS-DA) are the supervised classifiers which can realize the analysis of data with high accuracy (Figure 11.1) [10].

In the following parts, we will showcase the application of oncometallomics for cancer screening/diagnosis and drug development.

11.2 The Application of Oncometallomics in Cancer Studies

11.2.1 The Application of Oncometallomics in Cancer Diagnosis

Oncometallomics has been applied in the screening and diagnosis of several types of cancers like prostate cancer, breast cancer, lung cancer, gastric cancer, colorectal cancer, esophageal cancer, liver cancer, ovarian cancer, cervical cancer, and thyroid cancer.

11.2.1.1 Prostate Cancer

Oncometallomics studies were carried out by using blood or urine samples to help the diagnosis, pathological grading, and clinical staging of prostate cancer [15–17]. It was found that the exposure to trace metals had more significant effects on the occurrence of prostate disease (prostate cancer and benign prostatic hyperplasia) than the exposure to phthalates [18]. Martynko et al. [17] used ICP-OES and AAS to investigate 19 macro- and trace elements in urine samples from 34 prostate cancer patients and 32 controls. Machine learning algorithms such as PCA and logistic regression were applied for data modeling. They observed that the individual element concentrations and their ratios were significantly different between prostate cancer group and the control group. Moreover, classification models achieved 89% accuracy for predicting prostate cancer on the basis of elemental profile, suggesting that oncometallomics can be a simple and non-invasive method for screening prostate cancer.

11.2.1.2 Breast Cancer

Blood, tissue, and fingernail samples in oncometallomics were applied for diagnosing and monitoring radiotherapy/chemotherapy of breast cancer [19–27]. In addition, the regulation of trace elements levels to reduce the breast cancer risk was also proposed [19]. Choi et al. [23] used ICP-MS to study the levels of seven trace elements in the serum samples from female breast cancer patients and healthy controls. It was found that the Mn and Mo levels in the breast cancer group were significantly higher than those in the control group. Besides, the Cu content was positively correlated, while the Se content was negatively correlated with distant metastasis of breast cancer, indicating that trace elements played the potential roles and impacts in breast cancer. Topdagi et al. [24] used ICP-MS and ICP-OES to study the Al, Cu, Ca, Mg, Fe, Ni, and Pb concentrations in the blood samples from 40 female breast cancer patients and 40 healthy adult female controls. It was found that the Cu, Mg, and Al concentrations were statistically significantly different, whereas the Ca and Fe concentrations were not different between breast cancer group and the control group. Furthermore, a positive correlation existed between the concentrations of (Al–Cu), (Fe–Mg), (Cu–Ca), and (Al–Ca) in breast cancer group, which may be used for breast cancer screening.

11.2.1.3 Lung Cancer

Blood, pleural effusion (PE), and nail samples were studied through oncometallomics for the etiology, diagnosis, and prognosis judgement of lung cancer [28–37]. For example, Lee et al. [28] investigated 14 trace elements concentrations and biochemical indexes in the PE from 48 lung cancer patients (22 smokers and 26 non-smokers), and they found that Zn concentration was significantly higher in non-smokers than that in smokers. Zn was found to be negatively correlated with glucose concentration in all smokers, while it was positively correlated with lactate dehydrogenase concentration in current smokers and total protein concentration in non-smokers, indicating the relationship between Zn and certain biochemical indexes in lung cancer patients' PE. In addition, some other studies have explored the effect of individual metal element on lung cancer and found that the content of Mn was significantly lowered in the blood of lung cancer patient, indicating that the deficiency of Mn might play an important role in the occurrence of lung cancer [36]. It was also found that Cu ion could promote cell proliferation within a certain concentration range, while Cu chaperone genes might be used as biomarkers for predicting lung cancer prognosis, and the imbalance of Cu level in the body might lead to lung cancer occurrence and development [37].

11.2.1.4 Gastric Cancer

Blood, hair, nail, tissue, and urine sample had been used as samples to find an effective method for preventing, diagnosing and treating gastric cancer [38–48]. For example, Janbabai et al. [44] compared 11 trace elements levels in hair and nail samples from 73 gastric cancer patients and 83 healthy controls. They found that gastric cancer group had increased Cu, K, Li, P, and Se levels and decreased Mg and Sr levels in hair and nail samples while increased Fe level only in hair samples compared with the control group. Besides, the mean Se, Fe, and P concentrations in the hair and the mean K concentration in the nail were positively correlated with gastric cancer stage, indicating that the increase of some trace elements might be a potential diagnostic marker for predicting cancer progression and etiology. Other studies found that the contents of serum trace elements and superoxide dismutase

(SOD) decreased significantly in gastric cancer patients, and the changes of trace elements and SOD were positively correlated [42]. It was also recommended to regularly monitor some indicators such as trace elements and DNA in nuclei and mitochondria of gastric mucosal epithelium and C-erbB2, P21ras, P53, Ki-67 in gastric mucosa for the early detection of chronic gastritis cancer [46].

11.2.1.5 Colorectal Cancer

Blood and tissue samples had been applied in oncometallomics for the study of colorectal cancer to find effective methods for the pathogenesis, diagnosis, and treatment of colorectal cancer [49–58]. For example, Sohrabi et al. [53] used AAS to study the differences of Zn, Cr, Mn, Sn, Cu, Al, Pb, and Fe levels between colorectal cancer tissues and corresponding normal tissues. It was found that the median of Al, Cu, Zn, Cr, and Pb was significantly lower, while the median of Mn, Fe, and Sn was significantly higher in normal tissues than those in cancerous tissues, indicating that the difference in the above elements levels might play a role in the development of colorectal cancer. Wang et al. [57] used ICP-MS to study 13 trace elements levels in whole blood from 93 colorectal cancer patients and 48 healthy controls. It was noticed that the Mg, Fe, Zn, Se, Sr, Mo, and Ba levels were higher in the control group than those in colorectal cancer group, while the Mg, Fe, Zn, Se, and Ba levels decreased with the increase of TNM stage. Furthermore, the Cu and Ni levels were lower in the control group than those in colorectal cancer group while the Cu level increased with the increase of TNM stage. More interestingly, the Mg, Cr, Fe, Cu, Zn, Se, Sr, and Ba levels were associated with the number of circulating tumor cells, indicating that combined determination of trace elements levels and circulating tumor cells may help the early diagnosis, the staging, and the treatment of colorectal cancer.

11.2.1.6 Esophageal Cancer

Serum and tissue samples were applied in oncometallomics to investigate the diagnosis and etiology of esophageal cancer [59–62]. For example, Xie et al. [60] quantitatively analyzed 22 elements in esophageal squamous cell carcinoma tissues and paired non-cancer tissues. They found that the Mn, Se, Cu, Ti, Mg, Fe, Co, Zn, Sr, Ca concentrations were significantly different between cancer tissues and non-cancer tissues. Among them, Mn, Se, Cu, and Ti were the four most statistically significant elements which were higher in cancer tissues than those in the non-cancer tissues, suggesting that these elements may be applied for the screening of esophageal cancer.

11.2.1.7 Liver Cancer

Serum and tissue samples were applied in oncometallomics for the prognostic study of liver cancer [63] and identification of liver cancer risk [64]. For example, Fu et al. [64] used ICP-MS/MS to accurately determine the levels of Mn, Fe, Ni, Cu, Zn, and Se in serum samples of hepatocellular carcinoma (HCC) patients and healthy controls. They found that the Fe and Cu contents were higher while the Zn and Se contents were lower in HCC patients than those in healthy controls, indicating that the changes in trace elemental concentrations might be associated with the risk of HCC.

11.2.1.8 Ovarian Cancer

Tissue, blood, and ascitic fluid samples were used in oncometallomics for the study of ovarian cancer [65–70]. For example, Chmura et al. [66] used SRXRF to analyze neoplastic and normal ovarian tissues in order to identify elemental changes in spatial distribution and concentration. They confirmed the efficacy of XRF in distinguishing various types of ovarian tumors on the basis of elements and found that K, Fe, S, Cl, Zn, Br, and Rb were the most important elements for classifying neoplastic and normal ovarian tissues. Besides, Zhang et al. [69] found that Zn facilitated ovarian tumor metastasis, while Onuma et al. [70] clarified that elevated Cu in ascitic fluid was related with the progression of malignant ovarian tumor.

11.2.1.9 Cervical Cancer

Serum, plasma, and tissue samples had been used in the study of cervical cancer [71–73]. For example, Okunade et al. [72] compared serum Zn, Cu, and Se concentrations from 50 cervical cancer patients and 100 healthy volunteers. It was found that the Zn and Se levels were lower in cervical cancer patients, indicating that these changes in trace elemental levels might play an important role in cervical cancer pathogenesis. In addition, this study may offer a solution to improve therapeutic effect of cervical cancer through providing these trace elements routinely; however, this method is needed to be determined by robust prospective studies.

11.2.1.10 Thyroid Cancer

Blood, tissue, and fingernail samples had been applied in the oncometallomics study of thyroid cancer [74–76]. For example, Stojsavljevic et al. [76] used ICP-MS to investigate the levels of trace elements in blood samples from multinodular goiter (MNG) patients, thyroid adenoma (TA) patients, thyroid cancer (TC) patients, and healthy controls. It was found that the Mn, Co, Ni, Cu, Zn, Se, and Pb contents were lower, while the As, Cd, and U contents and the Cu/Zn and U/Se ratios were higher in pathological blood samples than those of the control samples. In addition, Co and Zn were most important for distinguishing MNG patients from TC patients while Co, Zn, and Mn influenced the identification of healthy controls from TA patients, which might lead to the establishment of new circulating screening markers for TC.

Besides the above examples, oncometallomics has also been used to investigate skin cancer [77, 78], bladder cancer [79, 80], brain tumor [81], vaginal cancer [82], and pediatric cancers [83, 84] etc. In summary, oncometallomics may provide new ideas and methods for early diagnosis of cancer, grading and staging of cancer, plan design of cancer treatment, evaluation of cancer prognosis, and the follow-up treatment of cancer.

11.2.2 The Application of Oncometallomics in Cancer Treatment

Cancer treatment includes surgery, chemotherapy, radiotherapy, biotherapy, and photodynamic therapy. Metallic drugs like platinum drugs have been used clinically for the chemotherapy of cancer. We will briefly introduce the application of oncometallomics in cancer chemotherapy taking platinum drugs as an example.

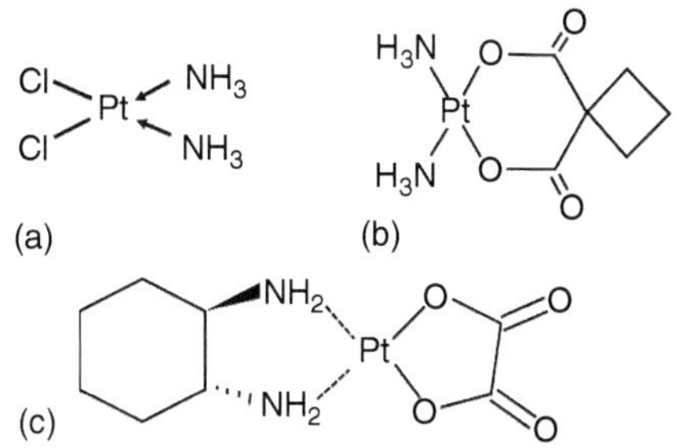

Figure 11.2 Chemical structures of (a) Cisplatin, (b) Carboplatin, and (c) Oxaliplatin. Source: Chao Li (author).

Cisplatin (Figure 11.2a) is one of the commonly used chemotherapy drugs clinically, which has been widely used for treating lung cancer, ovarian cancer, esophageal cancer, gastric cancer, and so on due to its strong anticancer effect [85]. However, cisplatin has nephrotoxicity and causes some complications such as electrolyte disturbance [86]. Carboplatin (Figure 11.2b) is one of the second-generation platinum drugs which is less nephrotoxic than cisplatin. Now carboplatin is being commonly used in the treatment of lung cancer, ovarian cancer, etc. [87]. However, the clinical application of carboplatin is limited due to its inhibitory action on bone marrow. Oxaliplatin (Figure 11.2c) is one of the third-generation platinum drugs. Compared with cisplatin and carboplatin, its nephrotoxicity and bone marrow toxicity are significantly reduced. Oxaliplatin is commonly used in the treatment of colorectal cancer and gastric cancer, etc. [88]. However, oxaliplatin can cause peripheral neurotoxicity, but the symptoms will gradually improve after drug withdrawal.

Oncometallomics has been used to reveal the anticancer mechanism of platinum drugs. For example, Hambley et al. used X-ray absorption near-edge spectroscopy and other techniques to study the distribution and metabolism in ovarian cancer cells after being treated by cisplatin for 24 hours, and Pt was found to mainly concentrate in the nucleus [89]. Lobinski et al. found that most of oxaliplatin gathered at the edge of cancer tissues, while cisplatin could penetrate deeply into cancer tissues by using laser ablation (LA) ICP MS and matrix-assisted laser desorption ionization imaging, suggesting that different action mechanism existed between oxaliplatin and cisplatin [90]. More works are still required for the study of new drugs, especially metallic drugs for cancer treatment through oncometallomics [91].

11.3 The Metallome that Involved in the Occurrence and Development of Cancer

Oncometallomics studies have revealed that trace elements such as Se, Fe, Mn, Zn, As, Cd, Ni, Cu, Pb, Co, and Mg play roles in the occurrence and development of cancer.

Se: Se can resist oxidation, block cancer cell cycle, inhibit cancer cell proliferation [92, 93]; hence, an appropriate amount of Se can reduce the risk of liver cancer, colon cancer, and so on [94, 95]. But excessive Se will lead to Se poisoning.

Fe: Excessive Fe can promote the occurrence and development of some cancers which will increase the risk of colorectal cancer and liver cancer by inducing oxidative damage [53, 66]. However, Fe deficiency may be a risk factor for gastric cancer and esophageal cancer [62].

Mn: Mn is an important component of SOD and it can scavenge free radicals. Low Mn content will increase the incidence rate of lung cancer, esophageal cancer, thyroid cancer, etc. [28, 60, 76] while high Mn content will lead to an increased risk of prostate cancer [96].

Zn: Zn deficiency can reduce immune function and promote the occurrence and development of cancer [97], and some researchers have shown the significant decrease of serum Zn in patients with lung cancer, colorectal cancer, liver cancer, etc. [28, 53, 64]. However, Zn excess will cause Zn poisoning and the inhibition of immune cell function [98].

As: As is mainly absorbed through the digestive tract and the respiratory tract, which can induce distortion, mutation, or canceration of normal cells [99]. In addition, As can increase the risk of lung cancer, bladder cancer, and so on [33, 100].

Cd: Cd acts on human body through genotoxic mechanism, including the inhibition of DNA repair, the activation of protooncogene, and the inhibition of tumor cell apoptosis [101]. In addition, Cd can increase the risk of lung cancer, prostate cancer, and so on [95, 102].

Ni: Ni promotes the formation of cancer by inhibiting cell signal transduction, inducing DNA chromosome deletion and aberration, inhibiting DNA repair, etc. [103] For example, Ni can increase the risk of lung cancer, prostate cancer, and so on [104, 105].

Cu: Cu plays a very important role in the human body, which is an important substance for constituting and synthesizing some enzymes [106]. However, high concentration of Cu may cause cell proliferation and even canceration, while Cu deficiency can cause decreased immune function [107].

Pb: Pb exposure can cause DNA and chromosome variation [108]. In addition, many epidemiological studies have shown that Pb exposure can increase the risk of several cancers such as lung cancer, laryngocarcinoma, and colorectal cancer [109].

Co: Proper amount of Co can prevent and treat thyroid tumors [110]. However, Co excess can cause respiratory disease, heart disease, cancer, and so on [111], while Co deficiency can cause macrocytic anemia, acute leukemia, and so on [110].

Mg: Mg is involved in DNA replication, DNA repair, and cell proliferation, and some studies have shown that long-term lack of Mg can increase the risk of liver cancer, colon cancer, and so on [112, 113].

11.4 Conclusions and Perspectives

As a branch of metallomics, oncometallomics is dedicated to study cancer in the following aspects as shown in Figure 11.3.

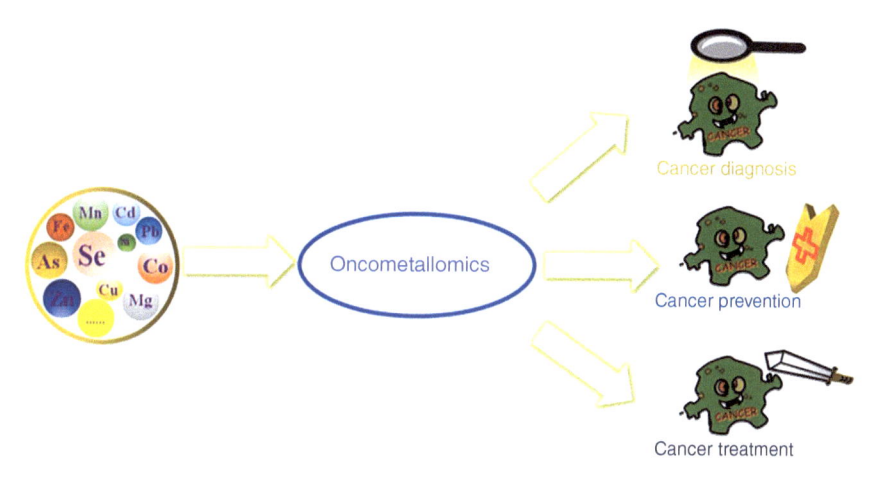

Figure 11.3 Oncometallomics as a tool for cancer diagnosis and treatment. Source: Xin Wang (author).

(1) *Cancer diagnosis.* The application of oncometallomics in cancer diagnosis has been gradually deepened, and new methodologies continue to emerge. For example, Li et al. proposed non-targeted metallomics method by comparing XRF spectra directly to find metallome differences without the transformation to quantitative data. This method was fast and accurate to distinguish cancer patients from healthy controls when combined with machine learning algorithms [10].

(2) *Cancer prevention.* It was found that the supplementation of some elements can effectively prevent cancer; for example, Se supplementation or Mn supplementation can reduce the risk of liver cancer [114, 115], while Zn or Fe supplementation can reduce the risk of esophageal cancer [116]. Hence the beneficial elements can be supplemented from agricultural production, eating habits, and food additives to prevent cancer. In addition, the reduced exposure to hazardous elements can also prevent cancer.

(3) *Cancer treatment.* Metallic drugs play a unique role in the treatment of cancer. For example, new platinum drugs such as oxaliplatin are used clinically with side effects significantly reduced compared with cisplatin. TOOKAD is used as a photosensitizer to treat prostate cancer, while Se compounds can enhance the sensitivity of cancer cells to radiotherapy and chemotherapy and reduce the side effects [117–119]. It is believed that there will be more metallic drugs for clinic use in the future.

Although oncometallomics has been applied in cancer study and it was found that some elements are highly connected to different cancers, more researches are required to understand the interlinks between different elements in cancer prevention, diagnosis, and treatment. Besides, with the fast development of artificial intelligence, it is highly desired to apply this emerging technology with oncometallomics to find novel prevention, diagnosis, and treatment methods for cancers.

Acknowledgments

The authors are grateful for the financial support from the National Natural Science Foundation of China (11975247), Natural Science Foundation of Anhui Province (No. 2108085MA26), Research Fund of Anhui Institute of translational medicine (No. 2022zhyx-C77), University Natural Science Research Project of Anhui Province (KJ2020A0196), and Discipline Construction Project of Anhui Medical University (No. ZCXK10). The authors also thank Jie Shi and Xinyao Jiang from Anhui Medical University for their kind help in drawing the figures.

List of Abbreviations

AAS	atomic absorption spectrophotometer
HCC	hepatocellular carcinoma
ICP-MS	inductive-coupled plasma mass spectrometry
ICP-OES	inductive-coupled plasma optical emission spectroscopy
LA	laser ablation
MNG	multinodular goiter
PCA	principal component analysis
PE	pleural effusion
PLS-DA	partial least squares discriminant analysis
SOD	superoxide dismutase
SRXRF	synchrotron radiation X-ray fluorescence analysis
SVM	support vector machine
TA	thyroid adenoma
TC	thyroid cancer
XRF	X-ray fluorescence analysis

References

1 Sung, H., Ferlay, J., Siegel, R.L. et al. (2021). *CA-Cancer J. Clin.* 71: 209.
2 Bleyer, A., Spreafico, F., and Barr, R. (2020). *Cancer* 126: 46.
3 Signal, V., Gurney, J., Inns, S. et al. (2020). *J. Roy. Soc. New Zeal.* 50: 397.
4 Hou, X., Shen, G.Y., Zhou, L.Q. et al. (2022). *Front. Oncol.* 12: 851367.
5 Yang, X.F., Wang, Y., Zhang, H.J. et al. (2022). *Biomed. Opt. Express* 13: 300.
6 Qu, C., Zeng, P.E., Wang, H.Y. et al. (2022). *J. Magn. Reson. Imaging* 55: 1625.
7 Bjerring, O.S., Fristrup, C.W., Pfeiffer, P. et al. (2019). *Brit. J. Surg.* 106: 1761.
8 Cheng, X.F., Tan, L., and Ming, F.P. (2021). *Math. Probl. Eng.* 2021: 7010438.
9 Li, Y.-F. and Sun, H.Z. (2021). *At. Spectrosc.* 42: 227.
10 He, L.N., Lu, Y., Li, C. et al. (2022). *Talanta* 245: 123486.
11 Ying, H.M. and Zhang, Y. (2018). *Acta Physiol. Sinica* 70: 413.
12 Zhang, Y., Xu, Y.Z., and Zheng, L. (2020). *Int. J. Mol. Sci.* 21: 8646.

13 Barman, I., Dingari, N.C., Singh, G.P. et al. (2012). *Anal. Bioanal. Chem.* 404: 3091.
14 Wu, Q.P., Li, C., Qi, Z.M. et al. (2020). *Infrared Phys. Techn.* 105: 103201.
15 Wang, Y.Y. and Zhao, T.J. (2018). *Modern Oncol.* 26: 2883.
16 Ke, Y.X., Gu, L., Xiang, M., and Yang, K. (2021). *Chinese J. Androl.* 35: 66.
17 Martynko, E., Oleneva, E., Andreev, E. et al. (2021). *Microchem. J.* 159: 105464.
18 Chang, W.H., Yen, C.C., Lee, Y.H., and Chen, H.L. (2018). *Environ. Int.* 221: 1179.
19 Ding, X., Jiang, M., Jing, H.Y. et al. (2015). *Environ. Sci. Pollut. Res.* 22: 7930.
20 Ahmadi, N., Mahjoub, S., Hosseini, R.H. et al. (2018). *Casp. J. Intern. Med.* 9: 134.
21 Cabré, N., Luciano-Mateo, F., Arenas, M. et al. (2018). *Breast* 42: 142.
22 O'Brien, K.M., White, A.J., Sandler, D.P. et al. (2019). *Epidemiology* 30: 112.
23 Choi, R., Kim, M.J., Sohn, I. et al. (2019). *Nutrients* 11: 37.
24 Topdagi, O., Toker, O., Bakirdere, S. et al. (2020). *Atomic Spectrosc.* 41: 29.
25 Huynh, P.T., Tran, T.P.N., Dinh, B.T. et al. (2020). *Nucl. Ch.* 324: 663.
26 Al-Dahhan, N.A.A., Al-hashemi, W.H.M., and Alkadhimi, B.J.H. (2020). AIP Conference Proceedings. Hashim, J.H., Joda, B.A., Aaber, Z. S. et al. (ed.). *International conference on applied science and technology*, Karbala, 030015.
27 Adeoti, M.L., Oguntola, A.S., Akanni, E.O. et al. (2022). *Indian J. Cancer* 52: 106.
28 Lee, K.Y., Feng, P.H., Chuang, H.C. et al. (2012). *Biol. Trace Elem. Res.* 182: 14.
29 Zhang, J.H., Wu, X.N., and Li, H.Z. (2013). *Gansu Med. J.* 32: 569.
30 Zhang, P., Fu, L., Huang, J.H., and Yang, H.J. (2014). *Chin. J. Pharm. Anal.* 34: 1348.
31 Dong, L., Zhang, X.X., Wang, J.L. et al. (2014). *Guangdong Trace Elem. Sci.* 21: 28.
32 Li, X.M., Chen, X.B., Li, G.J. et al. (2019). *J. Kunming Med. Univ.* 40: 29.
33 Unrine, J.M., Slone, S.A., Sanderson, W. et al. (2019). *PLoS One* 14: e0212340.
34 Li, W.J., Li, W.P., Jin, F.G. et al. (2019). *Lung Dis. (Electronic Edition)* 12: 335.
35 Saikawa, H., Nagashima, H., Cho, K. et al. (2021). *Medicina-Lithuania* 57: 209.
36 Yang, Q.B., Dong, L., Zhang, X.X. et al. (2014). *Anhui Med. J.* 35: 1644.
37 Wu, R., Wang, G.Z., Cheng, X., and Zhou, G.B. (2021). *China Oncol.* 31: 1072.
38 Yan, H.Y. and Dong, M.Q. (2010). *J. Qinghai Med. College* 31: 265.
39 Sun, Y.A., Xu, Z.B., Wang, G.Q. et al. (2011). *J. Chin. Spectrosc. Lab.* 28: 2815.
40 Liu, Y.Y., Hu, H.N., Chen, W. et al. (2011). *J. Chin. Pract. Diagn. Therapy* 25: 228.
41 Wang, L.L. and Yang, L.L. (2012). *Food and Drug* 14: 424.
42 Zhang, C.X., Wang, Y., Zhang, L.C. et al. (2014). *J. Chin. Front. Med. Sci. (E-version)* 6: 64.
43 Li, J.N., Feng, M., Zhao, X.Y. et al. (2015). *J. Shenyang Pharm. Univ.* 32: 271.
44 Janbabai, G., Alipour, A., Ehteshami, S. et al. (2018). *Ind. J. Clin. Biochem.* 33: 450.
45 Ji, J., Bai, X.F., and Liu, Y. (2019). *J. Baotou Med. College* 35: 1.

46 Sun, C.J., Yin, G.Y., Zhang, W.N., and Chen, Y. (2019). *J. Liaoning Trad. Chinese* 46: 238.
47 Lin, Y., Wu, C., Yan, W. et al. (2020). *Cancer Manag. Res.* 12: 4441.
48 Afzal, A., Qayyum, M.A., and Shah, M.H. (2021). *Biointerface Res. App.* 11: 10824.
49 Khoshdel, Z., Naghibalhossaini, F., Abdollahi, K. et al. (2016). *Biol. Trace Elem. Res.* 170: 294.
50 Ribeiro, S.M.D., Moya, A.M.T.M., Braga, C.B.M. et al. (2016). *Acta Cir. Bras.* 31: 24.
51 Ribeiro, S.M.D., Braga, C.B.M., Peria, F.M. et al. (2016). *Biol. Trace Elem. Res.* 169: 8.
52 Stepien, M., Jenab, M., Freisling, H. et al. (2017). *Carcinogenesis* 38: 699.
53 Sohrabi, M., Gholami, A., Azar, M.H. et al. (2018). *Biol. Trace Elem. Res.* 183: 1.
54 Juloski, J.T., Rakic, A., Cuk, V.V. et al. (2020). *J. Trace Elem. Med. Bio.* 59: 126451.
55 Nawi, A.M., Chin, S.F., Mazlan, L., and Jamal, R. (2020). *Sci. Rep.* 10: 18670.
56 Ranjbary, A.G., Mehrzad, J., Dehghani, H. et al. (2020). *Biol. Trace Elem. Res.* 194: 66.
57 Wang, H.T., Liu, H., Zhou, M.J. et al. (2020). *Biol. Trace Elem. Res.* 198: 58.
58 Cabral, M., Kuxhaus, O., Eichelmann, F. et al. (2021). *Eur. J. Nutr.* 60: 3267.
59 Joshaghani, H., Mirkarimi, H.S., Besharat, S. et al. (2017). *Middle East J. Digest. Dis.* 9: 81.
60 Xie, B.B., Lin, J.Q., Sui, K. et al. (2019). *Talanta* 196: 585.
61 Mirzaee, M., Semnani, S., Roshandel, G. et al. (2020). *J. Clin. Lab. Anal.* 34: e23269.
62 Sohrabi, M., Nikkhah, M., Sohrabi, M. et al. (2021). *J. Trace Elem. Med. Bio.* 68: 126761.
63 Udali, S., De Santis, D., Mazzi, F. et al. (2020). *Front. Oncol.* 10: 596040.
64 Fu, L., Xie, H.L., Huang, J.H., and Chen, L. (2020). *Anal. Chim. Acta* 1112: 1.
65 Canaz, E., Kilinc, M., Sayar, H. et al. (2017). *J. Trace Elem. Med. Biol.* 43: 217.
66 Chmura, L., Grzelak, M., Czyzycki, M. et al. (2017). *J Physiol. Pharmacol.* 68: 699.
67 Krauze, D.M., Grzelak, M., Wrobel, P. et al. (2018). *J. Anal. Atom. Spectrom.* 33: 1638.
68 Caglayan, A., Katlan, D.C., Tuncer, Z.S., and Yuce, K. (2019). *J. Trace Elem. Med. Biol.* 52: 254.
69 Zhang, R.T., Zhao, G.N., Shi, H.R. et al. (2020). *Free Radical Bio. Med.* 160: 775.
70 Onuma, T., Mizutani, T., Fujita, Y. et al. (2021). *J. Trace Elem. Med. Bio.* 68: 126865.
71 Ji, J., Liu, J., and Liu, H.J. Y. L(2014). *Biol. Trace Elem. Res.* 159: 346.
72 Okunade, K.S., Dawodu, O.O., Salako, O. et al. (2018). *Pan Afr. Med. J.* 31: 194.
73 Meghana, G.S., Kalyani, R., Sumathi, M.E., and Sheela, S.R. (2019). *J. Clin. Diagn. Res.* 13: EC17.
74 Baltaci, A.K., Dundar, T.K., Aksoy, F., and Mogulkoc, R. (2017). *Biol. Trace Elem. Res.* 175: 57.

75 Zidane, M., Ren, Y., Xhaard, C. et al. (2019). *Asian Pac. J. Cancer P.* 20: 355.
76 Stojsavljevic, A., Rovcanin, B., Jagodic, J. et al. (2021). *Biol. Trace Elem. Res.* 199: 4055.
77 Zhao, T.T., Gong, J.S., Li, J.F., and Liang, G. (2015). *J. Taiyuan Norm. Univ. (Nat. Sci. Edit.)* 14: 89.
78 Matthews, N.H., Koh, M., Li, W.Q. et al. (2019). *Cancer Epidem. Biomar.* 28: 1534.
79 Wach, S., Weigelt, K., Michalke, B. et al. (2017). *J. Trace Elem. Med. Bio.* 46: 150.
80 Moazed, V., Jafari, E., Ebadzadeh, M.R. et al. (2021). *Int. J. Cancer Manag.* 14: e106642.
81 Zhong, L.Y., Shan, X., Chen, B.S., and Jin, Q. (2019). *J. Med. Res.* 48: 27.
82 Dring, J.C., Forma, A., Chilimoniuk, Z. et al. (2022). *Nutrients* 14: 185.
83 Gokcebay, D.G., Emir, S., Bayhan, T. et al. (2018). *J. Pediat. Hematol. Oncol.* 40: e343.
84 Chung, J.H., Phalke, N., Hastings, C. et al. (2021). *Pediatr. Blood Cancer* 68: e29104.
85 Chattopadhyay, D. (2022). *Resonance* 27: 659.
86 Liu, S., Wen, X., Huang, Q.H. et al. (2022). *Antioxidants-Basel* 11: 1141.
87 Moore, K.N., Chambers, S.K., Hamilton, E.P. et al. (2022). *Clin. Cancer Res.* 28: 36.
88 Xu, Z.F., Wang, Z.G., Deng, Z.Q., and Zhu, G.Y. (2021). *Coordin. Chem. Rev.* 442: 213991.
89 Hall, M.D., Foran, G.J., Zhang, M. et al. (2003). *J. Am. Chem. Soc.* 125: 7524.
90 Bianga, J., Bouslimani, A., Bec, N. et al. (2014). *Metallomics* 6: 1382.
91 Wang, H.B., Wang, Y.C., and Sun, H.Z. (2016). *Metallomics* (ed. Y.F. Li, H.Z. Sun, C.Y. Chen, and Z.F. Chai). Beijing, China: Science Press Ch. 3.
92 Ip, C., Dong, Y., and Ganther, H.E. (2002). *Cancer Metast. Rev.* 21: 281.
93 Bera, S., De Rosa, V., Rachidi, W., and Diamond, A.M. (2013). *Mutagenesis* 28: 127.
94 Nawi, A.M., Chin, S.F., Azhar Shah, S., and Jamal, R. (2019). *Iran. J. Public Health* 48: 632.
95 Kazi, T.G., Kolachi, N.F., Afridi, H.I. et al. (2012). *Biol. Trace Elem. Res.* 150: 81.
96 Qayyum, M.A. and Shah, M.H. (2014). *Biol. Trace. Elem. Res.* 162: 46.
97 Rucker, D., Thadhani, R., and Tonelli, M. (2010). *Semin. Dial.* 23: 389.
98 Almutairi, R.M., Al-ayed, M.S., Abdelhalim, M.A., and Ghannam, M.M. (2022). *Egypt. J. Chem.* 65: 314.
99 El-Atta, H.M.A., El-Bakary, A.A., Attia, A.M. et al. (2014). *Biol. Trace Elem. Res.* 162: 95.
100 Karagas, M.R., Tosteson, T.D., Morris, J.S. et al. (2004). *Cancer Cause Control* 15: 465.
101 Chen, Z.M., Gu, D.W., Zhou, M. et al. (2016). *Chem-Biol. Interact.* 243: 35.
102 Cobanoglu, U., Demir, H., Sayir, F. et al. (2010). *Asian Pac. J. Cancer P.* 11: 1383.
103 Pavela, M., Uitti, J., and Pukkala, E. (2017). *Am. J. Ind. Med.* 60: 87.

104 Kang, Y.T., Hsu, W.C., Ou, C.C. et al. (2020). *Int. J. Mol. Sci.* 21: 619.
105 Guntupalli, J.N.R., Padala, S., Gummuluri, A.V.R.M. et al. (2007). *Eur. J. Cancer Prev.* 16: 108.
106 Cemek, M., Buyukokuroglu, M.E., Buyukben, A. et al. (2010). *Pediatr. Cardiol.* 31: 1002.
107 Scheiber, I., Dringen, R., and Mercer, J.F. (2013). *Met. Ions Life Sci.* 13: 359.
108 Duydu, Y., Dur, A., and Süzen, H.S. (2005). *Biol. Trace Elem. Res.* 104: 121.
109 Barry, V. and Steenland, K. (2019). *Environ. Res.* 177: 108625.
110 Li, Q.R., Su, B., and Li, S.C. (2008). *Guangdong Trace Elem. Sci.* 15: 66.
111 Turkdogan, M.K., Karapinar, H.S., and Kilicel, F. (2022). *J. Trace Elem. Med. Bio.* 72: 126978.
112 Tukiendorf, A. and Rybak, Z. (2004). *Magnes. Res.* 17: 46.
113 Polter, E.J., Onyeaghala, G., Lutsey, P.L. et al. (2019). *Cancer Epidemiol. Biomark. Prev.* 28: 1292.
114 Chen, J.G. (2001). *China Cancer* 10: 3.
115 Ma, X., Yang, Y., Li, H.L. et al. (2017). *Int. J. Cancer* 140: 1050.
116 Ma, J.F., Li, Q.W., Fang, X.X. et al. (2018). *Nutr. Res.* 59: 16.
117 Shah, N. and Dizon, D.S. (2009). *Future Oncol.* 5: 33.
118 Azzouzi, A.R., Lebdai, S., Benzaghou, F., and Stief, C. (2015). *World J. Urol.* 33: 937.
119 Handa, E., Puspitasari, I.M., Abdulah, R. et al. (2020). *J. Trace Elem. Med. Biol.* 62: 126653.

12

Bio-elementomics

Dongfang Wang[1], Jing Wu[2], Bing Cao[3], Lailai Yan[4], Qianqian Zhao[5], Tiebing Liu[6], and Jingyu Wang[4]

[1] *The First Affiliated Hospital of Chongqing Medical University, NHC Key Laboratory of Diagnosis and Treatment on Brain Functional Diseases, No. 1 Youyi Road, Yuzhong District, Chongqing 400016, China*
[2] *Peking University, Medical and Health Analysis Center, Biological Imaging and Analysis Laboratory, No. 38 Xueyuan Road, Haidian District, Beijing 100191, China*
[3] *Southwest University, Faculty of Psychology, No. 2 Tiansheng Road, Beibei District, Chongqing 400715, China*
[4] *Peking University, School of Public Health, Department of Laboratorial Science and Technology, No. 38 Xueyuan Road, Haidian District, Beijing 100191, China*
[5] *Beijing Shijingshan District Center for Disease Prevention and Control, Department of Epidemiology, Building 2, Courtyard No. 6, Stadium South Road, Beijing 100043, China*
[6] *Civil Aviation Administration of China (Civil Aviation General Hospital), Civil Aviation Medicine Center, Civil Aviation Public Health Emergency Management Office, No. 76 Chaoyang Road, Chaoyang District, Beijing 100123, China*

12.1 Introduction

12.1.1 The Concept of Bio-elementomics

Bio-elements are the collection of all inorganic elements that have biological significance and exist in various forms in living organisms, including biological macromolecules (such as metalloproteins and metalloenzymes), organic small molecules (such as metal compounds and complexes), and free ions. Bio-elementomics is a discipline that systematically studies the distribution, chemical species, content, structural characteristics, and functions of bio-elements in a specific system.

12.1.2 The Development History of Bio-elementomics

There are around 20 elements defined as essential, though the precise requirements can differ within different organisms, including the "organic" and "bulk" elements (H, C, N, and O), seven "macro-minerals" (Na, K, Mg, Ca, l, P, and S), and some "trace elements," namely Mn, Fe, Cu, Zn, Se, Co, Mo, and I. At present, some other elements are under discussion to be included as "essential," for instance V, Ni, Br, Si, and Sn. For As and Pb, which are considered to be toxic elements, positive effects have been discussed for certain organisms, indicating the uncertainties in classification of many trace elements [1]. In fact, the dose–response diagram shows that also

Applied Metallomics: From Life Sciences to Environmental Sciences, First Edition.
Edited by Yu-Feng Li and Hongzhe Sun.
© 2024 WILEY-VCH GmbH. Published 2024 by WILEY-VCH GmbH.

essential elements for human life can be toxic if the dose is high enough, pointing to the Paracelsus principle "the dose makes the poison" [2].

In addition to above essential and toxic elements, with the rapid development of analytical technology, the number of bio-elements in human organism will be further increased. Simultaneous multielement detection capability of analytical plasma spectrometry, such as ICP-AES and ICP-MS, allows the comprehensive analysis of almost all elements as a whole for the study on their biological functions and interactions. Genomics, proteomics, and metabolomics, as omics-sciences, have newly evolved in the life sciences and molecular cell biology since the late 1990s. Although genes and proteins are the key biomaterials in the construction, regulation, and maintenance of the animals, plants, and microorganisms, the biological functions and physiology of life systems also require the involvement of various bio-elements. Thus, it was highly desirable to establish a new scientific field for elements, which might be complementary to genomics, proteomics, and metabolomics. Considering these research trends in the life sciences, the concept of "bio-elementomics" came to mind.

In 2004, prior to the proposal of "bio-elementomics," the term "metallomics" has been coined to describe integrated biometal science [3], and great progress has been achieved as one of the omics-sciences. Not limited to metals, bio-elementomics also focuses on non-metal elements and metalloid elements, since these elements are also basic units making up cells, proteins, and small molecules of life.

Here, "bio-elementomics" is proposed as a new scientific field in order to integrate the research fields related to bio-elements. The study of bio-elements in biological systems is an increasingly important area of research. Bio-elementomics is receiving great attention as a new frontier in the study of various elements in life sciences.

12.1.3 Research Scope

The research scope of bio-elementomics covers investigations on the ions, valence, speciation, concentration, temporal and spatial distribution, biological function of inorganic elements of an entire biological system, and investigations on the interactions between bio-elements and the organism's other "omes," such as the genome, proteome, or metabolome. It is a global discipline encompassing many areas including biology, chemistry, geology, medicine, physics, and pharmacy.

12.2 Basic Laws of Bio-elementomics

12.2.1 Review of Bio-elementomics

Although the concentrations of bio-elements in the human body are low, they have extremely powerful biological effects. For example, the metal elements are the components or activators of enzymes, and they play unique roles in metabolic of hormones and vitamins and nucleic acid metabolism, etc. [2, 4]

12.2.2 Organizational Selectivity of Bio-elements

The distribution of bio-elements in human body is not only uneven, but also has a circadian cyclical difference. The uneven distribution of metal elements in the organ is related to its function. For example, different organs of animals selectively accumulate different trace elements, such as cobalt in the heart, iron in the liver, and calcium and magnesium in the bone marrow. Zinc is most abundant in retina, choroid (a thin film in the eyeball, composed of fibrous tissue, small blood vessels and capillaries, brown-red, between the sclera and retina) and prostate, suggesting that zinc may play an important role in vision and sexual organ development. In fact, zinc deficiency can cause visual disturbances, such as night blindness, and can also lead to male infertility. The uneven distribution of harmful elements in the human body reflects the selectivity of element toxicity. Organic mercury can accumulate in the brain, and its target organ is the brain; inorganic mercury is stored in the kidney, which mainly causes kidney disease.

12.2.3 Specific Correlation of Bio-elements

All bio-elements in the body are directly or indirectly connected, and they affect each other and are in a state of balance [5]. Once the balance of multiple elements is disrupted, pathological processes will develop, affecting health and even cause disease. For instance, there is usually a certain proportional relationship between zinc and copper in the body. Competitive inhibition can occur between the two in the process of absorption and transport. Studies have shown that the imbalance of zinc and copper metabolism is an important risk factor for arteriosclerosis. Short-term large or long-term small amounts of zinc can inhibit the body's absorption and utilization of copper. Excessive intake of copper can also inhibit zinc absorption and accelerate zinc excretion. In addition to zinc and copper, some other elements also have such mutual inhibition correlation, such as iron can increase the absorption of fluorine and inhibit the absorption of manganese and cobalt. An appropriate amount of iron supply can reduce the absorption of lead and prevent the adverse effects of lead. Calcium is the main component of bones and teeth and is closely related to its formation, growth, and maintenance. The precipitation and dissolution of calcium in the bone continue to maintain a dynamic balance. Anything that disrupts this homeostasis will affect bone and dental disease.

The bio-elements in the organism do not exist in isolation. There is a balance between elements, either synergy or antagonism. Changes in one element often affect one or more other elements. Therefore, when analyzing the effects of elements on the health of the body, it is necessary to comprehensively consider a variety of elements; otherwise, it is easy to ignore one and the other, and the conclusion is one-sided. In recent years, some scholars have found that the distribution of essential trace elements in the periodic table has certain rules, and they are also interconnected and restricted in the human body. Therefore, it is necessary to consider the elements as a whole.

12.2.4 Orderliness of Bio-elements

A large number of elements are mainly distributed in the first, second, and third cycles of the periodic table, among which hydrogen in the first cycle; carbon, nitrogen, and oxygen in the second cycle; sodium, magnesium, and oxygen in the third cycle; phosphorus, sulfur, potassium, and calcium in the fourth cycle. They account for more than 99.95% of the total mass of the living body. Except for boron in the second period and silicon in the third period, trace elements are mainly distributed in the fourth period of the periodic table, such as vanadium, chromium, manganese, iron, cobalt, nickel, copper, zinc, arsenic, and selenium. After the fifth cycle, there are basically no elements involved in the composition of life forms. If they appear, they are caused by pollution.

It should be pointed out that the concentrations of bio-elements of different organisms are different. For example, grasses contain more silicon, bird excrement on islands contains more phosphorus, and kelp contains more iodine, but the overall concentration trend of macro- and trace elements is consistent. Carbon, hydrogen, oxygen, nitrogen, sulfur, and phosphorus are the most important elements in life. The hardest metal element that constitutes life is calcium, which ranks 20th in the periodic table. The metals that appear later, especially heavy metal elements, are extremely small in living organisms, and many of them pose a threat to life.

12.2.5 Diversity of Bio-elements

The human body is composed of more than 80 elements. According to the different content of elements in the human body, they can be divided into two categories: macro-elements and trace elements. All elements that account for more than one ten thousandth of the total weight of the human body, such as carbon, hydrogen, oxygen, nitrogen, calcium, phosphorus, magnesium, and sodium, are called macro-elements; all elements that account for less than one ten thousandth of the total weight of the human body, such as iron, zinc, copper, manganese, chromium, selenium, molybdenum, cobalt, and fluorine, are called trace elements (iron is also called semi-trace elements).

Each trace element has its special physiological function. Although they are present in very small amounts in the human body, they are necessary to maintain some critical metabolism in the human body. Once these essential trace elements are lacking, the human body will suffer from disease and even life-threatening. Moreover, the trace elements in the human body not only maintain normal physiological functions, but also their content in the human body will also affect human intelligence, emotions, etc. which is the material basis for human mental health.

12.2.6 Biological Fractionation

The biological fractionation of lead isotopes was found when lead pollution sources were traced with lead isotopes. Usually, it was considered that lead isotope ratios did not change during physical, chemical, or biological processes. However, recent evidence has shown that the lead isotope ratios among different biological samples in human are not always identical from its lead origins in vitro. An animal experiment focusing on the respiratory tract of lead exposure was conducted to explore the biological fractionation of lead isotopes in biological systems [6]. In the study, there are significant differences in lead isotope ratios between blood, urine, and feces from the lead-poisoned SD rats. Moreover, a nonlinear relationship between the blood lead concentration and the blood lead isotope ratios was observed. There is also a threshold effect to the fractionation function. Therefore, any study that attempts to trace potential lead pollution sources from the environment must consider the blood lead level and the fractionation functional threshold of lead isotopes.

12.2.7 The Correlation Between the Bio-elementomes and Other "Omes"

In bio-elementomics, biological metals are bound with metalloproteins and metalloenzymes [7]. However, alkali and alkaline earth metal ions, which are mostly present in free form in body fluids, should also be included because they also play important roles in the existence of physiological functions in biological systems. Early research mainly focused on the distribution of elements in the environment and organisms and their effects to human body; thus, the total amount of elements was well studied [8]. Modern bio-elementomics mainly focus on the distribution and speication of metal ions and metalloids in cells and organs; as well as their association with genome, proteome, and metabolome [9]; and their complexes with biological macromolecules from the perspective of molecular biology. The analysis of structure, function, and activity of metalloproteins and metalloenzymes was also highlighted [10]. In bio-elementomics, in order to effectively describe the functions of proteins and enzymes that bind to metals, it is necessary to solve the full speciation analysis of metals.

12.3 Rare-Earth Elementome

12.3.1 Association of Rare-Earth Elements and Related Diseases

The rare-earth elements (REEs) are identified as a chemically uniform group which has similar physiochemical characters [11], including 15 active metals with similar physicochemical properties. Recently, excessive exploration of REEs has resulted in the contamination of soil or water, increasing environmental exposure to REEs in the general population [12]. A growing body of evidence reported that REEs generally accumulate in the blood, bones, and brain after entering the body through the food chain, which may be toxic to humans. Animal studies have indicated that REEs can cause fibrotic tissue injury, neurotoxicity, lung toxicity, renal toxicity, cytotoxicity, anti-testis effects, and male infertility [13]. Moreover, REEs can cross the blood–brain barrier and accumulate in the brain, which results in nerve damage ultimately. In addition, REEs can enter the infants' body through the placental barrier and have an effect on the growth and development of infants. In recent years, the environmental contamination and exposure caused by exploitation of REEs has improved risks of human health including birth defects, chronic disease, autism spectrum disorders (ASD), and schizophrenia.

Few epidemiological studies have investigated the adverse effects of REEs on perinatal outcomes. One previous study showed that concentrations of lanthanum (La), cerium (Ce), praseodymium (Pr), and neodymium (Nd) in scalp hair of pregnant women in the neural tube defects (NTDs) group were higher than those in the control group, but there were no statistically significances [13]. Researchers studied the associations between concentrations of 10 REEs in maternal serum and the risk for fetal NTDs in another study, including 200 pregnant women affected by NTDs and 400 pregnant women with healthy fetuses. Fifteen REEs in maternal serum were determined, and 10 of them were detectable in over 60% of samples and were included in statistical analyses, including La, Ce, Pr, Nd, samarium (Sm), europium (Eu), terbium (Tb), dysprosium (Dy), lutetium (Lu), and yttrium (Y). Results showed the risk for NTDs increased by 2.78-fold (1.25–6.17) and 4.31-fold (1.93–9.62) for La, and 1.52-fold (0.70–3.31) and 4.73-fold (2.08–10.76) for Ce, in the second and third tertiles, respectively, compared to the lowest concentration tertile by the use of logistic regression model. When Bayesian kernel machine regression was used to examine the joint effect of exposure to all 10 REEs, the risk for NTDs increased with overall levels of these REEs and the association between La and NTDs risk remained when other nine elements were taken into consideration simultaneously [13]. To sum it up, the study indicated that the risk for NTDs increased with La concentrations when single REE is considered and with concentrations of all 10 REEs when these REEs are considered as a co-exposure mixture. A similar phenomenon was found in schizophrenia. We recruited first-episode and drug-naive schizophrenic patients ($n = 96$) and age-sex-matched normal controls ($n = 96$) from Tangshan, Hebei Province, China, in a pre-experiment. Fifteen REEs in serum were determined, and six of them were included in statistical analyses finally, including La, Ce, Pr, Nd, Sm, and Y. Preliminary results showed that concentrations of above elements in serum were higher in cases than those in controls, and all these REEs had adjusted odds ratios (ORs) > 1.0 except for element Y.

Metals from food and drinking water are reported to be important contributors to regulating blood pressure. We recruited 398 housewives in Shanxi Province, China, consisting of 163 women with hypertension and 235 healthy women without hypertension, to investigate the relationship between REEs in hair and the risk of hypertension in housewives [14]. We used hair to assess long-term intake levels of various metals, because the special section of hair is expected to indicate the exposure levels, assuming a constant hair growth rate. We analyzed 15 REEs and calcium (Ca) accumulated in housewives' hair over a period of two years. The results showed that concentrations of the REEs in hair were higher in the cases than in the controls except Eu. The univariate ORs of the 14 REEs were >1.0, and four of the REEs (i.e. Dy, Y, thulium [Tm], ytterbium [Yb]) also had adjusted ORs > 1.0. The increasing dose–response trends of the four REEs further indicated the potential for increased hypertension risk (Figure 12.1). Previous studies have revealed that REEs can cause adverse effects by disturbing the balance of oxidative stress in the human body [15], which play a significant role in hypertension development. Our study population

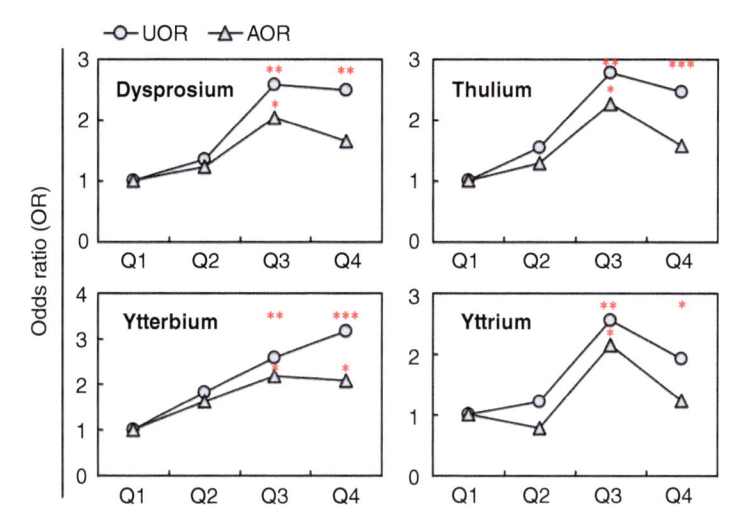

Figure 12.1 The dose–response effect of hair concentrations of the four rare-earth elements (REEs) on hypertension risk. Concentrations of REEs were classified into four levels by quartiles, i.e. <1st quartile (Q1), 1st quartile – 2nd quartile (Q2), 2nd quartile – 3rd quartile (Q3), >3rd quartile (Q4).

had higher hair REE contents than the general population, which might result in chronic oxidative stress damage to housewives, thereby increasing the risk of hypertension. Moreover, the REEs were negatively correlated with Ca content in hair, suggesting an antagonistic effect of REEs on Ca in the human body. It was concluded that high intake of REEs might increase the risk of hypertension among housewives. Most of the REEs ions were trivalent and had ionic radii similar to important nutritional ions such as Ca^{2+}, Mg^{2+}, Zn^{2+}, Mo^{2+}, and Cu^{2+}. Previous studies had supported the conclusion that a lower intake of Ca or Mg is associated with an elevated risk of hypertension.

In Wu's study, 34 elements including REEs in erythrocytes of autistic and typically developing children (TDC) were examined for the first time [6]. In the study, the levels of REEs (La, Ce, Gd, and Eu) in erythrocytes of children with and without ASD were examined. It was shown that the concentration of La had a dramatic decrease compared to the control level, and the concentrations of other three REEs (Ce, Gd, and Eu) were in the same range as that of the control group. In addition, the degree of the decrease of La was not only associated with the severity of the autistic syndrome as a whole, but also specifically with the impairment in three aspects, auditory response, taste, smell, and touch, and fear or nervousness (Figure 12.2). However, there were no other results reported so far, and more researches are needed to explore the possible roles of REEs in the development of ASD.

12.3.2 The Mechanism Studies of the Hormesis Effect of REEs Based on the Bio-elementomics

So far, the Hormesis effect of REEs has been found in many life phenomena [16], and there are many different hypotheses about the mechanism of this

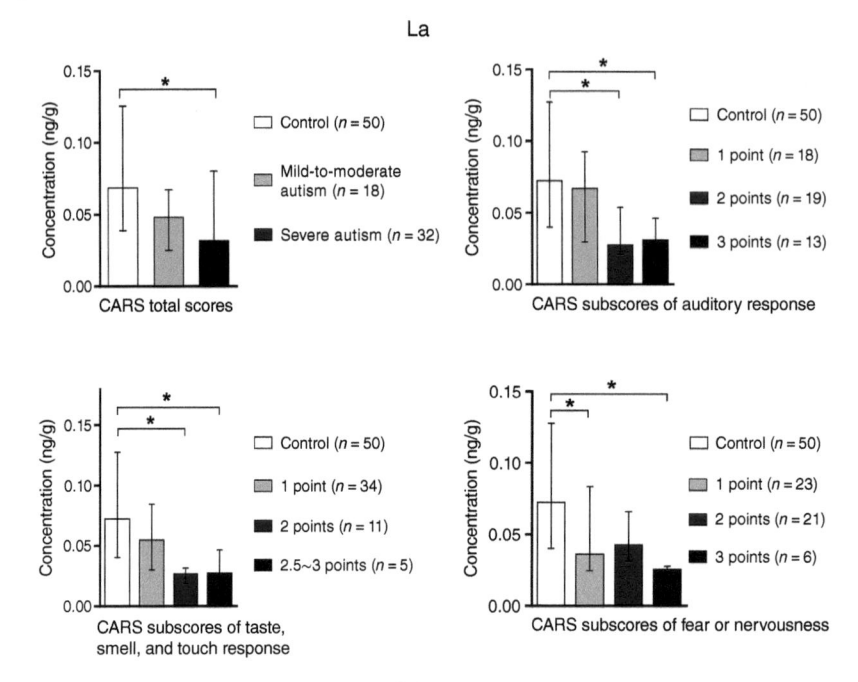

Figure 12.2 Relationship between CARS scores and La level in erythrocytes of control and autism.

effect [17]. Jingyu Wang's team at Peking University took the rare-earth element lanthanum-*Escherichia coli* as the research object and tried to explore the mechanism of the Hormesis effect from the perspective of bio-elementomics. The research results show that when the concentration of lanthanum in the culture solution is lower than 250 µg/ml, with the increase of the concentration of lanthanum, the stimulation response to the growth of *E. coli* gradually increases. At the same time, the content of other inorganic elements in *E. coli* bacteria will also change, among which the contents of Mn, Ba, Mo, and U have a good positive correlation with the content of lanthanum and the growth rate of *E. coli*.

Jingyu Wang's team conducted another research on the effect of Zn on the growth of *Lactobacillus acidophilus* [18]. The study found that adding an appropriate amount of Zn to the culture medium can not only up-regulate the prebiotic elements, such as Fe (453%) and Se (392%), but also down-regulate the absorption of harmful element Pb (1.41%) by *Lactobacillus acidophilus*. This interesting and important result suggests that we may be able to grow non-polluting agricultural products on polluted land with appropriate use of specific micro-fertilizers. We may also use appropriate amounts of specific auxiliary drugs to increase the efficacy of the original drug and reduce its side effects.

12.3.3 Beneficial Rebalancing Hypothesis for Hormesis Effect

Combining the results of the previous two experiments, the following hypothesis is put forward: the addition of an appropriate amount of lanthanum in the culture solution will break the inherent balance of the element content in *E. coli* and promote *E. coli* to absorb essential elements that are beneficial to its growth and form a new element content balance in *E. coli*. So far, there is no solid evidence proving that REEs are essential elements for life. Therefore, the increase in the content of lanthanum in the culture solution may not directly lead to the accelerated growth of E. coli, but it will up-regulate the absorption of essential elements by *E. coli.*, and then increase the growth rate of *E. coli* and/or improve its quality, thus producing the so-called hormesis effect. Here, La or Zn can be regarded as a "regulating substance (RS)," the essential element is regarded as "promoting substance (PS)," the hazardous element is considered as "hazardous substances (HS)," and *E. coli* is regarded as a "life form (LF)." RS may or may not be an essential element of life, its main function is to up-regulate the absorption of PS and down-regulate the absorption of HS by LF, thereby promoting the growth rate of LF and/or improving the quality of LF. In other words, RS broke the original substance balance in LF by up-regulating PS content and down-regulating HS content, and then formed a new substance balance that was beneficial to LF growth and/or its quality. This new balance may be called "beneficial rebalancing," which may be one of the potential mechanisms of the hormesis effect of REEs. The above mechanistic hypothesis has been further confirmed in subsequent experiments [19].

In fact, the PS found by the above method can be regarded as a new RS which can be used to further up-regulate other PSs and down-regulate HSs. Repeating this process can continuously approach the set target, e.g. increasing the growth rate of living organisms or improving their quality. In addition, RS or PS can be other substances besides inorganic elements, such as small organic molecules, short peptides, and protein macromolecules. Therefore, this method can also be used in other different research and clinical fields.

12.4 Limitations of Bio-elementomics

Many scholars believe that the content of trace elements in the human body will be affected by environmental and dietary factors. Therefore, the content of trace elements in various body fluids is difficult to be a reliable clinical indicator for human health or disease [20]. The above viewpoints limit the application of bio-elementomics in life sciences to some extent. To deal with the above limitations, we try to elaborate our views and solutions based on the experimental results.

12.4.1 Statistically Higher Level of Some Elements in the Patient's Body

Some researchers think that the disease will affect the patient's appetite, lead to malnutrition, and then affect, especially reduce, the level of certain trace elements in the patient's body. Therefore, the content of trace elements in human body fluids or cells should not be a biomarker for discriminating diseases. However, many experimental results have shown that the levels of some elements in the patient's body do decrease, but at the same time, the levels of some elements are statistically higher than those of healthy people. Therefore, the possibility of the content of trace elements in human body fluids or cells becoming a biomarker should not be simply denied since the decrease in the patient's appetite.

12.4.2 Environment-independent Biomarkers

To overcome the limitations of bio-elementomics as mentioned above, Wang Jingyu's team made an attempt to find trace elements that are not affected by the environment in living body fluids or cells. The results of rat experiments indicated that the content of elements in certain body fluids or cells has a very good correlation with the corresponding exposure of rats [21]. For example, when the intake of Sr in rats is 0–25 mg/(kg bw), the range of Sr content in rat plasma is 40–220 ng/ml, and there is a positive correlation between exposure and plasma Sr content ($r = 0.856$, $p < 0.001$). Another example, when the Mo intake of rats is 0–18 mg/(kg bw), the range of Mo content in rat plasma was 10–4300 ng/ml, and the correlation between exposure and plasma Mo content was also significant ($r = 0.872$, $p < 0.001$). The above results show that the contents of certain elements in the living body are indeed related to environmental exposure and cannot be directly used to evaluate the normal or abnormal of the living body.

However, the same rat experiment also showed different results. When plasma was replaced with erythrocytes, the Sr content in rat erythrocytes varied only by 20–30 ng/ml, and there was no longer any correlation between exposure and erythrocytes Sr content ($p = 0.166$); when plasma was replaced by PBMC, the range of Mo content in rat PBMC was only 0.55–1.25 ng/ml, and there was no correlation between exposure and PBMC Mo content and Mn in PBMC, and Mn in erythrocytes (all $p > 0.05$) exhibited the same characteristics [22]. It's entirely possible that we could find some elements in certain cells that are unaffected by environmental exposure. Such elements may be used as evaluation indicators for normal or abnormal living organisms, while those elements related to the environment can be used as environmental exposure assessment of living organisms.

12.4.3 Trace Elements in Immortalized Lymphocytes

Immortalized lymphocytes from 29 patients with schizophrenia and 12 normal persons were collected. The above-mentioned 41 people all come from families with a family history of schizophrenia. Using the same culture medium, each immortalized lymphocyte was expanded to 10^7 cells under the same culture conditions. Cells were washed with PBS solution, and the supernatant was discarded. The above operation was repeated three times to remove the elements contaminated outside the cells. The nutrient intake of the cells cultured in this way should be identical; if there is any difference in the element content in the cells between patients and normal people, it should not be related to the nutrient intake, but should be related to the cells themselves.

The above pre-experimental results showed that the content of 17 elements (B, Al, Ti, Cu, Zn, Se, Rb, Ba, Sr, Y, Bi, I, U, Mn, Mg, P, and S) was significantly different between schizophrenia patients and normal people. The content of most elements in patients with schizophrenia is lower than normal, only Ni and P are exceptions (unpublished data).

The above experimental method has the following advantages: (i) Cultivating immortalized lymphocytes under the same culture conditions can offset the interference of the environment, and then convincingly demonstrate the difference in element content between patients and normal people. (ii) Increasing the amount of the cells by one thousand or even ten thousand times by culturing can save lymphocytes which are extremely difficult to obtain, and at the same time greatly improve the detection accuracy of elements in cells. (iii) Optimizing the culture conditions of fresh lymphocytes, increasing their expansion, and using them to replace immortalized lymphocytes will further expand the scope of clinical and basic research applications of this method.

12.5 Perspectives

With the development of genomics and proteomics, the full investigation of bio-elementomics has received increasing attention. Early studies mainly focused on the distribution and effects of elements in the environment and organisms, and examined the total amount of elements. Future researches on the bio-elementomics may include the following.

12.5.1 Speciation Analysis of Elements

To obtain a good understanding of the toxicity and ecological effects of trace elements in the environment, it is necessary to determine not only the total amount, but also their existing species. Since the toxicity and bioavailability strongly depend on their chemical forms, obtaining the information of elements and their chemical species may contribute to understanding their physiological activities and toxicological properties, which is of great significance for the study of the metabolism, migration, and transformation of elements in human body.

12.5.2 Bio-elements and Their Interactions with Proteins, Genes, and Small Molecules

In the future, the bio-elementomics must be combined with other "omics" to deeply explore the interaction between the bio-elementome and other "omes" (genome, proteome as well as metabolome). Using the same set of biological samples to expand simultaneous multi-omics researches is particularly important. The experimental results obtained in this way can improve the comparability between multi-omics experimental data, making the research conclusions more comprehensive, scientific, and reliable.

12.5.3 Research Based on the Hormesis "Beneficial Rebalancing" Hypothesis

As shown before (Section 12.3.3), we may apply the Hormesis beneficial rebalancing hypothesis to different fields, such as safe and efficient anti-pollution micro-fertilizers, feed additives, development of biopharmaceuticals, and even personalized diagnosis and treatment.

12.5.4 Multi-element Analysis of Immortalized Lymphocytes

Multi-element analysis of immortalized lymphocytes (12.4.3) may have meaningful application prospects in basic medical and clinical medical researches, and it is worth trying.

12.5.5 Analysis of Bio-elements in Single Cell

Single-cell analysis has attracted growing attention from researchers in various fields of biology due to its effectiveness as a tool to study the precise mechanisms of cellular and molecular behavior. However, single-cell analysis can be very challenging due to the very small size of each cell as well as the large variety and extremely low concentrations of substances found in individual cells. As this technology continues developing and yields even higher spatial resolution and sensitivity, mass spectrometry will play an increasingly crucial role in single-cell analysis, and in turn it will also promote the development of biological research. This technology may provide a powerful method for achieving early detection of various diseases and will contribute to valuable insights about pathogenesis and cell metabolism.

References

1. Nielsen, F.H. (1991). *FASEB J.* 5: 2661.
2. Zoroddu, M.A., Aaseth, J., Crisponi, G. et al. (2019). *J. Inorg. Biochem.* 195: 120.
3. Haraguchi, H. (2017). *Metallomics* 9: 1001.
4. Jan, A.T., Azam, M., Siddiqui, K. et al. (2015). *Int. J. Mol. Sci.* 16: 29592.
5. Miao, F., Zhang, Y., Lu, S. et al. (2021). *Chemosphere* 277: 130353.
6. Wu, J., Liu, D.J., Shou, X.J. et al. (2018). *Autism Res.* 11: 834.
7. Chen, A.Y., Adamek, R.N., Dick, B.L. et al. (2019). *Chem. Rev.* 119: 1323.
8. Shi, L., Yuan, Y., Xiao, Y. et al. (2021). *Environ. Int.* 157: 106808.
9. Yuan, Y., Long, P., Liu, K. et al. (2020). *Redox Biol.* 29: 101404.
10. DiPrimio, D.J. and Holland, P.L. (2021). *J. Inorg. Biochem.* 219: 111430.
11. Ramos, S.J., Dinali, G.S., Oliveira, C. et al. (2016). *Curr. Pollut. Rep.* 2: 28.
12. Huo, W., Zhu, Y., Li, Z. et al. (2017). *Environ. Pollut.* 226: 89.
13. Wei, J., Wang, C., Yin, S. et al. (2020). *Environ. Int.* 137: 105542.
14. Wang, B., Yan, L., Huo, W. et al. (2017). *Environ. Pollut.* 220: 837.
15. Pagano, G., Aliberti, F., Guida, M. et al. (2015). *Environ. Res.* 142: 215.

16 Calabrese, E.J. and Baldwin, L.A. (2003). *Crit. Rev. Toxicol.* 33: 305.
17 a) Calabrese, E.J. (2004). *Int. J. Occup. Environ. Health* 10: 466. b) Nies, D.H. (2003). *FEMS Microbiol. Rev.* 27: 313.
18 Zhao, Q., Yan, L., Xie, Q. et al. (2020). *Spectrosc. Spectral Anal.* 40: 6.
19 Gao, X., Liu, Y., Yan, J. et al. (2020). *Chin. J. Rare Earth* 37: 11.
20 He, Z.L., Yang, X.E., and Stoffella, P.J. (2005). *J. Trace Elem. Med. Biol.* 19: 125.
21 Sun, X., Zhang, Y., Wen, H. et al. (2014). *J. Hyg. Res.* 43: 972.
22 Sun, X. (2014). *Environmental and Occupational Health*, Vol. Master. Beijing: Peking University.

13

Methodology and Tools for Metallomics

Xiaowen Yan[1], Ming Xu[2,3], and Qiuquan Wang[1]

[1] Xiamen University, College of Chemistry and Chemical Engineering, Department of Chemistry and the MOE Key Laboratory of Spectrochemical Analysis & Instrumentation, 422 South Siming Road, Xiamen 361005, China
[2] Chinese Academy of Sciences, Research Center for Eco-Environmental Sciences, State Key Laboratory of Environmental Chemistry and Ecotoxicology, No. 18, Shuangqing Road, Haidian District, Beijing 100085, China
[3] College of Resources and Environment, University of Chinese Academy of Sciences, No.1 Yanqihu East Rd, Huairou District, Beijing 100049, China

13.1 Brief Description of Metallomics

13.1.1 Why Do Research on Biometals?

All living matter is made of the elements as basic chemical components. Therefore, disclosing the elemental status in organisms is a key to understand the essence of life, especially for the trace metals/metalloids that considered to be essential or nonessential rather than limited to the bulk elements such as H, C, N, O, S, and P. In organisms, trace essential elements, such as Fe, Cu, and Se, participate in numerous biochemical reactions necessary to maintain a good health. Like the other side of the coin, widespread presence of heavy and/or noble metals, naturally occurred and released as a result of various anthropogenic activities or metallodrugs used in clinic, inevitably leads to their intake and brings unpredictable risks in our life. It has been a long and strong need to figure out what took place between the metals/metalloids and living things, as well as how they keep or interfere with a normal life activity.

Chemically, at the molecular level, the chemobiological property of the trace metals/metalloids is determined essentially by their atomic characteristics such as the nucleus and electrons, resulting in their benefits or hazards to life. However, the living organism and even a cell is a complex running machinery, and most of its parts have not been fully understood yet. It is still a daunting challenge to get a full picture of biological fate of the trace metals/metalloids because there must be certain unknown rules to fundamentally control the element–biomolecule interactions. To this end, around two decades ago, the terms "metallome" and "metallomics" were proposed to integrate the research field related to bio-trace elements as part of omics-science [1–4], in a similar manner to genome and genomics as well

Applied Metallomics: From Life Sciences to Environmental Sciences, First Edition.
Edited by Yu-Feng Li and Hongzhe Sun.
© 2024 WILEY-VCH GmbH. Published 2024 by WILEY-VCH GmbH.

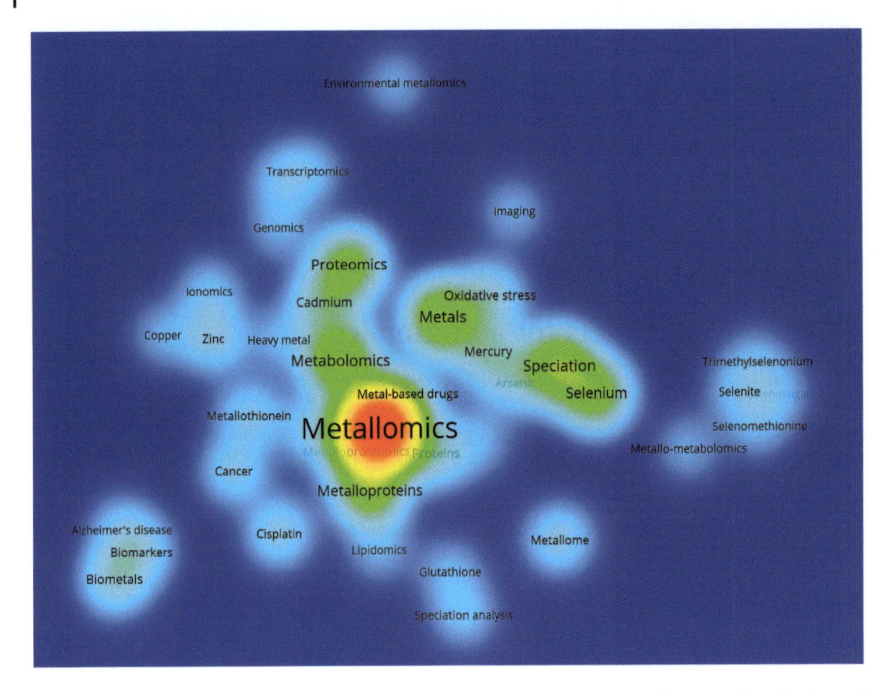

Figure 13.1 Density visualization map of co-occurrence keywords for metallomics based on total link strength from January 2004 to December 2021. Source: Adapted from Web of Science.

as proteome and proteomics (Figure 13.1). Meanwhile, strategic methodology and efficacious tools for exploring the roles of the trace metals/metalloids in life process have never stopped developing [5–8].

13.1.2 What's the Goal of Metallomics?

As the science of biometal/metalloids, metallomics aims to elaborate on the complex biological roles of the metal/metalloid-containing compounds in a metallome [9, 10]. In biologically relevant scenarios, once entering organisms, the metal/metalloid ion-variants will encounter diverse kinds of biomolecules such as nucleic acids, proteins, lipids, and polysaccharides, which may directly interact with the metal/metalloid ions via covalent or non-covalent bindings, forming multifarious metal/metalloid-containing species. Incorporation of the metal/metalloid into a biomolecule will bring conformation or functional transitions, thus giving rise to changes of molecular activity and enabling the biomolecule to perform its physiological roles. Taking Fe as an example, in normal physiological state, the concentration of liable Fe is tightly controlled by a series of Fe-binding proteins including hepcidin, transferrin, ferritin, and ferroportin, avoiding the generation of active oxygen free radical that can cause oxidative damages to cellular constituents via Fenton chemistry [11]. These biomolecules are the central regulators of Fe homeostasis, in charge of iron absorption, storage, recycling, and utilization in the body. Of course, Fe is an essential constituent of these biomolecules as well, determining their physiological function and expression level. Besides, metabolic transformation is sometimes necessary for the bioavailability of essential elements such as Se, which has to be metabolized to selenoamino acids for further synthesis of selenoproteins in vivo [12–14]. On the other hand, some nonessential metals can interact with biomolecules forming many unpredictable metal-containing compounds, most of which are still a mystery to be uncovered so far. According to the hard–soft–acid–base principle, soft acidic metal ions such as Hg(II) have a strong binding affinity to soft bases such as thiol and selenol groups in biomolecules, which commonly causes irreversible damages to the biomolecules, particularly for the thiol- or selenol-containing enzymes [15–17]. Nonetheless, not only what type of toxic metal-binding biomolecules forms within organisms/cells but also how the metallic compounds induce detrimental effects and undermine a biological process remain largely unclear at present. Collectively, not limited to the elements as mentioned above, all issues related to the trace metals/metalloids within a living system, e.g. structure, speciation, abundance, distribution, transformation, metabolism, function, toxicity as well as the underlying mechanisms, ought to be included in the scope of metallomic study. Ultimately, one should be able to understand how the metals/metalloids participate in life processes and work at the molecular, cellular, and tissue levels.

13.1.3 How to Perform a Metallomic Study?

Metallomics aims to uncover the interactions and functional connections between metals/metalloids and biomolecules or the role within a metal/metalloid-involved biomolecule. Not only the identification and quantification of the metals/metalloids present in biospecimens but also their global interaction network with biomolecules as well as their link to the functions of genome, transcriptome, proteome, and metabolome result in an increasing complexity and dynamics of metallomic study. Moreover, the widespread existence of various essential elements, possible metallodrugs, and heavy metals in our living environment [18], and their diverse properties such as stability, valence, redox, and coordination states might further increase the complexity of the metallome in different levels of dimension, leading to an enormous demand of strategic methodology and powerful analytical tools for metallomics research. Thanks to great advances of the technical tools used in elemental analysis, majority of the naturally occurring trace metals/metalloids in life can be well determined nowadays. In particular, interdisciplinary research is becoming more common, providing new opportunities for investigating the metallome in biological systems. From the point of methodology, we are able to study a metallome through an integration of complementary strategies and methods used in different fields in parallel for one research object. An ideal plan is to combine the physical, chemical, or biological protocols for element imaging, protein labeling, and gene editing together to examine the metallome in organisms in the manner of in vivo, ex vivo, in vitro, or in silico. In addition, the development and maturation of genomic, proteomic, and metabolomic methodologies offer very typical examples for metallomic researchers to design more effective platforms that facilitate the biometal analysis and their functions in a high-throughput manner, e.g. multi-omics analysis.

From a technical perspective, on the other hand, there are already many powerful instruments enabling the qualitative and quantitative analysis of metal/metalloid-containing compounds at the molecular, atomic, and sub-atomic levels. For example, when it needs to identify known or unknown metal/metalloid-binding proteins in biospecimens, elemental and molecular mass spectrometry (MS) can be used together because of their promising performance in the analysis of proteins, through coupling with gel electrophoresis or liquid chromatography [5, 19, 20]. Moreover, the three-dimensional structure of metal/metalloid-binding proteins can be assessed using spectroscopic, electron microscopic, and nuclear magnetic resonance (NMR) techniques. But even that does not fulfill the requirements of metallomic measurements, more technical tools with high detection sensitivity and spatial resolution are long-awaited to be developed, which will be discussed in more detail later.

13.2 Methodologic Strategy for Metallomic Research

In general, metallomics research may be carried out at different levels of complexity from molecule and cell to tissue and even intact organism (Figure 13.2). It requires strategic methodology for solving the scientific issues faced to researchers. For example, to explain why cisplatin induces nephrotoxicity in clinical treatment, potential protein targets of this Pt-containing chemotherapy drug can be first examined by taking advantage of mouse in vivo and ex vivo [21]. Then renal cells and organoids may be chosen to demonstrate the protein-cisplatin interactions in cellulo, and the underlying reaction mechanism can be further disclosed in vitro

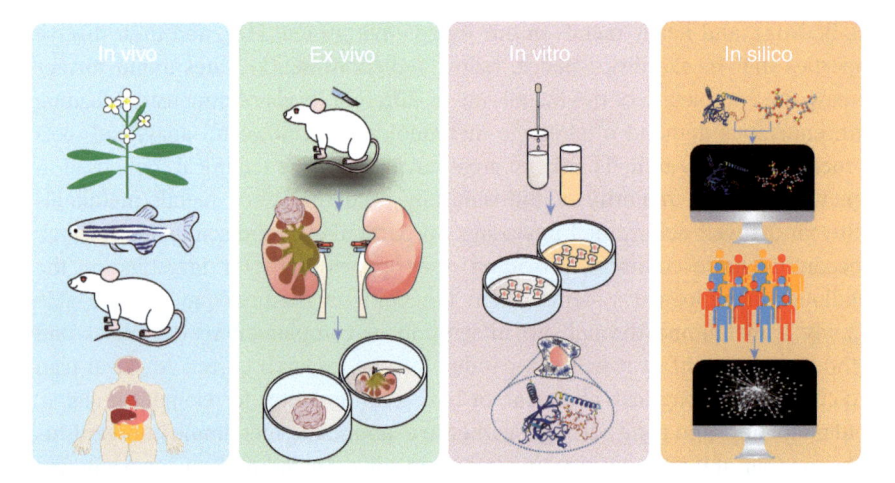

Figure 13.2 Summary of strategies used for the research on metallome in living organisms. Source: Ming Xu.

and in silico. Accordingly, the representative strategies for metallomic research will be briefly overviewed here.

13.2.1 In Vivo

Role of the metals/metalloids in vivo, i.e. within a living organism, is one of biggest concerns of the metallomics community, as essential and toxic metals/metalloids in body or metal/metalloid-containing chemicals administrated are tightly related to the health. Thus in vivo study of the metallome in human is the best choice; however, it almost can hardly be done due to the ethical issues, unless collected tissue samples are available. Alternatively, non-human mammal model such as mouse and macaque are frequently tested for studying bio-trace metals/metalloids in vivo, contributing greatly to our fundamental knowledge of the metals/metalloids' nature in human body. In case of some distinctive cases, besides wildtype animal, disease or gene knockout mammal models can be established in laboratory and used for metallomics research. Meanwhile, there is a wide variety of model plants, invertebrate and vertebrate animals that have long been used in the fields of agriculture, ecology, toxicology that may be employed for similar purposes.

In living organisms, the examination can be realized through imaging techniques such as magnetic resonance or computed tomography in combination with radioactive isotopes, or optical microscopy using molecular sensors, allowing a real-time analysis of in vivo fate of trace metals including absorption, distribution, metabolism, and transformation. However, in vivo analysis is always not an easy task for all kinds of metals/metalloids, due to the limitations of instruments and lack of available chemicals, such as reference materials and enriched isotopes. To supplement these, tissue or blood of human or model organisms can be collected through real-time sampling and examined with usable methods for fast analysis, reflecting the dynamic changes of biometals in vivo, as well as giving more detailed information about the speciation and activity of metal-containing compounds. In particular, there are many ready-to-use means allowing researchers to investigate the relationship between trace biometals and biomolecules. For instance, proteomic, transcriptomic, or metabolic analysis may be done together with metallomic analysis to spy upon the interaction networks between protein, gene, or metabolite and biometals, or temporal and spatial changes of metal-binding biomolecules within the body can be examined by elemental and molecular MS. In short, in vivo study can best uncover the true role of biometals related to the health, which is hardly forecasted.

13.2.2 Ex Vivo

Ex vivo culture represents a model between in vitro and in vivo, offering the advantages including tightly controlled experimental conditions and good compatibility with tools used in the routine examination of biospecimens, which is increasingly applied in life science. As described above, in vivo analysis of biometals has higher demands to experimental and technical conditions, especially in ethical issues for living subjects. Alternatively, freshly sampled organ, tissue, and cell can be cultured ex vivo in an artificial environment mimicking the natural condition, ensuring their metabolism and bioactivity. For example, ex vivo tests can been applied for evaluating the detrimental effects of heavy metals on the organ level such as liver, or assessing the therapeutic effects of metallodrugs on tumors, which are carried out outside organisms [22, 23]. Besides, cells can be isolated, for instance, from the lesions of patients, for ex vivo culture and further mechanistic study [24]. In this way, one is able to elucidate how biometals and metal-containing compounds damage the function of different organs and tissues, affect the progression of the disease or inhibit the growth of tumor, instead of using animals.

Meanwhile, there are more in situ methods that are technically feasible to determine the metals/metalloids within organs and tissues ex vivo, but difficult in vivo, because organs inside body are not easy to be directly observed without advanced tools unless sampled for in vitro examination. For instance, fluorescence imaging of metal/metalloid ions within organs requires a high reliability of fluorescent probes, providing the good signal penetration across skin and biological tissues. Furthermore, the chemical properties of biometals in organ, tissue, and cell cultured ex vivo is able to be examined using X-ray, and the metal–biomolecule interactions may be studied through any feasible biological means, like gene editing.

13.2.3 In Vitro

To date, in vitro testing provides a starting point for researchers to gain insights into how biometals act within cell and how cell responds to the metals/metalloids in a controlled environmental outside of a living organism. In comparison with in vivo and ex vivo studies, in vitro model is relatively low cost and easy to prepare, which can be conducted in most laboratories. Several strategies have been employed in vitro to better mimic the circumstances occurring in vivo [25]. For example, two-dimensional (2D) or three-dimensional (3D) cell culture systems are presentative in vitro platforms for metallomic research due to their evident advantages in providing biometal relevant information at the cellular and molecular levels. For 2D monolayer culture, all cells can directly come into contact with metals/metalloids from the medium during growth. Thus, the subcellular distribution of biometals can be examined within each cell, and cell characteristics such as morphology, growth, differentiation, and migration can be easily determined. When grown in 3D culture systems, cells form aggregates or spheroids that have tissue-like structures with more similarity to in vivo tissues, allowing researchers to study the spatial heterogeneity of biometals within tissue, as well as metal/metalloid–cell interactions. In addition, more types of in vitro testing models are being developed such as organoids and microfluidic chip, providing a wealth of possibilities to study the response of cells to biometals in culture [26].

More importantly, on the basis of in vitro models, the metallome within cell can be well studied. Taking Fe as an example, with the help of dyes, its cellular level can be examined by light microscopy after Prussian blue staining. Alternatively, molecule fluorescent probes can be used to detect labile $Fe(II)$ ion or $Fe(II)$-containing heme in living cells, giving subcellular dynamics of Fe homeostasis. At the same time, the level of Fe-binding proteins may be examined through the methods in molecular biology such as western blotting, or bioinorganic tools such as elemental MS. If one wants to know how Fe acts on specific protein, gene silencing or overexpression of target protein in cells is a good option. Furthermore, Fe-containing proteins may be separated and purified from cells for the identification of some unknown species. To elucidate the underlying mechanisms of Fe-protein interactions, thermodynamic and kinetic studies can be performed "in the glass" through circular dichroism spectroscopy or isothermal titration calorimetry. For structure analysis, cryo-electron microscopy even enables researchers to visualize individual atoms in metalloproteins. Collectively, in vitro study is a very effective alternative means for metallomic research.

13.2.4 In Silico

The use of computational methods and theoretical tools increases rapidly for the last few years, which predict how metals/metalloids interact with biomolecules or cells in silico, because in some cases, no effective experimental method is available for such kinds of measurements. For example, molecular dynamics simulation may be applied to study the binding domain and conformation stability of metalloproteins at the atomic level via software [27]. Moreover, the quick development of artificial intelligence and machine learning technologies is being expected to be widely applied for predictive modeling to biometals within living organisms. For example, machine learning analytics can be used to deal with multi-omics big data at different levels, enabling researchers to find out potential biomarkers and gain insights into how biometals are involved in disease [28]. Together, these computational methods are expected to be successful in describing metallome in all aspects of life in silico.

13.3 Tools for Metallomics

Modern analytical techniques have made remarkable progress during the past decades, allowing studying the identity, amount, uptake, binding, transport and storage of biometals in biological media, as well as the interaction with biomolecules and functions in life [8, 29]. Due to complexity of metallomics research, each analytical tool can achieve information on certain aspect of a full metallome analysis. Different analytical tools should be integrated to approach the comprehensive role and function of biometals in life [6]. The following subsection will give a brief overview of quantitative, qualitative, and imaging tools for metallomics study (Figure 13.3).

13.3.1 Tools for Quantitative Metallomics

To begin a metallomics study, one should first figure out the amount of biometals in a biological media. Atomic spectroscopy and elemental MS are the most used

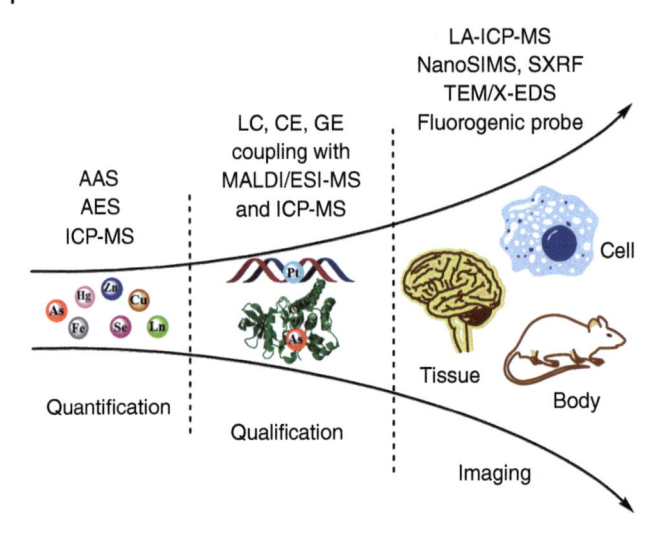

Figure 13.3 A brief overview of quantitative, qualitative and imaging tools used for metallomics study in live. Source: Xiaowen Yan.

techniques to accomplish this task. Because of its high selectivity and sensitivity and low cost, the well-developed atomic absorption spectroscopy (AAS) is widely used for metal quantification; when coupled with liquid chromatography (LC), it can be used for the separation and determination of metalloproteins. An obvious disadvantage of AAS is that it can only carry out single element determination at one time. From this point, atomic emission spectroscopy shows its advantages in multi-element determination. However, considering the comprehensive analytical performance, inductively coupled plasma–mass spectrometry (ICP-MS) is the most powerful tool for the identification of elements present in the sample and for the quantification of their total concentration. As a hard ionization source, the ICP is able to ionize most elements in the periodic table with high efficiency. On the other hand, the mass analyzers, either quadrupole or high-resolution time-of-flight and sector field, are able to resolve m/z of element isotopes generated in the ICP, allowing the isotope dilution strategies easily applied on ICP-MS. Taken together, ICP-MS is capable of detecting most elements below ppb level with broad dynamic range of 9 orders of magnitude, excellent mass resolution, and high throughput, which make it not only a primary quantitative tool for metallomics study [30], but also a powerful tool for quantitative proteomics through the metal-tagging strategies [31, 32].

13.3.2 Tools for Qualitative Metallomics

The species (chemical form and/or oxidation state) determines the bioavailability, mobility, and toxicity of biometals. Information on the stoichiometry and structure of biometals in the complex with biological molecules (metabolites, proteins and nucleic acids) is also crucial to understand the function and mechanism of biometals interaction with biological system. However, in the hard ICP ionization source, all molecule-specific information is lost due to the high-temperature plasma, which hampers the in-depth characterization of the chemical states of biometals using ICP-MS alone. To obviate this limitation, various separation approaches, such as liquid chromatography (LC), capillary electrophoresis (CE), or 2-dimension gel electrophoresis (2DGE), can be chosen to couple with ICP-MS through a specific interface. In these hyphenated techniques, the element-specific quantification information is obtained by ICP-MS, and thus chemical states of biometal can be validated with the biometal standards analyzed in the separation technique.

One should carefully select a suitable separation approach, based on not only the property (stability, hydrophobicity, charge, molecular weight, etc.) of the biometal species or the metal–biomolecule complex, but also the information expected to obtain. Special attention has to be paid to the mobile and stationary phases of the LC. Reversed-phase LC (RPLC) has good separation resolution for stable biometal species and complex; however the organic solvent used in the mobile phase could cause the denaturation or dissociation of the biometal–biomolecule complex and possible change of the plasma properties. Size-exclusion chromatography (SEC), which used buffer similar to physiological condition as mobile phase, is frequently applied for the separation of unstable biometal complex. Due to the relatively low chromatographic resolution, SEC merely achieves a fractionation of the sample rather than a separation of the individual compounds. CE, whose separation takes place in a fused-silica capillary filled with aqueous buffer, is able to separate analytes ranging from small molecules to large biological macromolecules with high resolution. A common feature of LC and CE is that both approaches can be coupled on-line with MS techniques. 2DGE represents a frequently used macromolecule separation technique in the field of genomics and proteomics. However, its application in metallomics is challenging because the denaturing conditions influence the integrity of the metal–proteins complex with weak affinity. Metal ion might dissociate from the complex during sample preparation or during electrophoretic separation process. Considering this limitation, 2DGE is mostly applied for selenoproteins and phosphorylated proteins, of which the elements remain covalently linked to the targeted molecules under denaturing conditions.

For unknown biometal species, the coupling of a separation technique to ICP-MS alone is not sufficient to reveal its chemical information due to lack of the species standards. This problem can be solved by the complementary application of soft ionization MS techniques, mainly electrospray ionization (ESI) and matrix-assisted laser desorption/ionization (MALDI) coupled with different mass analyzers. The application of molecular MS techniques for metallomic study allows the characterization of the stoichiometry and composition of the unknown biometal species from molecule weight of parent ion, but also the in-depth identification of the targeted molecules and even the possible modifications or other variations through the fragment ions generated in different MS/MS modes, such as collision-induced dissociation (CID), electron capture dissociation (ECD), electron transfer dissociation (ETD), and infrared multiphoton dissociation (IRMPD). The synergistic employment of coupling techniques and the integration of elemental as well as molecular MS techniques will significantly deepen our understanding of the diverse roles of biometals and their species in complex biological systems such as cells, tissues, and entire organisms.

13.3.3 Imaging Tools for Metallomics

Besides identification and quantification, another important analytical goal in metallomics is to achieve the in situ localization of biometal species in biological specimens, based on which the uptake, transport, storage, distribution information of biometals can be obtained. The advance of modern element-specific imaging techniques enables the localization of metals in biological tissue and cells from micro- to nanometer level [33, 34]. Mainly, laser ablation–inductively coupled plasma–mass spectrometry (LA-ICP-MS), synchrotron-based X-ray fluorescence (SXRF), and nano secondary ion MS (NanoSIMS) as well as transmission electron microscopy coupled with energy-dispersive X-ray spectroscopy (TEM/X-EDS) are the most powerful and frequently used element-specific bioimaging tools for metallomics. A common feature of these tools is that each pixel of the image is a sample spot of which the metal-specific ions generated by a high-energy beam scanned on the surface of the solid sample. The signal intensity for each metal is converted into a color or grayscale intensity plot, and the generated image represents the relative distribution of the metals in a cell or tissue. The high-energy beam of LA-ICP-MS/SXRF, NanoSIMS, and TEM/X-EDS consists of photons, ions, and electrons, respectively. Correspondingly, in LA-ICP-MS, the metal species are ablated from surface by the photons of laser beam, ionized in ICP, and then detected in MS. In SXRF, element-specific X-ray fluorescence spectrum is induced by the high-energy X-ray photons. Besides elemental imaging provided by SXRF, additional information about the chemical environment or neighboring atoms of metals, e.g. the coordination ligands, can be obtained by the extended X-ray absorption fine structure (EXAFS) or X-ray absorption near-edge structure (XANES) techniques. For NanoSIMS, a primary ion beam directly ionizes the biometal species, and the resulting secondary ions are latter detected by MS. In terms of TEM/X-EDS, the sample morphology is first imaged by electron beams, and then achieve the element compositions and distribution information by X-ray scanning. Due to their different ionization process and imaging mechanism, each tool should choose an appropriate sample preparation to producing high-quality images without artifacts. Particular attention should be paid to avoid the redistribution of biometals during sample preparation. Different to the aforementioned four tools which can only image fixed sample, metal-ion-responsive fluorogenic probe is capable to visualize metal ions in living cells [35]. The fluorescence of the probes turns on after specific chelation with the targeted metal ions or after chemical conversions of the probe molecules induced by the target. However, the disturbance on the real state of metal ions as a result of the interaction with the fluorogenic probes cannot be ignored.

Spatial resolution and element sensitivity are a trade-off in element-specific bioimaging [36]. Higher spatial resolution implies a smaller sample pixel, i.e. less signal generated, which calls for higher sensitivity of the detection techniques. LA-ICP-MS allows imaging of metals in biological samples with a resolution down to micrometer level. By using a microlensed fiber to focus the laser beam, a 3D distribution of specific molecules at the nanoscale level can be achieved by the so-called laser desorption post-ionization MS [37]. SXRF allows element imaging with resolution down to 50 nm using high flux X-ray beams provided by third-generation synchrotron facilities. NanoSIMS, another high-resolution element imaging technique, is suitable for metal imaging at subcellular level with resolution down to 50 nm. In general, the techniques described here have their own strengths and limitations, but they are excellent tools to investigate uptake, transport, and bioaccumulation processes of metals from tissue to subcellular level. Therefore, these element-imaging techniques should be applied in a complementary manner rather than competition in metallomic studies.

13.4 Concluding Remarks

Metallomics, a system biology discipline aiming to study the interactions and functional connections between biometals and DNA, RNA, proteins, and metabolites, is considered to be more complicated and challenging compared with genomic, transcriptomic, proteomic, and metabolomic studies. The advance of strategic methodology and modern analytical tools have made it possible to achieve the information in terms of the identity, amount, uptake, binding status, transport, and storage of metals in biological media, as well as the interactions with biomolecules and functions in life. However, due to complexity and dynamic nature of metallomic study, each analytical tool can only achieve information on certain aspect rather than a full picture in omics level. Consequently, metallomics study is conducted mostly on subgroups of the metallome [16, 38, 39]. Strategic methodology and complementary analytical tools should be wisely developed and reasonably integrated to approach the comprehensive roles and functions of biometals in life. We envision that there should be a long way to go to reach a real metallome picture. For the ultimate goal, we should not only make great and long-term efforts to setup biological models closer to the real state of biometals in biological system, but also develop revolutionary tools based on new physical, chemical, and biological principles in the future. Whether we would reach the real metallomic picture or not is a question that hard to answered on this moment, but to be sure, the tireless exploration of generations will deepen our understanding the roles of biometals in life.

List of Abbreviations

2D	two dimensional
3D	three dimensional
AAS	atomic absorption spectroscopy
CE	capillary electrophoresis
CID	collision-induced dissociation
ECD	electron capture dissociation

ESI	electrospray ionization
ETD	electron transfer dissociation
EXAFS	extended X-ray absorption fine structure
ICP-MS	inductively coupled plasma–mass spectrometry
IRMPD	infrared multiphoton dissociation
LA-ICP-MS	laser ablation–inductively coupled plasma–mass spectrometry
LC	liquid chromatography
MALDI	matrix-assisted laser desorption/ionization
MS	mass spectrometry
NanoSIMS	nano secondary ion mass spectrometry
NMR	nuclear magnetic resonance
RPLC	reversed-phase liquid chromatography
SEC	size-exclusion chromatography
SXRF	synchrotron-based X-ray fluorescence
TEM/X-EDS	transmission electron microscopy coupled with energy-dispersive X-ray spectroscopy
XANES	X-ray absorption near-edge structure

References

1 Williams, R.J.P. (2001). Chemical selection of elements by cells. *Coord. Chem. Rev.* 216–217: 583–595.
2 Haraguchi, H. (2004). Metallomics as integrated biometal science. *J. Anal. At. Spectrom.* 19: 5–14.
3 Wackett, L.P., Dodge, A.G., and Ellis, L.B.M. (2004). Microbial genomics and the periodic table. *Appl. Environ. Microbiol.* 70: 647–655.
4 Haraguchi, H. (2017). Metallomics: the History over the last decade and a future outlook. *Metallomics* 9: 1001–1013.
5 Szpunar, J. (2005). Advances in analytical methodology for bioinorganic speciation analysis: metallomics, metalloproteomics and heteroatom-tagged proteomics and metabolomics. *Analyst* 130: 442–465.
6 Mounicou, S., Szpunar, J., and Lobinski, R. (2009). Metallomics: the concept and methodology. *Chem. Soc. Rev.* 38: 1119–1138.
7 Lobinski, R., Becker, J.S., Haraguchi, H., and Sarkar, B. (2010). Metallomics: guidelines for terminology and critical evaluation of analytical chemistry approaches (IUPAC technical report). *Pure Appl. Chem.* 82: 493–504.
8 Michalke, B. (ed.) (2016). *Metallomics: Analytical Techniques and Speciation Methods*. Weinheim, Germany: Wiley-VCH Verlag GmbH & Co. KGaA, Boschtr. 12, 69469.
9 Li, Y.F., Sun, H.Z., Chen, C.Y., and Cai, Z.F. (2016). *Metallomics (in Chinese)*. Beijing: Science Press.
10 Maret, W. (2016). *Metallomics – A Primer of Integrated Biometal Sciences*. London: Imperial College Press.

11 Sheftel, A.D., Mason, A.B., and Ponka, P. (2012). The long history of iron in the Universe and in health and disease. *Biochim. Biophys. Acta* 1820: 161–187.
12 Reich, H.J. and Hondal, R.J. (2016). Why nature chose selenium. *ACS Chem. Biol.* 11: 821–841.
13 Xu, M., Yang, L.M., and Wang, Q.Q. (2008). Quantification of selenium-tagged proteins in human plasma using species-unspecific isotope dilution ICP-DRC-qMS coupled on-line with anion exchange chromatography. *J. Anal. At. Spectrom.* 23: 1545–1549.
14 Zhang, J.X., Zhou, Y., Zuo, D. et al. (2022). Quantification of active selenols in cells: a selenol-specific recognition europium-switched signal-amplification ICP-MS approach. *Anal. Bioanal.Chem.* 414: 257–263.
15 Xu, M., Yan, X.W., Xie, Q. et al. (2010). Dynamic labeling strategy with ^{204}Hg-isotopic methylmercurithiosalicylate for absolute peptide and protein quantification. *Anal. Chem.* 82: 1616–1620.
16 Xu, M., Yang, L.M., and Wang, Q.Q. (2013). Chemical interactions of mercury species and some transition and noble metals towards metallothionein (Zn7MT-2) evaluated using SEC/ICP-MS, RP-HPLC/ESI-MS and MALDI-TOF-MS. *Metallomics* 5: 855–860.
17 Xu, M., Yang, L.M., and Wang, Q.Q. (2015). Se-Hg dual-element labeling strategy for selectively recognizing selenoprotein and selenopeptide. *Chin. J. Anal. Chem.* 43: 1265–1271.
18 Chen, B.W., Hu, L.G., He, B. et al. (2020). Environmetallomics: systematically investigating metals in environmentally relevant media. *TRAC, Trend Anal. Chem.* 126: 1158752.
19 Mounicou, S. and Lobinski, R. (2008). Challenges to metallomics and analytical chemistry solutions. *Pure Appl. Chem.* 80: 2565–2575.
20 Yan, X.W., Li, Z.X., Liang, Y. et al. (2014). A chemical "hub" for absolute quantification of a targeted protein: orthogonal integration of elemental and molecular mass spectrometry. *Chem. Commun.* 50: 6578–6581.
21 Sun, X.S., Tsanga, C.N., and Sun, H.Z. (2009). Identification and characterization of metallodrug binding proteins by (metallo)proteomics. *Metallomics* 1: 25–31.
22 Kunst, R.F., Niemeijer, M., van der Laan, L.J.W. et al. (2020). From fatty hepatocytes to impaired bile flow: matching model systems for liver biology and disease. *Biochem. Pharmacol.* 180: 114173.
23 Pinto, C., Estrada, M.F., and Brito, C. (2020). In vitro and ex vivo models – the tumor microenvironment in a flask. *Adv. Exp. Med. Biol.* 1219: 431–443.
24 Zitter, R., Chugh, R.M., and Saha, S. (2022). Patient derived ex-vivo cancer models in drug development, personalized medicine, and radiotherapy. *Cancers* 14: 3006.
25 Duval, K., Grover, H., Han, L. et al. (2017). Modeling physiological events in 2D vs. 3D cell culture. *Physiology* 32: 266–277.
26 Huh, D., Hamilton, G.A., and Ingber, D.E. (2011). From 3D cell culture to organs-on-chips. *Trends Cell Biol.* 21: 745–754.
27 Banci, L. (2003). Molecular dynamics simulations of metalloproteins. *Curr. Opin. Chem. Biol.* 7: 143–149.

28 Reela, A.S., Reel, S., Pearson, E. et al. (2021). Using machine learning approaches for multi-omics data analysis: a review. *Biotechnol. Adv.* 49: 107739.

29 Li, Y.F., Chen, C.Y., Qu, Y. et al. (2008). Metallomics, elementomics, and analytical techniques. *Pure Appl. Chem.* 80: 2577–2594.

30 Vogiatzis, C.G. and Zachariadis, G.A. (2014). Tandem mass spectrometry in metallomics and the involving role of ICP-MS detection: a review. *Anal. Chim. Acta* 819: 1–14.

31 Yan, X.W., Yang, L.M., and Wang, Q.Q. (2013). Detection and quantification of proteins and cells by use of elemental mass spectrometry: progress and challenges. *Anal. Bioanal.Chem.* 405: 5663–5670.

32 Liang, Y., Yang, L.M., and Wang, Q.Q. (2016). An ongoing path of element-labeling/tagging strategies toward quantitative bioanalysis using ICP-MS. *Appl. Spectrosc. Rev.* 51: 117–128.

33 Dressler, V.L., Müller, E.I., and Pozebon, D. (2018). Bioimaging metallomics. *Adv. Exp. Med. Biol.* 1055: 139–181.

34 Becker, J.S., Matusch, A., and Wu, B. (2014). Bioimaging mass spectrometry of trace elements – recent advance and applications of LA-ICP-MS: a review. *Anal. Chim. Acta* 835: 1–18.

35 Gao, J., Chen, Y.C., Guo, Z.J., and He, W.J. (2020). Recent endeavors on molecular imaging for mapping metals in biology. *Biophys. Rep.* 6: 159–178.

36 Hare, D.J., New, E.J., de Jonge, M.D., and McColl, G. (2015). Imaging metals in biology: balancing sensitivity, selectivity and spatial resolution. *Chem. Soc. Rev.* 44: 5941–5958.

37 Li, X., Hang, L., Wang, T. et al. (2021). Nanoscale three-dimensional imaging of drug distributions in single cells via laser desorption post-ionization mass spectrometry. *J. Am. Chem. Soc.* 143: 21648–21656.

38 Chen, L.Q., Yang, L.M., and Wang, Q.Q. (2009). In vivo phytochelatins and Hg–phytochelatin complexes in Hg-stressed *Brassica chinensis* L. *Metallomics* 1: 101–106.

39 Yan, X.W., Li, J., Liu, Q. et al. (2016). p-Azidophenylarsenoxide: an arsenical "bait" for the in-situ capture and identification of cellular arsenic-binding proteins. *Angew. Chem. Int. Ed.* 55: 14051–14056.

14

ICP-MS for Single-Cell Analysis in Metallomics

Man He, Beibei Chen, and Bin Hu

Wuhan University, Department of Chemistry, 299 Bayi Road, Wuchang District, Wuhan, Hubei Province 430072, China

14.1 Introduction

Cells are the basic units of organisms, with the main components of water, metals/metalloids, sugars, lipids, nucleic acids, and proteins [1]. Trace metals/metalloids play an important role in the composition, structure, and physiological functions of cells [2]. The identification, quantification, and localization of specific elements in cells along with investigation of their physiological functions are of significant importance in biological, chemical, and clinical fields. Moreover, due to the random expression of genes and proteins in individual cells in the growth process and small variations in external conditions, there are huge differences among cells called cell heterogeneity. In contrast to stochastic average analysis by bulk measurements, the quantitative analysis of individual cells would provide more sensitive representation of cell-to-cell variations of elemental contents [3].

Metallome refers to the entirety of metal and metalloid species present in a cell or tissue type, their identity, quantity, and localization. Metallomics is the study of a metallome, interactions and functional connections of metal ions and their species with genes, proteins, metabolites, and other biomolecules within organisms and ecosystems [2]. It aims at understanding the biological functions of elements and their species in organisms and how their usage is finetuned in biological species and in populations of species with genetic variations [4]. Metallomics on the level of single cells are regarded as an ultimate goal in the development of the field [5].

Inductively coupled plasma-mass spectrometry (ICP-MS) is one indispensable technique in metallomics study with the highest sensitivity for elements among so many modern analytical techniques. There are mainly two data acquisition modes in ICP-MS analysis. One is the commonly used way in which continuous signals are collected and averaged or "bulk" information are obtained; in this way, the signal from individual cells would be covered or averaged. The other one is time-resolved analysis (TRA) mode, also known as single-event mode (e.g. single particle [sp] or single cell [sc], depending on the target entities). In this way, single

Applied Metallomics: From Life Sciences to Environmental Sciences, First Edition.
Edited by Yu-Feng Li and Hongzhe Sun.
© 2024 WILEY-VCH GmbH. Published 2024 by WILEY-VCH GmbH.

events resulting from the introduction of discrete entities are recorded one by one. The duration of individual event is generally less than 1 ms, and each event will be different and can appear randomly within the total acquisition time [6]. Houk's group [7] performed pioneering work by investigating the behavior of bacterium grown in a spiked uranium medium with TRA-ICP-MS. Positive U^+ spikes were observed in the presence of intact bacterium but not in aqueous U solution. Ho and Chan [8] obtained signal spikes of Mg, Mn, and Cu from single algal cells by TRA-ICP-MS, and the frequency of signal spikes was proportional to the number of cells. Furthermore, adsorption kinetics of Cr^{3+} on algal cells was investigated by this strategy. These pilot experiments demonstrated that TRA-ICP-MS was an emerging technique for counting cells and elemental quantification in single cells simultaneously.

While TRA-ICP-MS-based single-cell analysis for metallomics study is still a challenge for analysts, the problems exist in the extremely low concentration level of interest elemental species in one single cell which cannot be detected directly by ICP-MS, the identification of specific elemental species which involve unknown variation in living cells, separation of single cells and obtaining elemental information at single-cells level in large number of cells at the same time. So far, a great deal of effort has been made to improve the performance of TRA-ICP-MS in single-cell analysis. The efforts are focused on three main issues: (i) Improvement in ICP-MS instrumentation; (ii) Combination with microfluidic platform; (iii) Methodologies development and their applications.

14.2 ICP-MS Instrumental Optimization for Single-Cell Analysis

In order to fulfill single-cell analysis by TRA-ICP-MS, ultrafast signal acquisition is required and recording a sufficiently high number of events (for proper statistics) is mandatory, along with a prerequisite of one signal spike corresponding to one cell event. Table 14.1 lists ICP-MS instruments equipped with different sample introduction systems or mass analyzers for single-cell analysis, and relevant performance.

14.2.1 Sample Introduction System

14.2.1.1 Pneumatic Nebulization
With the use of pneumatic nebulization, ICP-MS is mainly used to analyze liquid samples. The sample introduction system plays an important role in ICP-MS analysis, by which liquid sample (e.g. cells solution) is transformed into tiny aerosols and ionized efficiently in ICP before detection. Detection efficiency (DE) and transport efficiency (TE) are two main factors for the evaluation of the performance of sample introduction system. DE of the cells is defined as the ratio of the number of cells arriving at the detector to the number of introduced cells [11]. TE is defined as the ratio of the difference between the waste volume exiting the spray chamber and the total sample uptake volume, also known as the waste collection method [60].

Table 14.1 Different sample introduction system combined with TRA-ICP-MS for single-cell analysis.

	Instruments	t_{dwell} (ms)	Nebulizer	Spray chamber	Sample flow rate (μL/min)	Cell DE (%)	Cell density (m/l)	The probability of single signal coming from two cells (%)	References
Microfluidic platform	SF (Element XR)	0.1	μDG		0.02	100	1.5×10^6	5×10^{-3}	[9]
	Q (ELAN6000)	10	LADE*		0.5	65.0 (droplet) 4.5	1×10^7	7.4	[10]
	Q (Thermo X2)	5	HPN (PFA microflow)	Impact bead	5	10.0** 3.0	5×10^5	3.4×10^{-3}	[11]
	Q (Agilent 7500a)	10	— (concentric 70 μm)	Scott-type double-pass	5	2.8	1×10^6	4.9×10^{-3}	[12]
	Q (Thermo X2)	5	HPN (Ari Mist HP)	Impact bead	0.1	24.0	5×10^5	1.2×10^{-3}	[13]
	Q (Thermo X2)	5	HPN (PFA microflow)	Impact bead	5	10.0** 3.0	5×10^5	3.4×10^{-3}	[14]
	QQQ (Agilent 8900)	1	HPN* (concentric 150 μm)	Mini coaxial*	200	42.1 ± 7.2 (Au NPs)	2.5×10^5	~0	[15]
	Q (NexIon 300X)	0.1	HPN* (concentric 50 μm)	Single-pass*	0.8	100	1.5×10^5	1.2×10^{-3}	[16]

(Continued)

Table 14.1 (Continued)

	Instruments	t_{dwell} (ms)	Nebulizer	Spray chamber	Sample flow rate (μL/min)	Cell DE (%)	Cell density (m/l)	The probability of single signal coming from two cells (%)	References
Direct injection	Q (Agilent 7700x)	10	HPN* (concentric 110 μm)	Low volume and sheath gas*	10	75	7.7×10^4	1.8	[17]
	Q (Agilent 7500a)	0.05 / — / 10	HPN* (concentric 150 μm)	Low volume and sheath gas*	10	86.0–100	1.7×10^6	1.8 (10 ms) 1.8×10^{-2} (1 ms) 1.8×10^{-4} (0.1 ms) 4×10^{-5} (0.05 ms)	[18]
	Q (NexION 300D)	4	HPN* (concentric 250 μm)	Heated single-pass*	10	2.0	1×10^5	—	[19]
	Q (Agilent 7700)	10	HPN (ENYA Mist)	Single-pass*	10	~48.0 (Au NPs) 25.0	5×10^4	—	[20]
	Q (NexION 350D)	0.05	HPN (HEN)	Asperon	15	31.3 (Au NPs)	2×10^5	—	[21]
	Q (Agilent 7700x)	5	HPN (X175)	Heated single-pass and sheath gas*	30	~1.0	1×10^6	—	[22]
	Q (NexION 300D)	0.1	HPN (Meinhard)	Asperon	21–22	45.6–63.7	5×10^5	—	[23]
	Q (NexION 300X)	0.05	HPN (PFA microflow)	Asperon	19.4	9.9 ± 0.9 (Au NPs)	1×10^5	5×10^{-2}	[24]

QQQ (ICAP TQ)	5	HPN (HPCN)	Low volume and sheath gas	10	50.0–55.0	2.5×10^4	—	[25]
SF (Element 1)	3	— (ES-20100)	—	200	—	1×10^8	—	[7]
Q (Agilent 7500a)	10	CN (V-groove)	Scott-type double-pass	400	0.5	1.2×10^6	8×10^{-2}	[8]
Q (Agilent 7500a)	10	— (microconcentric)	Single-pass*	20	~3.0	1×10^5	—	[26]
Q (Thermo X7)	5	CN (MicroMist)	Impact bead	250	3.0**	1×10^6	—	[27]
Q (NexION 300D)	5	CN (PFA-ST)	Baffled quartz cyclonic	320	~1.0** ~0.2	2×10^5	5×10^{-3}	[28]
Q (Thermo X7)	5	CN (MicroMist)	Impact bead	200	—	1×10^5	—	[29]
Q (Agilent 7500a)	10	CN (V-groove)	Scott-type double-pass	500	~1.0**	3×10^5	—	[30]
TOF (CyTOF2)	—	—	—	45	—	$0.5-1 \times 10^6$	—	[31]
Q (Agilent 7700)	5	— (microconcentric)	Single-pass*	30	1.0**	1×10^6	—	[32]
Q (Agilent 7500a)	10	CN (V-groove)	Scott-type double-pass	400	~0.6	1×10^5	—	[33]
TOF (CyTOF2)	0.013	—	—	45	—	$2-5 \times 10^5$	—	[34]
TOF (CyTOF)	—	—	—	45	—	3.3×10^5	—	[35]

(Continued)

Table 14.1 (Continued)

Instruments	t_{dwell} (ms)	Nebulizer	Spray chamber	Sample flow rate (μL/min)	Cell DE (%)	Cell density (m/l)	The probability of single signal coming from two cells (%)	References
QQQ (Agilent 8800)	3	CN (MicroMist)	Scott-type double-pass	166	7.5** 0.5	2.5×10^5	—	[36]
SF (Element XR/2)	0.1	CN (MicroMist)	Impact bead	250	2.0–3.0	1×10^5	—	[37]
Q (Agilent 7700x)	5	— (microconcentric)	Impact bead	30	—	1×10^6	—	[38]
Q (Agilent 7900)	0.1 — 10	— (concentric)	Scott-type double-pass	300	5.1**	2×10^5	—	[39]
Q (Agilent 7900)	10	CN (MicroMist)	Scott-type double-pass	300	0.02–0.03	1×10^6	—	[40]
Q (Agilent 7500a)	10	CN (concentric)	Scott-type double-pass	500	~1.0	3×10^5	—	[41]
Q (NexION 2000)	0.05 — 0.1	—	Asperon	—	—	1×10^4	—	[42]

	Instruments	t_{dwell} (ms)	LA system	Cell lines	Sample gas (l/min)	Mode	Scan speed (μm/s)	Laser spot size (μm)	References
LA	SF (Element XR)	10	NWR213	3T3	1	Imaging	5	4	[43]
	SF (Element XR)	10	NWR213	3T3	1	Imaging	5–8	4–8	[44]

SF (Element 2XR)	—	UP-213	PBMC	0.8	Single spot	—	25–30	[45]
Q (NexION 300D)	10	NWR213	Raw 264.7	0.8	Single spot	—	40	[46]
Q (NexION 300D)	10	NWR213	HEL	0.8	Single spot	—	20	[47]
Q (Thermo X2)	2–40	Analyte G2	*S. trochoidea* microalgae	1	Imaging	10	2	[48]
	1–5				Single spot	—	40	[49]
SF (Element XR)	3	NWR213	Neuro-2a	1	Imaging	5	4	
SF (Element XR)	—	NWR213	3T3	1	Imaging	4–25	4–25	[50]
SF (Element XR)	—	NWR213	3T3 and A549	1	Imaging	6–15	6–15	[51]
SF (Element XR)	—	NWR213	3T3	1	Imaging	25	30	[52]
					Single spot	—	110	
Q (Agilent 7900)	4	Analyte G2	HeLa	—	Imaging	250	1	[53]

(Continued)

Table 14.1 (Continued)

Instruments	t_{dwell} (ms)	Nebulizer	Spray chamber	Sample flow rate (μL/min)	Cell DE (%)	Cell density (m/l)	The probability of single signal coming from two cells (%)	References
Q (NexION 300D)	10	NWR213	16HBE	1	Single spot	—	60	[54]
TOF (icpTOF 2R)	—	Analyte Excite	Blood cells	1.1	Single spot	—	4	[55]
Q (Agilent 7900)	3–4	Analyte G2	MDA-MB-231 and MDA-MB-46	1.05	Imaging	100	2	[56]
					Single spot	—	30	
TOF (icpTOF 2R)	—	Analyte Excite	Fibroblast multicellular spheroid	1.1	Imaging	25	10	[57]
					Single spot	—	150	
TOF (icpTOF)	0.03	Analyte G2	THP-1	0.7	Single spot	—	<10	[58]

The asterisk "*" denotes the custom-built introduction system and "**" denotes the nebulization efficiency obtained by the specification or Waste Collection Method.
Source: Reproduced from Ref. [59] with permission from Elsevier B.V. (2020).

Conventional pneumatic nebulizer (PN) provides the DE of 1–3% under the uptake rate of 1 ml/min and can reach DE of 10–30% under 0.1 ml/min [7]. For single-cell analysis, a series of nebulizers fitting for microflow have been proposed with improved DE, such as PFA MicroFlow (Elemental Scientific), HEN/DIHEN (Meinhard), HPCN (AIST), and parallel path nebulizers, such as Ari Mist HP and ENYA Mist (Burgener) (Table 14.1). For example, cells DE of 25% was obtained by using ENYA Mist nebulizer and a 10-ml single-pass spray chamber [20]. A home-made high-efficiency cell introduction system (HECIS) consisting of high-performance nebulizer with a 110 μm i.d.-capillary, a 15-ml on-axis spray chamber and sheath gas was constructed [17]. The low-volume chamber was involved with a sheath gas flow to avoid deposits of cell samples at the coned exit. The cell DE of 1.8% was obtained by conventional PN system consisting of a concentric nebulizer and a cyclonic spray chamber. But it was increased to 9% by replacing the concentric nebulizer with HPCN and dramatically raised to 75% by using HECIS. Besides, the inner diameter of the capillary can be varied for improving DE. TE can be improved by desolvation. A HECIS was designed and constructed, consisting of a single-pass spray chamber wrapped with heating wires (80–200 °C) and a detachable microconcentric nebulizer [7]. The generated water vapor can be greatly reduced, and the TE was increased 10 times higher than that obtained in conventional nebulization system. Besides, the sensitivity of ^{115}In was approximately twofold higher at 140 °C than that obtained at 25 °C. Deng and coworkers [16] designed a single-cell sampling system consisting of a flow cell, a microscope to observe cell integrity, customized nebulizer, and a fabricated spray chamber. Cell suspension was introduced in the middle of the flow cell, and sheath flow was introduced on both sides to promote the formation of a single row of cell flow. The inner diameter of capillary tube was optimized, and the DE of cells reached 100%.

To avoid multiple cells contained in one aerosol and the resultant one signal spikes corresponding to multiple cells events, the cells density needs to be optimized before introducing into ICP-MS. Besides, some microfluidic platform or chips have been designed and constructed, endowing with functions of cells encapsulation/separation or droplets generation/splitting, and combined with TRA-ICP-MS for single-cell analysis. Relevant progress will be detailed in Section 14.3.

14.2.1.2 Laser Ablation

As a universal energy source, laser is one of the most commonly used sampling methods in mass spectrometry imaging. Laser ablation (LA) is an introduction technique for solid samples, in which high-energy laser irradiation is employed to make solid samples dissociated and the generated dry aerosols are transported to ICP by carrier gas (helium or argon) for subsequent dissociation, atomization, and ionization. Elemental quantification and distribution can be obtained by LA-TRA-ICP-MS imaging. In cells analysis, cells are generally fixed and dehydrated on a glass slide and subjected to the LA. At present, the spot size of the laser has reached sub-μm (7)-μm [61] level, meeting the requirements of single-cell analysis, while relatively large macrophage and fibroblast cells (>50 μm) are common experimental models [27] for a better resolution, considering more data points per cell available in large-size cells.

Presently, the major problems in single-cell analysis by LA-TRA-ICP-MS include lack of matrix-matching and cell-sized standards for elemental quantification, as well as low throughput due to manual movement for focusing on target cells. Jakubowski's group [50, 51] prepared matrix-matching calibration by doping nanoparticles (NPs) suspension or metal standard solution in nitrocellulose membrane for the quantification of target analyte in single cells. Moreover, a commercial inkjet printer was used to generate individual picoliter droplets, simulating matrix-matching standards and reducing the volume of calibration standard droplets [46]. High-density microarray plates filled with spiked gelatin were proposed by Malderen and coworkers and used as matrix-matching single-cell standards [48]. To improve the sample throughput, Zheng and coworkers [54] fabricated a polydimethylsiloxane (PDMS) microwell array. Benefiting from single cells regularly distributed in the grid pattern, the analytical throughput was increased at least fivefold. Löhr et al. [58] used a novel technique based on a piezo-acoustic dispenser to make single-cell arraying, avoiding cells overlapping and throughput of ~550 cells/h was achieved.

Jakubowski's and coworkers [62] obtained fast and highly spatially resolved (~µm) qualitative elemental distribution within single cells by an ultra-fast wash-out ablation chamber and a nanosecond laser. Ultra-fast wash-outs (<10 ms) were achieved reducing the aerosol mixing from consecutive laser shots even when operating the laser at high repetition rates (25–100 Hz). To realize subcellular spatial resolution imaging, Hang and coworkers [63] designed a LA system with a microlensed fiber and a "three-way" structure ablation chamber. By optimizing the laser energy and fiber-sample distance, ablation craters with flexible diameter ranging from 400 nm to 10 µm are obtained with ease. Detection limits of subfmol level can be achieved for various elements.

14.2.2 Mass Analyzer and Detector

Dwell time (t_{dwell}) in ICP-MS describes the actual scanning time the mass analyzer uses to collect signal for a specific isotope. Quadrupole (Q), sector field (SF), and time of flight (ToF) are three common mass analyzers for ICP-MS instruments. Among them, the t_{dwell} of ICP-SF-MS and ICP-ToF-MS is 100 and 10-µs level, respectively, and has been successfully applied to single-cell analysis. It is worth mentioning that the mass cytometry derived from ICP-ToF-MS can realize simultaneous detection of up to 50 elements in a single cell with the help of fast scanning speed. Therefore, using element tags instead of traditional fluorescent probes to analyze the immunophenotype of single cells has broad application prospects [64]. Bandura et al. [65] used 20 rare earth element polymer tags to simultaneously analyze 20 antigens on the surface of single cells. Bendall et al. [64] divided human bone marrow cell samples into 30 subtypes by analyzing 34 surface markers and intracellular signal proteins of single cells at the same time, and studied the corresponding variations of these subsets of cells under external stimulation, providing comprehensive information on the cell phenotype. The technical details and latest progress of mass cytometry can be found in reference [65], which is helpful to understand the physiological process and disease mechanism of cells more deeply.

The data acquisition ability of Q and SF mass analyzers is limited due to their sequentially read-out mode. To resolve this problem, Sharp's group improved the data acquisition characteristics of LA-ICP-SF-MS at 10-μs time resolution by using a "plug-and-play" multi-channel scaler board [66], and Engelhard's group developed a home-built data acquisition unit based on a programmable logic device for ICP-Q-MS with a time resolution of 5 μs [67, 68]. Miyashita et al. [18] used an external ion pulse counting unit and a function generator to directly read the ion pulse current with no dead time from ICP-Q-MS, and the time resolution was improved down to 0.1 ms. Compared to 10-ms dwell time, S/B of ^{31}P obtained with 0.1-ms dwell time was 13-fold higher, and the probability that a spike signal comes from multiple cells was also reduced. In addition, the DE of cells with a mean size of 6.4 μm was 86%, while could even reach c. 100% for cells of 2.0–3.0 μm size. In other words, cell size affects the DE of cells.

Many of the current generation ICP-Q-MS instruments have been designed with improved data acquisition capabilities to meet the demands of fast transient analysis. The t_{dwell} of commercial ICP-Q-MS can reach 100-μs level, e.g. Agilent 7900 and PlasmaQuant®MS of AnalytikJena can reach 50 μs and NexION 300/350/2000 of Perkin Elmer can reach 50 μs. To further reduce the spectral interference of coexisting ions, a series of ICP-MS was equipped with triple quadrupole, including Agilent 8800/8900 (t_{dwell} = 100 μs), iCAP Q™ of Thermo Scientific™ (t_{dwell} = 100 μs), and NexION®5000 of Perkin Elmer (t_{dwell} = 10 μs). Relevant details are presented in Table 14.1.

14.3 Microfluidic Platform for Single-Cells Analysis

Microfluidic platform generally has a microchannel structure of tens to hundreds of microns, suitable for manipulating a small amount of fluid, and can integrate multiple work units such as sample injection, reaction, separation, enrichment, and detection, rapidly realizing various functions of conventional chemical or biological laboratories [69]. Microfluidic platform includes microfluidic chips and microfluidic devices. Microfluidic chip refers to the fabrication of quasi 2D microchannels on a piece of several square centimeters of thin film made of glass, silicon, quartz, or polymer. Microfluidic devices usually have three-dimensional microstructure fabricated by using microelectromechanical process, 3D printing technology, or assembly of commercial components. These microfluidic platforms have outstanding advantages in the analysis of cell samples [70, 71].

(i) It can flexibly integrate or connect functional units, which is easy to be coupled with high sensitivity detection technology;

(ii) The microscale channel makes the heat and mass transfer fast, and the microscale sample pretreatment technology can improve the analysis performance of the macrodevice;
(iii) The micrometer channel matches the micrometer size of mammalian cells, which is convenient for controlling the culture medium of cells and reducing the damage and pollution to cells.
(iv) With its high spatial resolution, micromanipulation and analysis of cell samples can be realized, which is helpful for single-cell analysis;
(v) The microfluidic chip is portable and can realize "personalized" chemical analysis.

It is now an acknowledged ideal platform for the miniaturization of sample pretreatments and cell manipulation. The integration of appropriate sample pretreatment technology in the microfluidic platforms can realize the matrix removal of cell samples and the enrichment of target analytes without increasing the consumption of cell samples, so as to achieve highly sensitive, highly selective, and high-throughput trace elements and their speciation analysis in a small number of cells/single cells [72].

Single-cell analysis highly depends on the manipulation of cells, mainly including cell location and capture, cell sorting and fusion. The continuous development of modern technology has enabled the fabrication and application of various microscale structures on microfluidic chips/devices, which makes it possible to achieve single-cell analysis on the microfluidic platform [73–75]. The combination of microfluidic platform with TRA-ICP-MS is very promising to achieve the determination of trace elements and their corresponding species even in single cells. Several strategies are usually used to separate and obtain target cells on the microfluidic platform [76–78] (Figure 14.1).

(i) *Droplet-based single-cell separation* [79]. Specifically, single cells are isolated by liquid drops, merits of high throughput; but it is only applicable to purified cells of a single population.
(ii) *Hydrodynamic-method-based cells capture*. Target single cells can be separated from matrix, such as multiple cells, by size, deformability, and other characteristics; delicate chip design and fabrication are required.
(iii) *Magnetic separation*. Magnetic tags are used to capture target cells, merits of good specificity; additional antibodies and magnetic tags are necessary.
(iv) *Sound capture*. It is based on surface acoustic wave to control the movement of particle (or cells). Cells can be accurately located through particle size, but negative impact would occur on physiological properties such as cell vitality.
(v) *Dielectrophoresis capture*. Cell separation is based on the fact that cells (polarizable particles) have different dielectric constants in non-uniform electric fields, resulting in different motion characteristics. By regulating the electric fields, target cells can be simply selected, but it is difficult to avoid heat generation in long-term use.

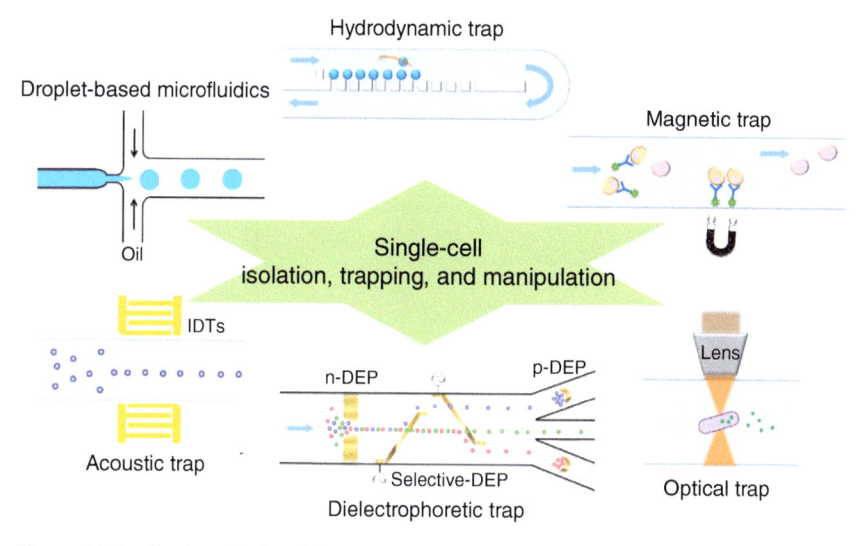

Figure 14.1 Design of microfluidic chip for single-cell separation, capture, and manipulation. Source: Reproduced from Ref. [76] with the permission from Multidisciplinary Digital Publishing Institute.

(vi) Optical capture [80], also known as optical tweezers. Optical potential wells generated by laser beams are used to control cells. This method is more applicable, but requires expensive optical systems.

At present, the most widely used methods for separating/capturing single cell are droplet encapsulation, hydrodynamic capture, and magnetic separation.

14.3.1 Droplet-Encapsulation-Based Single-Cell Separation

Based on Rayleigh principle, the jet obtained by microdroplet generator (μDG) will become unstable when it is subjected to the pressure transmitted by the mechanical vibration generated by the voltage, so as to overcome the surface tension and viscous force and spray droplets. Shigeta et al. [9, 81] used μDG (Figure 14.2) to generate droplets, which encapsulated single selenium-rich yeast cell. With He as the desolvent gas, the droplets were to introduced into ICP-SF-MS, and the DE of the cell reached ~100%. The spectral interference can be effectively reduced due to the small amount of matrix introduced into ICP. However, in the operation of μDG, the requirements for argon flow rate are more stringent, and the size of generated droplets is related to the nozzle. In particular, as a passive way to generate droplets, piezoelectric is very sensitive to the pH, viscosity, and salt content of the solution. Besides, the apparatus is expensive. Comparatively, self-made microfluidic chips/devices have greater application potential.

The droplet microfluidic chip generally generates W/O type droplets. In other words, the generation of droplets usually requires oil phase as the continuous phase to realize the separation of droplets. Verboket et al. [10] used perfluorohexane, a highly volatile organic phase, as the continuous phase of liquid drops in the

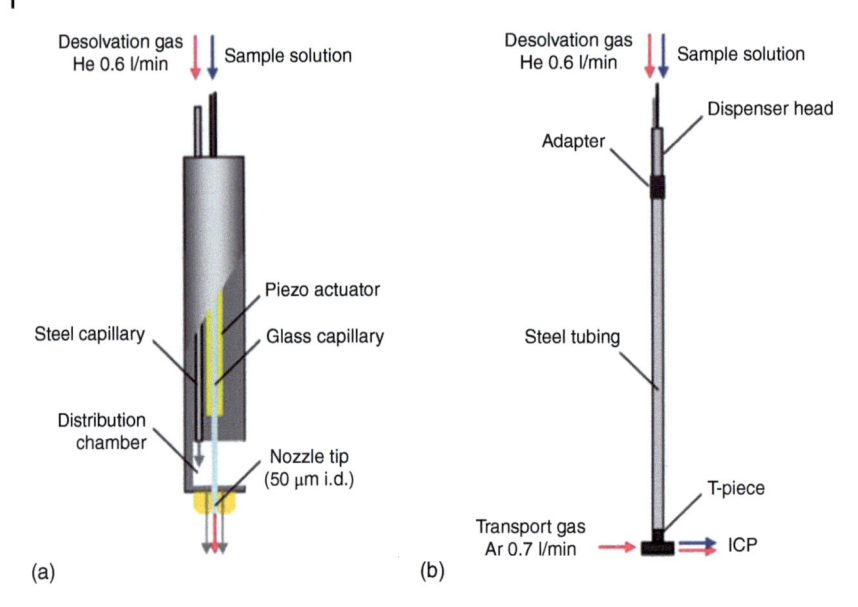

Figure 14.2 Microdroplet generator (a) and desolvation diagram (b). Source: Reproduced from Ref. [81] with permission from the Royal Society of Chemistry (2013).

cross focused droplet microfluidic chip. The organic solvents were removed by heating and a self-made membrane-assisted desolvent device after the droplets left the chip (Figure 14.3), effectively avoiding the instability of ICP and carbon deposition at the sampling cone caused by the organic solvents. The analytical system was used to detect iron in a single red blood cell. However, a large number of accessories in the system are self-made, and the composition of the analytical system is complex.

Taking full advantages of alcohols, e.g. high viscosity and low carbon content, Hu and coworkers [11] used n-hexanol as the organic phase, along with HPN and oxygenation in spray chamber, to reduce the adverse effects of the organic phase. With n-hexanol containing 1% Span80 as the organic phase and cell suspension as the aqueous phase, droplets of 25 μm encapsulating single HepG2 cell were achieved by a simple flow focusing structure (Figure 14.4). Through a glass capillary tube (75 μm i.d. × 365 μm o.d.), the droplets generated on the chip, wrapping single cells, were directly introduced into ICP-MS equipped with a microfluidic nebulizer. The droplet generation frequency on the chip is $3-6 \times 10^6$ min^{-1}, cells introduction frequency is 2500 min^{-1}, and the method has a high sample throughput. Conventional PDMS chips are generally quasi 2D chips. In microchannels with a height of tens of micrometers, droplets will inevitably contact the channel surface, so the surface properties of the channel need to be considered when generating droplets. Moreover, the pressure PDMS can withstand is often limited (0.3–0.5 MPa). To solve this problem, Hu and coworkers [13] built a 3D droplet microfluidic device with coaxial structure and visualization characteristics based on all commercial components such as PEEK head, PEEK tube, quartz capillary, and four-way valve (Figure 14.5). Compared with other 3D droplet microfluidic devices, the droplet generation process can be observed in

Figure 14.3 LADE chip-ICP-MS with membrane desolvention device. Source: Reprinted with permission from Ref. [10]. Copyright (2014) American Chemical Society.

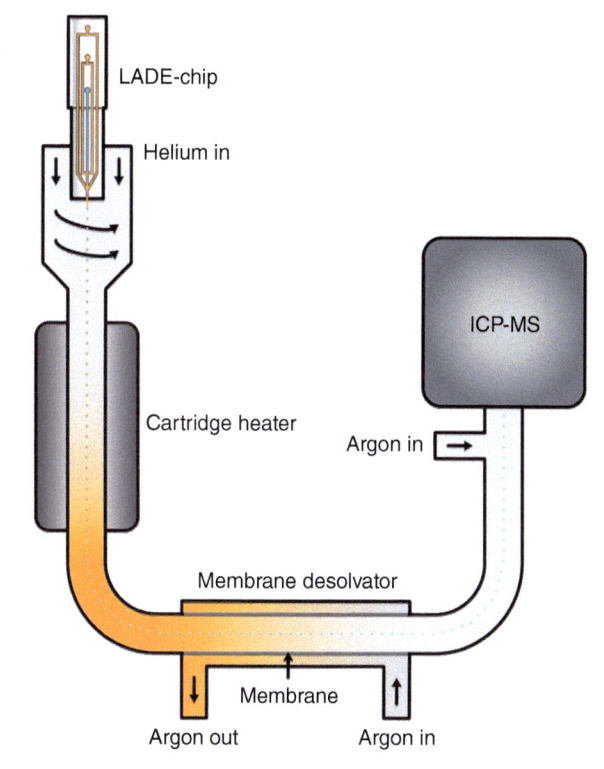

Figure 14.4 Design of droplet chip (a); droplet generation (b); schematic diagram of online droplet chip-ICP-MS (c). Source: Reprinted with permission from Ref. [11]. Copyright (2017) American Chemical Society.

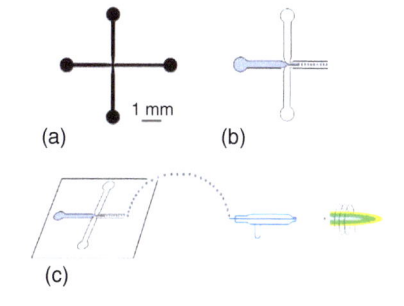

the droplet microfluidic device. Compared with the quasi 2D PDMS chip, the droplet microfluidic device can be directly combined with ICP-MS, and there is no need for aging of the bonded PDMS chip or bonding additional capillaries. It greatly simplifies the interface design and operation steps, with the advantages of simple device, low cost, convenient production, no secondary processing and high-pressure resistance. On this basis, through the hydrophobic modification of the glass capillary and the design of a controllable gas circuit system, the working gas (Ar) of ICP-MS was used as the continuous phase to build water in gas droplet microfluidic device that is friendlier to ICP-MS instrument, completely avoiding the instability of the ICP power caused by the organic phase and the carbon deposition at the sampling cone. No membrane desolvent device or expensive instruments such as ICP-MS and platinum cone with oxygenation technology are needed to form droplets encapsulating

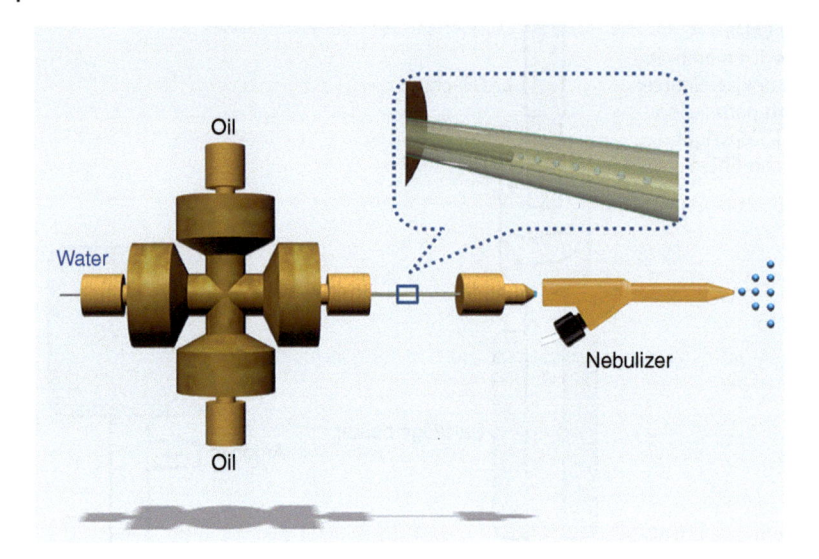

Figure 14.5 3D droplet microfluidic device with visual characteristics for single-cell analysis. Source: Reprinted with permission from Ref. [13]. Copyright (2019) American Chemical Society.

single cells, reducing the probability that one signal peak comes from multiple cells to 6.1×10^{-3}. The DE of cells is 17.6%, and the device has more universal applicability. Wang and coworkers [12] designed a four-way valve with a cross structure, which generates a droplet encapsulating single cells at the cross, and the probability that a droplet encapsulates one single cell is less than 0.005%.

The above works involve directly introducing single cell encapsulated in droplets into ICP-MS, which can only achieve the analysis of the total amount of trace elements in a single cell. In order to further obtain the information of elemental species in a single cell, Hu and coworkers [82] designed an integrated chip, consisting of a cross channel droplet generation area, a cell lysis area, and a T-channel droplet division area. An online analytical method based on droplet splitting microfluidic chip and TRA-ICP-MS detection was proposed to analyze the degradation behavior of FePt Cys NPs in a single cell. The design of T-shaped channel is used to divide the liquid droplets encapsulating single cells generated in the chip evenly. Under the effect of gradient magnetic field, the undifferentiated magnetic FePt@Cys is retained in the subdroplets close to the permanent magnet and discharged through the waste outlet when splitting. The subdroplets at the end far from the permanent magnet are homogeneous solutions, which contain free Fe and Pt formed in the single cells due to the degradation of FePt@Cys NPs. The outlet end far away from the permanent magnet is connected to the ICP-MS nebulizer to realize online analysis of free Fe and Pt in a single cell. In addition, if the permanent magnet is removed and one side of the outlet is blocked, the droplets will not split and will be directly introduced into ICP-MS, realizing online analysis of the total amount of FePt@Cys taken by a single cell (Figure 14.6).

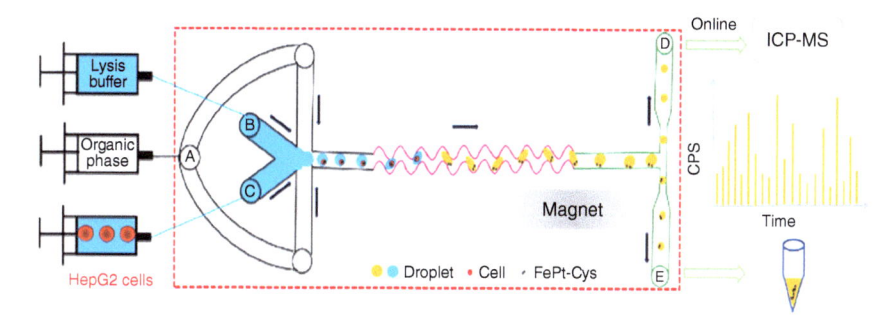

Figure 14.6 Droplet splitting microfluidic chip ICP-MS system. Source: Reprinted with permission from Ref. [82]. Copyright (2020) American Chemical Society.

14.3.2 Hydrodynamic-Capture-Based Single-Cell Separation

The flow field force can be used to make the cells be orderly arranged in the microchannel. Wang and coworkers [83] designed a spiral channel containing 104 periodic dimensional confinement micropillars. With the flow rate of 100–800 μl/min, single cells were focused into an aligned stream. This system is convenient to operate with a high sample throughput of 16 000 cells/min. In the subsequent work, the glass capillary is wound on the polyurethane cylinder, and Dean vortex is generated by using the curvature effect of the spiral structure to exert force on the velocity field and pressure field in the tube, so that the single cells are arranged in order (Figure 14.7). The DE of cell was 42.1% [15]. This method has a good spatial and temporal resolution (41.55 ± 17.46 μm, 0.97 ± 0.41 ms) and a high throughput (40 000 cells/min).

Wang and coworkers [84] designed a straight–curved–straight channel (Figure 14.8) to focus on single cells and used it with TRA-ICP-MS for single-cell analysis. The author used NH_4HCO_3 buffer with good thermal decomposition

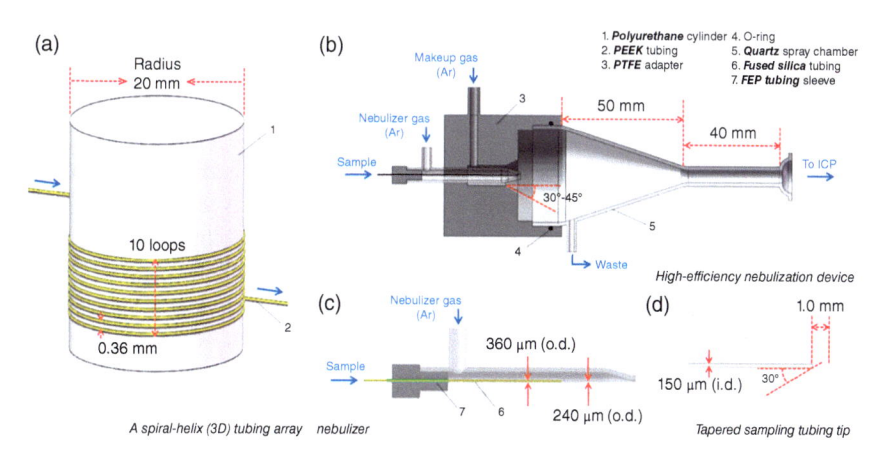

Figure 14.7 Single-cell focusing system of spiral channel; 3D spiral array (a); atomizer (b); concentric atomizer and coaxial atomization chamber (c); design drawing of atomizer nozzle tip (d). Source: Wei et al. [15]/American Chemical Society.

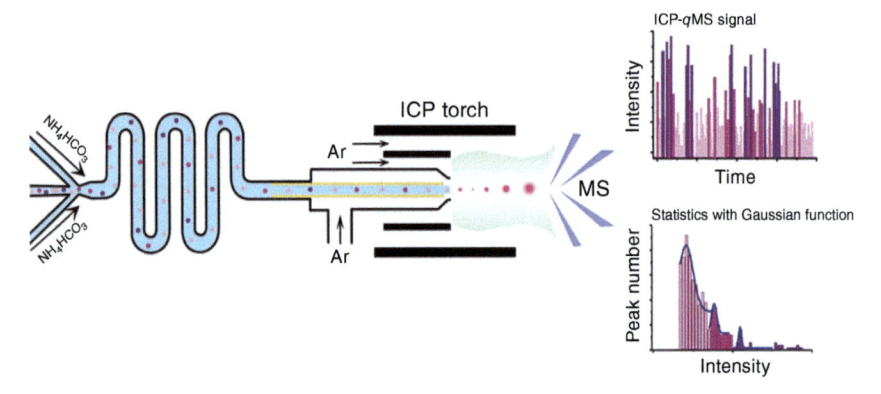

Figure 14.8 Single-cell sequencing microfluidic chip ICP-MS system [84].

performance as sheath flow to promote cell focusing, avoiding the use of oil phase. The DE of cells is as high as 70%, and the sample throughput is 25 000/min.

14.3.3 Magnetic-Separation-Based Single-Cell Capture

At present, magnetic separation is mainly applied to the screening of target cells in complex matrices and mixed samples of multiple cells; the application in single-cell analysis is scarce. It can be used to capture and analyze single cells by combining it with droplet single-cell separation. For example, Gu et al. [85] developed a microfluidic chip platform for single-cell DNA extraction by combining liquid droplets with magnetic solid-phase extraction. Chen et al. [86] fabricated a chip system in which magnetic tags were used to sort cells and droplets were used to wrap and separate single cells.

Although the magnetic capture strategy can effectively separate target cells from complex matrices and interfering cells quickly and easily by means of magnetic immunity, a single magnetic capture method cannot obtain a single target cell. This method needs to be combined with strategies of droplets or microgrooves capturing, etc. to achieve the separation and acquisition of single cells. However, such a system is often complex, and its application in ICP-MS-based single-cell analysis is greatly limited.

It can be seen that single-cell separation based on droplets is the main strategy for trace element analysis in single cells by TRA-ICP-MS. It has the advantages of simple operation, low difficulty in chip fabrication, and good stability. Although the introduction of organic phase will affect the stability of ICP-MS, it is still hopeful to establish simple, effective, reliable, and highly sensitive systems for single-cell analysis by optimizing the parameters of microfluidic platform and TRA-ICP-MS detection.

14.4 ICP-MS-Based Single-Cells Analysis in Metallomics

With the development and improvement of relevant methodologies, TRA-ICP-MS has been demonstrated with good potential in metallomics study at single-cells level.

The first step in this direction is the quantification of target elements in individual cells. It should be noted that NPs or standard solutions containing the same elements are generally used as the standards for the quantification of specific elements in a single cell by TRA-ICP-MS. When NPs are applied, the exact number of atoms of the specific elements in one NP is one prerequisite for accurate quantification. When standard solutions are applied, the TE of the system should be determined before quantification. The analysis of specific elemental species in individual cells is still a great challenge presently. Secondly, elemental distribution at subcellular level in individual cells can be obtained by analytical techniques with high sensitivity and spatial resolution (e.g. LA-ICP-MS). Moreover, to clarify the interactions and functional connections of elemental species with genes, proteins, metabolites, and other biomolecules within organisms and ecosystems, the identification and quantification of relevant molecules is necessary. TRA-ICP-MS has been applied to quantify biomolecules with the aid of elemental tagging strategy, and the binding ratio of element labels and target biomolecules should also be accurately determined before calculation. Based on these methodologies, the variation of specific element/species in terms of quantity and location can be obtained in organisms, along with the relationship between the element/species and relevant biomolecules. Table 14.2 presents the applications of TRA-ICP-MS in single-cell analysis in recent years.

14.4.1 Endogenous Elements in Single Cells

Droplet microfluidic chip online coupled with TRA-ICP-MS was used for the quantification of Zn in single HepG2 cell, with ZnO NPs suspension as the standard [11]. The results showed the heterogeneity of HepG2 cells. The distribution of zinc content in a single cell conforms to the Gaussian distribution, and the zinc content in one single cell is 21.7 fg, which is consistent with the result (25.0 fg) obtained by conventional acid digestion ICP-MS method. In addition, Deng and coworkers [16] proposed the formation of single-row cell flow by introducing sheath flow, and applied it to the analysis of Cu in red blood cells in combination with TRA-ICP-MS. The results of single-cell analysis showed that the content of Cu in a single red cell was 0.20–0.40 fg, which was consistent with the results of microwave digestion (0.266 fg/cell).

14.4.2 Exogenous Metal Exposure to Single Cells

Based on the combination of periodic microcolumn spiral channel chip and TRA-ICP-MS, Wang and coworkers [83] explored the antagonism of Cd^{2+} and Cu^{2+} at the single-cell level. An obvious antagonistic effect was observed for MCF-7 cell by culturing for 3, 6, 9, and 12 hours with 100 μg/l Cd^{2+} and 100 μg/l Cu^{2+}, and a rivalry rate of 12.8% was achieved at 12 hours. However, limited antagonistic effect was encountered for a bEnd3 cell within the same incubation time period, with a rivalry rate of 4.81%. On the contrary, an antagonistic effect was not observed for the HepG2 cell in six hours culturing process, while an obvious antagonistic effect was found by further culturing to 12 hours, with a rivalry rate of 10.43%. For all three cell lines, significant heterogeneity was observed among individual cells.

Table 14.2 Applications of TRA-ICP-MS in single-cell analysis.

	Application	Elements	Highlights	Quantification approach	References
Difference analysis	Dead cells versus live cells	$^{102/104}$Ru, $^{107/109}$Ag, ^{195}Pt	Different mass of silver in dead and live bacterial cells	No quantification	[35]
	Cell lines	^{31}P^{16}O, ^{32}S^{16}O, ^{55}Mn, ^{56}Fe, ^{63}Cu, ^{68}Zn	Different masses and distributions of essential elements in cancer and normal cells	Standard solution	[28]
	Cell lines	^{31}P^{16}O, ^{32}S^{16}O, ^{55}Mn, ^{56}Fe, ^{59}Co, ^{65}Cu, ^{66}Zn	Different masses and distributions of essential elements in cancer and normal cells	Standard solution	[19]
	Cell lines	^{157}Gd, ^{195}Pt	Different uptaken kinetics of anti-tumor agents in cancer and normal cells	Standard solution	[29]
	Cell lines	^{59}Co	Different uptake kinetics of ([Co(tpa)(cur)](ClO$_4$)$_2$) in HepG2 cells and MCF-7 cells	Standard solution	[41]
	Cell subsets	^{159}Tb, ^{195}Pt	Different masses of Pt in cisplatin sensitive and resistant cells	Standard solution	[20]
	Cell subsets	^{75}As, ^{146}Nd, ^{163}Dy, ^{195}Pt	Different cytotoxicity of As$_2$O$_3$ in NB4 and HL60 cells from the perspective of viability, apoptosis, and differentiation	Standard solution (no validation by acid-digestion)	[32]
	Cell subsets	^{31}P^{16}O, ^{142}Nd	Different masses of TfR1 in MCF7 and MDA-MB 231 cells	Standard solution	[25]
	Cell subsets	^{107}Ag	Different uptaken Ag NPs numbers in THP-1 monocytes and their partially differentiated macrophages	NPs	[37]

Cell subsets	^{142}Nd, $^{147/152}$Sm, ^{148}Nd, ^{151}Eu, ^{159}Tb, ^{165}Ho, $^{168/170}$Er, $^{172/176}$Yb, ^{174}Y, ^{197}Au	Different masses of AuNPs in alveolar macrophages, dendritic cells, and B/T-cells from the lung tissue of mice	No quantification	[31]
Cell cycle phases	^{75}As	Different uptake behaviors of As-based drugs in different cell cycle phases of NB4 and HL60 cells	No quantification	[38]
Chemical valence	^{53}Cr	Different uptake behaviors of Cr(VI) and Cr(III) in HeLa cells	Standard solution	[32]
Ions versus nanoparticles	^{66}Zn, ^{107}Ag	Different uptake behaviors of dissolved Ag and Ag NPs in HepG2 cells	NPs	[13]
Ions versus nanoparticles	^{197}Au	Different uptake behaviors of dissolved Au and Au NPs in alga cells	Standard solution	[21]
Nanoparticles of different sizes and surface coatings	^{66}Zn, ^{197}Au	Different uptake kinetics of Au NPs@citric acid and Au NPs@DNA with different sizes in Hela cells	NPs	[14]
Metal- and metalloid-containing drugs	^{75}As	Different uptake kinetics of ZIO-101 and ATO (arsenic-based drugs) in NB4 and HL60 cells	No quantification	[38]
Metal- and metalloid-containing drugs	^{24}Mg, ^{65}Cu	Different uptake kinetics of cupric sulfate and EarthTec® (copper-based algaecides) in *Microcystis aeruginosa* cells	Standard solution (no validation by acid-digestion)	[23]

(continued)

Table 14.2 (Continued)

Application		Elements	Highlights	Quantification approach	References
Correlation analysis	Endogenous elements versus cell sizes	^{25}Mg, ^{31}P, ^{44}Ca, ^{55}Mn, ^{56}Fe, ^{63}Cu, ^{66}Zn	Interrelation of Mg and cell size	Only by acid-digestion	[17]
	Endogenous elements versus cell sizes	^{24}Mg	Interrelation of Mg and cell size	No quantification	[22]
	Endogenous elements versus cell sizes	^{13}C, ^{25}Mg, ^{27}Al, ^{31}P, ^{34}S, ^{39}K, ^{44}Ca, ^{52}Cr, ^{54}Fe, ^{55}Mn, ^{63}Cu, $^{64/66}$Zn	Interrelation of P and cell size	No quantification	[18]
	Interrelation of endogenous elements	^{25}Mg, ^{31}P, ^{44}Ca, ^{55}Mn, ^{56}Fe, ^{63}Cu, ^{66}Zn	Interrelation of P and Zn, P and Mg, as well as Mg and Zn	Only by acid-digestion	[17]
	Interrelation of endogenous elements	^{25}Mg, ^{55}Mn, ^{107}Ag	The ratio of Mn to Mg after Ag NPs exposure	Standard solution (no validation by acid-digestion)	[40]
	Interrelation of endogenous elements	^{23}Na, ^{24}Mg, ^{27}Al, ^{31}P, ^{32}S, ^{44}Ca, ^{55}Mn, ^{56}Fe, ^{65}Cu, ^{66}Zn, ^{75}As, ^{88}Sr, ^{138}Ba, ^{165}Ho, $^{191/193}$Ir	Interrelation of P, Zn, and Ir-DNA	Microdroplets of matrix-matched standard solution (LA system)	[58]
	The effect of exogenous elements on intracellular elements and components	^{24}Mg, ^{65}Cu	The effect of copper-based algaecides on intracellular Mg	Standard solution (no validation by acid-digestion)	[23]
	The effect of exogenous elements on intracellular elements and components	^{25}Mg, ^{31}P, ^{39}K, ^{53}Cr, ^{55}Mn, ^{63}Cu, ^{64}Zn	The effect of toxic Cr(VI) on Mg	No quantification	[33]
	The effect of exogenous elements on intracellular elements and components	^{75}As	The effect of As on the membrane lipids of cells	Standard solution (lower)	[24]

	The effect of exogenous elements on intracellular elements and components	^{26}Mg, ^{31}P, ^{39}K, ^{55}Mn, ^{56}Fe, ^{59}Co, ^{63}Cu, ^{66}Zn, ^{109}Ag	Ag NPs and Fe were colocalized in the outer rim	Nitrocellulose membrane doped with standard solution (LA system)	[57]
	The effect of exogenous elements on intracellular elements and components	^{31}P, ^{32}S, ^{57}Fe, ^{65}Cu, ^{66}Zn, ^{81}Br, ^{195}Pt	The preferential accumulation of Pt in red and white blood cells	Only by acid-digestion	[55]
	Interrelation of exogenous elements	^{24}Mg, ^{209}Bi	The effect of ferric ions on the uptake of CBS in single Helicobacter pylori cells	Only by acid-digestion	[26]
Interaction behaviors	Uptake kinetics of stimulus	^{111}Cd	Uptake kinetics of CdSeS QDs	Standard solution	[27]
	Uptake kinetics of stimulus	^{31}P^{16}O, ^{32}S^{16}O, ^{75}As^{16}O	Uptake kinetics of arsenite	Standard solution	[36]
	Action mechanisms	^{107}Ag, $^{191/193}$Ir	Distinguish strong adhesion on the cell membrane and internalization by using chemical etching	Standard solution	[34]
	Action mechanisms	^{66}Zn, ^{197}Au	Distinguish strong adhesion on the cell membrane and internalization by using low-temperature method	NPs	[14]
	Action mechanisms	^{139}La, ^{141}Pr, ^{142}Nd, $^{151/153}$Eu, ^{152}Sm, ^{160}Gd	The antibacterial mechanisms of vancomycin and Ag NPs	Isotope dilution method	[42]

(continued)

Table 14.2 (Continued)

Application	Elements	Highlights	Quantification approach	References
Others	^{44}Ca, ^{24}Mg, ^{238}U	Single-cell spikes in the spectrogram of TRA-ICP-MS	Standard solution (lower)	[7]
	^{25}Mg, ^{53}Cr, ^{55}Mn, ^{65}Cu	Quantify intrinsic elements	Standard solution (lower) NPs	[8]
	^{24}Mg, ^{39}K, ^{45}Sc, ^{55}Mn, ^{63}Cu, ^{66}Zn, ^{72}Ge	Quantify intrinsic elements	Standard solution	[39]
	$^{24/25}$Mg, ^{57}Fe, $^{63/65}$Cu, $^{64/66}$Zn, $^{74/77/82}$Se	Quantify intrinsic elements	Standard solution (lower)	[9]
	^{56}Fe	Quantify intrinsic elements	Monodisperse microdroplets of standard solution (lower)	[10]
	^{63}Cu	Quantify intrinsic elements	Standard solution	[16]
	^{66}Zn	Quantify exogenous ZnO NPs	NPs	[11]
	^{89}Y, ^{197}Au	Quantify exogenous Au NPs	NPs	[12]
	^{197}Au	Quantify exogenous Au NPs	NPs	[15]

Source: Reproduced from Ref. [59] with permission from Elsevier B.V. (2020).

14.4.3 Nanoparticles Uptake by Single Cells

Single HepG2 cell analysis after ZnO NPs incubation showed that some cells will not ingest/adsorb ZnO NPs, some cells will ingest/adsorb one ZnO NP, some cells will ingest/adsorb two ZnO NPs, and some cells will ingest/adsorb three ZnO NPs [11]. This result also shows the heterogeneity for HepG2 cells in ingesting/adsorbing ZnO NPs.

Wang et al. [12] conducted single-cell analysis on MCF-7 cells incubated with Au NPs. With the increase of incubation concentration of Au NPs, the intensity of Au signal peak in a single cell increased. Gaussian fitting of single-cell peaks of two different samples showed that the content of Au NPs in a single cell was 2.27 fg and 3.77 fg, respectively. The results of acid digestion showed that the content of Au NPs in a single cell was 2.38 fg, which further explained the heterogeneity of cells. Hu and coworkers [14] employed droplet-chip-TRA-ICP-MS single-cell analysis system to study the uptake behavior of HeLa cells for Au NPs with different modification. The statistical results demonstrated that though the average uptake amount of Au NPs@citric acid was higher than that of Au NPs@ssDNA in the cell population, the percentage of cells uptaking Au NPs@ssDNA was higher than that uptaking Au NPs@citric acid. The distribution of the number of uptaken Au NPs in single HeLa cells shows a great difference when HeLa cells are incubated with different Au NPs. The percentage of cells uptaking Au NPs in the cell population also reveals a difference in cellular uptake between Au NPs@citric acid and Au NPs@DNA at the single-cell level. To explain the abovementioned phenomenon, the endocytosis mechanisms of Au NPs are investigated. Clathrin-mediated endocytosis is found to be the major internalization pathway for 15 and 30 nm Au NPs@DNA. In addition, based on the effect of Dean force in the spiral channel to promote the orderly arrangement of single cells, Wang and coworkers [15] analyzed the K562 cells incubated by Au NPs. With the increase of incubation concentration of Au NPs, the peak jumping intensity of K562 cells increased, and the intensity distribution was more diversified. In order to explore the differences in the uptake behavior of Au NPs by different kinds of cells, Wang and coworkers [84] studied the differences in the uptake of Au NPs by Hela and macrophages based on a straight–curved–straight channels on chip and TRA-ICP-MS. The results showed that the uptake of Au NPs by the two kinds of cells showed the heterogeneity between cells, and the difference of uptake behavior of Hela single cells was more obvious, while the uptake of Au NPs by macrophages was about 20 times that of Hela cells.

Hu and coworkers [13] used a visualized 3D droplet microfluidic device based on coaxial structure and TRA-ICP-MS to analyze Ag NPs in HepG2 cells. The results showed that the cellular heterogeneity of Ag NPs uptake by cells was more obvious than that of Ag^+, which might be related to the fact that the medium used to incubate Ag NPs was a heterogeneous solution. After incubating HepG2 cells with Ag^+ for six hours, almost all cells in the cell population uptake Ag^+. After HepG2 cells were incubated with Ag NPs for 12 hours, almost all cells were able to ingest Ag NPs. After incubation of Ag^+ or Ag NPs, the content of Ag uptake by cells increased with the increase of incubation time, and did not reach the absorption platform at 24 hours, and the content of Ag$^+$ uptake by cells was slightly less than that of Ag NPs.

In order to explore the degradation behavior of NPs in a single cell, Hu and coworkers [82] constructed a droplet splitting microfluidic chip to study the behavior of cells in uptake and degradation of FePt@Cys magnetic NPs. With the increase of incubation time, the proportion of FePt@Cys uptake and degraded by cells increased. When the cells were incubated for six hours, almost all cells uptake FePt@Cys NPs. However, FePt@Cys NPs were degraded only in 60% of cells incubated for six hours. When the incubation time reached 18 hours, FePt@Cys NPs were degraded in almost all cells. In addition, the release rate of Pt in cells is higher than that of Fe. It is speculated that the released Pt is more toxic to cells and causes apoptosis.

Fast and highly spatially resolved images of elemental distribution within mouse embryonic fibroblast cells (NIH/3T3 fibroblast cells) and human cervical carcinoma cells (HeLa cells), incubated with Au NPs and Cd-based QDs, respectively, are obtained by a ns-LA-ICP-SFMS equipped with an ultra-fast wash-out ablation chamber [62]. Elemental distribution of Au and Cd in single cells was achieved using a high scanning speed (50 µm/s) and high repetition rate (100 Hz). The results obtained for the distribution of fluorescent Cd-based QDs within the HeLa cells are in good agreement with those obtained by confocal microscopy.

14.4.4 Metal-containing Drugs Uptake by Single Cells

After an exposure to various concentrations of cisplatin as a chemotherapeutic drug, Vanhaecke et al. [87] determined Pt and five endogenous elements (P, S, Fe, Cu, and Zn) in individual human cells (Raji, Jurkat, and Y79) by TRA-ICP-MS. The content of Pt in single cells exhibited a concentration and time-dependent way. Besides, a higher Pt uptake was found for Jurkat cells than other two cell types. It is consistent to the difference in chemosensitivity to cisplatin for these three cell types. The increasing amount of ingested Pt hardly affects the contents of endogenous elements in Raji and Y79 cells, while for Jurkat cells, the content of P and S in single cell was found to decrease with the increase of cisplatin exposure concentration. It is probably due to the stress induced by the chemotherapeutic treatment in cells showing chemosensitivity toward cisplatin. Ruprecht and coworkers [88] quantified cisPt in single cancer cells and in isolated nuclei. A comparison of cisPt uptake was carried out between a wild-type (wt) cancer cell line and related resistant sublines. The amount of cisPt was lower in resistant cell lines and their nuclei than that in wt cells. Moreover, the abundance of internalized cisPt decreased with increasing resistance, and the concentration of cisPt within the nuclei was higher than cellular concentrations.

Hang and coworkers [63] fabricated a LA system with an adjustable spatial resolution down to 400 nm. In combination with TRA-ICP-MS, the distribution of various photodynamic therapy drugs in the intestine of mouse can be clearly observed. The comparison imaging results showed that the drug distribution in tissue slice could be identified at the subcellular level with the high-resolution mode. More valuably, gold nanorods (GNRs) and carboplatin in a single cell are able to be visualized at organelle level due to the nanoscale resolution, which is able to reveal the mechanism of cell apoptosis.

14.4.5 Biomolecular Quantification at Single-Cell Level

The elemental labeling TRA-ICP-MS strategy is still in its infancy for the quantification of biomolecules in single cells. Montes Bayón et al. [25] used Maxpar X8 antibody labeling kit to modify a polymer label containing 21 Nd atoms (average number) on the antibody of transferrin receptor 1 (TfR1), and labeled the TfR1 on the surface of breast cancer cells with the Nd-labeled antibody. The quantification of Nd labeled on single cells was based on Nd standard solution. By using 30 nm AuNPs and cell/Eu-doped microspheres, TE of aqueous solution and cell in the applied system was measured to be 70% and 55%, respectively. Along with the measured stoichiometry of Nd and TfR1 antibodies, the number of TfR1 molecules on the surface of single cells was calculated for two cell models of breast cancer with different malignancy (MCF7 and MDA-MB 231). Compared with MCF-7 cells (6.4×10^3 TfR1/cell), more invasive MDA-MB-231 cells expressed more TfR1 (2.3×10^4 TfR1/cell). Wang and coworkers [42] labeled five bacteria with *alkynyl-D-alanine* (aDA) firstly. After labeling, aDA would be assembled into the peptidoglycan layer of the bacterial cell wall. Then, N_3-DOTA-Eu was grafted to aDA in the peptidoglycan layer through a click chemical reaction between alkynyl and azide, resulting in labeling of all aDA on the bacteria with Eu. Five antibodies modified with different rare earth elements (La, Pr, Nd, Sm, and Gd) were used to label corresponding bacteria, respectively, and the five bacterial mixtures were subjected to TRA-ICP-MS analysis simultaneously. The type of bacteria was judged according to the existence of the corresponding rare earth jump peak, and the aDA on the surface of a single bacteria was quantified according to the strength of the Eu jump peak. The vancomycin and AgNPs treatment were found to cause a down-regulation of bacterial aDA expression, with obvious cell heterogeneity.

Ogra and coworkers [89] transformed plasmid vector containing enhanced green fluorescent protein (EGFP) or red fluorescent protein (mCherry) gene fused with His-tag into *Escherichia coli* (*E. coli*). After labeling the His-tag with Ni/Co, *E. coli* was subjected to TRA-ICP-MS, and the amount of EGFP or mCherry protein was obtained according to the signal spikes of Ni/Co. The specific binding of Co over Ni to His-tag of the protein was demonstrated. The content of Co was found to increase after protein induction for six hours, while it showed a decrease for EGFP and level off for mCherry from 6 to 24 hours. The two proteins were mainly recovered in the insoluble fraction 24 hours after the induction. It suggests that the overexpressed fluorescent proteins with His-tag are transferred into inclusion bodies, hampering the reaction between the proteins and Co ions.

When NPs are used as the elemental tags in elemental labeling TRA-ICP-MS strategy, an enhanced sensitivity can be achieved due to plenty of atoms contained in one NP. They can be detected by ICP-MS after dissolution with conventional integral mode or detected directly with the acquisition mode of TRA. In TRA-ICP-MS analysis of NPs, elemental signals originating in dissolved state can be distinguished from NPs or aggregation form. That means the separation of excess NPs tags by washing, magnetic separation, or ultrafiltration can be avoided if the pulse signals in TRA-ICP-MS are attributed to the variation in aggregation state of the NPs tags rather than the NPs themselves. In this strategy, target molecular can be used to trigger the aggregation of NPs tags directly, or initiate signal amplification process firstly and then aggregate a large number of NPs into a whole, followed by TRA-ICP-MS detection of high-intensity signal obviously different from the background of dispersed NPs.

Zhang and coworkers [90] prepared two AuNP probes modified with different single stranded (ss)DNA sequences, which would hybridize with target DNA. The existence of target DNA would make these two kinds of AuNPs probes cross-linked into dimer, trimer, and other aggregation structures. The cross-linking state of AuNPs in solution depended on the content of target DNA. Based on the signal frequency and average intensity of Au, the quantification of target DNA was achieved. Jiang and coworkers [91] utilized hybridization chain reaction (HCR) to mediate AuNP assembly. The large AuNP aggregates provide a high signal-to-background ratio for single-particle counting in TRA-ICP-MS. Based on the statistics of the peak hopping frequency of the assembly and the concentration of target nucleic acid, ultra-sensitive quantification of target nucleic acid was realized with the LOD of 3 fmol/l. Hu and coworkers [92] employed small-particle-size Au NPs as the elemental tags to lower the blank in TRA-ICP-MS detection. With the presence of target DNA, long ssDNA possessing a large number of repeating sequence units was generated by rolling circle amplification (RCA). After the addition of spermidine, AuNP probes agglomerated, resulting in easily distinguishable pulse signal of Au. LOD of 5.1 fmol/l was obtained for target DNA. By grafting two complementary chains of SARS CoV-2 nucleic acid on AuNPs and two complementary chains of influenza A H3N2 nucleic acid on AgNPs, respectively, simultaneous homogeneous detection of the two nucleic acids was achieved [93]. Lv and coworkers [94] replaced the complementary nucleic acid with antigen antibody pairs, and realized the quantification of carcinoembryonic antigen (CEA) by TRA-ICP-MS based on the aggregation of AgNPs in the presence of CEA. This method was then applied for simultaneous quantification of three antigens (CEA, CA125, and CA199) with the use of AuNPs, AgNPs, and PtNPs tags [95]. Jiang and coworkers [96] proposed a TRA-ICP-MS system for a sensitive uracil-DNA glycosylase (UDG) activity detection, which relied on the specific recognition and enzymatic reaction of UDG to remove uracil and induce the cleavage of the DNA probe. The cleavage of the DNA probe could induce the release of AuNPs from the composites of AuNPs@magnetic beads, resulting in obvious signal spikes of Au.

14.4.6 Other Applications

Combined with toxicity assessment, the cytotoxicity of ZIF-8 NPs was comprehensively investigated with the aid of TRA-ICP-MS, along with a comparison of ZnO NPs with similar particle size and Zn^{2+} under the same concentration [97]. It was found that the cytotoxicity of ZIF-8 and ZnO NPs depended on their sizes and concentrations. Due to the higher degradation rate of ZIF-8 NPs in the medium than that of ZnO NPs, the cellular uptake and elimination behaviors of ZIF-8 NPs were more similar to those of Zn^{2+}. It is speculated that ZIF-8 NPs have higher bioavailability than ZnO NPs, enabling more cells to absorb more Zn. This would lead to elevated intracellular Zn accumulation, which in turn caused elevated ROS levels and cellular inflammation, ultimately leading to cell necrosis. Although ZIF-8 NPs could be metabolized and/or eliminated, the fraction of ZIF-8 NPs retaining in the cell may still pose a potential risk.

The combined effect of zinc-based NPs (e.g. ZnO, ZIF-8) and Cd^{2+} on HepG2 cells was investigated by combining biological indicator detection methods with TRA-ICP-MS-based single-cell analysis [98]. Specifically, Zn and Cd amount in cells exposed to ZnO/ZIF-8 NPs and Cd^{2+} was determined at single-cell level. Simultaneously, the co-exposure of Zn^{2+} and Cd^{2+} at the same concentration was evaluated for a comparison. The combination of quantitative data with toxicity assessment showed that high dose of ZnO/ZIF-8 NPs exposed with Cd^{2+} exhibited synergistic toxicity, and low dose of ZnO/ZIF-8 NPs exposed with Cd^{2+} could protect cells from Cd^{2+}-induced damage. Low doses of ZnO/ZIF-8 NPs can activate the Nrf2 pathway, increase GSH production to counteract Cd^{2+}-induced ROS elevation, and decrease metallothionein concentrations. Then, ZIP8 and ZIP14 transporters that can transport Cd were downregulated, reducing the cellular Cd amount, further reducing ROS and subsequent DNA damage and apoptosis. In the co-exposure group of Zn^{2+} and Cd^{2+}, the detoxification effect of Zn^{2+} mainly depended on the increase of cellular Zn amount absorbed through the increase of ZIP8 and ZIP14 transporters, which improved the antioxidant capacity of cells.

14.5 Summary and Perspectives

Trace elements play an important role in the physiological activities of cells. The study of metallomics at single-cells level is of great significance to reveal the mechanism of trace elements in the physiological activities of cells. This chapter focuses on the combination of different sample introduction system (e.g. pneumatic nebulization, microfluidic devices, LA) with TRA-ICP-MS for single-cell analysis and the application in metallomics study. As can be seen, TRA-ICP-MS can provide comprehensive information that cannot be obtained by conventional cell population measurements. With the help of microfluidic platform/device, the probability of multiple events (e.g. cells, NPs) existing in one signal spike can be reduced significantly, ensuring accurate information of cells heterogeneity. The quantification of interest elements in single cells has been realized. By using elemental labeling strategy, interest biomolecules can also be sensitively quantified. With the aid of LA, the distribution and transportation of specific elements or molecules in single cell or even cell organelle can be further "visualized" by TRA-ICP-MS. The obtained differences in absolute amount of endogenous elements for individual cell between different cell types demonstrate the potential of TRA-ICP-MS as a "metallo-fingerprinting" tool.

Although a variety of TRA-ICP-MS-based methodologies have been developed for single-cell analysis, paving the way for metallomics study at single-cell level, there are still some technical difficulties in the quantification, sensitivity, separation/identification of specific species, special resolution, standardization of the analytical process, and accuracy validation. Much more efforts are expected on these issues. Simultaneously, with the help of other single-cell detection technologies (such as flow cytometry), further exploration of the relationship between the spatial distribution of elements/elemental species and various physiological indicators (e.g. endogenous elements, DNA, lipids, proteins) is necessary. These would help to draw element fingerprint identification spectrum library and understand the function mechanism of elements in cells, promoting relevant progress in metallomics study at single-cell level.

List of Abbreviations

CEA	carcinoembryonic antigen
DE	detection efficiency
DNA	deoxyribonucleic acid
EGFP	enhanced green fluorescent protein
HCR	hybridization chain reaction
HECIS	high-efficiency cell introduction system
ICP-MS	Inductively coupled plasma mass spectrometry
LA	laser ablation
NPs	nanoparticles
PDMS	polydimethylsiloxane
PN	pneumatic nebulizer
Q	quadrupole
sc	single cell
SF	sector field
sp	single particle
t_{dwell}	dwell time
TE	transport efficiency
ToF	time of flight
TRA	time-resolved analysis
UDG	uracil-DNA glycosylase
μDG	microdroplet generator

References

1 Schmid, A., Kortmann, H., Dittrich, P.S., and Blank, L.M. (2010). Chemical and biological single cell analysis. *Curr. Opin. Biotechnol.* 21 (1): 12–20.
2 Mounicou, S., Szpunar, J., and Lobinski, R. (2009). Metallomics: the concept and methodology. *Chem. Soc. Rev.* 38 (4): 1119–1138.

3 Myashita, S., Fujii, S., and Inagaki, K. (2017). Single-particle/cell analysis by highly time-resolved ICP-MS using a high-efficiency sample introduction system. *Bunseki Kagaku* 66 (9): 663–676.

4 Maret, W. (2016). The metals in the biological periodic system of the elements: concepts and conjectures. *Int. J. Mol. Sci.* 17 (1): 66.

5 Jimenez-Lamana, J., Szpunar, J., and Lobinski, R. (2018). New frontiers of metallomics: elemental and species-specific analysis and imaging of single cells. *Adv. Exp. Med. Biol.* 1055: 245–270.

6 Resano, M., Aramendia, M., Garcia-Ruiz, E. et al. (2022). Living in a transient world: ICP-MS reinvented via time-resolved analysis for monitoring single events. *Chem. Sci.* 13 (16): 4436–4473.

7 Li, F., Armstrong, D.W., and Houk, R.S. (2005). Behavior of bacteria in the inductively coupled plasma: atomization and production of atomic ions for mass spectrometry. *Anal. Chem.* 77 (5): 1407–1413.

8 Ho, K.-S. and Chan, W.-T. (2010). Time-resolved ICP-MS measurement for single-cell analysis and on-line cytometry. *J. Anal. At. Spectrom.* 25 (7): 1114–1122.

9 Shigeta, K., Koellensperger, G., Rampler, E. et al. (2013). Sample introduction of single selenized yeast cells (*Saccharomyces cerevisiae*) by micro droplet generation into an ICP-sector field mass spectrometer for label-free detection of trace elements. *J. Anal. At. Spectrom.* 28 (5): 637–645.

10 Verboket, P.E., Borovinskaya, O., Meyer, N. et al. (2014). A new microfluidics-based droplet dispenser for ICPMS. *Anal. Chem.* 86 (12): 6012–6018.

11 Wang, H., Chen, B., He, M., and Hu, B. (2017). A facile droplet-chip-time-resolved inductively coupled plasma mass spectrometry online system for determination of zinc in single cell. *Anal. Chem.* 89 (9): 4931–4938.

12 Wei, X., Zheng, D.H., Cai, Y. et al. (2018). High-throughput/high-precision sampling of single cells into ICP-MS for elucidating cellular nanoparticles. *Anal. Chem.* 90 (24): 14543–14550.

13 Yu, X., Chen, B., He, M. et al. (2019). 3D droplet-based microfluidic device easily assembled from commercially available modules online coupled with ICPMS for determination of silver in single cell. *Anal. Chem.* 91 (4): 2869–2875.

14 Wang, H., Chen, B., He, M. et al. (2019). Study on uptake of gold nanoparticles by single cells using droplet microfluidic chip-inductively coupled plasma mass spectrometry. *Talanta* 200: 398–407.

15 Wei, X., Zhang, X., Guo, R. et al. (2019). A spiral-helix (3D) tubing array that ensures ultrahigh-throughput single-cell sampling. *Anal. Chem.* 91 (24): 15826–15832.

16 Cao, Y.P., Feng, J.S., Tang, L.F. et al. (2020). A highly efficient introduction system for single cell-ICP-MS and its application to detection of copper in single human red blood cells. *Talanta* 206: 120174.

17 Groombridge, A.S., Miyashita, S., Fujii, S. et al. (2013). High sensitive elemental analysis of single yeast cells (*Saccharomyces cerevisiae*) by time-resolved

inductively-coupled plasma mass spectrometry using a high efficiency cell introduction system. *Anal. Sci.* 29 (6): 597–603.

18 Miyashita, S., Groombridge, A.S., Fujii, S. et al. (2014). Highly efficient single-cell analysis of microbial cells by time-resolved inductively coupled plasma mass spectrometry. *J. Anal. At. Spectrom.* 29 (9): 1598–1606.

19 Wang, H.L., Wang, M., Wang, B. et al. (2017). Interrogating the variation of element masses and distribution patterns in single cells using ICP-MS with a high efficiency cell introduction system. *Anal. Bioanal. Chem.* 409 (5): 1415–1423.

20 Corte Rodriguez, M., Alvarez-Fernandez Garcia, R., Blanco, E. et al. (2017). Quantitative evaluation of cisplatin uptake in sensitive and resistant individual cells by single-cell ICP-MS (SC-ICP-MS). *Anal. Chem.* 89 (21): 11491–11497.

21 Merrifield, R.C., Stephan, C., and Lead, J.R. (2018). Quantification of Au nanoparticle biouptake and freshwater algae using single cell – ICP-MS. *Environ. Sci. Technol.* 52 (4): 2271–2277.

22 von der Au, M., Schwinn, M., Kuhlmeier, K. et al. (2019). Development of an automated on-line purification HPLC single cell-ICP-MS approach for fast diatom analysis. *Anal. Chim. Acta* 1077: 87–94.

23 Shen, X., Zhang, H.T., He, X.L. et al. (2019). Evaluating the treatment effectiveness of copper-based algaecides on toxic algae *Microcystis aeruginosa* using single cell-inductively coupled plasma-mass spectrometry. *Anal. Bioanal. Chem.* 411 (21): 5531–5543.

24 Mavrakis, E., Mavroudakis, L., Lydakis-Simantiris, N., and Pergantis, S.A. (2019). Investigating the uptake of arsenate by *Chlamydomonas reinhardtii* cells and its effect on their lipid profile using single cell ICP-MS and easy ambient sonic-spray ionization-MS. *Anal. Chem.* 91 (15): 9590–9598.

25 Corte-Rodriguez, M., Blanco-Gonzalez, E., Bettmer, J., and Montes-Bayon, M. (2019). Quantitative analysis of transferrin receptor 1 (TfR1) in individual breast cancer cells by means of labeled antibodies and elemental (ICP-MS) detection. *Anal. Chem.* 91 (24): 15532–15538.

26 Tsang, C.N., Ho, K.S., Sun, H., and Chan, W.T. (2011). Tracking bismuth antiulcer drug uptake in single *Helicobacter pylori* cells. *J. Am. Chem. Soc.* 133 (19): 7355–7357.

27 Zheng, L.N., Wang, M., Wang, B. et al. (2013). Determination of quantum dots in single cells by inductively coupled plasma mass spectrometry. *Talanta* 116: 782–787.

28 Wang, H.L., Wang, B., Wang, M. et al. (2015). Time-resolved ICP-MS analysis of mineral element contents and distribution patterns in single cells. *Analyst* 140 (2): 523–531.

29 Zheng, L.N., Wang, M., Zhao, L.C. et al. (2015). Quantitative analysis of Gd@C-82(OH)(22) and cisplatin uptake in single cells by inductively coupled plasma mass spectrometry. *Anal. Bioanal. Chem.* 407 (9): 2383–2391.

30 Wei, X., Hu, L.L., Chen, M.L. et al. (2016). Analysis of the distribution pattern of chromium species in single cells. *Anal. Chem.* 88 (24): 12437–12444.

31 Yang, Y.S.S., Atukorale, P.U., Moynihan, K.D. et al. (2017). High-throughput quantitation of inorganic nanoparticle biodistribution at the single-cell level using mass cytometry. *Nat. Commun.* 8: 14069.

32 Zhou, Y., Li, H.Y., and Sun, H.Z. (2017). Cytotoxicity of arsenic trioxide in single leukemia cells by time-resolved ICP-MS together with lanthanide tags. *Chem. Commun.* 53 (20): 2970–2973.

33 Lau, W.Y., Chun, K.H., and Chan, W.T. (2017). Correlation of single-cell ICP-MS intensity distributions for the study of heterogeneous cellular responses to environmental stresses. *J. Anal. At. Spectrom.* 32 (4): 807–815.

34 Ivask, A., Mitchell, A.J., Hope, C.M. et al. (2017). Single cell level quantification of nanoparticle-cell interactions using mass cytometry. *Anal. Chem.* 89 (16): 8228–8232.

35 Guo, Y.T., Baumgart, S., Stark, H.J. et al. (2017). Mass cytometry for detection of silver at the bacterial single cell level. *Front. Microbiol.* 8: 1326.

36 Meyer, S., Lopez-Serrano, A., Mitze, H. et al. (2018). Single-cell analysis by ICP-MS/MS as a fast tool for cellular bioavailability studies of arsenite. *Metallomics* 10 (1): 73–76.

37 Oliver, A.L.S., Baumgart, S., Bremser, W. et al. (2018). Quantification of silver nanoparticles taken up by single cells using inductively coupled plasma mass spectrometry in the single cell measurement mode. *J. Anal. At. Spectrom.* 33 (7): 1256–1263.

38 Zhou, Y., Wang, H.B., Tse, E. et al. (2018). Cell cycle-dependent uptake and cytotoxicity of arsenic-based drugs in single leukemia cells. *Anal. Chem.* 90 (17): 10465–10471.

39 Liu, Z.H., Xue, A.F., Chen, H., and Li, S.Q. (2019). Quantitative determination of trace metals in single yeast cells by time-resolved ICP-MS using dissolved standards for calibration. *Appl. Microbiol. Biotechnol.* 103 (3): 1475–1483.

40 Lum, J.T.S. and Leung, K.S.Y. (2019). Quantifying silver nanoparticle association and elemental content in single cells using dual mass mode in quadrupole-based inductively coupled plasma-mass spectrometry. *Anal. Chim. Acta* 1061: 50–59.

41 Sun, Q.X., Wei, X., Zhang, S.Q. et al. (2019). Single cell analysis for elucidating cellular uptake and transport of cobalt curcumin complex with detection by time-resolved ICPMS. *Anal. Chim. Acta* 1066: 13–20.

42 Liang, Y., Liu, Q., Zhou, Y. et al. (2019). Counting and recognizing single bacterial cells by a lanthanide-encoding inductively coupled plasma mass spectrometric approach. *Anal. Chem.* 91 (13): 8341–8349.

43 Giesen, C., Waentig, L., Mairinger, T. et al. (2011). Iodine as an elemental marker for imaging of single cells and tissue sections by laser ablation inductively coupled plasma mass spectrometry. *J. Anal. At. Spectrom.* 26 (11): 2160–2165.

44 Drescher, D., Giesen, C., Traub, H. et al. (2012). Quantitative imaging of gold and silver nanoparticles in single eukaryotic cells by laser ablation ICP-MS. *Anal. Chem.* 84 (22): 9684–9688.

45 Managh, A.J., Edwards, S.L., Bushell, A. et al. (2013). Single cell tracking of gadolinium labeled CD4(+) T cells by laser ablation inductively coupled plasma mass spectrometry. *Anal. Chem.* 85 (22): 10627–10634.

46 Wang, M., Zheng, L.N., Wang, B. et al. (2014). Quantitative analysis of gold nanoparticles in single cells by laser ablation inductively coupled plasma-mass spectrometry. *Anal. Chem.* 86 (20): 10252–10256.

47 Zhai, J., Wang, Y., Xu, C. et al. (2015). Facile approach to observe and quantify the alpha(IIb)beta3 integrin on a single-cell. *Anal. Chem.* 87 (5): 2546–2549.

48 Van Malderen, S.J.M., Vergucht, E., De Rijcke, M. et al. (2016). Quantitative determination and subcellular imaging of Cu in single cells via laser ablation-ICP-mass spectrometry using high-density microarray gelatin standards. *Anal. Chem.* 88 (11): 5783–5789.

49 Hsiao, I.L., Bierkandt, F.S., Reichardt, P. et al. (2016). Quantification and visualization of cellular uptake of TiO_2 and Ag nanoparticles: comparison of different ICP-MS techniques. *J. Nanobiotechnol.* 14: 50.

50 Mueller, L., Herrmann, A.J., Techritz, S. et al. (2017). Quantitative characterization of single cells by use of immunocytochemistry combined with multiplex LA-ICP-MS. *Anal. Bioanal. Chem.* 409 (14): 3667–3676.

51 Herrmann, A.J., Techritz, S., Jakubowski, N. et al. (2017). A simple metal staining procedure for identification and visualization of single cells by LA-ICP-MS. *Analyst* 142 (10): 1703–1710.

52 Lohr, K., Traub, H., Wanka, A.J. et al. (2018). Quantification of metals in single cells by LA-ICP-MS: comparison of single spot analysis and imaging. *J. Anal. At. Spectrom.* 33 (9): 1579–1587.

53 Van Malderen, S.J.M., Van Acker, T., Laforce, B. et al. (2019). Three-dimensional reconstruction of the distribution of elemental tags in single cells using laser ablation ICP-mass spectrometry via registration approaches. *Anal. Bioanal. Chem.* 411 (19): 4849–4859.

54 Zheng, L.N., Sang, Y.B., Luo, R.P. et al. (2019). Determination of silver nanoparticles in single cells by microwell trapping and laser ablation ICP-MS determination. *J. Anal. At. Spectrom.* 34 (5): 915–921.

55 Theiner, S., Schweikert, A., Van Malderen, S.J.M. et al. (2019). Laser ablation-inductively coupled plasma time-of-flight mass spectrometry imaging of trace elements at the single-cell level for clinical practice. *Anal. Chem.* 91 (13): 8207–8212.

56 Van Acker, T., Buckle, T., Van Malderen, S.J.M. et al. (2019). High-resolution imaging and single-cell analysis via laser ablation-inductively coupled plasma-mass spectrometry for the determination of membranous receptor expression levels in breast cancer cell lines using receptor-specific hybrid tracers. *Anal. Chim. Acta* 1074: 43–53.

57 Arakawa, A., Jakubowski, N., Koellensperger, G. et al. (2019). Quantitative imaging of silver nanoparticles and essential elements in thin sections of fibroblast multicellular spheroids by high resolution laser ablation inductively coupled plasma time-of-flight mass spectrometry. *Anal. Chem.* 91 (15): 10197–10203.

58 Lohr, K., Borovinskaya, O., Tourniaire, G. et al. (2019). Arraying of single cells for quantitative high throughput laser ablation ICP-TOF-MS. *Anal. Chem.* 91 (18): 11520–11528.

59 Yu, X., He, M., Chen, B., and Hu, B. (2020). Recent advances in single-cell analysis by inductively coupled plasma-mass spectrometry: a review. *Anal. Chim. Acta* 1137: 191–207.

60 Pace, H.E., Rogers, N.J., Jarolimek, C. et al. (2011). Determining transport efficiency for the purpose of counting and sizing nanoparticles via single particle inductively coupled plasma mass spectrometry. *Anal. Chem.* 83 (24): 9361–9369.

61 Wu, B., Niehren, S., and Becker, J.S. (2011). Mass spectrometric imaging of elements in biological tissues by new BrainMet technique-laser microdissection inductively coupled plasma mass spectrometry (LMD-ICP-MS). *J. Anal. At. Spectrom.* 26 (8): 1653–1659.

62 Pisonero, J., Bouzas-Ramos, D., Traub, H. et al. (2019). Critical evaluation of fast and highly resolved elemental distribution in single cells using LA-ICP-SFMS. *J. Anal. At. Spectrom.* 34 (4): 655–663.

63 Meng, Y., Gao, C., Lu, Q. et al. (2021). Single-cell mass spectrometry imaging of multiple drugs and nanomaterials at organelle level. *ACS Nano* 15 (8): 13220–13229.

64 Bendall, S.C., Simonds, E.F., Qiu, P. et al. (2011). Single-cell mass cytometry of differential immune and drug responses across a human hematopoietic continuum. *Science* 332 (6030): 687–696.

65 Bandura, D.R., Baranov, V.I., Ornatsky, O.I. et al. (2009). Mass cytometry: technique for real time single cell multitarget immunoassay based on inductively coupled plasma time-of-flight mass spectrometry. *Anal. Chem.* 81 (16): 6813–6822.

66 Managh, A.J., Douglas, D.N., Makella Cowen, K. et al. (2016). Acquisition of fast transient signals in ICP-MS with enhanced time resolution. *J. Anal. At. Spectrom.* 31 (8): 1688–1692.

67 Strenge, I. and Engelhard, C. (2016). Capabilities of fast data acquisition with microsecond time resolution in inductively coupled plasma mass spectrometry and identification of signal artifacts from millisecond dwell times during detection of single gold nanoparticles. *J. Anal. At. Spectrom.* 31 (1): 135–144.

68 Strenge, I. and Engelhard, C. (2020). Single particle inductively coupled plasma mass spectrometry: investigating nonlinear response observed in pulse counting mode and extending the linear dynamic range by compensating for dead time related count losses on a microsecond timescale. *J. Anal. At. Spectrom.* 35 (1): 84–99.

69 Manz, A., Graber, N., and Widmer, H.M. (1990). Miniaturized total chemical-analysis systems – a novel concept for chemical sensing. *Sens. Actuators B Chem.* 1 (1–6): 244–248.

70 Bhagat, A.A.S., Bow, H., Hou, H.W. et al. (2010). Microfluidics for cell separation. *Med. Biol. Eng. Comput.* 48 (10): 999–1014.

71 Chao, T.-C. and Ros, A. (2008). Microfluidic single-cell analysis of intracellular compounds. *J. R. Soc. Interface* 5: S139–S150.

72 Zhao, B., He, M., Chen, B., and Hu, B. (2019). Fe_3O_4 nanoparticles coated with double imprinted polymers for magnetic solid phase extraction of lead(II) from biological and environmental samples. *Microchim. Acta* 186 (12): 775.

73 Yin, H. and Marshall, D. (2012). Microfluidics for single cell analysis. *Curr. Opin. Biotechnol.* 23 (1): 110–119.

74 White, A.K., Heyries, K.A., Doolin, C. et al. (2013). High-throughput microfluidic single-cell digital polymerase chain reaction. *Anal. Chem.* 85 (15): 7182–7190.

75 Zhang, X., Wei, X., Wei, Y. et al. (2020). The up-to-date strategies for the isolation and manipulation of single cells. *Talanta* 218: 121147.

76 Lo, S.-J. and Yao, D.-J. (2015). Get to understand more from single-cells: current studies of microfluidic-based techniques for single-cell analysis. *Int. J. Mol. Sci.* 16 (8): 16763–16777.

77 Valizadeh, A. and Khosroushahi, A.Y. (2015). Single-cell analysis based on lab on a chip fluidic system. *Anal. Methods* 7 (20): 8524–8533.

78 Shields, C.W., Reyes, C.D., and Lopez, G.P. (2015). Microfluidic cell sorting: a review of the advances in the separation of cells from debulking to rare cell isolation. *Lab Chip* 15 (5): 1230–1249.

79 Kang, D.-K., Ali, M.M., Zhang, K. et al. (2014). Droplet microfluidics for single-molecule and single-cell analysis in cancer research, diagnosis and therapy. *Trends Anal. Chem.* 58: 145–153.

80 Huang, N.-T., Zhang, H.-l., Chung, M.-T. et al. (2014). Recent advancements in optofluidics-based single-cell analysis: optical on-chip cellular manipulation, treatment, and property detection. *Lab Chip* 14 (7): 1230–1245.

81 Shigeta, K., Traub, H., Panne, U. et al. (2013). Application of a micro-droplet generator for an ICP-sector field mass spectrometer – optimization and analytical characterization. *J. Anal. At. Spectrom.* 28 (5): 646–656.

82 Chen, Z., Chen, B., He, M., and Hu, B. (2020). A droplet-splitting microchip online coupled with time-resolved-ICPMS for analysis of released Fe and Pt in single cells treated with FePt nanoparticles. *Anal. Chem.* 92 (18): 12208–12215.

83 Zhang, X., Wei, X., Men, X. et al. (2020). Inertial-force-assisted, high-throughput, droplet-free, single-cell sampling coupled with ICP-MS for real-time cell analysis. *Anal. Chem.* 92 (9): 6604–6612.

84 Zhou, Y., Chen, Z., Zeng, J. et al. (2020). Direct infusion ICP-qMS of lined-up single-cell using an oil-free passive microfluidic system. *Anal. Chem.* 92 (7): 5286–5293.

85 Gu, S.Q., Zhang, Y.X., Zhu, Y. et al. (2011). Multifunctional picoliter droplet manipulation platform and its application in single cell analysis. *Anal. Chem.* 83 (19): 7570–7576.

86 Chen, A., Byvank, T., Chang, W.J. et al. (2013). On-chip magnetic separation and encapsulation of cells in droplets. *Lab Chip* 13 (6): 1172–1181.

87 Liu, T., Bolea-Fernandez, E., Mangodt, C. et al. (2021). Single-event tandem ICP-mass spectrometry for the quantification of chemotherapeutic drug-derived Pt and endogenous elements in individual human cells. *Anal. Chim. Acta* 1177: 338797.

88 Gale, A., Hofmann, L., Ludi, N. et al. (2021). Beyond single-cell analysis of metallodrugs by ICP-MS: targeting cellular substructures. *Int. J. Mol. Sci.* 22 (17): 9468.
89 Tanaka, Y.K., Shimazaki, S., Fukumoto, Y., and Ogra, Y. (2022). Detection of histidine-tagged protein in *Escherichia coli* by single-cell inductively coupled plasma-mass spectrometry. *Anal. Chem.* 94 (22): 7952–7959.
90 Han, G., Xing, Z., Dong, Y. et al. (2011). One-step homogeneous DNA assay with single-nanoparticle detection. *Angew. Chem. Int. Ed.* 50 (15): 3462–3465.
91 Li, B.-R., Tang, H., Yu, R.-Q., and Jiang, J.-H. (2020). Single-nanoparticle ICPMS DNA assay based on hybridization-chain-reaction-mediated spherical nucleic acid assembly. *Anal. Chem.* 92 (3): 2379–2382.
92 Xu, Y., Xiao, G.Y., Chen, B.B. et al. (2022). Single particle inductively coupled plasma mass spectrometry-based homogeneous detection of HBV DNA with rolling circle amplification-induced gold nanoparticle agglomeration. *Anal. Chem.* 94 (28): 10011–10018.
93 Xu, Y., Chen, B.B., He, M., and Hu, B. (2021). A homogeneous nucleic acid assay for simultaneous detection of SARS-CoV-2 and influenza A (H3N2) by single-particle inductively coupled plasma mass spectrometry. *Anal. Chim. Acta* 1186: 339134.
94 Huang, Z., Wang, C., Liu, R. et al. (2020). Self-validated homogeneous immunoassay by single nanoparticle in-depth scrutinization. *Anal. Chem.* 92 (3): 2876–2881.
95 Huang, Z., Li, Z., Jiang, M. et al. (2020). Homogeneous multiplex immunoassay for one-step pancreatic cancer biomarker evaluation. *Anal. Chem.* 92 (24): 16105–16112.
96 Li, B.R., Tang, H., Yu, R.Q., and Jiang, J.H. (2021). Single-nanoparticle ICP-MS for sensitive detection of uracil-DNA glycosylase activity. *Anal. Chem.* 93 (24): 8381–8385.
97 Chen, P., He, M., Chen, B., and Hu, B. (2020). Size- and dose-dependent cytotoxicity of ZIF-8 based on single cell analysis. *Ecotoxicol. Environ. Saf.* 205: 111110.
98 Chen, P., Chen, B., He, M., and Hu, B. (2021). Combined effects of different sizes of ZnO and ZIF-8 nanoparticles co-exposure with Cd^{2+} on HepG2 cells. *Sci. Total Environ.* 786: 147402.

15

Novel ICP-MS-based Techniques for Metallomics*

Panpan Chang and Meng Wang

Chinese Academy of Sciences, Institute of High Energy Physics, CAS Key Laboratory for Biomedical Effects of Nanomaterials and Nanosafety, & CAS-HKU Joint Laboratory of Metallomics on Health and Environment, & Beijing Metallomics Facility, & National Consortium for Excellence in Metallomics, 19B Yuquan Road, Beijing 100049, China

15.1 Introduction

Metals play important roles in cells and organisms by participating in fundamental processes such as cell signaling, gene expression, and enzyme catalysis [1]. Accordingly, a cell (or an organism) must be characterized not only by its nucleic acids (genome), proteins (proteome), and metabolites (metabolome), but also by its metals, i.e. metallome [2–5]. As an emerging interdisciplinary science, metallomics aims to integrate the research fields related to metallome in biological systems from a systematic perspective [3, 6]. In this context, metallomics is extremely complicated. On the other hand, the metallome is highly dynamic because many metal complexes and intermediates are thermodynamically unstable and easily change [7]. This complexity and variation pose huge challenges to analytical methods because many techniques that are successfully used in other omics are not suitable for metallomics.

The analytical techniques for metallomics must have specific responses to metals and metalloids, a high-throughput capability, and sufficient sensitivity with a limited sample amount, such as single cells [3, 8]. The techniques should allow not only the quantitative analysis of multiple elements but also the determination of their species in the samples. To completely characterize the metallome, it is also necessary to understand the localization, speciation, and metabolism of the metals and metalloids at a subcellular level and study their connections and interactions with the genes, proteins, or metabolites in the spatial dimension, which is a new topic in metallomics and is regarded as *spatial metallomics*.

New analytical tools and strategies are necessary to address these challenges. Inductively coupled plasma–mass spectrometry (ICP-MS) is considered one

---

* This chapter has been modified to feature as Review: Chang, P., Zheng, L., Wang, B., et al. (2022). ICP-MS-based methodology in metallomics: towards single particle analysis single cell analysis, and spatial metallomics. *At. Spectrosc.* 43:255–265.

Applied Metallomics: From Life Sciences to Environmental Sciences, First Edition.
Edited by Yu-Feng Li and Hongzhe Sun.
© 2024 WILEY-VCH GmbH. Published 2024 by WILEY-VCH GmbH.

of the most versatile tools for metallomics research because of its outstanding characteristics, including high analytical throughput, excellent detection limits for most elements, minimal matrix effects, wide linear dynamic range, and simple coupling to other analytical methods (e.g. laser ablation or high-performance liquid chromatography [HPLC]) [9, 10]. However, ICP-MS-based methods can only provide qualitative and quantitative information on elements and must be integrated with other analytical tools for metallomics research. For example, in parallel with the detection of metal species, metal ligands or metal-binding molecules have been identified using biological mass spectrometric methods coupled with a separation method such as two-dimensional gel electrophoresis or HPLC [11].

This chapter provides a brief overview of ICP-MS and describes recent advances in ICP-MS instrumentation. Then, ICP-MS-based methods and applications in metallomics are discussed, focusing on single-particle analysis, single-cell analysis, and spatial metallomics. Lastly, the conclusion and the perspecive of ICP-MS-based methodology in metallomics are presented.

15.2 ICP-MS: A Powerful Method in Metallomics

ICP-MS instrumentation was introduced to the commercial market in the early 1980s [12] and has since become the most powerful tool for ultratrace element analysis. In ICP-MS, the high-temperature plasma (6000–10 000 K) is used as an ion source; the energy is sufficient for desolvation, atomization, and ionization of the analyte [13]. Different mass analyzers are utilized in ICP-MS, such as the quadrupole, magnetic sector analyzer, and ion trap [14]. Besides solution analysis, in situ solid sampling can be achieved when laser ablation (LA) is used as the sample introduction system. In LA-ICP-MS analysis, solid samples are ablated with high-power laser shots, and the resulting aerosol is transported and analyzed by ICP-MS. [15] LA-ICP-MS has the unique advantages of high spatial resolution (down to ~1 μm), minimal sample preparation, and high sensitivity (<1 fg for many elements). Recent advances in ICP-MS instrumentation are discussed in the following section.

15.2.1 Solution Introduction System and Plasma Source

Currently, the main application of ICP-MS involves the analysis of liquid samples. The standard sample introduction systems in ICP-MS suffer from low transport efficiencies. For example, the typical transport efficiency of single cells using ICP-MS is reported to be less than 1% [16]. In addition, polydisperse aerosols from standard systems induce a sampling bias and adversely affect the accuracy of the results of single-particle analysis [17]. New introduction systems have been developed to improve the transport efficiency and reduce the sampling bias. Miyashita et al. achieved approximately 100% transport efficiency of single cells using a microflow concentric nebulizer coupled to a total consumption spray chamber [18]. A heated single-pass spray chamber with a sheath gas was reported to improve the transport efficiency and reduce the sampling bias in single-cell analysis [19]. Zhou et al. developed a novel oil-free passive microfluidic system coupled to ICP-MS with a direct infusion micronebuliser and achieved >70% transport efficiency of single cells [20].

Monodisperse droplets generated by either a commercial piezoelectric dispenser [21] or a microfluidics-based droplet dispenser [22] have proven to be ideal calibrations for single-cell/particle analysis. Monodisperse droplets exhibit almost the same behavior in the plasma (e.g. trajectory, ionization efficiency, and transport efficiency) so that the sampling bias in standard sample introduction systems is reduced [23]. Using the sample introduction system with the monodisperse droplets resulted in more accurate distributions in single cells or single particles [24].

Better detection limits are continuously pursued in metallomics studies. The detection efficiency of current ICP-MS instruments ranges from 10^{-4} to 10^{-6} counts per atom [25]. Achieving the better detection limit is hindered by several conventional ICP-MS designs. A new conical torch with a reduction of argon consumption and power density was developed to realize a stable plasma with a 1000–1700 K higher excitation temperature and a fivefold increase in the electron number, compared with conventional torches [26]. The ICP-MS equipped with the conical torch has a higher sensitivity, better signal-to-background ratios, and better trajectories, thus achieving a better performance in single-particle analysis [27].

Almost all commercial ICP-MS systems have a horizontal orientation of the plasma in which the trajectories of sample droplets in different sizes are affected by gravity and transport efficiencies of the droplets become mass-dependent, particularly for droplets tens of micrometers in diameter. Vonderach et al. designed a vertical downward orientation of the plasma, allowing sample introduction from the top [28]. This gravity-assisted sampling approach is beneficial for transporting large-sized droplets that are difficult to transport in a horizontally oriented plasma. In addition, the transport efficiency of the droplets was independent of the droplet size [28]. This new plasma source design has the potential to be widely applied in single-cell/particle analysis.

15.2.2 Time-of-flight Mass Analyzer

Most commercial ICP-MS instruments use a scanning analyzer (e.g. a quadrupole filter); thus, only ions of a certain mass-to-charge ratio (m/z) can be determined at every moment. However, in many metallomics applications, multiple elements must be determined in a short period. For example, a typical temporal duration of ion clouds from a single cell is between 200 and 500 μs [29]. Scanning analyzers fail to determine more than one isotope at such a short pulse. Time-of-flight (TOF) technology offers unique advantages over scanning analyzers [30]. The ions are sampled simultaneously, and a full mass spectrum can be obtained in tens of microseconds. ICP-TOF-MS also offers a higher mass resolving power, minimizing polyatomic interferences in the mass spectrum [31]. In addition, ICP-TOF-MS offers improved precision in isotope ratio analysis owing to its quasi-simultaneous characteristics [31].

Table 15.1 Comparison of three modern ICP-TOF-MS instruments.

	icpTOF 2R	Vitesse	CyTOF Helios
Mass range (m/z)	6–270	6–280	75–209
Mass resolution (FWHM)	6000	4500	900
Minimum integration time (μs)	46	25.5	13
Sensitivity (kcps/ppb)	30 (^{238}U)	40 (^{238}U)	N/A
Abundance sensitivity (ppm)	100	30	3000
Linear dynamic range	10^6	10^7	$10^{4.5}$
Time resolution (continuous acquisition, μs)	2000	77	13
Ion blanking	Notch filter	Bradbury–Nielsen gate	Quadrupole high-pass filter

TOF was first introduced in ICP-MS in the 1990s [32]. However, the last generation of ICP-TOF-MS suffered from a limited sensitivity and dynamic range compared with other types of ICP-MS instruments, such as the quadrupole ICP-MS. With breakthroughs in high-speed digitization electronics and related techniques [33], ICP-TOF-MS has evolved into a powerful tool for metallomics studies. New generation of ICP-TOF-MS instruments are currently available in the market, including the icpTOF series from TOFWERK, Vitesse from Nu Instruments, and CyTOF from Fluidigm. These instruments use the same orthogonal design and single-pass reflectron TOF design. The CyTOF was designed to analyze samples labeled by lanthanide stable isotopes, and it has a mass range of 75–209 m/z [34]. The other two ICP-TOF-MS instruments cover a wider mass spectrum and thus are more versatile in metallomics studies [31, 35], particularly for the study of important trace elements in biological systems (e.g. iron, copper, and zinc). A comparison and the specifications of the ICP-TOF-MS instruments are presented in Table 15.1.

As an emerging technique, ICP-TOF-MS is expected to be the instrument of choice in metallomics studies that require high-throughput and high-resolution analyses, such as single-cell/particle analysis [36] and high-resolution elemental bio-imaging [37].

15.2.3 Laser Ablation Systems

Laser ablation coupled with ICP-MS (LA-ICP-MS) enables in situ quantitative analysis and imaging of metals and metalloids in sample sections. However, the slow analytical speed and limited spatial resolution of traditional instruments hinder the wider application of LA-ICP-MS.

Many commercial ablation cells produce a signal response to a single laser shot, which is also called a single-pulse response (SPR), in the range of a second. In the past several decades, numerous efforts have been made to optimize ablation cell geometries to improve the aerosol transport efficiency, reduce the signal duration, and decrease the aerosol dispersion [38]. Tanner et al. designed an in-torch laser ablation cell that greatly reduced the cell volume and achieved an SPR of 4 ms. [33] The signal-to-noise ratios were also significantly improved because the ablated aerosols were less diluted during the transport [33]. However, bio-imaging analysis was not possible because the cell was not able to sample a normal specimen. Liu et al. developed a two-volume cell by placing a movable inner cell inside an external cell, which was sufficiently large to accommodate large samples. The cell also greatly reduced the memory effects and the SPR [39]. Different designs of fast and low-dispersion ablation cells have since been developed and commercialized [40, 41]. These cells all achieved a single-pulse response of less than 10 ms at FW0.01M (e.g. a 1% height of the maximum peak) and a higher sensitivity compared to the traditional ablation cells [40]. The stable distance between the cell inlet and sample surface in these low-dispersion cells is reported to be key for a shorter SPR and improved data precision [42], Thus, some of the latest laser ablation systems contain a z-axis drive to control the distance [43].

In addition to the ablation cell, the laser performance must also be considered. Selective removal of biological tissues from glass substrates can be achieved after careful control of the laser fluence using an energy meter [44]. Compared to lasers with a low repetition rate (e.g. ~20 Hz), an excimer laser can operate at 200–1000 Hz, enabling fast analysis and imaging. Van Malderen et al. achieved submicrometer LA-ICP-MS imaging at a pixel acquisition rate above 250 Hz with an ArF excimer laser [43]. A laser with a short wavelength also produces smaller thermal effects during the ablation, a more controlled ablation process, and a better distribution of ablated particles, which are beneficial to a successful LA-ICP-MS analysis [45]. The sizes of the laser spots in most modern laser systems are down to a few microns, making subcellular imaging possible. However, the lateral resolutions of commercial systems are limited to ~1 μm, owing to the diffraction limit.

15.3 Recent Advances in ICP-MS-based Metallomics

15.3.1 Single-particle Analysis

Single-particle (SP)-ICP-MS is a new technique that utilizes the excellent detection capabilities of ICP-MS. spICP-MS can provide the elemental composition, number concentration, and size distribution of nanoparticles. In addition, ions dissolved from nanoparticles can be determined under the optimal conditions. This section discusses the basic theory and the application of spICP-MS in metallomics.

The spICP-MS method was developed by C. Degueldre and P.Y. Favarger [46]. The theory of spICP-MS can be found in many excellent reviews [29, 47]; therefore, only a brief overview is given here. spICP-MS requires a sufficiently diluted nanoparticle solution at a constant flow rate. Under these conditions, only a single particle is statistically introduced into the ICP-MS instrument at a time. After being atomized and ionized, the nanoparticle is detected as a signal pulse in ICP-MS. The frequency of the nanoparticle pulses is directly proportional to the number concentration of the nanoparticle solution, and the intensity of the pulse is a function of the element mass of the nanoparticle. After the correction of transport efficiencies, standard solutions can be used for calibrations of single particles. spICP-MS can provide valuable information about the nanoparticles and reach extremely low number detection limits between 10^3 and 10^5 ml^{-1} [48].

Despite the remarkable progress made to date, spICP-MS is still a developing technique that faces many challenges. The current size detection limits for metal nanoparticles, most of which are in the range of 10–80 nm in diameter [49], often fail to meet the requirements of metallomics studies. Accordingly, more sensitive ICP-MS instruments and improved sample introduction systems are required. Hadioui et al. used a high-sensitivity sector field ICP-MS and a dry aerosol sample introduction system to successfully lower the size detection limits of Ag and TiO_2 nanoparticles to 3.5 and 12.1 nm, respectively, by spICP-MS. [50]

Multi-isotope analysis of a single particle is another challenge for spICP-MS, particularly for instruments equipped with a scanning analyzer, such as a quadrupole. ICP-MS instruments with a simultaneous analyzer can circumvent this challenge. Multi-collector (MC)-ICP-MS has been applied to multi-isotope analysis in single particles [51], but the determined isotopes must be close to each other due to the instrument limitations. As mentioned in Section 15.2.2, TOF is considered a better choice for multi-isotope analysis of single particles. Praetorius et al. employed SP-ICP-TOF-MS to analyze multi-isotopes in single particles [52]. In this way, the engineered CeO_2 nanoparticles were distinguished from the natural Ce-containing particles in soils by multi-isotope fingerprints that were classified using a machine learning method [52].

Sample preparation is crucial for spICP-MS analyses. The unique physical and chemical characteristics of nanoparticles should be considered to avoid changes in the form, size distribution, or aggregation state. The dilution of samples is always necessary in spICP-MS, and for real samples, further extraction and separation are also necessary to remove the matrix and minimize interference [53]. The nanoparticles from biological matrices have been successfully extracted from biological matrices using acids, alkalis, or enzymes, as reported in the literature [54]. However, standard protocols cannot be applied to nanoparticles in different biological matrices. Reliable and standard protocols for spICP-MS sample preparation are urgently needed, for example, the development of certified reference materials for nanoparticles in biological matrices for method validation [55].

In combination with nanoparticle labeling, spICP-MS offers a new opportunity for highly sensitive bioassays. Hu et al. developed a competitive heterogeneous immunoassay for the determination of α-fetoprotein in serum with an AuNP-tagged antibody and spICP-MS analysis. The quantification limit was 0.016 μg/l with a relative standard deviation of 4.2% for α-fetoprotein [56]. Later, Han et al. reported a novel method of a one-step homogeneous DNA assay using spICP-MS. [57] The hybridization of DNA targets with probes immobilized on the AuNP surfaces resulted in the formation of dimers, trimers, or even large aggregates of AuNPs. These changes were detected and the DNA concentration was obtained with spICP-MS. After the above pioneer works, spICP-MS based methods were successfully applied for the analysis of DNAs [58], rRNAs [59], and carcinoembryonic antigens [60]. Table 15.2 shows the selected applications of spICP-MS.

15.3.2 Single-cell Analysis

Cell heterogeneity is always present in all cell populations. Therefore, single-cell analysis can provide valuable insights that are often covered in cell populations [71]. For the analysis of metals and metalloids in single cells, ICP-MS is considered to be the first choice [72]. Two main ICP-MS-based methods have been applied for single-cell analysis: single-cell ICP-MS (scICP-MS) for the analysis of cells in a suspension and LA-ICP-MS for the analysis of cells on a substrate.

scICP-MS shares a similar theory and technique with spICP-MS. The single cells were sprayed sequentially into the high-temperature plasma, where the constituents in each cell were atomized, ionized, and detected by ICP-MS in the time-resolved mode. In the mass spectra, the intensity of each transient signal is related to the atomic constituents in a single cell, and the frequency of the transient signals is directly proportional to the number of cells. scICP-MS has been utilized for the analysis of metals in bacteria [73], trace elements in algae [74, 75] and in mammalian cells [19, 76–79], metal medicines in human cells [80, 81], and nanoparticles in *Tetrahymena* [82]. Using a new chemical labeling strategy, the fucose contents in single cells were determined via europium using scICP-MS, achieving a detection limit of 4.2 zmol fucose [83]. Single bacteria metabolically labeled by a lanthanide-encoding were counted and recognized with scICP-MS, providing a new way to identify bacteria and study variability at a single bacterium level [84]. Table 15.3 shows the recent applications of scICP-MS analysis in single-cell suspensions.

Mass cytometry, a specific scICP-MS using a TOF analyzer, was originally designed to analyze single cells labeled with lanthanide isotopes [94]. Instead of the fluorophores in traditional flow cytometry analysis, enriched isotopes are labeled with single cells as reporters and are analyzed by mass cytometry. More than 40 parameters can be determined in single cells using mass cytometry because many stable isotopes are available as reporters without any spectral overlap [95]. This post-fluorescence technique is believed to be the next platform for flow cytometry, allowing simultaneous analysis of the cell surface and intracellular proteins, signaling components, cell cycle state, cell viability, and nucleic acids (mRNA and DNA) in a single cell [95].

The discussions in Section 15.3.1 on the challenges and solutions in spICP-MS analysis can also be applied to scICP-MS. In addition, other issues should be considered in scICP-MS analysis. First, the sample preparation of single cells should be performed with additional caution to maintain the cell integrity and avoid elemental losses. The buffer solutions used for the cell sample preparation should be compatible with ICP-MS because organic solvent or inorganic salts often have adverse effects. In mass cytometry, fixation and permeabilization are usually required. However, these treatments may lead to the loss of naturally occurring metals in the cells, which restricts their application in metallomics.

Table 15.2 Applications of spICP-MS analysis.

Instrument	Nebulizer	Spray chamber	Dwell time (ms)	Transport efficiency (%)	Size detection limit (nm)	Application	References
ICP-QMS Agilent 4500	Babington	—	10	—	Aluminum particles: 30	Colloids in water	[46]
ICP-QMS Thermo X2	Concentric	Conical with impact bead	10	—	AuNPs: 15	Immunoassay with AuNP tags	[56]
ICP-QMS Agilent 7500	Concentric	Scott double pass	10	~9	—	AgNPs and AuNPs in water	[48]
ICP-QMS Thermo X2	Concentric	Conical with impact bead	0.5	0.75	AuNPs: 15	DNA assay with AuNP probes	[57]
ICP-QMS PerkinElmer NexION 300Q	Concentric (glass)	Cyclonic	10	4–6	15–20	AgNPs and AuNPs in biological tissues	[61]
MC-ICP-MS Nu Plasma HR	Concentric	Cyclonic	200	—	130	ErNPs in water	[51]
ICP-QMS Perkin Elmer NexION 350D	Concentric	Cyclonic	0.1	—	20	AuNPs in tomato plants	[62]
ICP-QMS Thermo iCAP	Microflow concentric	Cyclonic	5	—	56	Lead nanoparticles in game meat	[63]
ICP-TOFMS TOFWERK icpTOF	Concentric	Cyclonic (quartz)	0.3	—	180	CeO_2 nanoparticles in soils	[52]

Instrument	Nebulizer	Spray chamber				Application	Ref.
ICP-QMS PerkinElmer NexION 300D	Concentric (PFA-ST)	Cyclonic (glass)	0.05	~9	19	AgNPs in soils	[64]
ICP-QMS Agilent 7900	Concentric	Cyclonic (quartz)	0.1	3.5	18	SeNPs in yeast	[65]
ICP-QMS PerkinElmer NexION 300X	Concentric (PFA-ST)	Cyclonic (PC³ glass)	0.05	3–5	~14	AgNPs in bivalve mollusks	[66]
ICP-SFMS Nu AttoM ES	Microflow concentric	Desolvation system	0.05	16	AgNPs: 3.5 TiO$_2$NPs: 12.1	TiO$_2$ NPs in sunscreen lotions, rainwater, and swimming pool water	[50]
ICP-QMS PerkinElmer NexION 350	Concentric	Cyclonic	0.05	15	—	Protein analysis with AuNP tags	[60]
ICP-QMS Agilent 7700x	Concentric	Scott double pass	3	7.8	AgNPs: 23 AuNPs: 16	AgNPs and AuNPs in sewage sludge	[67]
ICP-TOFMS TOFWERK icpTOF 2R	Concentric	Cyclonic (quartz)	3	7–15	6–311 Dependence on isotopes	Particles in rivers	[68]
ICP-TOFMS TOFWERK icpTOF 2R	Concentric	APEX Ω desolvation system	0.3	11.6	40 to several hundred Dependence on isotopes	Gunshot residues	[69]
ICP-TOFMS TOFWERK icpTOF	Concentric	Cyclonic	2	5–7	15–307 Dependence on isotopes	Anthropogenic nanomaterials in urban rain and runoff	[70]

Table 15.3 Applications of scICP-MS analysis.

Instrument	Nebulizer	Spray chamber	Dwell time (ms)	Transport efficiency (%)	Sample flow rate (µl min^{-1})	Application	References
ICP-SFMS Thermo Element 1	Microflow concentric	Scott double pass	4	—	200	Uranium in bacteria	[73]
ICP-QMS Agilent 7500a	V-groove	Scott double pass	10	0.6	400	Heterogeneous cellular responses to external stresses	[74]
ICP-QMS Agilent 7500a	Microflow concentric	Custom-made single pass	10	3.0	20	Determination of Bi drugs in single *Helicobacter pylori* cells	[81]
ICP-QMS Thermo X7	Concentric	Conical with impact bead	5	3.1	250	Determination of quantum dots in single cells	[85]
ICP-QMS Agilent 7500a	Concentric	Single pass	0.05–0.1	~100	10	Highly efficient single-cell analysis of microbial cells	[18]
ICP-QMS Thermo X7	Microflow concentric (glass)	Conical with impact bead	5	3.1	200	Quantification of Gd@C$_{82}$(OH)$_{22}$ and cisplatin in HeLa and 16HBE cells	[80]
ICP-QMS PerkinElmer NexION 300D	Concentric (PFA-ST)	Cyclonic (quartz)	5	1	320	Determination of Fe, Cu, Zn, Mn, P, and S in HeLa and A549 and 16HBE	[16]

ICP-TOFMS Fluidigm CyToF2	Concentric	Single pass	0.013	—	45	Quantification of AgNPs in human T-lymphocytes	[86]
ICP-QMS Agilent 7700	Parallel path (EnyaMist)	Single pass	0.010	25	10	Cisplatin in human ovarian carcinoma cells (A2780)	[87]
ICP-QMS Thermo XII	Microflow concentric	—	5	2.96	30	Quantification of Fe, Pt in HepG2 cells with a droplet-splitting microchip	[76]
ICP-QMS PerkinElmer NexION 300D	Microflow concentric	Single pass (heated)	4	2	10	Quantification of Mn, Fe, Co, Cu, Zn, P, and S in HeLa and A549 and 16HBE cells.	[19]
ICP-QMS PerkinElmer NexION 350D	Concentric (Meinhard)	Single pass (Asperon)	0.05	31.33	283	Quantification of Au NPs in freshwater algae	[75]
ICP-QMS Agilent 7500a	Concentric (quartz)	Scott double pass	10	4	5	Quantification of Au NPs in MCF-7 cells	[88]
ICP-QMS Agilent 7900	Concentric (MicroMist)	Scott double pass	10	0.02–0.03	300	Quantification of Ag NPs in single cells	[89]
ICP-MS/MS Thermo iCAP-TQ	Concentric (MicroMist)	Cyclonic	1	25	10	Quantification of Cu in individual spores	[90]
ICP-QMS PerkinElmer NexION 300X	Concentric (PFA)	Single pass (Asperon)	0.05	9.9	19.4	Evaluation of As uptake and lipid profile changes from cells	[91]

(Continued)

Table 15.3 (Continued)

Instrument	Nebulizer	Spray chamber	Dwell time (ms)	Transport efficiency (%)	Sample flow rate (µl min^{-1})	Application	References
ICP-TOFMS Fluidigm CyTOF	Concentric	Single pass	—	—	30	Quantification of Au NPs in *Tetrahymena thermophila*	[82]
ICP-MS/MS Agilent 8900	Concentric (Quartz)	Scott double pass	1	—	200	Intracellular antagonism of Cu^{2+} against Cd^{2+} in different cells	[79]
ICP-QMS PerkinElmer NexION 300D	Microflow concentric	Single pass	0.1	12	40	Quantification of AuNPs in individual HepG2 cells	[92]
ICP-TOFMS Fluidigm CyTOF	Concentric	Single pass	—	—	30	Quantification of Pb in human erythrocytes	[78]
ICP-QMS PerkinElmer NexION 300D	Concentric (Meinhard)	Single pass (Asperon)	0.05	30–40	15–20	Quantification of AgNPs in yeast cells	[93]
ICP-QMS PerkinElmer NexION 2000	Concentric	Single pass (Asperon)	0.10	15–36	20	Counting and recognizing single bacteria via lanthanide encoding	[84]
PerkinElmer ELAN DRC-II	micronebulizer (Homemade)	None (Direct infusion)	10	>70	13	An oil-free passive microfluidic system for high transport efficiency of single-cell analysis	[20]

Second, the behaviors of single cells in the plasma, such as transport efficiency and atomization–ionization efficiencies, are different from those of standard solutions. Therefore, the intracellular elements usually exhibit different behaviors from the calibration solutions. For example, Li et al. found a 30% lower sensitivity for uranium in single bacteria, compared to the uranium solution [73]. The development of calibration methods and certified standard materials of single cells will ensure that scICP-MS is an accurately quantitative method.

Unlike solution-based scICP-MS, LA-ICP-MS enables in situ analysis of single cells. A single cell is ablated by laser shots, and the resulting aerosol is introduced and analyzed by ICP-MS. The transport efficiency of the aerosols is high, nearly 100% under the optimal conditions, and is independent of the cell size. In addition, aerosols are more easily ionized compared with intact cells, making quantitative analysis of single cells possible. However, some obstacles must be overcome to enable wider applications of LA-ICP-MS for single-cell analysis. The first obstacle is the low analytical throughput. In many applications reported in the literature, single cells are manually targeted under a microscope and then analyzed by LA-ICP-MS. [96–98] The entire process is typically slow and time consuming. To improve the analytical throughput, Zheng et al. designed a microwell array to trap single cells, which were analyzed by LA-ICP-MS using a grid pattern, greatly improving the analytical throughput [99]. Löhr et al. used a piezo-acoustic spotter to produce a single-cell array for high-throughput LA-ICP-MS analysis. Under the optimal parameters, a single-cell occupancy of >99%, high throughput of up to 550 cells per hour, and high cell recovery (>66%) were achieved [100].

The second obstacle is the lack of commercially available reference materials for single cells; thus, it is difficult to achieve accurate and reliable quantification with LA-ICP-MS and to compare the results with those from different laboratories. Many in-house standards have been developed to overcome this obstacle. Wang et al. used the dried residues of picoliter droplets as the single-cell standards for LA-ICP-MS analysis [98]. Van Malderen et al. added a range of concentrated Cu standard solutions to gelatin and used microfabrication techniques to prepare a high-density microarray gelatin [101]. Besides the in-house standards for external calibration, isotope dilution calibration was developed in which each cell is dispensed with a known picoliter droplet of an enriched isotope solution using an inkjet printer and then analyzed using isotope dilution LA-ICP-MS [102].

15.3.3 Spatial Metallomics

The spatial organization of cells and tissues is closely related to biological functions, and understanding the spatial context is critical for life sciences research [103]. When cells are dissociated from tissues, the spatial context is lost. To address this matter, techniques for spatially resolved multi-omics have been rapidly developing in recent years. For example, spatial transcriptomics was selected by *Nature Methods* as the Method of the Year 2020 [104]; a high-spatial-resolution multi-omics sequencing technique named DBiT-seq was developed for co-mapping mRNAs and proteins in tissue sections [105]. In terms of metallomics, it is also necessary to understand the localization, speciation, and metabolism of the metals and metalloids at a subcellular level and study their connections and interactions with the genes, proteins, or metabolites in the spatial dimension, i.e. spatial metallomics. In combination with other spatial omics, spatial metallomics will provide profound insights into not only cellular phenotypes but also the basic chemicals underlying these cellular properties.

Although there is no specific technique for spatial metallomics thus far, LA-ICP-MS is one of the most promising tools owing to its unique characteristics, as described in Section 15.2. LA-ICP-MS was used for the elemental imaging of biological tissues in 1994 [106]. Since then, LA-ICP-MS has been applied in the imaging of metal contrast agents [107] such as manganese [108]; however, most of these are limited to a few metals or metalloids and are not combined with other spatial omics techniques. With rapid advances in laser ablation systems and ICP-TOF-MS instruments, the new generation of LA-ICP-MS can provide the capabilities of fast imaging speed, high spatial resolution, and full mass spectral scan [40], making it more suitable for spatial metallomics. Quantitative imaging can also be achieved using the appropriate matrix-matched standard materials, either certified standard materials [109] or in-house standards [110, 111]. Finally, many open-source or commercial software packages have been developed to process and reconstruct the enormous spatial imaging data from LA-ICP-MS. [112, 113]

Owing to these advances, new methods have been developed. Three-dimensional (3D) quantitative imaging of metals and metalloids has been realized by imaging successive slices of a sample with LA-ICP-MS and reconstructing the images into three dimensions. This method has been applied to the 3D modeling of metallome in a wide range of biological samples, such as single cells [114] and mouse brains [115]. LA-ICP-MS can image biomolecules in tissues in combination with immunohistochemistry methods, which is usually called imaging mass cytometry (IMC) by life scientists [116]. Biomolecules in tissue sections are stained with metal probes and then imaged via the probes using LA-ICP-MS [116]. Giessen et al. simultaneously imaged 32 proteins and protein modifications in tissue sections at a subcellular resolution with LA-ICP-TOF-MS [117]. The success of IMC demonstrates that LA-ICP-MS could become a spatial multi-omics platform. The distributions of metals and proteins in the same tissue section were obtained using a strategy of two consecutive imaging acquisitions with LA-ICP-MS analysis followed by tissue staining and a standard IMC analysis (Figure 15.1) [118]. The LA-ICP-MS technique may be further expanded to image mRNAs or DNAs once suitable metal probes are developed, realizing spatial multi-omics analysis using the same LA-ICP-MS instrument.

15.4 Conclusions

ICP-MS was commercialized approximately 40 years ago and has become a versatile tool for both routine analysis and pioneering research. The development of ICP-MS is progressing toward higher sensitivity, better detection limits, faster acquisition speed, in situ analyses, and automatic data processing. The ICP-MS application also

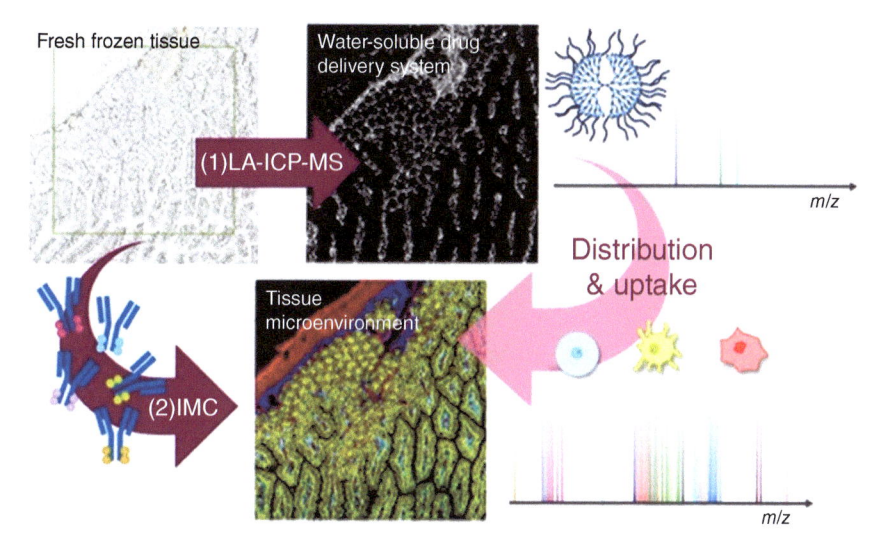

Figure 15.1 A new strategy for imaging both metals and proteins in tissue sections. The strategy involves two consecutive imaging acquisitions from the same tissue section with LA-ICP-MS analysis followed by tissue staining and a standard IMC analysis. Source: Strittmatter et al. [118], © 2021/Reproduced with permission from American Chemical Society.

greatly expands to many aspects of metallomics, such as single-particle analysis, single-cell analysis, and spatial metallomics. It must be emphasized that a single technique cannot meet all the challenges of metallomics; thus, comprehensive strategies should be developed, such as the integration of metal determination with ICP-MS, metal–complex identification with biological mass spectrometry, and speciation characterization with synchrotron X-ray absorption spectroscopy. As with other omics research, metallomics is both a method-driven and data-driven science. An enormous amount of data acquired from various instruments must be handled and interpreted. Thus, a bioinformatics approach is crucial and urgently required in the future. With the rapid development of instrumentation and methodology, ICP-MS-based techniques will evolve further and play a dominant role in metallomics research.

Acknowledgment

This work was supported by the National Natural Science Foundation of China (Nos. 11975251, 11875268) and Guangdong Basic and Applied Basic Research Fund (No. DG2231351B).

List of Abbreviations

ICP-MS   inductively coupled plasma-mass spectrometry
HPLC   high performance liquid chromatography

LA-ICP-MS	laser ablation-inductively coupled plasma-mass spectrometry
ICP-TOF-MS	inductively coupled plasma time-of-flight mass spectrometry
SPR	single pulse response
spICP-MS	single particle ICP-MS
scICP-MS	single cell ICP-MS
MC-ICP-MS	multi-collector ICP-MS

References

1 Chen, C., Chai, Z., and Gao, Y. (2010). *Nuclear Analytical Techniques for Metallomics and Metalloproteomics*. Royal Society of Chemistry.
2 Haraguchi, H. (2004). *J. Anal. At. Spectrom.* 19: 5.
3 Mounicou, S., Szpunar, J., and Lobinski, R. (2009). *Chem. Soc. Rev.* 38: 1119.
4 Lobinski, R., Becker, J.S., Haraguchi, H., and Sarkar, B. (2010). *Pure Appl. Chem.* 82: 493.
5 Haraguchi, H. (2017). *Metallomics* 9: 1001.
6 Li, Y.-F. and Sun, H. (2021). *At. Spectrosc.* 42: 227.
7 Kim, B.-E., Nevitt, T., and Thiele, D.J. (2008). *Nat. Chem. Biol.* 4: 176.
8 Li, Y.-F., Chen, C., Qu, Y. et al. (2008). *Pure Appl. Chem.* 80: 2577.
9 Wang, M., Feng, W.-Y., Zhao, Y.-L., and Chai, Z.-F. (2010). *Mass Spectrom. Rev.* 29: 326.
10 Feng, L. (2021). *At. Spectrosc.* 42: 262.
11 Montes-Bayón, M., Sharar, M., and Corte-Rodriguez, M. (2018). *Trends Anal. Chem.* 104: 4.
12 Houk, R.S. (1986). *Anal. Chem.* 58: 97A.
13 Ammann, A.A. (2007). *J. Mass Spectrom.* 42: 419.
14 Koppenaal, D.W., Eiden, G.C., and Baringa, C.J. (2004). *J. Anal. At. Spectrom.* 19: 561.
15 Lin, X., Guo, W.J., Jin, L., and Hu, S. (2020). *At. Spectrosc.* 41: 1.
16 Wang, H., Wang, B., Wang, M. et al. (2015). *Analyst* 140: 523.
17 Franze, B., Strenge, I., and Engelhard, C. (2012). *J. Anal. At. Spectrom.* 27: 1074.
18 Miyashita, S.I., Groombridge, A.S., Fujii, S.I. et al. (2014). *J. Anal. At. Spectrom.* 29: 1598.
19 Wang, H., Wang, M., Wang, B. et al. (2017). *Anal. Bioanal. Chem.* 409: 1415.
20 Zhou, Y., Chen, Z., Zeng, J. et al. (2020). *Anal. Chem.* 92: 5286.
21 Gundlach-Graham, A. and Mehrabi, K. (2020). *J. Anal. At. Spectrom.* 35: 1727.
22 Verboket, P.E., Borovinskaya, O., Meyer, N. et al. (2014). *Anal. Chem.* 86: 6012.
23 Niemax, K. (2012). *Spectrochim. Acta Part B At. Spectrosc.* 76: 65.
24 Hendriks, L., Ramkorun-Schmidt, B., Gundlach-Graham, A. et al. (2019). *J. Anal. At. Spectrom.* 34: 716.
25 Gschwind, S., Flamigni, L., Koch, J. et al. (2011). *J. Anal. At. Spectrom.* 1166–1174.
26 Alavi, S., Khayamian, T., and Mostaghimi, J. (2018). *Anal. Chem.* 90: 3036.
27 Guo, X., Alavi, S., Dalir, E. et al. (2019). *J. Anal. At. Spectrom.* 34: 469.

28 Vonderach, T., Hattendorf, B., and Günther, D. (2021). *Anal. Chem.* 93: 1001.
29 Montaño, M.D., Olesik, J.W., Barber, A.G. et al. (2016). *Anal. Bioanal. Chem.* 408: 5053.
30 Guilhaus, M. (2000). *Spectrochim. Acta Part B At. Spectrosc.* 55: 1511.
31 Hendriks, L., Gundlach-Graham, A., Hattendorf, B., and Günther, D. (2017). *J. Anal. At. Spectrom.* 32: 548.
32 Myers, D.P., Li, G., Yang, P., and Hieftje, G.M. (1994). *J. Am. Soc. Mass Spectrom.* 5: 1008.
33 Tanner, M. and Günther, D. (2008). *Anal. Bioanal. Chem.* 391: 1211.
34 Bandura, D.R., Baranov, V.I., Ornatsky, O.I. et al. (2009). *Anal. Chem.* 81: 6813.
35 Azimzada, A., Farner, J.M., Jreije, I. et al. (2020). *Front. Environ. Sci.* 8: 1.
36 Gundlach-Graham, A. (2021). *Compr. Anal. Chem.* 93: 69–101.
37 Burger, M., Schwarz, G., Gundlach-Graham, A. et al. (1946). *J. Anal. At. Spectrom.* 2017: 32.
38 Van Malderen, S.J.M., Managh, A.J., Sharp, B.L., and Vanhaecke, F. (2016). *J. Anal. At. Spectrom.* 31: 423.
39 Liu, Y., Hu, Z., Yuan, H. et al. (2007). *J. Anal. At. Spectrom.* 22: 582.
40 Gundlach-Graham, A. and Günther, D. (2016). *Anal. Bioanal. Chem.* 408: 2687.
41 Wang, H.A.O., Grolimund, D., Giesen, C. et al. (2013). *Anal. Chem.* 85: 10107.
42 Pisonero, J., Bouzas-Ramos, D., Traub, H. et al. (2019). *J. Anal. At. Spectrom.* 34: 655.
43 Van Malderen, S.J.M., Van Acker, T., and Vanhaecke, F. (2020). *Anal. Chem.* 92: 5756.
44 Van Acker, T., Van Malderen, S.J.M., Colina-Vegas, L. et al. (1957). *J. Anal. At. Spectrom.* 2019: 34.
45 Guillong, M., Horn, I., and Günther, D. (2003). *J. Anal. At. Spectrom.* 18: 1224.
46 Degueldre, C. and Favarger, P.-Y. (2003). *Colloids Surf., A* 217: 137.
47 Meermann, B. and Nischwitz, V. (2018). *J. Anal. At. Spectrom.* 33: 1432.
48 Pace, H.E., Rogers, N.J., Jarolimek, C. et al. (2011). *Anal. Chem.* 83: 9361.
49 Lee, S., Bi, X., Reed, R.B. et al. (2014). *Environ. Sci. Technol.* 48: 10291.
50 Hadioui, M., Knapp, G., Azimzada, A. et al. (2019). *Anal. Chem.* 91: 13275.
51 Su, Y., Wang, W., Li, Z. et al. (2015). *J. Anal. At. Spectrom.* 30: 1184.
52 Praetorius, A., Gundlach-Graham, A., Goldberg, E. et al. (2017). *Environ. Sci.: Nano* 4: 307.
53 Mozhayeva, D. and Engelhard, C. (2020). *J. Anal. At. Spectrom.* 35: 1740.
54 Laycock, A., Clark, N.J., Clough, R. et al. (2022). *Environ. Sci.: Nano* 9: 420.
55 Gajdosechova, Z. and Mester, Z. (2019). *Anal. Bioanal. Chem.* 411: 4277.
56 Hu, S., Liu, R., Zhang, S. et al. (2009). *J. Am. Soc. Mass Spectrom.* 20: 1096.
57 Han, G., Xing, Z., Dong, Y. et al. (2011). *Angew. Chem. Int. Ed.* 50: 3462.
58 Li, B.-R., Tang, H., Yu, R.-Q., and Jiang, J.-H. (2020). *Anal. Chem.* 92: 2379.
59 Xu, X., Chen, J., Li, B. et al. (2019). *Analyst* 144: 1725.
60 Huang, Z., Wang, C., Liu, R. et al. (2020). *Anal. Chem.* 92: 2876.
61 Gray, E.P., Coleman, J.G., Bednar, A.J. et al. (2013). *Environ. Sci. Technol.* 47: 14315.
62 Dan, Y., Zhang, W., Xue, R. et al. (2015). *Environ. Sci. Technol.* 49: 3007.

63 Kollander, B., Widemo, F., Ågren, E. et al. (1877). *Anal. Bioanal. Chem.* 2017: 409.

64 Schwertfeger, D.M., Velicogna, J.R., Jesmer, A.H. et al. (2017). *Anal. Chem.* 89: 2505.

65 Jiménez-Lamana, J., Abad-Álvaro, I., Bierla, K. et al. (2018). *J. Anal. At. Spectrom.* 33: 452.

66 Taboada-López, M.V., Alonso-Seijo, N., Herbello-Hermelo, P. et al. (2019). *Microchem. J.* 148: 652.

67 Moreno-Martín, G., Gómez-Gómez, B., León-González, M.E., and Madrid, Y. (2022). *Talanta* 238: 123033.

68 Montaño, M.D., Cuss, C.W., Holliday, H.M. et al. (2022). *ACS Earth Sp. Chem.* 6: 943.

69 Brünjes, R., Schüürman, J., von der Kammer, F., and Hofmann, T. (2022). *Forensic Sci. Int.* 332: 111202.

70 Wang, J., Nabi, M.M., Erfani, M. et al. (2022). *Environ. Sci.: Nano* 9: 714.

71 Oomen, P.E., Aref, M.A., Kaya, I. et al. (2019). *Anal. Chem.* 91: 588.

72 Yu, X., He, M., Chen, B., and Hu, B. (2020). *Anal. Chim. Acta* 1137: 191.

73 Li, F.M., Armstrong, D.W., and Houk, R.S. (2005). *Anal. Chem.* 77: 1407.

74 Ho, K.-S. and Chan, W.-T. (2010). *J. Anal. At. Spectrom.* 25: 1114.

75 Merrifield, R.C., Stephan, C., and Lead, J.R. (2018). *Environ. Sci. Technol.* 52: 2271.

76 Wang, H., Chen, B., He, M., and Hu, B. (2017). *Anal. Chem.* 89: 4931.

77 Wei, X., Hu, L.L., Chen, M.L. et al. (2016). *Anal. Chem.* 88: 12437.

78 Liu, N., Huang, Y., Zhang, H. et al. (2021). *Environ. Sci. Technol.* 55: 3819.

79 Zhang, X., Wei, X., Men, X. et al. (2020). *Anal. Chem.* 92: 6604.

80 Zheng, L.N., Wang, M., Zhao, L.C. et al. (2015). *Anal. Bioanal. Chem.* 407: 2383.

81 Tsang, C.N., Ho, K.S., Sun, H., and Chan, W.T. (2011). *J. Am. Chem. Soc.* 133: 7355.

82 Wu, Q., Shi, J., Ji, X. et al. (2020). *ACS Nano* 14: 12828.

83 Liu, Z., Liang, Y., Zhou, Y. et al. (2021). *iScience* 24: 102397.

84 Liang, Y., Liu, Q., Zhou, Y. et al. (2019). *Anal. Chem.* 91: 8341.

85 Zheng, L.-N., Wang, M., Wang, B. et al. (2013). *Talanta* 116: 782.

86 Ivask, A., Mitchell, A.J., Hope, C.M. et al. (2017). *Anal. Chem.* 89: 8228.

87 Corte Rodríguez, M., Álvarez-Fernández García, R., Blanco, E. et al. (2017). *Anal. Chem.* 89: 11491.

88 Wei, X., Zheng, D.H., Cai, Y. et al. (2018). *Anal. Chem.* 90: 14543.

89 Lum, J.T.S. and Leung, K.S.Y. (2019). *Anal. Chim. Acta* 1061: 50.

90 González-Quiñónez, N., Corte-Rodríguez, M., Álvarez-Fernández-García, R. et al. (2019). *Sci. Rep.* 9: 4214.

91 Mavrakis, E., Mavroudakis, L., Lydakis-Simantiris, N., and Pergantis, S.A. (2019). *Anal. Chem.* 91: 9590.

92 Liu, J., Zheng, L., Shi, J. et al. (2021). *At. Spectrosc.* 42: 114.

93 Rasmussen, L., Shi, H., Liu, W., and Shannon, K.B. (2022). *Anal. Bioanal. Chem.* 414: 3077.

94 Bendall, S.C., Simonds, E.F., Qiu, P. et al. (2011). *Science* 332: 687.

95 Spitzer, M.H. and Nolan, G.P. (2016). *Cell* 165: 780.
96 Drescher, D., Giesen, C., Traub, H. et al. (2012). *Anal. Chem.* 84: 9684.
97 Managh, A.J., Edwards, S.L., Bushell, A. et al. (2013). *Anal. Chem.* 85: 10627.
98 Wang, M., Zheng, L.-N., Wang, B. et al. (2014). *Anal. Chem.* 86: 10252.
99 Zheng, L.N., Sang, Y.B., Luo, R.P. et al. (2019). *J. Anal. At. Spectrom.* 34: 915.
100 Löhr, K., Borovinskaya, O., Tourniaire, G. et al. (2019). *Anal. Chem.* 91: 11520.
101 Van Malderen, S.J.M., Vergucht, E., De Rijcke, M. et al. (2016). *Anal. Chem.* 88: 5783.
102 Zheng, L., Feng, L.-X., Shi, J. et al. (2020). *Anal. Chem.* 92: 14339.
103 Zhuang, X. (2021). *Nat. Methods* 18: 18.
104 Marx, V. (2021). *Nat. Methods* 18: 9.
105 Liu, Y., Yang, M., Deng, Y. et al. (2020). *Cell* 183: 1665.
106 Wang, S., Brown, R., and Gray, D.J. (1994). *Appl. Spectrosc.* 48: 1321.
107 Becker, J.S., Matusch, A., and Wu, B. (2014). *Anal. Chim. Acta* 835: 1.
108 Doble, P.A. and Miklos, G.L.G. (2018). *Metallomics* 10: 1191.
109 Jackson, B., Harper, S., Smith, L., and Flinn, J. (2006). *Anal. Bioanal. Chem.* 384: 951.
110 Kysenius, K., Paul, B., Hilton, J.B. et al. (2019). *Anal. Bioanal. Chem.* 411: 603.
111 Liu, J., Zheng, L., Wei, X. et al. (2022). *Microchem. J.* 172: 106912.
112 Halbach, K., Holbrook, T., Reemtsma, T., and Wagner, S. (2021). *Anal. Bioanal. Chem.* 413: 1675.
113 Paul, B., Kysenius, K., Hilton, J.B. et al. (2021). *Chem. Sci.* 12: 10321.
114 Van Malderen, S.J.M., Van Acker, T., Laforce, B. et al. (2019). *Anal. Bioanal. Chem.* 411: 4849.
115 Hare, D.J., George, J.L., Grimm, R. et al. (2010). *Metallomics* 2: 745.
116 Chang, Q., Ornatsky, O.I., Siddiqui, I. et al. (2017). *Cytom. Part A* 91: 160.
117 Giesen, C., Wang, H.A.O., Schapiro, D. et al. (2014). *Nat. Methods* 11: 417.
118 Strittmatter, N., England, R.M., Race, A.M. et al. (2021). *Anal. Chem.* 93: 3742.

16

Machine Learning for Data Mining in Metallomics

Wei Wang[1] and Xin Wang[2]

[1]China Agricultural University, College of Engineering, Beijing Key Laboratory of Optimization Design for Modern Agricultural Equipment, No. 17 Qinghua East Road, Haidian District, Beijing 100083, China
[2]Anhui Medical University, & National Consortium for Excellence in Metallomics, No. 81 Meishan Road, Shushan District, Hefei, 230032, Anhui, China

Metallomics is the integrated understanding of the quantification, distribution, chemical species, and functions of metals/metalloids in biological systems. High-throughput techniques are applied in metallomics. For example, for the quantification of metallome, inductively coupled plasma mass spectrometry (ICP-MS), flame atomic absorption spectroscopy (FAAS), and neutron activation analysis (NAA) can be applied. For the spatial distribution analysis, synchrotron radiation X-ray fluorescence analysis (SR-XRF) and laser ablation inductance coupled plasma mass spectrometry (LA-ICP-MS) are suitable techniques. High-performance liquid chromatography inductively coupled plasma mass spectrometry (HPLC-ICP-MS), X-ray absorption spectroscopy (XAS), etc. are techniques that can be used for chemical speciation study.

The high-throughput techniques applied in metallomics brings high volume of data like in many other –omics studies, which requires dedicated tools for data mining. By automatically altering the weights of numerous components and digesting enormous volumes of data, machine learning algorithms bear the capability for data mining in metallomics. Data types can be categorized as one-dimensional vector data and two-dimensional matrix data from the view of data dimension. The data mining tool set for machine learning includes statistics, pattern recognition, visualization, and classification models.

This chapter will first introduce the data mining methods including data preprocessing, dimensionality reduction, model building, model evaluation, and other processing processes, then introduce the application of data mining methods in metallomics, especially in medical, agricultural, and environmental sciences.

Applied Metallomics: From Life Sciences to Environmental Sciences, First Edition.
Edited by Yu-Feng Li and Hongzhe Sun.
© 2024 WILEY-VCH GmbH. Published 2024 by WILEY-VCH GmbH.

16.1 Data Mining Methods in Metallomics

Data preprocessing, data dimensionality reduction, feature information extraction, data set partitioning, building of the machine learning model, and model assessment are the essential processes in the data mining in metallomics.

16.1.1 Data Preprocessing

The initial data is required to be preprocessed and be modified to meet the analysis or mining requirements. A series of operations on the data known as data preprocessing are performed throughout the entire process, from the initial data to the analysis or mining results [1]. The surrounding environmental factors, parameter settings, non-standard measurement tools, and technical procedures might affect the data collection, resulting in received vector data and matrix data with significant noise. Highlighting the feature information of the heterogeneous samples and enhancing the precision of the qualitative and quantitative model are crucial to the preprocessing of the data in order to lessen the influence of the noise. Choosing the best preprocessing method is crucial because, despite the advantages of preprocessing, it is possible that it could introduce erroneous data that will cause loss of crucial information.

16.1.1.1 Smoothing Process

Suppressing noise while preserving as many data details as possible is an essential operation in data pre-processing. The effectiveness and dependability of subsequent data processing and analysis will be significantly impacted by the processing effect's quality [2]. While the information of interest is frequently masked by noise in the higher-frequency bands of the amplitude spectrum, the majority of the energy of the data is concentrated in the low- and middle-frequency bands. Therefore, a filter that could lower the high-frequency components' amplitude can lessen the effect of noise. In smoothing processing, mean filters, Gaussian filters, and median filters are frequently employed. The mean filter is a linear filter that determines the average value of the values in the window area and then sets the value of the anchor point to the average value determined in the window. In order to suppress noise with a normal distribution, a one-dimensional or two-dimensional convolution operator (linear filter) known as "Gauss filtering" is used. The median filter is a nonlinear filter that reduces noise by using the window area's median.

16.1.1.2 Normalization

Normalization is a common data preprocessing technique used in machine learning and data analysis to unify the numerical range of different features or samples for better comparison and analysis [3]. Normalizing the data to a suitable range during the training of machine learning models can improve stability, convergence speed, and ensure a balanced influence of different features on the model. Maximum–minimum normalization, Z-score normalization, normalization by decimal scaling are examples of common normalizing.

Maximum-Minimum normalization rescales the data to a specific range, typically between 0 and 1. By subtracting the minimum value and dividing by the range (maximum minus minimum), the data is transformed to a normalized range where the minimum value maps to 0 and the maximum value maps to 1. The values in between are linearly scaled accordingly. The maximum–minimum normalization is as follows:

$$x' = \frac{x - \min}{\max - \min} \tag{16.1}$$

where x is the original data, x' is the processed data, min is the minimum value of the data, and max is the maximum value of the data.

Z-score standardization is the standardization of data based on the mean and standard deviation of the original data. The formula for conversion is:

$$x' = \frac{x - \mu}{\delta} \tag{16.2}$$

where x is the original data, x' is the processed data, μ is the mean value of the original data, δ is the standard deviation of the original data.

By shifting the attribute value's decimal places, normalization by decimal scaling transforms the attribute value to $[-1, 1]$. The maximum absolute value of the property value determines the number of decimal places that are moved. This method is applicable to scenarios with large data analysis. The formula for conversion is:

$$x' = \frac{x}{10^k} \tag{16.3}$$

where k depends on the size of the data.

16.1.1.3 Fourier Transform

Fourier transform is a term used to describe the linear combination of trigonometric functions (such as the sine and/or cosine functions) or their integrals, which may be used to define certain functions under specific circumstances [4].

In vector data and matrix data, the Fourier transform may distinguish between high frequency and low frequency. When comparing high-frequency and low-dimension signals, high-frequency signals include mostly edge information (details). Numerous disciplines, including medicine, data science, physics, acoustics, optics, communications, finance, and other fields, use the Fourier transform extensively. For instance, the Fourier transform is frequently used in signal processing to break down complicated signals into single-frequency components of varying amplitudes in order to perform filtering and other operations. In addition, the wave function in the momentum space is the Fourier transform of the quantum mechanical wave function in the position space.

16.1.1.4 Wavelet Transform

A recent technique for transform analysis is the wavelet transform (WT) [5]. The concept of short-time Fourier transform localization is inherited and expanded upon, and it resolves the issue of the window size not changing with frequency by being able to give a "time-frequency" window that varies with frequency. The use of WT for signal time-frequency analysis and processing is highly recommended. Its primary characteristic is its ability to fully highlight the characteristics of some aspects of the issue through transformation, analyze the localization of time (space) frequency, and gradually refine the signal (function) at multiple scales through expansion and translation operations, in order to finally achieve time subdivision at high frequencies and frequency subdivision at low frequencies, as well as its ability to automatically adapt to the requirements of time–frequency signal analysis.

The value of each function point affects the outcome of the Fourier transform since it is a global transformation. Only the points on the support of the basis can have an impact on a basis's coefficients. As a result, the wavelet includes time information (reflected by the local time axis position of the base) in addition to frequency information (reflected by the frequency of the base).

16.1.1.5 Convolution Operation

Convolution operation is the process of opening an active window that is the same size as the template from the matrix's upper left corner, multiplying and averaging the data from the window and the template, and then replacing the data value in the window's middle with the outcome of the calculation. Then the window is gradually moved and the same operation is performed. For example, a new matrix may be created by moving from top to bottom and from left to right. Spatial domain filtering is a technique for convolutionally filtering vector or matrix data based on the spatial connection between values and their immediate surroundings. Frequency domain filtering is the process of changing the frequency characteristics of data by applying the Fourier transform to vector or matrix data, converting the matrix from plane to frequency domain space, and analyzing and processing the plane spectrum in the latter space.

It may be separated into single-channel convolution and multi-channel convolution depending on the quantity of data channels. Each convolution kernel's channel count needs to match the number of input channels. Padding and stride are key variables in convolution operations. Padding refers to filling elements (usually 0 elements) on both sides of the input height and width. The stride in convolution operations refers to the length of the convolution kernel's movement on the data.

16.1.2 Data Dimensionality Reduction

There is a correlation between the data in the actual world, which is redundant and connected. High-dimensional data provides several benefits, including a lot of information and additional tools for making decisions. The downside of high-dimensional data is that it consumes a lot of computer resources and takes a long time to process. However, Data with more dimensions is not beneficial since it also has to take into account the real computing power. The experimental findings will be impacted by redundant and linked data at the same time, and it may even result in a "dimensional disaster." In order to increase the accuracy of models and guarantee that the feature attributes are independent of one another, it is necessary to decrease the number of feature attributes, remove redundant or unnecessary features, and minimize the number of features.

Principal Component Analysis, Independent Component Analysis, Multi-Dimensional Scaling, Local Preserving Projection, T-Stochastic Neighbor Embedding, etc. are common techniques for reducing the dimensionality of data.

16.1.2.1 Principal Component Analysis

The main analysis process of the principal component analysis (PCA) [6] is described in the following procedure. Firstly, the m data $x_1, x_2, \cdots, x_m$ are processed by center zeroing, that is

$$\sum_{i=1}^{m} x_i = 0 \tag{16.4}$$

Secondly, define the covariance matrix

$$C = \frac{1}{m} \sum_{i=1}^{m} x_i x_i^T \tag{16.5}$$

Since C is a positive semidefinite matrix, it is diagonalized by:

$$C = V \Lambda V^T \tag{16.6}$$

among, $\Lambda = \text{diag}(\lambda_1, \lambda_2, \cdots, \lambda_n)$, $VV^T = 1$.

If the rank of C is p, namely rank $(C) = p$, then C has p non-zero eigenvalues, recorded as $\lambda_1 \geq \lambda_2 \geq \cdots \geq \lambda_p \geq 0$.

Finally, taking the eigenvectors $v_1, v_2, \cdots, v_k$ corresponding to the first k eigenvalues as the direction of projection, then for any sample x, we can calculate $y_i = x^T v_i$, $i = 1, 2, \cdots, k$, that is, we can use $[y_1, y_2, \cdots, y_k]^T$ to represent sample x. Thus, the data is reduced from n dimension to k dimension.

16.1.2.2 Independent Component Analysis

Data in the original feature space is really linearly transformed by Independent Component Analysis (ICA). It may be used as a compression to dimension reduction approach, as opposed to PCA's rank reduction process. The degree of data dispersion in this direction, which is how much variation there is in different directions, is not used by ICA to determine which components are the primary ones. The so-called main component and minor component are not established by ICA. According to ICA, each component is equally significant, and finding a linear transformation that would provide the changed data the most independence is our aim rather than isolating significant characteristics. However, the uncorrelation in PCA is too weak; therefore, we employ statistics bigger than 2 to define the data in the hopes that all levels of statistics may be utilized. Furthermore, orthogonality of the features is not necessary for ICA.

Since the observed signal $x(k)$ in the independent component analysis model is a linear mixture of m different independent source signals, it tends to have a Gaussian distribution more than any other source signal. One of the source signals will be acquired if a vector w can be discovered to operate on the mixed signal $x(k)$, causing the processed output to diverge from the Gaussian distribution as much as feasible, or non-Gaussian maximizing. The remaining $m - 1$ source signals may also be isolated in a similar manner.

16.1.2.3 Multidimensional Scaling

One of the traditional techniques for dimensionally reducing linear data is multidimensional scaling (MDS) [7]. Its fundamental concept is to locate a collection of data in low-dimensional space that corresponds to a set of high-dimensional data points given the non-similarity matrix, with the goal of maintaining consistency between the actual distance between these points and the distance between high-dimensional data. The MDS is compatible with the low-dimensional outcomes produced by PCA if a particular phase specificity measure is similar to the Euclidean distance.

When high-dimensional data is supplied in vector form for use in practical issues, both the MDS and the PCA are equally suitable.

16.1.2.4 Local Preserving Projection

He et al. developed the Local Preserving Projection (LPP) [8], a linear approximation of the nonlinear Laplacian feature graphs, as a dimensionality reduction technique for linear projection mapping in 2005. LPP is a linear dimensionality reduction technique that may be defined anywhere in the surrounding space and is not limited to the training dataset, even if it shares many non-linear data to represent qualities with other techniques like Laplace feature mapping and local linear embedding. The LPP dimension reduction algorithm's calculating procedure is as follows: adjacency diagram construction, weight matrix construction, and feature mapping construction. The suggested approach has a lengthy running time and needs a lot of memory for data with a sizable amount of data.

16.1.2.5 T-Stochastic Neighbor Embedding

For high-dimensional data reduction to 2D or 3D for display, the nonlinear dimension reduction method t-Stochastic Neighbor Embedding (t-SNE) is a great choice [9]. In real-world applications, t-SNE may be utilized for visualization in addition to dimensionality reduction. The fundamental idea is to transform Stochastic Neighbor Embedding (SNE) into symmetric SNE, which increases computation efficiency and marginally enhances impact. The original Gaussian distribution is swapped out for the t distribution in the low-dimensional space in order to address the congestion issue. The optimization SNE overvalues local features while undervaluing global features.

The drawback of t-SNE is that it is difficult to employ for other applications and is mostly used for visualization. t-SNE does not have a single optimal solution and cannot be used for prediction, for instance, if the dimension is lowered to 10 dimensions, the t distribution of one degree of freedom may need to be set to a larger degree of freedom in order to retain local characteristics.

16.1.3 Sample Set Division

Choosing the representative training set and the test set samples from the original spectral samples is very important for the establishment and evaluation of the discriminative models. The current common training set division algorithms include Random Sampling, Kennard-Stone Sampling, and Sample Set Partitioning Based on Joint x-y Distances.

16.1.3.1 Random Sampling

Using the Random Sampling (RS) method of randomly downsampling, K points are evenly chosen from N input points, each having an equal chance of selection. The RS has a constant time complexity, meaning that the amount of computation does depend only on the number of points K after downsampling and is not at all dependent on the total number of points in the input point cloud. As a result, it has excellent scaling capabilities and a very high efficiency, but it does not guarantee that the training set comprises extreme sample data.

16.1.3.2 Kennard–Stone Sampling

All samples were taken into account as potential members of the training set, and a set of samples was chosen. First, from the training set, the two samples with the greatest Euclidean distance are chosen. The chosen sample with the least distance is then identified and added to the training set until the requisite number of samples is obtained. This is done by computing the Euclidean distance of each remaining sample to each known sample in the training set. This technique has the benefit of uniformly distributing the samples in the training set based on spatial distance. The drawback is that a lot of work is required to convert the data and determine the sample pairwise spatial distance.

The Euclidean metric, a widely used definition of distance, determines the actual separation between any two points in an m-dimensional space, or the vector's natural length (i.e. the distance from that point to the origin). The Euclidean distance in two-and three-dimensional space is the actual distance between two points.

$$d_x(p,q) = \sqrt{\sum_{j=1}^{N} [x_p(j) - x_q(j)]^2}; p, q \in [1, N] \quad (16.7)$$

x_p and x_q represents two different samples, and N represents the number of spectral wave points of the sample.

16.1.3.3 Sample Set Partitioning Based on Joint x−y Distances

Based on Kennard–Stone Sampling (KS) method, Galvo et al. [10] proposed Sample Set Partitioning Based on Joint x−y Distances (SPXY) method that can consider the sample spectral information and the measured index parameter. The principles and steps of the SPXY method are similar to the KS method, except in the different calculation methods of the Euclidean distance between the two samples. The formula

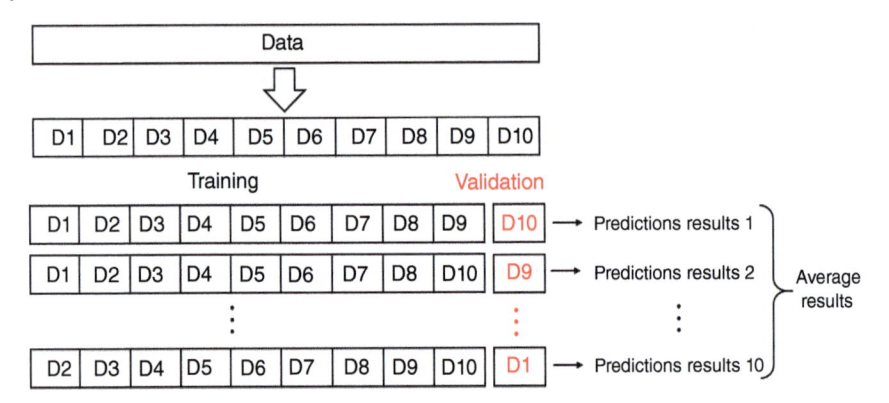

16 Machine Learning for Data Mining in Metallomics

Figure 16.1 Principle of 10-fold Cross-validation. Source: Wei Wang.

for calculating the sample Euclidean distance is shown in Eq. (16.8).

$$d_{xy}(a,b) = \frac{d_x(a,b)}{\max_{a,b\in[1,n]} d_x(a,b)} + \frac{d_y(a,b)}{\max_{a,b\in[1,n]} d_y(a,b)} \tag{16.8}$$

In formula (16.8), $d_x(a,b) = \sqrt{\sum_{i=1}^{m}[x_a(i) - x_b(i)]^2}$, $x_a(i)$ and $x_b(i)$ is the spectral information value of the i-th band of samples a and b, respectively, $d_y(a,b) = |y_a - y_b|$, y_a and y_b is the measured index parameter of samples a and b, respectively, $d_x(a,b)$ is the Euclidean distance between samples a and b in spectral space, $\max_{a,b\in[1,n]} d_x(a,b)$ is the maximum Euclidean distance between two samples in spectral space, $\max_{a,b\in[1,n]} d_y(a,b)$ is the maximum Euclidean distance between two samples in the measured index parameter space, $d_{xy}(a,b)$ refers to the distance between samples a and b taking into account the sample spectral information and the measured index parameters, and the total number of samples is expressed in n.

16.1.3.4 Cross-Validation

The sample data set is split into two complementary subsets using a cross-validation. The training set is the subset that the classifier or model is trained on. A testing set is another subset that is used to confirm the accuracy of an analysis (of a classifier or model). The trained classifier or model, which serves as the classifier or model's performance index, is put to the test using the test set. High prediction accuracy and low prediction error are desired outcomes of study. A sample dataset was partitioned into several complementary subsets for numerous cross-validations in order to decrease the unpredictability of the cross-validation findings. As the verification result, use the average of several validations. 10-fold Cross-validation is shown in Figure 16.1.

Remove the majority of the samples from the given modeling samples so that the model can be built. Then, leave a small portion of the samples so that the newly established model can be used to make predictions. Finally, calculate the prediction error of the small portion of the samples and record their sum of squares. Once and only once, this method is repeated until all samples are predicted.

16.1.3.5 Leave-One-Out Cross Validation

A large data set is divided into k small data sets, of which $k-1$ is the training set, the remaining one is the test set, and then the next one is selected as the test set, the remaining $k-1$ is the training set, and so on. Among them, the value of k is more important. It is mentioned that 10 is generally taken as the value of k. This method is also called "k-fold cross validation." Only one is used as the test set each time, and the rest is used as the training set. The result obtained by this method is the closest to the expected value of training the entire test set, but the cost is too large.

16.1.4 Predictive Model Building Method

16.1.4.1 Partial Least Squares Regression

In various scientific domains, including nature, economy, and society, partial least squares regression (PLSR) is a cutting-edge multiple statistical data analysis technique that is often utilized. The technique creates a single or multiple dependent variable Y model for numerous independent variables. Principal components analysis (PCA) is used in the modeling process of the X regression model to extract the principal components of Y and X and to take into account the correlation between X and Y to optimize the Canonical Correlation Analysis (CCA) concept. The three analysis techniques PCA, CCA, and multiple linear regression analysis are therefore combined in partial least squares regression. When there are multiple correlations between variables, numerous variables but a small sample size, heteroskedasticity, and other issues, PLSR is primarily used to solve problems involving many to many linear regression analyses. This method has unmatched advantages over traditional linear regression.

The following traits apply to PLSR: it is capable of doing regression modeling if the independent variables exhibit strong multiple correlations; regression modeling is permitted when there are fewer sample points than variables; the final model will incorporate all the initial independent variables using PLSR; PLSR model makes system information and noise (including some non-random sounds) simpler to recognize; the regression coefficients of each independent variable in the PLSR model will be simpler to understand.

16.1.4.2 Support Vector Machine

Data is binary classified using supervised learning with SVM, a family of generalized linear classifiers. The suggested approach uses nonlinear mapping and the structural risk reduction principle as its fundamental concepts to convert a low-dimensional space nonlinear problem into a high-dimensional space linear problem. Its decision boundary, the maximum margin hyperplane for learning samples, can convert the issue into a convex quadratic programming issue. SVM is a more powerful and understandable method of learning complicated nonlinear equations than logical regression and neural networks. In particular, when two types of data are linearly separable, the ideal classification hyperplane is discovered in the original space. Add relaxation variables when the linearity is indivisible, and then use nonlinear mapping to convert the samples from the low-dimension input space to the high-dimension space to make them linearly separable so that the best classification hyperplane may be determined in the feature space.

The advantages of SVM include its ability to handle small sample sizes while solving machine learning challenges and its ability to streamline common classification and regression issues; when mapping to high-dimensional space, the complexity of the computation is not increased since the problem of dimension disaster and nonlinear separability is avoided by employing the kernel function approach. In other words, the difficulty of the calculation depends on the number of support vectors, not the size of the sample space, as the final decision function of the support vector calculation technique is controlled by just a small number of support vectors. By adding relaxation variables, the SVM technique may allow some points' distance from the classification plane to deviate from the initial specifications, avoiding the influence of these points on model learning.

16.1.4.3 Decision Tree

A widely popular method of categorization is the decision tree. It is a kind of supervised learning algorithm. Given a number of samples, so-called supervised learning is the process of assigning characteristics and categories to each sample. These categories are predetermined, and a classifier is created through learning that can correctly classify newly discovered items. Decision Tree is a graphical method to intuitively use probability analysis. It determines the probability that the expected value of the net present value is greater than or equal to zero, assesses project risk, and determines feasibility based on the known probability of occurrence of various situations. Because it resembles a tree trunk, this object is known as a decision tree. Decision trees are prediction models used in machine learning that show the mapping between object properties and object values. The decision tree is a tree structure where each leaf node represents a category, each branch represents an output from a test, and each internal node represents a test on an attribute.

The benefit of decision tree is that it may quickly and efficiently produce results for big data sources that are viable and effective by processing both data type and regular type properties. Disadvantages: it's challenging to anticipate continuous fields; there must be extensive preparation for chronological data; too many categories might lead to rapid increases in mistakes.

16.1.4.4 K-means Clustering

Unsupervised learning is a type of clustering. Finding the connections between data items in a set of data without prior "tags" is referred to as clustering. "A cluster" is another name for a collection. The clustering effect is enhanced by increasing similarity within a group and decreasing difference across groups. In other words, the more similarity of an object within a cluster, and the lower the similarity of an object between clusters, the better the clustering effect. K-means is a clustering algorithm.

In the K-means clustering technique, "K" stands for grouping data into K clusters, and "means" is the center of each cluster, also known as the center of mass, which is determined by taking the mean value of the data in that cluster. K-means clustering employs a computational approach to assess the similarity or dissimilarity between objects, aiming to group similar items together and dissimilar items apart. This distance-based clustering technique commonly utilizes Euclidean distance as the default measure of similarity, considering proximity as an indicator of similarity. The closer two items are in distance, the more they are deemed similar.

The key differentiation between clustering and classification lies in their respective targets. In classification, the target variable is predefined, whereas clustering deals with variable targets. Unlike classification, where categories are predetermined, clustering operates without prior knowledge of the target variable. As a result, clustering is primarily employed for knowledge discovery rather than prediction and is commonly referred to as unsupervised learning. The beauty of clustering lies in its ability to uncover patterns and groupings without the need for prior instructions, as sometimes we may not even know what we are seeking.

16.1.4.5 Deep Learning

Deep Learning (DL) is a data-driven method that extracts end-to-end features instead of artificially extracting implicit complex features of the raw data [11]. Convolutional neural network (CNN) is a deep neural network containing the convolutional layer, including Alexnet, VGG, and other famous networks, which is one of the representative algorithms of deep learning. The basic CNN framework is presented in the Figure 16.2a.

Convolutional neural network contains the input layer, convolutional layer, pooling layer, and fully connected layer. Each layer has multiple feature maps, each extracting the input features via a single convolution filter. Among them, the role of the input layer is to input data into the network and extract data features. The important feature of the convolutional layer is to enhance the signal and reduce noise through the convolution operation. Convolution operation is presented in the Figure 16.2b. The pooling layer can reduce the model size, improve the calculation speed, while maintaining the robustness extracting features. The maximum pooling layer is one of the common down sampling methods. It can optimize the spatial hierarchy of learning features, maintain the overall input information, and reduce the number of feature graphic elements. Pooling operation is presented in Figure 16.2c. A reasonable strategy to alternately use a convolutional layer with pooling layer is to generate dense feature maps, achieve maximum activation on each block of the feature layer, and maximize the feature information. The fully connected layer can learn the feature information of the output data of the convolution layer, and integrate the local features to form the overall features, so that the network can be applied to the data classification and regression problems.

The process of constructing a convolutional neural network model to analyze the one-dimensional spectral information generally includes: model structure design, hyperparameter optimization, and model training and testing. Among them, the model structure design mainly includes the number of convolution layers, the number of convolution kernel, the number of layers of the fully connected layers, and the number of neuron nodes. The more layers of the network model, the

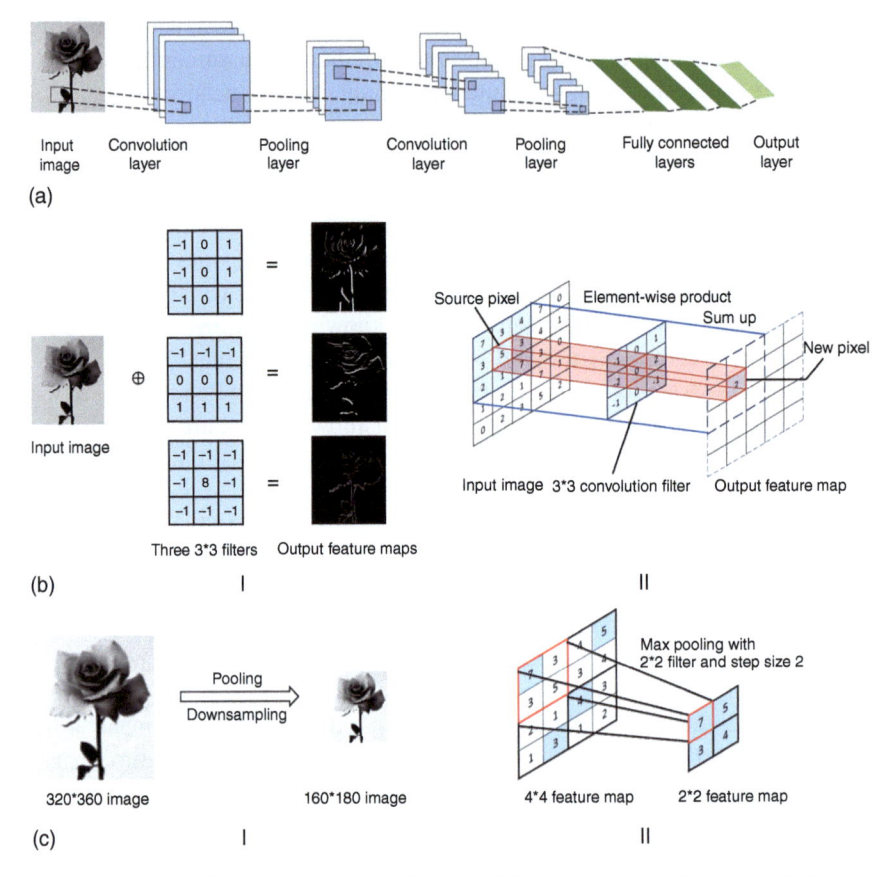

Figure 16.2 The CNN architecture. (a) The basic CNN framework. (b) Examples of three feature maps developed by using three different convolution filters, where ⊕ represents the convolution operation; (c) Pooling operation to generate a feature map with a filter size of 3*3. Source: Yu et al. [11]/Reproduced with permission from Elsevier.

complex extracting features, the more time taken, the more overfitting. Therefore, in the network design process, we can simplify the model as much as possible while ensuring the accuracy. The hyperparameter optimization of the model mainly includes the convolution kernel size, learning rate, regularized system, and the dropout coefficient, which is mainly selected through the automatic selection method of grid search.

In addition, in the network model, a nonlinear activation function is superimposed after each linear change, the network has more powerful fitting and learning ability, and Rectified Linear Unit (ReLU) has sparse activation, unilateral inhibition and relatively wide excitation boundary, zero all negative inputs, cause some neurons to die, improve training speed, as shown in Eq. (16.9), where x is the nonlinear activation input on the channel.

$$\text{ReLU}(x) = \max(0, x) \tag{16.9}$$

In the neural network, the output of the last layer is classified by the Softmax function, distributing the probability of each category to 0–1 and with a sum of 1. The position of the maximum is the category corresponding to the spectral curve. Softmax is as follows:

$$\text{Softmax}(x_i) = \frac{\exp(x_i)}{\sum_{k=1}^{k} \exp(x_k)} \quad (16.10)$$

where x_i is the size of the i-th output value; K is the total number of categories.

The loss function and the objective function of the convolutional neural networks, used to evaluate the deviation between the model predicted value and the true value, are the evaluation criteria for the model training results. In the process of qualitative discrimination analysis, the cross-entropy function is used as the loss function. The smaller the value of the loss function is, the smaller the distance between the prediction result and the realism category. The cross-entropy function in this study is shown in Eq. 16.11.

$$\text{Loss} = -\frac{1}{N}\sum_{n=1}^{N}[y_n \log \hat{y}_n + (1-y_n)\log(1-\hat{y}_n)] + \alpha\|\omega\|^2 \quad (16.11)$$

where N is the total number of training samples; y_n is the actual classification category of the sample; $\hat{y}_n$ is the prediction classification category of the sample; α is the regularization coefficient, which is 0–1; ω is the weight to be regularized.

16.1.5 Model Evaluation

16.1.5.1 Evaluation Index of the Quantitative Model

Determination Coefficient The determination coefficient (R^2) is also known as the goodness of fit. It represents the degree to which the independent variable interprets the dependent variable in the model, with values ranging from 0 to 1. The closer the value of R^2 is to 1, the better fit of the model is. The calculation formula is as follows:

$$R^2 = 1 - \frac{\sum_{i=1}^{n}(\hat{y}_i - y_i)^2}{\sum_{i=1}^{n}(\hat{y}_i - \bar{y})^2} \quad (16.12)$$

where $\hat{y}_i$ and y_i represents the predicted value and actual value of the ith sample, respectively, $\bar{y}$ is the average of the actual values of all samples.

Root Mean Square Error of Calibration Set The correction set root mean square error (RMSEC) is the root mean square error of the model prediction on the correction set sample which is used to evaluate the prediction ability of the model. The calculation formula is as follows:

$$\text{RMSEC} = \sqrt{\frac{\sum_{i=1}^{n}(\hat{y}_i - y_i)^2}{n}} \quad (16.13)$$

where $\hat{y}_i$ and y_i represent the predicted value and actual value of the i-th sample, respectively, and n represents the number of samples in the correction set.

Root Mean Square Error of Cross-Validation The root mean square error of cross validation (RMSECV) is used to evaluate the prediction ability of the obtained model by calculating the root mean square error in the process of cross validation. The interactive verification steps are as follows: (i) Select one or a group of samples i from the calibration set, and remove the spectral matrix x_i and concentration matrix y_i of the sample from the calibration set; (ii) The new correction set composed of the remaining samples is used to establish the model; (iii) Use the model to predict the rejected samples and obtain the predicted value $\hat{y}_i$; (iv) Remove another sample or group of samples from the original correction set, and go back to step (ii) for circular calculation until the prediction of all samples in the correction set is completed. The calculation formula of RMSECV is as follows:

$$\text{RMSECV} = \sqrt{\frac{\sum_{i=1}^{n}(\hat{y}_i - y_i)^2}{n}} \tag{16.14}$$

Root Mean Square Error of the Validation Set The root mean square error of prediction set (RMSEP) of the validation set is the RMSEP error of the model prediction to the validation set sample, which is used to evaluate the prediction ability of the model to the external samples. The calculation formula is as follows:

$$\text{RMSEP} = \sqrt{\frac{\sum_{i=1}^{n}(\hat{y}_i - y_i)^2}{n}} \tag{16.15}$$

where $\hat{y}_i$ and y_i represent the predicted value and actual value of the i-th sample, respectively, and n represents the number of samples in the correction set.

Bias Bias is the error between the predicted value of the prediction model and the true value of the sample, reflecting the accuracy of the model and the fighting ability of the algorithm. The calculation formula is as follows:

$$\text{Bias} = \frac{\sum_{i=1}^{n}(\hat{y}_i - y_i)}{n} \tag{16.16}$$

16.1.5.2 Evaluation Indicators of the Qualitative Model

The main evaluation indicators of the classification models include the confusion matrix, specificity, sensitivity, accuracy, and overall accuracy. Among them, the confusion matrix and the overall accuracy are the evaluation indicators of the classification model population. Accuracy, sensitivity, and specificity are the evaluation indicators of the classification effect of the classification model on each class of the sample.

Confusion Matrix Confusion matrix is the most basic method to evaluate the classification models. The confusion matrix displays the classification results of each class sample in the same table as the actual category, which can most intuitively reflect the prediction results. Table 16.1 presents the format of the confusion matrix with a 3-classification result:

Table 16.1 The confusion matrix.

Confusion matrix		Predicted value		
		Class 1	Class 2	Class 3
True value	Class 1	a	b	c
	Class 2	d	e	f
	Class 3	g	h	i

Specificity Taking the specificity of category 1 as an example, the specificity represents the proportion of all outcomes of non-category 1 results, which the model predicts as non-category 1 results, reflecting the model's ability to test negative samples. The calculation formula is as follows:

$$\text{Specificity}_{class1} = \frac{e+f+h+i}{d+e+f+g+h+i} \tag{16.17}$$

Sensitivity Take the sensitivity of category 1 as an example, where the sensitivity indicates that the true value is among all the results of category 1, and the model predicts the correct proportion, reflecting the ability of the model to detect positive samples. The calculation formula is as follows:

$$\text{Sensitivity}_{class1} = \frac{a}{a+b+c} \tag{16.18}$$

Precision With the precision of category 1, the precision represents the proportion of the correct prediction of all results that the model predicts as category 1. The calculation formula is as follows:

$$\text{Precision}_{class1} = \frac{a}{a+d+g} \tag{16.19}$$

Accuracy Accuracy represents the overall accuracy of all categories in the sample, and the calculation formula is as follows:

$$\text{Accuracy} = \frac{a+e+i}{a+b+c+d+e+f+g+h+i} \tag{16.20}$$

16.2 Application of Machine Learning for Data Mining in Metallomics

16.2.1 Applications in Medical Science

Elemental homeostasis is maintained in healthy people while diseases may cause or be caused by elemental dyshomeostasis. Therefore, studying the elemental homeostasis through metallomics may help understand the diseases and find way to treat them.

He et al. [12] distinguished healthy people from cancer patients using serum samples with non-targeted metallomics through X-ray fluorescence spectroscopy (XRF) combining machine learning algorithms. In order to reduce the unnecessary noise interference of the XRF spectra, eliminate the error caused by optical path difference and scattering, and highlight the difference between spectral lines, smoothing, standard normal variable transformation (SNV), and derivative processing were used as pretreatment means. The data dimensionality reduction method (PCA) was used to reduce the dimension of the original data, retained the first three principal components to express more than 90% of the original spectral information, and formed two-dimensional and three-dimensional scatter plots to analyze the classification of healthy people and cancer patients. In order to further improve the classification accuracy, a linear model (PLS-DA) and a nonlinear model (SVM) were established with full spectra and the first three principal components as model inputs to distinguish healthy people from cancer patients. Because there was redundant information in the full spectral information, the successive projection algorithm (SPA) was used to select XRF feature information bands, and a simplified machine learning model was established. Through analyzing the selected feature information bands, it was found that the fluorescence channels correspond to Ca, Cu, and Zn levels in the serum, which is important information about the difference between the serum components of healthy people and patients, and is consistent with previous studies. It demonstrated that the non-targeted metallomics approach through SRXRF with machine learning algorithms is fast (5s for data collection for one sample) and accurate (over 96% accuracy) for the screening of cancers using blood samples (Figure 16.3). The combination of machine learning algorithm and XRF spectral data is feasible to quickly distinguish healthy people from cancer patients.

In addition, Chen et al. [13] classified normal and cardiovascular disease patients by studying the metal concentration in blood/urine samples with machine learning algorithms. Fisher linear discriminant analysis (FLDA), support vector machine

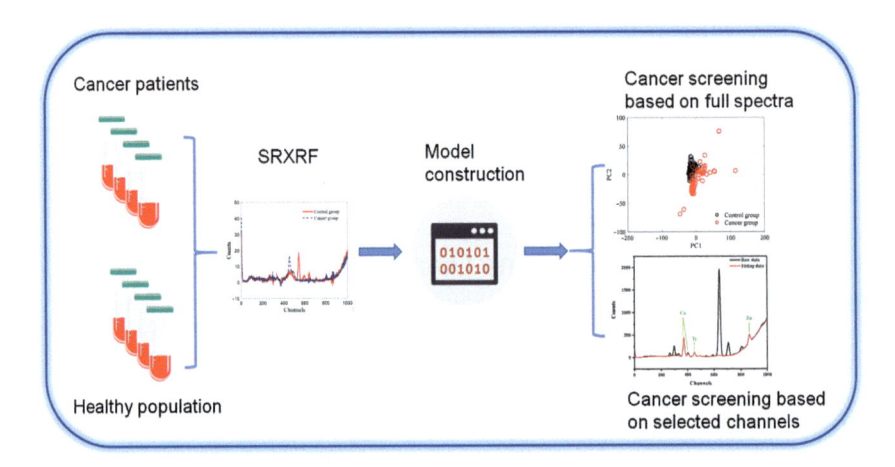

Figure 16.3 The combination of SR-XRF and machine learning was used for cancer screening. Source: Reproduced with permission from Elsevier.

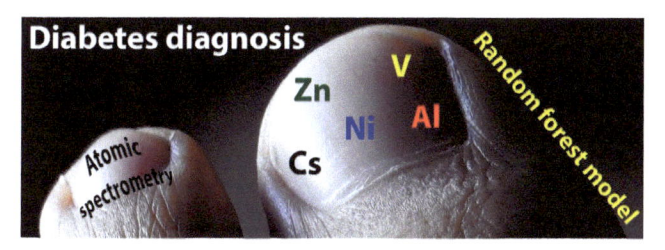

Figure 16.4 Differences in nail elemental composition between healthy and type 2 diabetes patients. Source: Jake et al. 2019/Reproduced with permission from Elsevier.

(SVM), and decision tree (D-Tree) were used to analyze the concentrations of nine elements (i.e. chromium, iron, manganese, aluminum, cadmium, copper, zinc, nickel, and selenium). Principal component analysis (PCA) was used for preliminary classification analysis. Then, a series of classifiers were constructed and compared. In terms of accuracy, sensitivity, and specificity, D-Tree classifier hold the best overall performance, followed by SVM and FLDA. In addition, the analysis of blood samples was superior to that of urine samples. The combination of D-Tree classifiers with blood sample element analysis can be used for cardiovascular disease screening, especially in routine physical examination. Carter et al. [14] studied the elemental levels in the toenails of diabetes patients with machine learning algorithms for the robust classification of type 2 diabetes. It was found that the concentrations of aluminum, cesium, nickel, vanadium, and zinc in the toenails of healthy volunteers and patients with type 2 diabetes were significantly different. Through the fusion of sample concentration and different machine learning algorithms, data mining is carried out to distinguish diabetes patients (Figure 16.4).

Liu et al. [15] studied seven machine learning algorithms for blood cell classification using single-cell Raman spectroscopy. With the Laser Target Recognition System (LTRS) in Raman spectroscopy, seven species were collected without disrupting red blood cells (RBCs). Seventy candidate features were extracted using the Boruta algorithm, and the effects of the top 30 features were selected using the seven classifiers based on cross-validation. These features and the corresponding information, such as chemical functional groups and chemical bonds, can illustrate which specific vibration patterns affect the performance of the classification. Finally, the best classification model was identified for the validation of the test set. Classification metrics showed that SVM and ANN ranked highest in the evaluation data. In addition to the machine learning algorithms described here, deep learning algorithms can handle more data and the classification performance can be further improved. For example, it was noticed that the neural network algorithm is the most accurate method. Figure 16.5 shows single-cell Raman spectroscopy for blood classification using different machine learning and deep learning algorithms.

The above examples show the data mining can be effectively applied in one-dimensional vector like the elemental levels in biological samples. Machine learning algorithms are also applicable for data mining tasks involving two-dimensional matrices. For example, with the development of medical

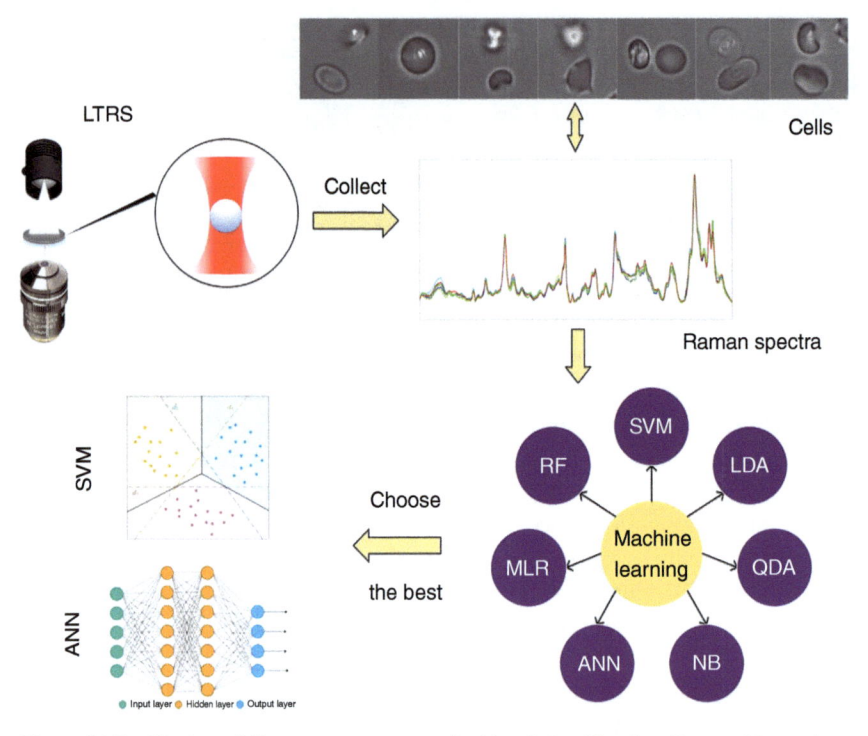

Figure 16.5 Single-cell Raman spectroscopy for blood classification. Source: Liu et al. [15]/Reproduced with permission from Elsevier.

imaging technology, more and more medical images are obtained while data mining with machine learning algorithms can help risk assessment, detection, diagnosis, prognosis, and treatment response through medical images. Kline et al. [16] summarized the application of artificial neural networks, k-close neighbor, SVM, D-tree, and Navier–Stokes for medical imaging. Park et al. [17] used machine learning and radiological features extracted from preoperative Magnetic Resonance Imaging (MRI) in patients with Oropharyngeal Squamous Cell Carcinoma (OPSCC) to predict pathological factors and treatment outcomes. The area under the subject operating characteristic curve of the logistic regression model for predicting human papillomavirus (HPV) was 0.792. The Light-GBM model predicted the HPV with the Area Under Curve (AUC) of 0.8333. It showed that machine learning models using MRI radio mics perform well in predicting pathological factors and treatment effects in patients with OPSCC. However, the study on the spatial distribution of elements in medical samples with machine learning and deep learning algorithms has not been rarely reported, which deserves further study.

16.2.2 Applications in Agricultural Science

Cd, As, Pb, Cr, and Hg are typical heavy metals in soils, which can enter the human body through food chain and cause a variety of health problems. At the same time, farmland heavy metal pollution will also lead to declined crop yield and quality.

Enzyme-linked immunosorbent assay (ELISA) and Gold Immune Colloidal Technique (GICT) are two relatively mature and widely used immunoassay methods, which use antigen antibodies to specifically identify the measured substances for qualitative and quantitative screening of heavy metals/metalloids. Ouyang [18] developed a Cu chemiluminescence immunosensor based on the dual specificity of antibody recognition mechanism and the sensitive mechanism of luminol hydrogen peroxide system to heavy metals, which has realized the detection of Cu in Chinese herbal medicine, with the detection range of 1.0~1000 µg/kg, detection limit of 0.33 µg/kg. However, it is worth noting that because the sensors need bioactive substances and other antibodies to participate in the recognition work, most sensors have problems such as short life and strict working conditions. Besides, for some heavy metals or metalloids, such as Cr and As, their chemical species have different toxicological implications [19, 20]. Therefore, in the quality and safety assessment of agricultural products and related research, the chemical form of heavy metals is more significant than the total amount of heavy metals [21].

X-ray fluorescence spectroscopy (XRF) with high-energy X-ray excites sample and qualitatively and quantitatively studies the metal element content in the sample according to the fluorescence spectral characteristics and intensity emitted by the sample. Romero et al. [22, 23] used XRF to analyze the elements in the samples of Mexican dry pepper varieties. The samples of 9 pepper varieties were analyzed and the contents of 12 elements (P, S, Cl, K, Ca, Mn, Fe, Cu, Zn, Br, Rb, and Sr) were determined. This work identified the differences in element concentrations for each type and compared them with previously analyzed pepper samples. XRF analysis was performed with Rh X-ray tube spectrometer and Si PIN detector. The differences of different pepper variety were extracted by cluster analysis and principal component analysis.

Anode dissolution voltammetry is an effective method to measure trace heavy metal ions. Ye et al. [24] used machine learning technology to learn and model the voltammetric peak shapes of different heavy metal ions, which improved the selectivity of anodic stripping voltammetry in detecting various heavy metal ions in complex water samples. Leon-Medina et al. [25] developed a technology for rapid detection of As, Pb, and Cd by using pulse voltammetry combined with machine learning with the accuracy of 98.31%.

The rapid detection of metal in soil and agricultural products plays an important role in the quality and safety monitoring of agricultural products and the safe utilization of farmland. Besides, with the more data in agricultural science, machine learning algorithms for data mining are also highly required. More works are desired for data mining with machine learning, deep learning, and also artificial intelligence in metallomics for agricultural science.

16.2.3 Applications in the Environmental Science

Environmental pollution, especially the heavy metal pollution, is a serious issue that affects the environmental health and human health. Heavy metals refer to those density greater than or equal to $4.5\,g/cm^3$, coming from energy, mining, chemical,

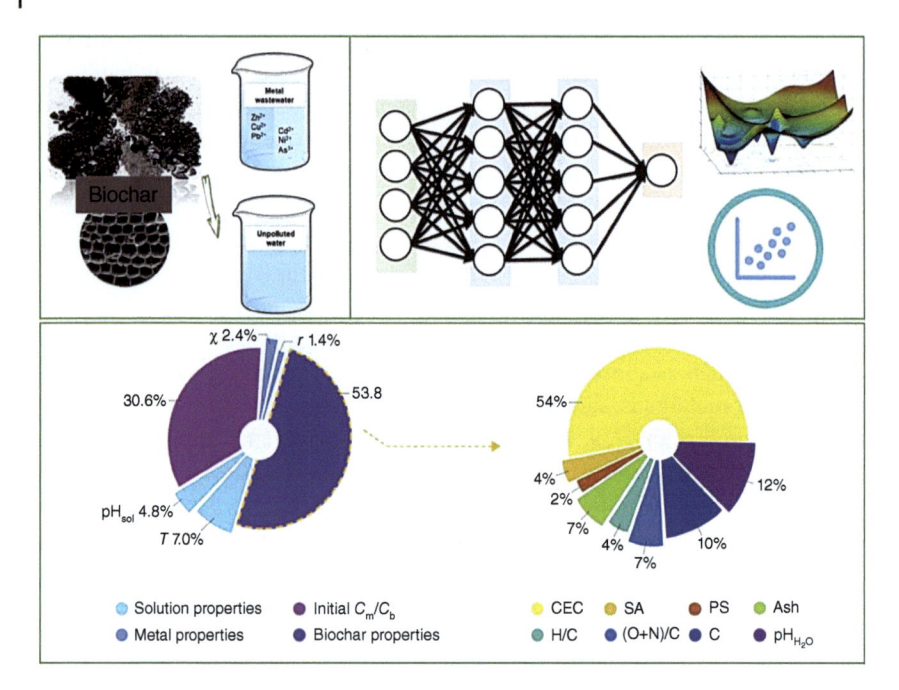

Figure 16.6 Flow chart for prediction the biochar adsorption capacity for heavy metals. Source: Zhu et al. [26]/Reproduced with permission from Elsevier.

electroplating, medical, printing and dyeing, papermaking, pesticides, fertilizers, pigments, and metal smelting and processing industries. In addition, rock weathering and riverbank erosion in nature may also cause some heavy metals into the water.

Biochar was applied to remove heavy metals in water. Zhu et al. [26] used RF and ANN models to predict the adsorption efficiency of biochar for heavy metal ions based on the properties of biochar and solution environment. It was showed that due to the small number of samples and the relatively high-dimensional characteristics, RF was slightly better than ANN, and the prediction accuracy was $R^2 = 0.973$. Figure 16.6 shows the flow chart for prediction the biochar adsorption capacity for heavy metals. Lv et al. [27] used ICP-MS and deep learning algorithms to trace the source of heavy metals in soils. The accuracy of 100% and 97%, respectively, was achieved through heavy metal content in combination with SVM algorithm and naive Bayesian discriminant model.

Dai et al. [28] provided a scheme for the content of ore-forming elements in large-scale surface soil based on existing data, taking 80% of samples as the training set and 20% as the test set, in view of the important indicative role of soil ore-forming elements in mineral resources exploration. Eight elements (K, B, Ni, V, Zn, As, Co, Cu) were selected as predictors by the combination method of variable importance measure ranking and learning curve construction. The goodness-of-fit R^2 of the model to the training data and test data reached 0.9832 and 0.8956. This study shows that it is feasible to introduce random forest algorithm into spatial quantitative prediction of surface soil geochemical element content, which can further expand the service and application dimension of land quality geochemical data.

Wang et al. [29] took farmland soils around two typical industrial and mining plants in Jiangxi Province and collected 19 complete soil profile samples with a depth of about 100 cm in order to explore the potential of visible-near infrared (VNIR) spectroscopy in predicting the content of heavy metals in undisturbed soil profiles. The VNIR data and Cu content of soil profile samples were measured. The Partial Least Squares Regression (PLSR), Cubist Regression Tree, Gaussian Process Regression (GPR), and Support Vector Machine Regression (SVR) were used to study the influence of different spectral pretreatment methods on the prediction accuracy of soil Cu content. It was showed that the prediction accuracy of Cubist, GPR, and SVM was generally higher than that of PLSR, and the First-Order Derivative had the highest prediction accuracy ($R^2 = 0.95$, root mean square error of 7.94 mg/kg, and relative analysis error of 4.34). This shows that using VNIR and machine learning can effectively predict the Cu content in native soil profiles, providing a reference for the rapid monitoring of Cu and other heavy metals.

In all, machine learning algorithms can be used for data mining in metallomics in different research field, which can mine system features through limited experimental data. Besides, the cutting-edge artificial intelligence represented by deep learning algorithms can also be applied to achieve the best prediction accuracy.

References

1 Famili, F., Shen, W.M., Weber, R., and Simoudis, E. (1997). *Intell. Data Anal.* 1: 3.
2 Eilers, P. (2003). *Anal. Chem.* 75: 3631.
3 Bolstad, B.M., Irizarry, R.A., Astrand, M., and Speed, T.P. (2003). *Bioinformatics* 19: 185.
4 Bailey, D.H. and Swarztrauber, P.N. (1994). *SIAM J. Sci. Comput.* 15: 1105.
5 Akansu, A.N. and Haddad, R.A. (2001). Chapter 6 - Wavelet Transform. In: *Multiresolution Signal Decomposition*, 2e (ed. A.N. Akansu and R.A. Haddad), 391. San Diego: Academic Press.
6 Jolliffe, I.T. (2002). *J. Mark. Res.* 87: 513.
7 Davison, M.L. (1983). *Multidimensional Scaling*, vol. 34 (ed. M. Lovric), 1463. https://doi.org/10.2307/2988171.
8 He, X. (2003). *Adv. Neural Inf. Process. Syst.* 16: 186.
9 Laurens, V.D.M. and Hinton, G. (2008). *J. Mach. Learn Res.* 9: 2579.
10 Galvão, R.K.H., Araujo, M.C.U., José, G.E. et al. (2005). *Talanta* 67: 736.
11 Yu, J., Dai, Q., and Li, G. (2022). *Drug Discovery Today* 27: 1796.
12 He, L., Lu, Y., Li, C. et al. (2022). *Talanta* 245: 123486.
13 Chen, H., Tan, C., Lin, Z. et al. (2013). *Comput. Biol. Med.* 43: 865.
14 Tepanosyan, G., Sahakyan, L., Maghakyan, N., and Saghatelyan, A. (2020). *Environ. Pollut.* 261.

15 Liu, Y.M., Wang, Z.Q., Zhou, Z.H., and Xiong, T. (2022). *Spectrochim Acta, Part A* 277.
16 Erickson, B.J., Korfiatis, P., Akkus, Z., and Kline, T.L. (2017). *RadioGraphics* 37: 505.
17 Park, Y.M., Lim, J.Y., Koh, Y.W. et al. (2022). *Head Neck-J. Sci. Spec.* 44: 897.
18 Ouyang, H., Shu, Q., Wang, W.W. et al. (2016). *Biosens. Bioelect.* 85: 157.
19 Sall, M.L., Diaw, A., Gningue-Sall, D. et al. (2020). *Environ. Sci. Pollut. Res.* 27: 29927.
20 Yang, Y., Peng, Y.M., Ma, Y.B. et al. (2022). *Environ. Pollut.* 296.
21 Luvonga, C., Rimmer, C.A., Yu, L.L., and Lee, S.B. (2020). *J. Agri. Food Chem.* 68: 943.
22 Romero-Davila, E., Miranda, J., and Pineda, J.C. (2020). *J. Food Compos. Anal.* 93.
23 Romero-Davila, E., Miranda, J., and Pineda, J. C. (2015). *Radiation physics: XI International Symposium On Radiation Physics*, Vol. 1671 (Eds: G. Espinosa, C. V. Lopez, J. A. Lopez), 11th International Symposium on Radiation Physics (ISRP).
24 Ye, J., Lin, C., and Huang, X. (2020). *J. Electroanal. Chem.* 872: 113934.
25 Leon-Medina, J.X., Tibaduiza, D.A., Burgos, J.C. et al. (2022). *IEEE Access* 10: 7684.
26 Zhu, X., Wang, X., and Ok, Y.S. (2019). *J. Hazard. Mater.* 378: 120727.
27 Lv, R.L., Jia, Z., He, H.Y. et al. (2021). *Chem. Res. Appl.* 33: 1776.
28 Dai, L., Nie, X., Guo, J. et al. (2022). *Geoscience* 36: 972.
29 Wang, Y., Xu, S., Zhao, Y., and Shi, X. (2022). *Trans. Chin. Soc. Agric. Eng.* 38: 336.

Index

a

accuracy, qualitative model   463
adsorption chromatography   12
affinity chromatography   13, 42, 313
agricultural animal and derived food
   elemental speciation and state analysis   135–137
   inductively coupled plasma–mass spectrometry (ICP-MS)   132–133
   inductively coupled plasma optical emission spectroscopy (ICP/OES)   134
   instrumental neutron activation analysis (INAA)   135
   laser ablation inductively coupled plasma mass spectrometry (LA-ICP-MS)   137–138
   laser-induced breakdown spectroscopy (LIBS)   139
   ultrasonic slurry sampling electrothermal vaporization inductively coupled plasma mass spectrometry (USS-ETV-ICP-MS)   138
   X-ray fluorescence (XRF) technique   134–135
agrimetallome   51
agrometallomics   2
   agricultural animal and derived food   131–139
   agricultural plants and fungi and derived food   127–131
   atomic spectrometry
      low temperature plasma atomic spectrometry   119–120
      optical emission spectrometry   119
   chromatographic hyphenation for atomic spectrometry/mass spectrometry   121–122
   concept   51–52
   electrothermal vaporization hyphenation technique   125
   energy spectroscopy based on X-ray   123–124
   instrumental neutron activation analysis (INAA)   120
   laser ablation inductively coupled plasma mass spectrometry (LA-ICP-MS)   124–125
   laser-induced breakdown spectroscopy (LIBS)   125–126
   logical diagram of agricultural system   49–50
   mass spectrometry
      glow discharge mass spectrometry (GD-MS)   118
      inductively coupled plasma mass spectrometry (ICP-MS)   52–117
      laser ionization mass spectrometry (LIMS)   118
      secondary ion mass spectrometry (SIMS)   118–119
      thermal ionization mass spectrometry (TIMS)   118
   single cell and micro particle analysis   126–127
   soil, water, fertilizer   139–143
   summarization and comparison of analytical methodologies   52–53
   synchrotron radiation (SR) analysis   122–123
   X-ray fluorescence spectrometry (XRF)   120–121
alpha-fetoprotein (AFP)   165
Alzheimer's disease (AD)
   application of metallomics in the predication and diagnosis of   321–322
   dysregulation of metal homeostasis in   315–320
   metal-associated dysfunction in   320–321
   metal ionophores as AD therapeutics   322

Applied Metallomics: From Life Sciences to Environmental Sciences, First Edition.
Edited by Yu-Feng Li and Hongzhe Sun.
© 2024 WILEY-VCH GmbH. Published 2024 by WILEY-VCH GmbH.

amyotrophic lateral sclerosis (ALS) 209
  dysregulation of metal homeostasis in 333–334
  metal-associated dysfunction in 334–336
anode dissolution voltammetry 467
anodic stripping method 198
archaeometallomics 2
  atomic absorption spectroscopy (AAS) 267
  in studying the aging mechanism of painting cultural relics 276–278
  in studying the authenticity identification of Painting cultural relics 278–279
  in studying the color mechanism and firing technology of ancient ceramics 271–272
  in studying the corrosion of metal cultural relics 275
  in studying the manufacturing technology of metal cultural relics 274–275
  in studying the origin and dating of ancient ceramics 269–271
  in studying the origin of metal cultural relics 273–274
  laser ablation inductively coupled plasma mass spectrometry (LA-ICP-MS) 267
  laser-induced breakdown spectroscopy (LIBS) 267
  neutron activation analysis (NAA) 266
  neutron diffraction 268–269
  X-ray absorption fine structure spectroscopy (XAFS) 267–268
  X-ray diffraction (XRD) 268
  X-ray fluorescence analysis (XRF) 266–267
arsenic and lung cancer 291–292
artificial crystal materials 257–258
As and related diseases 222
asymmetric flow-field flow fractionation (AF4) 13, 17
atomic absorption spectrophotometer/spectroscopy (AAS) 120–122, 125, 127, 130, 132, 196, 198, 200, 201, 266, 267, 350, 351, 353, 384
atomic emission spectrometry 125, 131, 197, 200, 250–251
atomic fluorescence spectrometry 120, 135, 197, 250
atomic spectrophotometry 196
atomic spectroscopy detection technology 196–197, 200–201
autism spectrum disorders (ASD) 312, 336–338, 368

b

beneficial rebalancing hypothesis for hormesis effect 370–371
bias 205, 206, 430, 431, 462
bimetallic/polymetallic doping 239, 241
bio-elementomics
  biological fractionation 366–367
  concept 363
  development history of 363–364
  diversity of 366
  Hormesis beneficial rebalancing hypothesis 374
  interactions with proteins, genes, and small molecules 373
  limitations of 371–373
  multi-element analysis of immortalized lymphocytes 374
  omes 367
  orderliness of 366
  organizational selectivity of 365
  rare earth elementome 367–371
  research scope 364
  review of 364
  single-cell analysis 374
  speciation analysis of elements 373
  specific correlation of 365
biological fractionation 366–367
biometals 285–287, 377–378, 381–387
  research on 377–378
biomolecular quantification at single cell level 417–418
bonded phase chromatography 12–13
breast cancer 162, 214, 351–352, 417

c

cadmium (Cd), epigenetic effects of 292–294
calcium-based-magnetic biochar (Ca-MBC) 141
Canadian Health Measures Survey (CHMS) 205
cancer
  clinimetallomics 213–214
  diagnostic methods 349
capillary electrophoresis (CE) 13, 40, 122, 128, 141, 200, 201, 313, 385
carbon-containing gaseous species (CCGS) 171
carbon containing particles (CCP) 171
Carboplatin 226, 355, 417
cardiovascular disease 217–219, 223, 293, 464, 465
case-control medimetallomics 205
Cd and related diseases 222–223
CdSe@ZnS core/shell QDs 19
cell heterogeneity 391, 417, 435
certified reference material (CRM) 125, 153, 159, 168, 246, 434
ceruloplasmin 331, 332, 337
cervical cancer 351, 354

chemotoxicity   294–296
chromatography   12
  adsorption   12
  affinity   13, 42, 313
  hyphenation for atomic spectrometry/
    mass spectrometry   121–122
chronic kidney disease(CKD)   216, 217
circular dichroism (CD) spectroscopy   16, 17,
    221
cisplatin   225, 226, 355, 357, 380, 416
clinical element morphology and valence
    analysis technology   199–203
clinimetallomics   195
  atomic spectroscopy detection technology
    196–197
  cancer   213–214
  cardiovascular disease   217–218
  combined toxicity of multiple heavy metal
    mixtures   223–224
  electrochemical analysis   198
  genetic diseases associated with metallomics
    224
  ischemic heart disease   221
  kidney disease   216–217
  liver diseases   215–216
  mass detection technology   197–198
  metabolic diseases   211–212
  neurodegenerative diseases   209–211
  obesity   214–215
  toxic element related diseases   221–222
Clioquinol   322–323
cobalt nanoparticle (CoNP)   14, 18
colorectal cancer   214, 351, 353, 355
comparative nanometallomics   11, 12, 20–21
confusion matrix   462–463
convolution operation   452, 459, 460
copper, Parkinson's disease (PD)   330–331
Cr and related diseases   223
cross-validation   456, 462, 465
Cu specific chelators   323
cytometry time of flight (CyTOF)   36, 37, 432

d

data dimensionality reduction
  independent component analysis (ICA)
    453–454
  local preserving projection (LPP)   454
  multidimensional scaling (MDS)   454
  principal component analysis (PCA)   453
  T-Stochastic Neighbor Embedding (T-SNE)
    454
data mining methods in metallomics
  data dimensionality reduction   452–453
  data preprocessing   450
  model evaluation   461–463
  predictive model building method   457–461

  sample set division   455–457
data preprocessing
  convolution operation   452
  Fourier transform   451
  normalization   450–451
  smoothing process   450
  wavelet transform   451–452
DBiT-seq   441
Decision Tree   458, 465
deep learning (DL)   351, 459–461, 465–469
deferiprone (DFP)   322, 333
deferoxamine (DFO)   322, 333
desorption electrospray ionization (DESI)
    185, 186
determination coefficient   461
dielectrophoresis   402
direct current plasma (DCP)   251
droplet based single cell separation   402
droplet encapsulation based single cell
    separation   403–406
droplet-chip-TRA-ICP-MS single cell analysis
    system   415
dynamic light scattering (DLS)   13, 14
dynamic reaction cell (DRC) system   171, 251,
    252
dyslipidemia   219–221

e

electrochemical analysis   196, 198, 200
electron-probe micro-analysis (EPMA)   247,
    253
electron spectroscopy for chemical analysis
    (ESCA)   123, 129, 142
electrophoresis   12, 13
  capillary   13
  gel   13
electrospray ionization mass spectrometry
    (ESI-MS)   40–42, 128, 135–137, 154,
    163, 164, 186, 200, 202, 203
electrothermal atomic absorption spectrometry
    (ETAAS)   196, 197
electrothermal vaporization (ETV)   119, 120,
    125, 130, 138–140, 142
  hyphenation technique   125
elemental homeostasis   12, 20, 349, 463
elemental speciation and state analysis
    121–124, 135–137
element coded affinity tags (ECAT) method
    163
endogenous elements in single cells   409
energy dispersive spectrometer (EDS)   15, 17
energy dispersive X-ray fluorescence (EDXRF)
    39, 121, 269, 271
energy spectroscopy based on X-ray   123–124
enhanced green fluorescent protein (EGFP)
    417

environmental protection policy 194
environmentallomics 2, 194
environment-independent biomarkers 372
environmetallome 34–35, 51
　defined 34–35
environmetallomics
　concept of 33–34
　environmental science and ecotoxicological science 43–44
　metal distribution and mapping for 37–39
　metalloprotein analysis 41–43
　metal speciation 39–41
　quantitative analysis 35–37
　requirements for 34–35
　scope of 34, 35
enzyme-linked immunosorbent assay (ELISA) 164, 467
esophageal cancer 214, 351, 353, 355, 357
ex vivo culture 381
exogenous metal exposure to single cells 409–414
external calibration 125, 159, 166, 167, 171, 174–182, 185, 257, 441

f

α-fetoprotein 434
field-flow fractionation (FFF) 12, 13, 40
flame atomic absorption spectroscopy (FAAS) 119, 140, 196, 197, 250, 449
Fourier transform 278, 451, 452

g

gallbladder cancer (GBC) 214
gas exchange device (GED) 171
gastric cancer 214, 351–353, 355
Gauss filtering 450
GD mass spectrometry (GD-MS) 118, 140, 253–255
GD optical emission spectroscopy (GD-OES) 253–256
gel electrophoresis (GE) 13, 40, 42, 135, 155, 184, 200, 313, 380, 385, 430
genetic diseases associated with metallomics 224
glow discharge mass spectrometry (GD-MS) 118, 140, 253–255
Gold Immune Colloidal Technique (GICT) 467
gold nanoparticles (AuNPs) 13, 38, 165, 166
graphite furnace atomic absorption spectrophotometer (GF-AAS) 198

h

hepatocellular carcinoma (HCC) 353
Hg and related diseases 222
high energy resolution fluorescence detected (HERFD) XAS 16, 17
high-performance liquid chromatography isotope dilution inductively coupled plasma mass spectrometry (HPLC-ID-ICP-MS) 168
high-resolution double-focusing sector-field ICP-MS (HR-ICP-SF-MS) 252
high-temperature furnace atomic absorption spectrometry 196
Hormesis beneficial rebalancing hypothesis 374
Hormesis effect of REEs 369–371
Huntington's disease (HD) 209, 311
hydrodynamic capture based single cell separation 407–408
hydrodynamic chromatography (HDC) 12, 13, 17, 40
hydrodynamic method-based cells capture 402
hypertension 33, 215, 217–219, 293, 368, 369
hyperthyroidism 212
hypothyroidism 212

i

imaging mass cytometry (IMC) 442, 443
imaging tools for metallomics 383, 386–387
immobilized metal affinity chromatography (IMAC) 42, 313
immortalized lymphocytes 372–374
Independent Component Analysis (ICA) 453–454
inductively coupled plasma (ICP), elemental or molecular ions 154
inductively coupled plasma atomic emission spectrometry (ICP-AES) 2, 134, 140, 197, 198, 200, 201, 250–251, 257, 273, 364
inductive coupled plasma optical emission spectroscopy (ICP-OES) 2, 35, 36, 119, 121, 125–128, 130, 132–134, 139, 140, 200, 350–352
inductively coupled plasma mass spectrometry (ICP-MS) 2, 13, 35, 52, 132, 139, 156, 198, 251, 312, 350
　applications of 409, 410
　based metallomics
　　single cell analysis 435–441
　　single particle analysis 433–435
　　spatial metallomics 441–442
　based method 164, 167, 420, 430, 435
　biomolecular quantification at single cell level 417–418
　biomolecules analysis 160, 161
　endogenous elements in single cells 409
　exogenous metal exposure to single cells 409
　laser ablation (LA) 399–400, 432–433
　mass analyzer and detector 400–401

metal-containing drugs uptake by single cells 416–417
metalloproteins 167–168
microfluidic platform for single cell analysis 401–408
nanoparticles uptake by single cells 415–416
other applications 418–419
pneumatic nebulization 392, 399
solution introduction system and plasma source 430–431
time-of-flight mass analyzer 431–432
instrumental neutron activation analysis (INAA) 120, 132, 134–135, 139, 199
integrated biometal science 1, 50, 311, 364
internal standardization
  calcium 172
  carbon 170–171
  sulfur 171–172
ion chromatography 12, 121
iron, Parkinson's disease (PD) 324, 330
ischemic heart disease (IHD) 221
isothermal circular strand-displacement polymerization reaction (ICSDPR) 165
isotope dilution 163, 166–168, 182–185, 384, 441
isotope-dilution analysis (IDA) 155
isotope dilution mass spectrometry (IDMS) 125, 138, 155, 159, 166, 182, 184

k
Kennard-Stone sampling 455
kidney disease 216–217, 293, 365
K-means clustering technique 458–459

l
large research infrastructures (LRIs) 2, 3
laser ablation (LA) 124, 155, 399, 432–433, 442
  system 155
  LA-ICP-MS 13, 36
laser ablation inductively coupled plasma mass spectrometry (LA-ICP MS) 17, 37, 124–125, 137, 142, 155, 158, 165, 243, 246–247, 266, 267, 386
laser-induced breakdown spectroscopy (LIBS) 21, 125–126, 131, 138, 139, 142, 243, 247, 255, 266, 267, 270
laser ionization mass spectrometry (LIMS) 118, 124, 140
Leave-One-Out Cross Validation 457
limit of detection (LOD) 36, 37, 118, 119, 123, 126, 129, 142, 160, 164, 165, 173, 179, 418

liquid chromatography (LC) 12, 40–42, 119, 121–122, 126–128, 140, 155, 313, 380, 384–385
liver cancer 222, 351, 353, 355–357
liver diseases 215–216
Local Preserving Projection (LPP) 453, 454
low temperature plasma atomic spectrometry 119–120
lung cancer (LC) 207, 213, 222, 291–292, 351, 352, 355–356

m
machine learning for data mining in metallomics
  agricultural science 466–467
  environmental science 467–469
  medical science 463–466
magnetic separation 402–403, 418
  based single cell capture 408
mass analyzer and detector 197, 400–401
mass detection technology 197–198
mass spectrometry 13
  based technique 12, 13, 36, 37
  detection technology 201–203
Materials Genome Initiative (MGI) 237–238, 242
matermetallomics 2
  artificial crystal materials 257–258
  GD mass spectrometry (GD-MS) 253–254
  GD optical emission spectroscopy (GD-OES) 253
  Inductively Coupled Plasma Atomic Emission Spectrometry (ICP-AES) 250–251
  Inductively Coupled Plasma Mass Spectrometry (ICP-MS) 251–252
  Laser Ablation Inductively Coupled Plasma Mass Spectrometry (LA-ICP-MS) 246–247
  laser-induced breakdown spectroscopy (LIBS) 247
  metallic elements as crosslinkers 242–243
  metallic elements as dopant 239–241
  metallic elements as impurities 241–242
  Raman spectroscopy 254–255
  secondary ion mass spectrometry (SIMS) 247–248
  semiconductor materials 256–257
  Synchrotron Radiation X-ray Fluorescence Spectrometry (SR-XRF) 249
  techniques providing depth information 255–256
  TEM/X-EDS 248–249
  X-ray fluorescence (XRF) spectrometry 252–253
  X-ray photo electron spectroscopy (XPS) 255

medimetallomics 195
  atomic spectroscopy detection technology 196–197
  Canadian Health Measures Survey (CHMS) 205
  clinical element morphology and valence analysis technology 199–203
  electrochemical analysis 198
  longitudinal study 208
  mass detection technology 197–198
  National Health and Nutrition Examination Survey (NHANES) 204–205
  neutron activation analysis (NAA) 198–199
  speciation analysis of trace metal element 208
  study design 205–206
  study population/recruitment criteria 206
  susceptible population 207–208
metabolic diseases 211–212
metal-associated-proteins, Parkinson's disease (PD) 331
metal-binding protein profiles 41
metal-containing drugs uptake by single cells 416–417
metallic elements
  as crosslinkers 242–243
  as dopant 239–241
  as impurities 241–242
metallome, defined 1
metallome signature 20
metallomics 1, 193
  application of metal-containing clinical drugs 225–226
  different branches of 5
  exploiting intermetallic interactions 224–225
  ex vivo culture 381–382
  goal of 378–379
  in silico 383
  in vitro testing 382–383
  in vivo 381
  key issues and challenges in 3–4
  performing 379–380
  perspectives 226
  research on biometals 377–378
  tools for 383–387
metallomics in toxicology
  arsenic and lung cancer 291–292
  biometals 285–287
  cadmium (Cd) 292–294
  knowledge gaps, challenges and perspectives 297–298
  mercury, oxidative stress and cell death 287–291
  nephrotoxicity of uranium in drinking water 294–297
metalloproteins 323
  absolute quantification of 168

analysis 41–43
  directly protein tagging 162–164
  elemental labeling 160–161
  ICP-MS 167–168
  immunological tagging 164–165
  LA-ICP-MS 165–166
  naturally present elements 159–160
metal-related nanomaterials 11, 12, 19, 21
metrometallomics 153
  CRMs 174, 177
  external calibration 174–176
  in-house prepared standard 177–180
  isotope dilution 182–185
  LA-ICP-MS 168–170
  metalloproteins in
    absolute quantification of 168
      directly protein tagging 162–164
      ICP-MS 167–168
      LA-ICP-MS 165–166
      naturally present elements 159–160
      elemental labeling 160–161
      immunological tagging 164–165
  online addition standard 181–182
  protein quantification in 154–155
  quantitative in-situ analysis in 155–159, 185–186
    internal standardization 168–170
micro droplet generator (μDG) 403, 404
microfluidic platform for single cell analysis
  droplet encapsulation based single cell separation 403–407
  hydrodynamic capture based single cell separation 407–408
  magnetic separation based single cell capture 408
microwave induced plasma (MIP) 251
monodisperse droplets 431
monomethylarsonous acid (MMA^{III}) 41
multi-collector (MC)-ICP-MS 16, 52, 252, 434
multidimensional scaling (MDS) 454
multi-element analysis of immortalized lymphocytes 374
multiple-reaction monitoring (MRM) 154

n
nanometallomics 2
  application in nanotoxicology 17–21
  concept of 11–12
  high-throughput quantification in biological system 14
  metabolism of nanomaterials in biological system 16–17
  size characterization of nanomaterials in biological system 12–14
  spatial distribution of nanomaterials in the biological system 15
nanoparticles uptake by single cells 415–416

nanoparticle tracking analysis (NTA)   13, 17
nanosafety evaluation   11, 19–21
nano secondary ion MS (NanoSIMS)   15, 119, 386, 387
National Health and Nutrition Examination Survey (NHANES)   204–206, 215–217, 220
nephrotoxicity of uranium in drinking water   294–297
neurodegenerative diseases   209–211, 288, 311, 312, 314–338
neutron activation analysis (NAA) technique   14, 17, 35, 40, 198, 199, 266, 274, 449
neutron diffraction   266, 268–269, 274, 275
non-flammable electrothermal atomic absorption spectrometry (ETAAS)   196, 197
non-human mammal model   381
normalization   170–173, 450–451

o

obesity   211, 214–215, 220
oncometallomics   349–351
   application of   351
   breast cancer   351–352
   cancer treatment   354–355
   cervical cancer   354
   colorectal cancer   353
   esophageal cancer   353
   gastric cancer   352–353
   liver cancer   353
   lung cancer   352
   occurrence and development   355–356
   ovarian cancer   354
   prostate cancer   351
   thyroid cancer   354
optical capture   403
optical emission spectrometry   35, 119, 255
optical tweezers   403
ovarian cancer   351, 354–355
oxaliplatin   226, 355, 357
oxidative stress, mercury-binding sites in   288, 290

p

pancreatic cancer (PaC)   213–214
Parkinson disease (PD)   209
   application of metallodrugs and metalloproteins in the treatment of   333
   application of metallomics in the predication of   332–333
   and metallomics   324–333
partial least squares discriminant analysis (PLS-DA)   351, 464
partial least squares regression   457, 469
partition chromatography   12

pathometallomics
   Alzheimer's disease (AD)   314–324
   in amyotrophic lateral sclerosis (ALS)   333–336
   autism spectrum disorders (ASD)   336–338
   concept and scope of   311–312
   introduction to methodologies for   312–314
   perspectives of   338
Pb and related diseases   222
pleural effusion (PE)   213, 352
pneumatic nebulization   181, 392, 399, 419
polymethylmethacrylate (PMMA)   172
positron emission tomography (PET)   15
precision, qualitative model   463
predictive model building method   457–461
principal component analysis (PCA)   351, 453, 454, 457, 464–465, 467
prostate cancer (PC)   213, 351, 356, 357
proteolysis targeting chimera (PROTAC)   43
proton-induced X-ray emission (PIXE)   15, 253, 266, 315, 330

q

quadrupole ICP-MS   251–252, 432, 434
qualitative metallomics, tools for   384–386
quantitative metallomics, tools for   383–384

r

radiometallomics   2
Raman spectroscopy   254–255, 270, 278, 465, 466
Random Sampling (RS)   205, 455
rare earth elementome   367–371
rare earth elements (REEs)   3, 121, 127, 133, 172, 239, 274, 367–371, 400, 417
red fluorescent protein (mCherry) gene   417
relative standard deviation (RSD)   139, 165, 171, 173, 434
reversed-phase high-performance liquid chromatography (RP-HPLC)   40, 127
rhodamine-based mercury probe   39
root mean square error of calibration set (RMSEC)   461
root mean square error of cross validation (RMSECV)   462
root mean square error of prediction set (RMSEP) of the validation set   462

s

sample introduction system combined with TRA-ICP-MS for single-cell analysis   393–398
Sample Set Partitioning Based on Joint x-y Distances (SPXY) method   455–456
scanning electron microscope (SEM)   12, 17, 124, 255
scattered light imaging (SLi) technique   38

secondary ion mass spectrometry (SIMS) 15, 37, 118–119, 243, 247–248
sensitivity, qualitative model 463
separation neutron activation analysis 199
silicon nanoparticles (SiNPs) 20
single cell analysis 374, 435
  droplet-chip-TRA-ICP-MS 415
  and micro particle analysis 126
single-cell ICP-MS (SC-ICP-MS/scICP-MS) 36, 126, 435, 438, 441
single particle analysis 430–431, 433–435, 443
single-photon emission computed tomography (SPECT) 15
single-pulse response (SPR) 432–433
size characterization of nanomaterials in biological system 12–14
size exclusion chromatography (SEC) 12–13, 40, 164, 385
small-angle neutron scattering (SANS) 14, 18
smoothing process 450
solution introduction system and plasma source 430–431
sound capture 402
spatial metallomics 4, 429, 430, 441–443
species-specific isotope dilution mass spectrometry (SS-IDMS) 166
specificity, qualitative model 463
SR X-ray fluorescence (SRXRF) 128
stable isotope probing (SIP) 140
stable isotope tracing 16
superoxide dismutase (SOD) 224, 225, 296, 332, 352–353
support vector machine (SVM) 351, 457–458, 464–466, 468, 469
synchrotron-based X-ray fluorescence (SXRF) 243, 249, 255, 386, 387
synchrotron radiation (SR)
  analysis 122–123
  techniques 128
synchrotron radiation based CD (SRCD) 15–17, 37, 129, 135, 249, 449, 464
synchrotron radiation-based SAXS 14, 18
synchrotron radiation based XRF (SR-XRF) 15, 313
synchrotron radiation X-ray absorption spectroscopy (SR-XAS) 37
synchrotron radiation X-ray fluorescence (SR-XRF) 37, 249, 266, 351, 449
  analysis 351
  spectrometry 249

t

TEM/X-EDS 243, 248–249, 386
tetraazacyclododecane tetraacetic acid (DOTA) 163, 164, 417

thermal ionization mass spectrometry (TIMS) 118, 140
three-dimensional (3D) cell culture systems 382
thyroid cancer 212, 351, 354
time-of-flight mass analyzer 37, 431–432
time-of-flight SIMS (TOF-SIMS) 38, 119, 140
time-resolved analysis (TRA) mode 13, 391
trace metals/metalloids 377–379, 381, 391
transmission electron microscope (TEM) 12, 13, 17, 18, 243, 248–249
transmission electron microscopy coupled with energy-dispersive X-ray spectroscopy (TEM/X-EDS) 243, 386
triple-quadrupole ICP-MS 252
T-Stochastic Neighbor Embedding (T-SNE) 453, 454
two-dimensional (2D) cell culture system 382
tyramide signal amplification (TSA) 165

u

ultrasonic slurry sampling electrothermal vaporization inductively coupled plasma mass spectrometry (USS-ETV-ICP-MS) 138
uranium, nephrotoxicity in drinking water 294–297

v

visualized 3D droplet microfluidic device 415
volatile species generation (VSG) 197, 201

w

wavelet transform 451–452
weak anion exchange chromatography (WAX) 42

x

X-ray absorption fine structure spectroscopy (XAFS) 266–268, 271, 276
X-ray absorption spectroscopy (XAS) 16, 17, 37, 41, 123, 128, 141, 249, 314, 443, 449
X-ray diffraction (XRD) 249, 266, 268
X-ray fluorescence (XRF) 14
  analysis 266, 350
  based technique 37
  spectrometry 120–121, 252
  technique 134
X-ray photo electron spectroscopy (XPS) 123, 124, 129, 142, 254, 255

z

zinc, Parkinson's disease (PD) 331